First published 1995

ISBN 0 9523448 1 5

Published by DTS Publishing
PO Box 105, Croydon, Surrey

Printed by the KPC Group, Ashford, Kent.

© Michael C. Davis 1995

British Library Cataloguing in Publication Data. A catalogue record for this book is available from the British Library.

FRONT COVER: The Daimler CVG5 was the first type of double-deck bus to operate in Hong Kong, four examples having entered service with Kowloon Motor Bus in April 1949. Here the traditional radiator-shell is combined with the yellow bands across the front to indicate that the bus is operated by a single, seated, conductor and that intending passengers should board by the front doorway. *(John Shearman*

TITLEPAGE: The Kowloon-side "Star"Ferry bus terminal was, for years, a central point in KMB bus operations and remained so until the advent of fixed-link harbour crossings, in the form of both road tunnels and rail links, the first of which, the Cross-Harbour Tunnel opened in 1972. .Since that time the ferry lost much of its importance as a major commuter crossing point between Kowloon and Hong Kong Island. In this 1978 view of the bus terminal, examples of Daimler CVG5, CVG6, CRG6 (Fleetline) and AEC Regent types can be seen. *(Mike Davis)*

BELOW: The Dennis Dragon in its eleven-metre, three-axle, air-conditioned, form. AD128 in Canton Road in September 1994 en-route for Yuen Long on the 68X, 39.7 kilometers from Jordan Road Ferry bus station. *(Mike Davis*

BACK COVER: The modern air-conditioned three-axle bus epitomises KMB's fleet in the mid-1990's. Here Leyland Olympian, AL33 poses for the camera in 1991. *(Clement Lau.*

Hong Kong Buses
Volume Two

The Kowloon Motor Bus Company (1933) Limited
From 1933

Mike Davis

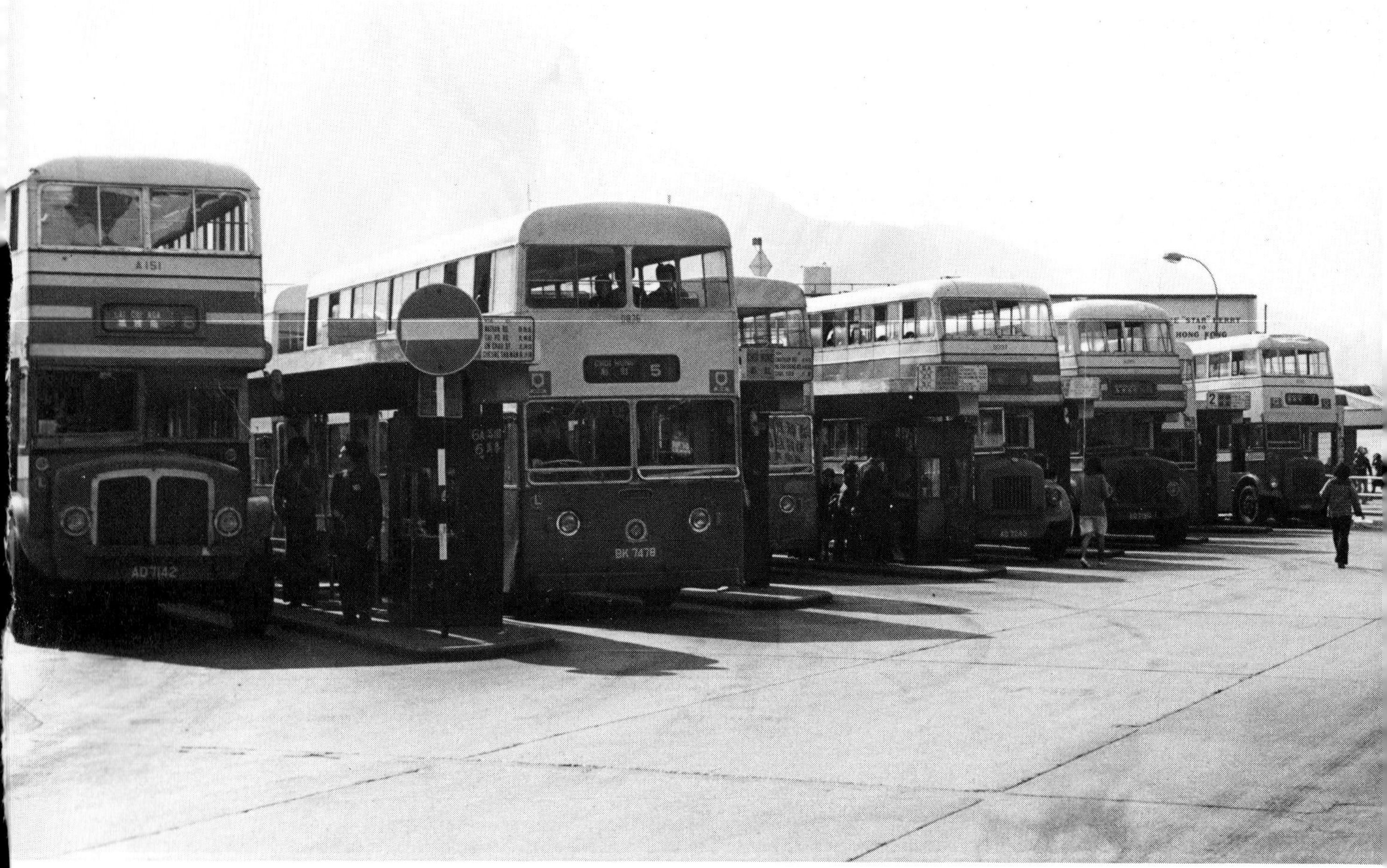

DTS
Publishing

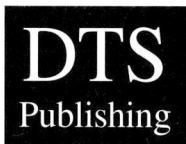

Hong Kong Buses
Volume Two
The
Kowloon Motor Bus
Company (1933) Limited

Contents

The Development of the Bus in Hong Kong from 1921 to 1933.	1
An Explanation of Body Layout Codes.	13
Buses of the Kowloon Motor Bus Co. (1933) Ltd.	14
Part One:	
From 1933 to 1941.	14
Part Two:	
Single-deck Buses from 1945.	19
Part Three:	
Double-deck Buses Purchased New 1949-1980.	79
Fleet number prefixes.	80
KMB type-classifications.	80
Part Four:	
Buses Purchased Second-hand.	161
Part Five:	
KMB Double-deck Buses 1980 to 1994:	
Section A - The Development of 12-metre, 3-axle Buses for Hong Kong	205
Section B - Prototype Buses for Evaluation.	209
Section C - Production Double-deckers buses from 1980.	223
Section D - Air-conditioned Double-deck buses.	251
Part Six:	
Other Aspects of KMB:	
Franchise.	260
KMB Fleet Livery.	270
KMB Insignia.	271
KMB Registration Numbers.	271
KMB Fleet Numbering.	272
KMB Depots.	272
KMB Fares, Tickets and Fare Collection.	273
KMB Route Development Since 1945 and Route Lists for 1977 and 1994.	276
Part Seven:	
Fleet List - Kowloon Motor Bus Co. (1933) Ltd..	290
Annex: Bigger Bus Development	330

Introduction

The Kowloon Motor Bus Company (1933) Limited is the largest bus operator in Hong Kong with over three thousand buses and it claims, almost certainly correctly, to be the "worlds's largest privately owned bus company operating in a single city".

The present-day KMB - it has been popularly known by its initials for as long as it has existed - has origins that extend back to about 1921, when it operated a small number of Model 'T' Ford buses on routes centred on the "Star" Ferry pier. Others, including the original China Motor Bus and Kai Tack Motor Bus Companies ("Tack" being the official spelling), set-up additional services, each keeping to their own routes. Smaller operators ran rural services with rudimentary vehicles and, at weekends, there were trips to the beach.

By 1933, the Hong Kong Government felt that bus services would better serve the public if they were controlled and so it sought tenders from interested parties to operate bus routes under exclusive franchises, either separately on Hong Kong Island and in Kowloon (including the New Territories) or for the entire Colony.

Two successful tenderers emerged with China Motor Bus moving across the harbour to take up its exclusive franchise on Hong Kong Island, leaving KMB to absorb the mainland operators and to change its registered name to 'Kowloon Motor Bus Company (1933) Limited'.

It should be noted that, although Kowloon and the New Territories are vastly larger in area than Hong Kong Island, the N. T. was then largely rural with few services while 'the Island' was heavily populated and bus ridership was high, redressing any perceived geographical imbalance.

Kowloon continued to grow through the 1930's and KMB's fleet also grew from 115 buses in 1933 to 140 buses by the day of the Japanese invasion on 8th December 1941. During the subsequent occupation of Hong Kong, early in 1942, bus services were restarted at the suggestion of the Japanese but a joint 'Hong Kong Motor Transport Company' was soon formed, including lorries and taxis on both the Island and in Kowloon and the buses of the two franchised companies KMB and CMB. The Managing Director of CMB was compelled to act as manager of this combine and, having no desire to co-operate, tried to resign but it was not until late in 1943, on grounds of ill health, that he was able to terminate his service and took refuge in nearby neutral Portuguese Macau.

Following the return of peace, in August 1945, the (British) Military Administration permitted the same organisation to continue and had the name changed to 'Hongkong & Kowloon Transport Service'. It ran bus services both on the Island and in Kowloon, although its management was largely in the hands of KMB who had two directors on the board, while CMB was not represented.

An application from CMB to resume their own individual operations was received by the Military Administration late in October 1945 but it has not, as yet, become clear when separate operations were resumed and KMB, probably gladly, was once again able to give its whole attention to restoring its own devastated and depleted organisation and fleet.

From 1946, KMB was, by almost daily increment, able to improve its services as rehabilitated buses were joined by converted military lorries and new buses arrived from Britain in the form of Bedford OB's, probably in partially knocked-down form. Many of the trucks were subsequently rebodied as small buses. Full-sized single-deck Tilling-Stevens were followed, in 1949, by the Colony's first ever double-deckers in the shape of Daimler CVG5's and other single-decks of varied make and model. Since that time, KMB's fleet has steadily grown in size with improving standards of comfort and safety. Today, with the application of advanced technology, KMB's newest buses, both single and double-deck, have evolved as the world leaders that they are.

The author first experienced KMB buses in 1964, just as the first 'jumbo' buses (in relative terms), in the form of 34ft long AEC Regent MkV's were taking to the road alongside 1949 vintage, 26ft long Daimlers. Between those were Daimler CVG6's, 30ft long and 8ft wide. All had bodies assembled locally from ckd (completely knocked down) kits of parts and they rattled to varying degrees according to age and how long since they had been overhauled. Since that time, both Hong Kong and the KMB bus fleet have changed enormously; Hong Kong has cast off much of its Colonial character and become an advanced city-state, playing on the world stage, while KMB buses have kept in step, even setting the pace with respect to urban bus development.

The bus service now offered reaches into the far flung corners of the Colony; to places that in 1964 the author could only reach as part of a joint Army/Police 'penetration patrol'. That was when the only urban areas were on Hong Kong Island and Kowloon Peninsula; Kwun Tong and Tsuen Wan were just emerging as industrial and residential centres. The New Towns, cross-harbour tunnels, underground railways, etc., were unheard of.

Transport on the mainland was once centred the Kowloon-side "Star" Ferry (traditionally the word "Star" has been written within double inverted commas), the most popular means of crossing the harbour before the days of fixed links. Here, as one emerged from the covered pier, there was an amazing transport interchange. To ones left was a series of piers where the ocean liners of the world tied-up; P & O, American President Lines, Messageries Maritimes, etc., the more prestigious of which berthed at the Sea Terminal - this was before the Ocean Terminal was built. Turning to one's right, there was the terminus of the Kowloon-Canton Railway - its clock tower remains, preserved as a landmark - from where diesel-electric locomotives hauled passenger trains, on the then single-track, only as far as the frontier with China at Lo Wu. This was in the days of Mao Tse Tung's turbulent regime and through trains were a thing of the future. For bus interest, one had only to look straight ahead to see the KMB terminus with its four lines of buses, mostly double-deck but with a few single-deckers on more lightly used routes such as the 7 and 8 - the latter was a short-lived route around the narrow streets of Tsim Sha Tsui, whereas the former is now run by three-axle, air-conditioned, double-deck buses.

Another focal point of KMB's bus operations was the Jordan Road Vehicular Ferry Pier, about one mile north of the "Star" Ferry, which was the terminus for other urban routes and several rural routes to the New Territories, as well as some new services to Government Resettlement Housing Estates, provided to rehouse the hundreds of thousands of squatters who had arrived homeless in the Colony as refugees from China's political and economic situation. These people pushed the population of the Colony to over four million by the mid-1960's but, despite the overcrowding and often sub-standard housing conditions, almost overwhelming the already overcrowded city, they continued to flood-in, until today there are about six million within the same boundaries.

It has been in order to accommodate this mass of people that Hong Kong's urban areas have been extended and the New Towns built. Naturally, to serve such a number of widely dispersed localities, KMB has been active in providing ever expanding bus services. In 1973, double-deckers were introduced onto the trunk routes in the New Territories using imported second-hand buses from Britain. Later, by ordering new buses in large numbers, the British bus manufacturing industry may well have been prevented from collapse during recent otherwise lean times on the home market.

During 1993, KMB's 3000-plus buses, the vast majority of which are high-capacity double-deckers, carried 2.65 million passengers per day on 342 routes - many in the comfort of air-conditioning.

Acknowledgements.

Having resided in Hong Kong over a twenty year span (1964-84) and subsequently visited the Colony on business trips, I have been able to keep as up to date as possible in Hong Kong's fast-moving environment, from which developments in the bus industry do not escape. However, it would have been impossible, even whilst there, for me to have gathered the detailed information for this book without the willing help offered by many individuals and organisations. Whilst this is in no way an official KMB publication, , the Directors and Managers, past and present, have shown every sympathy with my project and made documents and data available without which there would have been many gaps. Sadly, many are no longer living but my thanks, nevertheless, go out to the late Mr. Ronnie Lui, the late Mr. Lawrence Louey, the late Mr. Humberto Wu, the late Mrs. Lorna Everest and the late Mr. Trevor Williams, all of whom made my 1978 investigations so easy by their enthusiasm. More recently, my thanks go to Mr. Charles Lui, General Manager, Mr. John Choi, Mr. Tim Ip, and Ms. Winnie Ng.

Thanks go also to three former members of the Hong Kong Government Transport Department, Mr. Peter Leeds, Commissioner for Transport, Mr. John Shearman (who, in turn, thanks Mr. Mark Houghton and the late Mr. Bob Everest for providing facilities and making information available to him which he has since been able to pass on to the author) and Mr. Tim Runnacles.

Help also came from other friends, amongst whom Lyndon Rees, Derek Nicholls, Clement Lau and Tim Phillips were particularly helpful and, very recently, members of both Bus Fan World and the Transport Society have offered and given assistance. Over many years, information has been provided by Leyland, in all its corporate forms, Metal Sections, British Aluminium, Hestair-Dennis and W. Alexander's Ray Braithwaite. Doug Jack helped check early drafts and Jim Wilkinson, who served in the RAF at Kai Tak from 1945-47, provided some invaluable first-hand knowledge of that period, while other support and constructive criticism has been contributed by John Shearman, Geoff Morant and Julian Osborne, who have each checked different sections of the text. Many others have added small but valuable 'snippets' here and there. Nigel Eadon-Clarke has been a valuable source of photographs taken during his many visits to Hong Kong in recent years and Ron Phillips of the PSV Circle has shared information on a reciprocal basis. The usual, but sincere, thanks are extended to anyone else whose name I have omitted.

MIKE DAVIS. CROYDON, OCTOBER 1994.

The Development of the Bus in Hong Kong from 1921 to 1933.

For the benefit of new readers, this section on the early development of the bus in Hong Kong is repeated from Volume One - China Motor Bus 1933-1993.

It is not clear from Government reports exactly when bus services were first operated in Hong Kong but it was probably about 1921 when the first Kowloon Motor Bus Company was established. In that year, legislation permitting their operation was finalized. Four routes were authorized in Kowloon and, in a bid to avoid the free-for-all situation to be found at that time in Britain, specifications were issued, covering vehicles, fares to be charged and recognized stopping places. From contemporary reports it would appear that further regulations, based on those laid down by Scotland Yard for buses operating in London, were introduced circa 1925 or 1926.

During the 1920's dissatisfaction with the bus service was expressed but Government consideration of the matter did little to improve things. It was not until September 1932 that tenders were invited for the operation of bus services covering either the whole Colony, or Hong Kong Island and Kowloon as two separate operations for a period of 15 years from 11 June 1933. The Kowloon operation was to include the New Territories and any outlying islands as necessary. In January 1933, Government announced the successful tenderers as being the China Motor Bus Co. Ltd. for Hong Kong Island and the Kowloon Motor Bus Co. Ltd. for the mainland. Unfortunately there is no authoritative list of the unsuccessful tenderers.

Due to the destruction of most, but not all, company records during the Japanese occupation of Hong Kong from Christmas Day 1941 until 14 August 1945, the details of vehicles set out in this section have been gleaned from many sources. In the main, contemporary periodicals have been consulted, principally 'Commercial Motor' (CM) and 'The Tramway and Railway World' (T&RW). In addition, the personal records and recollections of individuals have provided much useful information as have other publications and therefore where the pieces fit, one can be fairly sure of an accurate account. Even so, the following pre-war information cannot be considered as being indisputable.

1927 Vehicle Data.

Under the heading 'British Vehicles in Hong Kong', a short article in 'Commercial Motor' for March 22, 1927 reads as follows:

"According to a recent return, there are now 116 motorbuses in use in Hong Kong, 106 of the evhicles being American and the others British. So far as motorvans and lorries are concerned, there are 220 in use in the city, of which 157 are American (mainly Fords), 58 British, three German and two Italian. Of the British vehicles, the most numerous are those of Dennis and Thornycroft makes." This last applies equally to the ten British-built buses.

From other sources we know that many buses had Ford Model 'T' and White chassis as well as a few Studebakers, all three of American origin. Later legislation required a 15% import duty on buses of non-British origin, and thus was established the tradition, seldom broken, that KMB buses have been British built.

BELOW: One of the earliest buses of KMB was this Model T Ford bodied by Hongkong & Whampoa Dock Co. circa 1923, from 'Whampoa - Ships on the Shore' by Austin Coates. *(H&W Dock Co.*

Early Kowloon Operators.

A report in the March/April 1978 edition of the Hong Kong Automobile Association magazine "Auto Gazette", records the following;

"... In Kowloon.... a few rickerty buses, all Model 'T' Fords, began operation years before the Penninsula Hotel was built, all without radiator caps, in front of the "Star" Ferry pier; the first bus driver, behind whom three or four buses were waiting, would start the engine though steam continued to come out of the radiator, and wait until he had collected five passengers. Those obsolete buses would go as far as Kowloon Tong."

It would appear that there was an element of dramatization about this uncredited piece, but there is more than a ring of truth and one is reminded somewhat of the latter day Public Light Buses (minibuses) as they refuse to commence a journey until they have at least half a load!

ABOVE: Detail from a photograph showing a KMB 'Model T' Ford at the Kowloon-side "Star" Ferry pier circa 1925.

NOTE: The term 'Hong Kong' can cause confusion to those unfamiliar with the local vernacular and a quick explanation here may be of help.
Hong Kong: Usually refers to the entire Territory of 404 sq miles.
Hong Kong Island (29 sq miles): Refers to the original Colony of 1841, situated just under one mile off the tip of the Kowloon peninsular, across 'the Harbour'. Referred to locally as "Hong Kong-side".
Kowloon (4.3 sq miles): That part of the mainland Colony added in 1860; lies between the Harbour and Boundary Street (about 1 miles to the north). Local people refer to "Kowloon-side".
New Territories (370 sq miles): The major land-mass of the Colony extending north from Boundary Street to the present frontier with China and added in 1898, includes the Outlying Islands, including Lantau Island, the largest and more extensive than Hong Kong Island.
New Kowloon: The urban area of the Kowloon peninsula north of Boundary Street; this is the southernmost part of the New Territories south of the range of hills which divides the mainland roughly east to west. Kwun Tong is the principal centre within New Kowloon.

BELOW: Kowloon Motor Bus Co. Thornycroft, photographed at the maker's Basingstoke factory. The 18-seat body was built at Basingstoke and the complete vehicle was shipped to Hong Kong in 1924. *(Tramway & Railway World)*

The Kowloon Motor Bus Company Limited.

Probably from the time of its formation, the Kowloon Motor Bus Co. Ltd. has been known by its initials, KMB. This designation is used widely in the text and applies equally, where appropriate, to the post-1933 Kowloon Motor Bus Company (1933) Limited.

Vehicles.

Two of the routes authorized in 1921 were taken-up by the first Kowloon Motor Bus Company Limited which commenced operations in that year, using nine Ford lorry-buses probably based on the ubiquitous Model 'T', incorporating a rear chassis extension of 3ft 6ins which was, at that time the most usual way of making this famous chassis suitable for bus work. In 1921 buses operated two services, one from Sham Shui Po to Kowloon-side "Star" Ferry pier - north-south route - and a second service from Kowloon City to Yau Ma Tei - an east-west route. Thus a large proportion of the then urban area was provided with a bus service.

Thornycroft:
Thornycroft 18-seaters - 'BT' or 'BX' type.

During 1924 the Company acquired an unknown number of Thornycroft single-deck buses which were most probably of the 'BT' or 'BX' type with bodies built by the chassis manufacturer and having provision for 18-seats; 6 in first-class and 12 in third-class. One source states that the order was for five such buses but this remains unconfirmed.

Thornycroft 'Cygnet' - 32-seaters.

KMB's next delivery appears to have been of ten Thornycroft half-cab 32-seaters which were supplied in 1928/29. These were again of two-class layout, having both front and rear entrances. Roof route boards were carried, these being lettered bi-lingually in Chinese and English. A repeat order for four similar vehicles was delivered in 1929.

Vulcan:
'Duke'.

Commercial for 10th April 1928 illustrated a Vulcan normal control single-decker for Kowloon Motor Bus. It was reported that they had teak bodies to withstand the climatic conditions but the seating capacity was not specified. The then usual two-class accommodation was entered through two separate doorways, one behind each axle.

Commer:
6TK 'Invader' - 20-seaters.

1930 saw the addition of a somewhat smaller type in the form of ten Commer 6TK 'Invaders' with 20-seat bodies (CM Dec. 30, 1930). These presumably also had two-class accommodation, but this is not certain.

BELOW: The rear view of the 1924 Thornycroft 18-seater, showing the rear door. It is unclear as to what function these somewhat luxurious-looking buses performed but from the upholstered seat beside the driver and roof luggage rack, they may have been for the transfer of passengers between ocean liners and hotels. *(Manufacturer's photograph*

ABOVE: A Kowloon Motor Bus Vulcan with "teak wood body to withstand the climate" according to the original text that accompanied this photograph in a contemporary magazine. *(Commercial Motor*

BELOW: A commercial post-card of circa-1926, showing the Kowloon-side "Star" Ferry concourse and bus terminal, with Leyland, Thornycroft and Dennis buses and, across the harbour, in the background, the famous 'Peak' on Hong Kong Island. The Rickshaw - or 'jin-riksha' was still a significant means of transport. *(Fred York collection*

ABOVE: KMB Thornycroft half-cab 32-seater. Reproduced from a photograph published in Commercial Motor magazine in February 1929.

LEFT: A reproduction of a Vickers advertisement circa 1926/7. *(Commercial Motor*

ABOVE: Vickers-bodied Leyland Lion PLSC1 of 1926 was destined for KMB when photographed in England prior to shipping. Two-class accommodation was provided within the dual-doorway bodies with a total seated capacity of thirty. *(Leyland Vehicles Ltd.)*

BELOW: A Vickers photograph used to illustrate an article in Commercial Motor magazine. The caption helped to perpetuate the myth of 'Hong Kong, China' - not to become fact until 1997! *(Commercial Motor, courtesy KMB.)*

30-Seater Double Entrance Saloon Bus on Leyland "Lion" chassis. Built for the Kowloon Motor Bus Co., China.

Leyland.

Leyland 'Lion' Model PLSC 1 - (1926)

After the Thornycroft 18 seaters, KMB turned to larger vehicles in order to meet its growing traffic commitments and so, in 1926, they purchased three Leyland PLSC1 Lions, with dual entrance by Vickers of Crayford, Kent to a standard Leyland design. The saloon was partitioned off to provide eight first-class seats behind the driver and 22 third-class seats to the rear. A sliding door in the partition allowed the conductor access to the third, or in the vernacular of the day, 'Native-class'. The first-class passengers were provided with fully upholstered seats but those in third-class were of the wooden-type. The PLSC1 'Lion' was powered by the Leyland 5.1 litre petrol engine and has a 4-speed sliding mesh gearbox. The wheelbase was 14ft 6in and the overall length 25ft.

Leyland 'A13' Model - Normal Control - (1927)

The following year, 1927, saw the arrival of ten Leyland 'A13' normal-control chassis (driver behind the engine) with chassis numbers 37024-33. This was a two-ton model with a wheelbase of 14ft 3ins and a body length of 21ft 10in. They were powered by Leyland's 30-32hp petrol engine, driving the rear wheels through a four-speed sliding-mesh gearbox. At that time Leyland were regularly constructing their own bodywork, but it is not certain whether these were shipped in chassis form for local bodying or as completely built up vehicles.

Leyland 'Lion' Model 'LT1' - (1929).

In addition to the 1929 Thornycrofts, two further Leyland 'Lions' arrived that year, but this time they were 'LT1' models, with chassis numbers 50264/5, having a wheelbase of 16ft 7ins and an overall length of 27ft 6ins: two feet six inches longer that the PLSC1 'Lion'. They were powered by a 5.1 litre, 4-cylinder, petrol engine, driving through a four-speed sliding-mesh gearbox. Again there is uncertainty as to the make of the bodywork on these buses but it is known that they were of the two-class type with dual entrances seating 36-passengers.

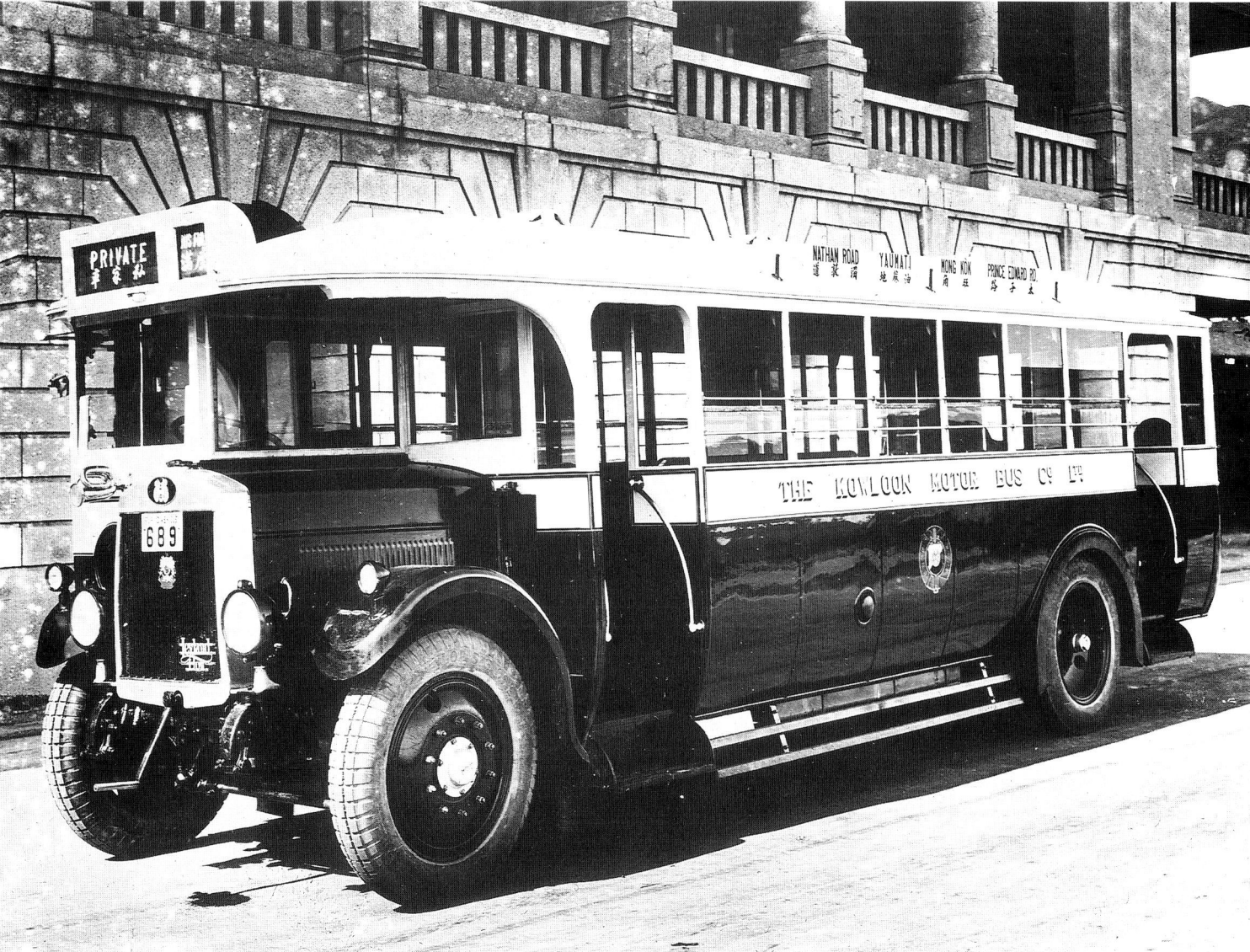

BELOW: KMB purchased further Leyland Lions in 1929, in the form of two Model LT1's which had a longer wheelbase than the earlier PLSC models. *(Leyland Vehicles Ltd.*

The Kai Tack Motor Bus Company.

The Kai Tack Motor Bus Company was formed in 1923 and it is interesting to note the spelling of the Chinese name 'Tack' which today is invariably spelt 'Tak'. The Company operated two services, one of which ran from Kowloon "Star" Ferry pier to Kowloon City, using a fleet of 20 lorry-buses. The Company was re-formed three years later to become the Kai Tack Motor Bus Company (1926) Ltd. In 1928 it was purchased outright by Hongkong Tramways Ltd. (qv), however the day to day operations remained entirely in the hands of the Kai Tack company's administrative Chinese staff.

It is, perhaps, of interest to learn that the origin of the name of both Kai Tak Airport and the bus company of that name, stems from the names of the two gentlemen who founded the airport, a Mr. Kai and a Mr Tak.

Vehicles.

At the time of the Hongkong Tramways takeover, Kai Tack owned 16 Dennis single-deck buses of two types.

Dennis 15ft wheelbase - 27-seaters.

At least three were Dennis 15ft wheelbase models, bodied in England by Strachan and having two-class seating for a total of 27; ie 10 first-class and 17 third-class. There was a separate off-side door to the driver's compartment, unusual for a bonnetted type, and two nearside doors to the passenger accommodation, front for first-class and rear for third-class travellers. All the seats faced forward with those in first-class having sprung backs and being upholstered in antique leather. Third-class had wooden-slatted seats with no frills. The rear door was of interest by reason of its oblique arrangement. On the nearside of the body there were four drop windows, with five on the off-side. Pivoted ventilators were fitted above the main windows and the ventilation was further assisted by two 'Airvac' vents in the roof. Mechanical features of the chassis included a 4-cylinder petrol engine developing up to 55hp, a cone clutch, four-speed gearbox and worm-driven rear-axle. A single line roller-blind type destination indicator was provided at the top of the nearside windscreen. Livery was described by Commercial Motor as follows: white roof and window frames, a black waist-band and grey panels below the waist. The operator's name appeared in both English and Chinese with the former in gold shaded black and the latter in plain black. In order to comply with police requirements, lifeguards were fitted on both the near and off sides, whilst a driver's mirror was required on the off-side.

Dennis 'G-type' - 20-seaters.

The second model operated was the smaller Dennis 'G' type, with eight first-class and twelve third-class seats. They were a low load-line model with the chassis cranked over the rear axle to provide a much lower floor level. Entry to the first-class compartment was by way of the wide front door which was hinged, car style, while the third-class passengers entered through the inset rear door which opened inwards. Inside, the two compartments were separated by a glazed partition in which was set a sliding door. The first few buses were bodied in England by Strachans but later bodies were locally constructed and varied in appearance from the original. Another four Dennis 'G' type were ordered by Kai Tack but were not delivered until after the Tramway takeover. Dennis records show a total of 15 'G' type supplied between 1927 and 1929. A further four Dennis bus chassis, this time of the longer 15ft wheelbase model, were ordered in 1929 for local bodying as 20 and 28 seaters respectively.

Dennis 15ft wheelbase - 28-seaters.

Shortly after its being taken over by Hongkong Tramways, Kai Tak ordered four additional 15ft wheelbase model buses with 28-seat bodies; 8 first and 20 second-class. These were required when two additional routes were introduced, including one from "Star" Ferry Pier to Tai Wan beach.

A photograph from 1929 shows a line-up of three different Dennis models, but none of them have bodywork with the same window arrangements as those shown in 1926 illustrations, leading one to believe that either, the bodies were rebuilt in the intervening months or, that there were other types so far unrecorded.

Diesel Engines.

Increased revenues, following a fare increase in 1931 enabled the Company to look ahead and an experimental diesel engined bus was purchased - possibly Crossley powered. At the same time a number of buses were rebuilt but to what extent is unclear. The entire fleet of 25 buses was sold to KMB in 1933 but spares were reported to have been sold separately; the trade press considered this to be an unusual arrangement.

ABOVE: A 1929 line-up of three types of Dennis buses owned by the Kai Tack Motor Bus Co. (1926) Ltd. Which model is which remains unclear. *(Tramway & Railway World, October 1929)*

ABOVE: To fulfil its requirements for a smaller bus on lightly loaded routes, the Kai Tack company purchased a number of Dennis 'G-types'. The early example illustrated here was, reportedly, bodied in the UK by Strachans but some later bodied were locally constructed, or assembled, in Hong Kong. This vehicle had not been provided with destination-blind equipment but later in-service photos indicate that these were fitted in due course. *(Commercial Motor, courtesy KMB*

BELOW: The original caption, reproduced here, tells most of the story. The Dennis 15ft wheelbase chassis was a very high model - years later the Dennis Jubilant owed much to this aspect of the earlier design! Reproduced from a 1926 issue of Commercial Motor that was kindly made available by KMB in 1978.

BUILT FOR SERVICE IN HONG-KONG. Local requirements have been studied in the 15-ft. wheelbase Dennis omnibuses which are now on their way to the Kai Tack Motor Bus Co. (1926), Ltd. The 27-seated body is divided into first and second class compartments, the former, at the front, holding 10 passengers on upholstered seats. Doors to each compartment are provided on the near side.

ABOVE: A front nearside view of the Dennis 15ft wheelbase type as illustrated in a 1926 edition of Commercial Motor. A copy of the magazine was kindly made available by Leyland Vehicles in 1981.

Miscellaneous Operators.

Nam Hing Bus Company.

In the New Territories, the 'Nam Hing Bus Company' started a service between Yuen Long and Sheung Shui in 1921 or 1922, using four lorry buses.

Cheung Mei Bus Company.

Also in the New Territories, the 'Cheung Mei Bus Company' was founded in 1925, operating six lorry-buses

Chun Hing Motor Bus & Co.

A photograph exists of a very spartan looking bus bearing the title 'Chun Hing Motor Bus & Co'. This bus, it is reported, operated during the 1920's and, on Sundays, took passengers to the beach. The vehicle appears to be a Ford Model 'T', with rear chassis extension.

Hong Kong and Cowloon Taxicab Co.

In 1924, the 'Hong Kong and Cowloon (sic) Taxicab Co.' cabled an order to Messrs. Henry Garner Ltd., vehicle manufacturers of Mosely, for three Garner chassis; two of 'standard' length and one bus chassis of longer length. Both were quickly prepared for shipping and subsequent enquiries revealed that the operator ran a fleet of 125 'Maxwell' taxicabs and, at the time of the Garner order, claimed to have a further 75 on order.

They had an expanding bus operating business, already owning a 'White' omnibus of American origin and one 'Dennis' of unspecified model. (From a report in 'Commercial Motor' magazine for May 27th, 1924).

The China Motor Bus Company Limited.

The third major operator in Kowloon prior to 1933 was the China Motor Bus Company Limited (CMB) which commenced its operations in 1924. CMB ran its Kowloon services alongside those of KMB until 1933 and has been fully described in Volume One.

A brief summary of its bus fleet pre-1933 would include Leyland PLSC and LT Lions, similar to those of KMB, together with various Thornycrofts. Many larger CMB buses, including the Leylands, were the subject of negotiations between KMB and CMB and were transferred to the newly re-formed Kowloon operator in 1933, where they continued to work alongside KMB's own Lions.

A China Motor Bus Leyland PLSC 1 'Lion' (right) and a Thornycroft 'Boadicea' (below), both photographed in England prior to shipping. At least the Leyland was transferred to KMB in 1933, possibly together with the Thornycroft.

ABOVE: Probably a Model T Ford with rear chassis extension, this rugged-looking vehicle was described by the 'South China Morning Post' as being used to take parties to the beach at weekends. *(South China Morning Post)*

BELOW: Illustration and caption from 'Commercial Motor' for 27th May, 1924. It is quite likely that the spelling of the word 'Cowloon', ie commencing with a 'C', was as per the registered name of the company.

Two 2-ton Garner chassis on pneumatic tyres, shipped last week in part fulfilment of a cabled order from the Hong Kong and Cowloong Taxicab Co., referred to in a paragraph on this page

Franchise and Monopoly.

Public dissatisfaction with the quality of bus services, in a 'free-for-all' situation of open competition, grew during the late 1920's and early 1930's, until Government action was deemed to bring about some degree of regulation. In a move, somewhat the antithesis of the usual 'laissez faire' approach to business practice in Hong Kong, it was decided that Government should award either a franchise, or franchises, for monopoly operation of bus services within geographical areas. This, it was agreed, would offer better control over fares, vehicles and standards generally. It was also decided that the proposed franchise holder(s) should pay Government a royalty, calculated on a percentage of revenue taken.

In September 1932, tenders were invited for bus services covering the whole Colony, or the whole island of Hong Kong, or Kowloon plus the New Territories, for a period of 15-years from 11th June 1933.

In January 1933 the successful tenderers for the new bus services were announced, in the Government 'Gazette' and the event was reported in the 'South China Morning Post' on Monday, January 16th, 1933.

Hong Kong Island services, covering 18 routes, 11 of which had second-class fares, were awarded to the 'China Motor Bus Company Limited', a title under which buses had previously been run in Kowloon.

The 18 routes for Kowloon and the New Territories were awarded to the Kowloon Motor Bus Company Limited, but there was no mention of second-class fares.

During the few months up to the vesting day of the new services, a series of mergers took place. The style of the Kowloon operator was amended to 'Kowloon Motor Bus Company (1933) Limited', on 13th April 1933 and, when the new services commenced on 11th June of that year, the new 'KMB', as the company is popularly known, was able to field 115 buses.

BUS CHASSIS EXPORTED TO UNSPECIFIED BUYERS.

These chassis were consigned to the Hong Kong agents, or branch, without mention of the operator who had placed the order.

THORNYCROFT:

Chassis Nos.	Quantity.	Type/Model.	Date.
18232/5	2 chassis,	JJ/MB4,	sent 6.2.29
18461/2	2 chassis	A2-long,	sent 14.9.29
18474/5/6	3 chassis	A2-long/FB4,	sent 4.10.29
20886/7/8	3 chassis	A2-long/FB4,	sent 18.10.29
18975	1 only	EC/WB6,	sent 13.5.32
18709-16/19/20	10 chassis	A2-long	Date unknown.

DENNIS.

Chassis Nos.	Quantity.	Type/Model.	Date.
45453-5	3 chassis	'2 -ton'	1926
70018/9/32/45/63/79	6 chassis	'G-type'	1928
80151	4 chassis	'FS-type'	1929
70400/12/41/8	4 chassis	'G-type'	1929

KOWLOON ROUTES

Immediately prior to the amalgamations of 1933, the routes operated by Kowloon Motor Bus and Kai Tack in Kowloon, according to long-serving members of the Company were as follows: *(CMB routes in italics).*

Route:	Operator:	From/to:
1	CMB	*"Star" Ferry to Lai Chi Kok, via Canton Road.*
2	KMB	?
3/3A	Kai Tack	Kowloon City to Yaumati Ferry - unconfirmed
4	Kai Tack	"Star" Ferry to Kowloon City - unconfirmed
5	Kai Tack	?
6	KMB	?
7	CMB	*"Star" Ferry to Kowloon Tong via Waterloo Road*
8	KMB	?
9	KMB	?
10	KMB	?
11	CMB	*Sham Shui Po to To Kwa Wan*
12	CMB	*"Star" Ferry to Lai Chi Kok, via Nathan Road.*
?	Kai Tack	"Star" Ferry to Tai Wan beach (Summer only).

By the time of the commencement of the new franchised operations, lorry-buses had gone and all buses were conventional in construction and appearance. It is of interest that the terms of the new franchisees laid-down that KMB would, upon specific application, be permitted to use double-deck buses on certain approved routes, while CMB were specifically debarred from double-deck operation altogether.

A view of the Kowloon-side "Star" Ferry concourse shortly before the amalgamations of 1933. On the left is the entrance to the pier and nearest to it are two CMB buses on the 1 and the 7; the former, No 666, being a small 20-seat Thornycroft, while the latter is a Leyland PLSC1 'Lion'. In the second line from the left are two Kowloon Motor Bus vehicles. Nearest to the camera is No 94, a half-cab Thornycroft 'Cygnet' on Route 6 and in front of that is a Leyland LT1 'Lion', with the driver's cab door swung open. In the third row are two Dennis buses of the Kai Tack company; a 'G-type' on Route 5, fleet number 12, registered 542, while in front is a Dennis 15ft wheelbase model on Route 3. The KMB bus departing, far right, is probable a normal-control Thornycroft or Vulcan 'Duke'

Commer.

Commer B3 - 1936.

During 1936, a batch of fourteen Commer B3 chassis arrived with either Dorman-Ricardo or J.U.R. diesel engines; the former was more usually associated with the contemporary products of the Associated Equipment Company (AEC). No details have come to light regarding the bodies carried or mechanical specification, neither is there any inforamtion regarding the type of body, seating or doorway arrangements.

VEHICLE SPECIFICATION:	
Chassis:	Commer B3
Chassis numbers:	B3-59020-33.
Engine:	Dorman-Ricardo or J.U.R. diesel
Body make:	?
Body layout:	B ??
Date introduced:	1934

Thornycroft single-deckers.

Thornycroft 'Cygnet' CD4LW - 1934.

As passenger traffic increased, the Company sought to rationalize its fleet, larger buses were required and the 1934 order was for ten Thornycroft CD4 'Cygnets' which were powered by Gardner 4LW diesel engines. These buses were despatched from the Basingstoke, England, factory of J. I. Thornycroft Ltd., during June and July 1934 and, upon arrival in Kowloon, were fitted with 37-seat bodies of local construction.

Hitherto, bodies had either been fully constructed in England or were shipped in large pre-fabricated segments, packed flat on the chassis, for final assembly by local craftsmen.

This was also the first large order from KMB, or, indeed, the Colony, to specify the Gardner diesel engine; a make that remains a favourite with KMB, CMB and other Hong Kong operators into the 1990's.

VEHICLE SPECIFICATION:	
Chassis:	Thornycroft CD4LW
Chassis numbers:	24794-803.
Engine:	Gardner 4LW
Gearbox:	Thornycroft 4-speed
Body make:	KMB contractor
Body layout:	B37FEX, REX; 2-class.
Date introduced:	1934
Length:	?
Width:	?
Wheelbase:	?

BELOW: A Kowloon Motor Bus (1933) Ltd. Thornycroft 'Cygnet, CD4LW, registered 779, posed in pre-war sunshine and probably brand new, circa 1938. *(Author's collection*

Thornycroft 'Cygnet' CD4LW - 1938-39.

A further seventeen Thornycroft 'Cygnet' chassis were added to the fleet in 1938/9, having been shipped from the UK in August and October 1938. These vehicles were fitted with Gardner 4LW diesel engines, as the chassis designation suggests.

LEFT: HK5551 was a pre-1941 Thornycroft, similar to 779, above, cut-down as a crane truck. It has, since this 1978 photograph, lost its Thornycroft radiator but remains in stock and is probably the oldest example of its kind still working for its original owner. *(Mike Davis*

VEHICLE SPECIFICATION:	
Chassis:	Thornycroft CD4LW
Chassis numbers:	29703-19
Engine:	Gardner 4LW
Gearbox:	Thornycroft 4-speed
Body make:	Unknown; possibly KMB
Body layout:	B37FEX, REX; 2-class.
Date introduced:	1938/9
Length:	?
Width:	?
Height:	?
Wheelbase:	?

ABOVE: Kowloon-side "Star"Ferry concourse in 1936, showing the KMB bus station, together with examples of Thornycroft, long-Thornycroft and Leyland LT1 Lions. The line of taxis in the foreground are all Hillmans. *(Commercial Motor 21st Feb., 1936*

RIGHT: This 'normal-control' Thornycroft remains unidentified but it is believed to have been taken in the late 1930's although the actual vehicle is possibly a pre-1933 vehicle. The bodywork mof these vehicles shows similarities with contemporary Vickers products on this make of chassis; this may me idle speculation. *(Jim Wilkinson collection*

Dennis single-deckers.

It is believed that these, the first Dennis chassis purchased by KMB since its first post-1933 orders, were all diverted from the nearby Chinese City of Canton, more usually referred to today as Guangzhou, due to the occupation of South China by the Japanese during the late 1930's. Some contemporary Leyland trolleybus chassis for Canton were diverted for similar reasons and found their way to Spain and Australia.

Dennis 'Ace' - 1938.

Prior to the fall of South China to the Japanese in 1938, it would appear that the City of Canton had ordered a number of small Dennis 'Ace', forward-control (bonnetted) single-deckers but, following the fall of that city, they were diverted to Hong Kong where they were purchased by KMB. Dennis records show that eleven 'Ace' bus chassis were supplied to Hong Kong in 1938. Of these eleven Aces, ten had Gardner 4LK engines while the odd one had a Perkins P4 unit. 4-speed gearboxes were fitted to both types.

(It should be noted that, according to 'Registration Plates of the World' (pub. Europlates), until midnight on 12th July 1946, Canton, together with Shanghai and Kunming drove on the left, the remainder of China having changed to the right on 1st January that year.)

Dennis 'Falcon' - 1939.

Fifteen Dennis Falcons were ordered in 1940 but, due to the constraints placed upon the manufacturers by the war in Europe, only twelve were built, although the chassis numbers were allocated. Dennis records show that these were, unusaully, forward control (driver beside the engine), they had Perkins 6-cylinder P6 engines, 4-speed gearboxes and 14ft wheelbase chassis. The first Falcon for KMB left the factory on 15th May 1940 and the twelth on 26th February 1941.

VEHICLE SPECIFICATION:	
Chassis:	Dennis Ace, **forward** control
Chassis numbers:	200592/3/5-603.
Engine:	Gardner 4LK
	(200600; Perkins P4).
Gearbox:	4-speed.
Body make:	?
Body layout:	?
Date built:	14-4-38 to 8-6-38;
(ex-works):	200600: 25-7-38
Length:	?
Wheelbase:	?

VEHICLE SPECIFICATION:	
Chassis:	Dennis Falcon **normal** control
Chassis numbers:	280034-7/9-46.
Engine:	Perkins P6
Gearbox:	4-speed
Body make:	?
Body layout:	B ?? ?
Introduced:	1940/41
Length:	?
Width:	?
Height:	?
Wheelbase:	14ft 0in
Order for 15 curtailed - only 12 being built due to war conditions in Europe	

Dennis 'Lancet II'.

Ordered 1937.

Through the Dennis Agent in Hong Kong, Alex Ross (Hong Kong) Ltd, KMB placed an order on 8th April 1937 for eleven Dennis Lancet II, half-cab, single-deck chassis for bodying in Hong Kong. Of these, ten had Gardner 5LW engines and the eleventh had a Dennis 117.47mm x 150mm engine but all had 5-speed gearboxes.

Ordered 1940.

A further twelve Dennis Lancet II chassis were ordered in 1939 with the first leaving the Dennis factory at Guildford on the 15th November 1939 and the eleventh on 11th March 1940, reflecting the shortages of wartime. The final chassis was not completed until 28th September 1940.

Further difficulties were encountered in obtaining engines which were required for the war effort and only the first five were shipped with engines. An exception was the tenth chassis (175753) which, somehow, was shipped with its engine fitted. The lower powered 4LW

Presumably, KMB fitted reconditioned engines or units obtained elsewhere; it is known that exemption was granted from the terms of their franchises allowing the two bus companies to obtain vehicles and supplies from non-British sources under the exceptional circumstances of the war.

Following the liberation of Hong Kong in August 1945, a number of chassis were recovered, although most had been cut down to make flat-bed lorries.

A December 1945 photograph shows a survivor of the occupation in Kowloon with bodywork of a most un-KMB-like appearance, with downward swept style-lines over the rear wheel-arches, flared skirt panels and a half-canopy in addition to the half-cab. A destination box was fitted in the front of the nearside roof and the rear window appears to be a large oval. Nothing further has so far been brought to light about these vehicles.

A 1945 photograph of this type of Dennis Lancet is reproduced on page 20.

VEHICLE SPECIFICATION: 1937 type.

Chassis:	Dennis Lancet II.
Chassis numbers:	175330/3/41/4/6/7/59/60/6/9/70
Engine:	Gardner 5LW; or 175344: Dennis 5-cyl.
Gearbox:	5-speed, manual
Body make:	Locally built.
Body layout:	B ?? ?
Introduced:	1937
Length:	27ft 6in
Width:	7ft 6in
Height:	?
Wheelbase:	17ft 5in

NOTE: The Dennis records describe their engine as 117.47 x150mm

VEHICLE SPECIFICATION: 1939/40 type.

Chassis:	Dennis Lancet II.
Chassis numbers:	175730/8/45-9/51-55.
Engine:	Gardner 4LW
Gearbox:	4-speed, manual
Body make:	Locally built.
Body layout:	B ?? ?
Introduced:	1939/40
Length:	27ft 6in
Width:	7ft 6in
Height:	?
Wheelbase:	17ft 5in

Daimler.

Double-deck chassis for trial with KMB.

Daimler COG5DD - 1939.

Two Daimler COG5DD chassis, designed for double-deck bodywork were ordered by KMB and shipped in 1939. Upon arrival, these double-deck chassis, like those of CMB, received single-deck bodies which, it is believed, were built by Leung Kam Kee, a contractor for bus body assembly into at least the late 1980's. It is reported that these two Daimlers carried registration numbers 375 and 376 but this remains unconfirmed. One long serving employee in the Service Manager's office at KMB recalled, in 1978, having seen the two Daimlers in Canton shortly after the war, possibly 1946, but they were not recovered to Hong Kong.

In passing, it may be of interest to learn, from local newspaper reports, that the reason the use of double-deckers was not approved by the Authorities was that, pre-1941, the streets of Kowloon were lined with large, mature trees, whose branches would have had to be pruned to permit the passage of double-deckers. This was unacceptable at the time as the Government was proud of its leafy urban areas but, when the question was resurrected after the war, most of the trees had been felled by the Japanese for use as fuel and double-deckers were accepted.

VEHICLE SPECIFICATION: Daimler COG5.

Chassis:	Daimler COG5DD
Chassis numbers:	10365 & 10366
Engine:	Gardner 5LW
Gearbox:	?
Body make:	?
Body layout:	?
Date introduced:	1939
Length:	26ft
Width:	7ft 6in
Height:	?
Wheelbase:	16ft 3in

Summary.

It is not possible to say if this is the complete pre-1941 Kowloon bus fleet as the Company records were lost during the Japanese occupation. While figures suggest that the pre-war fleet could have included only those buses described above, it seems likely that, with the reported rationalization and scrapping of unsuitable buses after 1933, no additional buses were acquired. It is known from late 1930's photographs, that some of the pre-1933 vehicles remained in service, particularly normal-control Thornycrofts.

Occupation.

Shortly after commencement the occupation, the Japanese confiscated all the Company's buses and then loaned them back to KMB to operate when not required for military purposes. As outlined in the Introduction, early in 1942, the enemy re-introduced a bus serviceen but later forced the amalgamation of all bus, trucking and taxi operations into a single operation, thus KMB joined CMB in the joint operation. Later in the war, the operation of buses became almost impossible and many vehicles were appropriated, officially by the occupier and unofficially by dubious characters who took engines to power fishing junks. Those taken by the Japanese were dispersed within the adjacent areas of occupied China and many were lost forever. Others that remained in Hong Kong were cut down to become chassis/cab flat-bed lorries but, somehow, a few survived to return to service soon after Liberation in August 1945.

BELOW: The rear of an unidentified single-decker in Nathan Road. Although minute, a Dennis Lancet with half-canopy can be seen approaching in the far distance to the immediate right of the retreating car ahead of the rear of the bus. *(Jim Wilkinson)*

Examination of a pre-1941 photograph showing the interior of KMB's Leyland Lion, No 689, reveals a notice on the front bulkhead showing the Company's routes and fares which were as follows:

Route	From & To
1	"Star" Ferry and Sham Shui Po
2	"Star" Ferry and Lai Chi Kok
3	"Star" Ferry and Kowloon City via Hung Hom
4	Yaumati and Kowloon City via Boundary Street
5	Yaumati and Hung Hom
6	"Star" Ferry and Kowloon City via Mongkok
7	"Star" Ferry and Ho Man Tin
8	"Star" Ferry and Kowloon Tong
9	Blank
10	"Star" Ferry and Ngau Chi Wan via Shanghai Street

Known Dennis chassis.

Known Dennis chassis exported to Hong Kong and/or China and believed diverted to KMB due to the Japanese occupation of South China:

DENNIS LANCET II - 1937

Chassis number	Date left works	Engine number	Order number
175330	24.06.37	37232(G)	34719
175333	24.06.37	37215(G)	34720
175341	24.06.37	37085(G)	34721
175344	29.06.37	100191	34730
175346	25.05.37	38356(G)	34722
175347	25.06.37	38287(G)	34723
175359	16.07.37	38395(G)	34724
175360	16.07.37	38463(G)	34725
175266	30.07.37	39409(G)	34726
175369	30.07.37	39516(G)	34727
175370	10.08.37	39515(G)	34728

(G) indicates a Gardner 5LW engine. The odd chassis was fitted with a Dennis 5-cyl. engine. All had 5-speed gearboxes.

DENNIS LANCET II - 1940/41.

Chassis number	Date left works	Engine number	Order number
175730	15.11.39	47560	34815
175738	15.11.39	47686	34816
175745	26.11.39	47742	34817
175746	05.12.39	47810	34818
175747	13.12.39	47809	34819
175748	28.12.39	Less engine	34829
175749	16.12.39	Less engine	34828
175751	10.01.40	Less engine	34830
175753	29.02.40	48251	34820
175754	11.03.40	Less engine	34832
175755	28.09.40	Less engine	34833

Where fitted, engine was Gardner 4LW with 4-speed gearbox.

DENNIS ACE forward control - 1938.

Chassis number	Date left works	Engine number	Order number
200592	14.04.38	41746(G)	
200593	14.04.38	41747(G)	
200595	14.04.38	41838(G)	
200596	03.05.38	42001(G)	
200597	17.05.41	41978(G)	
200598	21.05.38	41979(G)	
200599	21.05.38	42160(G)	
200600	25.07.38	10093(P4)	
200601	24.05.38	42161(G)	
200602	08.06.38	42231(G)	
200603	30.05.38	42209(G)	

(G) Gardner 4LK engine.
(P4) Perkins P4 engine.

DENNIS FALCON forward control - 1940/41

Chassis number	Date left works	Engine number	Order number
280034	23.07.40	11947(P6)	34854
280035	15.05.40	11942(P6)	34853
280036	31.07.40	10779(P6)	34855
280037	06.11.40	11949(P6)	34856
280039	?	10621(P6)	34857
280040	05.12.40	11955(P6)	34858
280041	05.12.40	11937(P6)	34859
280042	27.12.40	12082(P6)	34860
280043	03.01.41	11951(P6)	33861
280044	20.01.41	11950(P6)	34862
280045	04.02.41	10787(P6)	34863
280046	26.02.41	11945(P6)	34864
280047	Not built.		
280048	Not built.		
280049	Not built.		

(P6): Perkins P6-diesel engine.

ABOVE: One of the pre-1941 Bedford airport buses seen here in late 1945 having passed to the custody of the Royal Air Force at Kai Tak where it continued its function, commenced by the occupying Japanese forces, of transferring VIP's between the airfield and Kowloon hotels. The registration number, L167 was issued by the Military Administration in lieu of lost pre-occupation registration numbers. *(Jim Wilkinson*

LEFT: From a photograph showing two pre-1941 buses in KMB's pre-war premises in Nathan Road, Yau Ma Ti. *(Kowloon Motor Bus*

Part Two:
Single-deck Buses From 1945.

The compilation of the foregoing general description of Hong Kong's pre-war buses was a challenge, due to little detailed, documented, information surviving the Japanese occupation and therefore there are few local records upon which to base any degree of an accurate account. The post-1945 period has far more to offer but even then the records of early post-war vehicles are scant. No doubt rebuilding their shattered operations was of more concern to the owners than recording details of their bus fleet.

It is known that the Japanese had formed a single road transport 'company', including taxis and lorries as well as the buses on both sides of the harbour. After the liberation, a British 'Military Administration' assumed government of the Colony until such time as a Civil Administration could be re-established (in mid 1946). The Military Administration retained the Japanese combined transport operation, under the new title of 'Hongkong & Kowloon Transport Services'. The joint bus operations were under the control of two KMB Directors; no representatives of CMB were included. After applications to return to the pre-1941 franchise operation, CMB were given back control of the Island operations, either late in 1945 or early 1946 and KMB could attend to their own rehabilitation.

There remained in 1978 a handful of KMB employees who had joined the company shortly after the end of the Pacific War, one of whom, Superintendent (Mechanical) Mr. Wan Kwai, had joined as a boy in the mid-1930's and had risen to foreman by the time of the Occupation in December 1941.

When, in September 1945, the Company received its orders from the Military Administration to recommence operations, the pre-war staff were located and told to report for duty. Mr. Wan Kai remembered being sent to collect the remnants of the bus fleet from Boundary Street playing fields where, for some unknown reason, the Japanese had seen fit to park them.

Of the seventeen, or so, buses recovered, only six were capable of moving under their own power and none was roadworthy. Of the six that functioned, only two could be made sufficiently safe to carry members of the public so that, when the service was recommenced between "Star" Ferry and St. Theresa's Church, in Prince Edward Road, only two buses were available. Within ten days the other four mobile buses were repaired and put on the same route.

Of the remaining pre-war buses, all of which needed much more attention, few had engines. This was due in part to the popularity of the Gardner engines with the local fishermen who, it is alleged, stole them from abandoned vehicles and installed them in their fishing junks. KMB mechanics, armed with registration books showing the engine numbers, searched the anchorages and successfully located many of their pre-war engines. Restoration work was, however, not merely a case of replacing engines for other major components had been either looted or damaged thus further delaying the re-entry into service of many battered survivors. Some were never returned to service and many of their components were used to keep other less damaged buses on the road. The bodywork was, without exception, in a terrible state and, during 1946/47, most pre-war buses were either rebuilt or rebodied. KMB buses were located all over neighbouring South China but few, if any, were recovered.

Survivors recalled.

During 1978, the then Service Manager, Mr. Poon Keung, who joined the Company as an apprentice in 1949, recalled some "quite old" Dennis buses which were probably Dennis 'Lancets' or the mysterious diverted Canton Dennis 'Aces'. There were also some Thornycrofts amongst the surviving pre-1941 vehicles, one of which survived in 1984 as a workshop crane truck inside Lai Chi Kok depot. This vehicle, a 'Cygnet' CD4LW, which was re-registered as a goods vehicle in 1950, still had the forward part of its single-deck bus bodywork of a design similar to that built for the Tilling-Stevens in 1947. This was probably the oldest Thornycroft half-cab ex-bus still used by its original owner, assuming a chassis vintage of circa 1938. When examined in 1975, the engine was still a Gardner 4LW, driving through a four-speed crash gearbox. The original radiator shell was removed late in 1977 and replaced by a type which is usually concealed behind a full-fronted grille, resulting in a very ugly appearance. In 1992 the remains of the vehicle still existed on KMB premises, although no longer in use. During its long life this vehicle carried three different registrations; a three-figure pre-war number; a four figure 4xxx post-war 'Public Omnibus' number from 1946 to 1950, finally becoming 'Goods Vehicle' HK5551.

BELOW: A 1947 view of the Tsimshatsui Piers, "Star" Ferry pier and bus terminus in which can be seen parts of six or seven Dodge G5 lorry buses and two long wheelbase single-deckers, both of which appear to have half-canopies. The bus nearest the camera would seem to have similarities with that in the 1946 photograph, above. The wide rear platform is more akin to that of a double-decker than a single-deck bus and the rear window is unusually segmented. The steps in the rear of the lorry-buses can be appreciated from this view. *(Kowloon Motor Bus.)*

Reinstated pre-war and unidentified buses.

It is understood from conversations with long-serving KMB staff, kindly made possible by KMB's Management in 1978, that only seventeen semi-serviceable buses remained at the time of Liberation in August 1945.

Another very reliable source, then a member of the Royal Air Force who arrived on a ship of the liberating fleet recalls that many buses had been cut down to become flat-bed lorries but still with their half-cabs. As the serviceable buses returned to service had mostly lost their pre-occupation registration numbers, they were allocated new numbers in a temporary series initiated by the interim Military Administration until these were in turn replaced by numbers in the permanent post-war series which was introduced in mid-1945 where all public carriage vehicles received registrations which all fell between 4001-4999. Blocks of numbers were allocated to KMB, as well as CMB, and these were allocated to vehicles by the operators themselves.

There are a large number of registration numbers against which no vehicle can be satisfactorily placed, although the blocks of numbers were issued to KMB in 1946/47. It is believed that these vehicles were the ex-military lorry-buses and reinstated pre-war buses, many of which were relatively quickly withdrawn and their registration numbers transferred to newly introduced types from as early as 1948.

Dennis Lancet II.

Already referred to in the pre-1941 section and identified from a photograph by Jim Wilkinson, is a Dennis Lancet single-decker with a refined body-style, including flared-out lower panels and a style-line, curving from the moulding beneath the lower-deck side windows to the top of the rear wheel-arch. Both this chassis type and the superior bodywork are unlike any previously identified as having been sent to Hong Kong immediately pre-1941. The half-cab vehicle also had a half-canopy with two-part destination/route-number box in the front canopy on the passenger-entrance side. While Lancets *were* sent to Hong Kong pre-war, the body appears quite different from contemporary local practice. The history of this vehicle would be interesting to trace; it *may* have originally been a pre-war private-hire vehicle as it was described as having a similar colour scheme to that of the Bedford airport buses which are reported as having been an orange shade on the lower body panels, a white waist rail and the roof a light lime green.

LEFT: A rare 1946 photograph of a Dennis 'Lancet' single-decker with a half-canopy/half-cab body. The registration number appears to have been painted directly on the front panel and cannot be fully made-out but *could* have been 2198; probably in a temporary series issued by the Military Administration as pre-war registration numbers were all three-figures, while post-war bus numbers were all in the 4000 series. The body style and livery certainly do not conform to the usual KMB pattern, either pre- or post-war. There appear to have been side-lights but the headlamp recess looks empty, probably awaiting spares. Note the flared skirt panels. Whatever its origin, teak-framed, drop-windows are fitted and there is some evidence of an oval rear window. (Jim Wilkinson.

Thornycroft CD4LW 'Cygnet'.

Also identified from the 1947 photograph on page 19, are two long single-deckers, both also with unusual half-canopy, half-cab, bodies. The radiator shape of one suggests that it may have been a Thornycroft, presumably from pre-1941 days. The rest of the body shows unusual features; the rear entrance was wide, as on a double-decker, with what *appears* to be a flat platform, suggesting a drop-frame to the rear chassis extension, also as found on double-deckers. There was also a half-width front entrance and an almost full-width rear window was formed of eight separate openings. The actual back of the bus was quite curved from skirt to dome; more so than was usual at the time. Between the two entrances were six side windows with half-drop windows. Little else can be seen as the bus is obscured by a canopy structure.

The second, longer, bus is seen in the same photograph, from the off-side but is partly obscured by a Dodge lorry-bus. Sufficient is visible to deduce that the bodywork was more angular than that described above, and the cab front was vertical and fitted with a rear-hinged cab door. Again the canopy only covered the half-cab. In this bus, the side windows appear to have been full-depth sliders, with a stout safety rail along the length of the vehicle.

BELOW: Detail, enlarged from the illustration on page 19, showing a single-decker which displays most of the characteristics of a Thornycroft Cygnet, partly hidden by a Dodge lorry-bus when almost new.. (KMB

Lorry-buses based on military truck chassis.

KMB made a remarkable recovery from the run-down state that existed on 30th August 1945, when the Colony was liberated by Rear Admiral Harcourt, RN, commanding units of the British Fleet.

From the seventeen semi-serviceable buses that remained, the fleet strength was supplemented by the provision of lorry-buses.

The first lorry-buses which were put on the streets shortly after the recommencement of operations in 1945 were ex-military and naval lorries with the absolute minimum in the way of fittings and just a canvas roof to keep out the elements.

Dodge (Canada) G5; 3-ton.

During 1946, the civilian authorities provided the bus companies with a number of Canadian Dodge 3-ton G5 truck chassis (one source describes the chassis as 1½-ton) which were modified to have wooden seats and steps fitted in the centre of the tail-board. Standard canvas roofs were provided over a tubular metal frame and roll-down canvas sheets took the place of 'windows' in wet or cold weather. To fit them for their bus role, a small glazed window and a roller-blind destination box were fitted across the front, above and behind the cab..

These Dodges were of the standard type, typical of the time, with pressed steel cabs, a long rounded bonnet with front mudguards, or wings, also of a rounded pattern, integrated with the front grille but not incorporating the headlights which were mounted on top of the wings; contemporary photographs show them to have been relatively smart and at least six appear in one picture taken at the "Star" Ferry.

Although 1946 saw the arrival of thirty new Bedford 'OB' type bus chassis from England, via Crown Agents, traffic was expanding far too rapidly for the vehicles available and so, once again, the Military Administration allocated a selection of military chassis to KMB which were duly converted for passenger use as before. Amongst these were Fords, Chevrolets and more Dodges.

These converted lorries reached a probable maximum of about 75 by 1947 but it was not until about 1955 that the last of these makeshift buses were withdrawn from routes in the rural New Territories. By this time, however, many had received conventional timber-framed bus bodies with perimeter seating on wooden benches. It is most probable that most, if not all, these buses with American lorry chassis were of left-hand-drive and remained so until withdrawn.

VEHICLE SPECIFICATION:	Dodge G5
Chassis:	Dodge G5 1½/3ton
Engine:	Dodge 4-cyl, petrol
Transmission:	Manual fout-speed gearbox
Body make:	KMB (originally truck body)
Body layout:	approx. B25F (perimeter seats.)
Date introduced:	1946
Total:	?

Thanks are extended to Mr. Peter Leeds who furnished much of the information on these ex-military chassis from personal observations during his early days as a Civil Servant in Hong Kong; he retired having reached the office of Commissioner for Transport.

Other valuable contributions were made by Jim Wilkinson who, as a member of 201 Staging Post, RAF, was among the first twenty RAF personnel to disembark at Kowloon Docks, having been part of Tiger Force, diverted to Hong Kong following the surrender of the Japanese in August 1945. He helped assemble the first Bedford OWB's to arrive and took a number of transport related photographs, some of which were published in Truck & Driver Magazine, to whom thanks is extended for assisting in locating Jim.

ABOVE & BELOW: Two photographs of the same vehicle; before and after the lorry-bus body of Dodge 4216 had been replaced by a bus-type body. The lorry-bus photograph dates from 1947, the other from 1952. *(KMB [above] and Derek Parsons [below].*

Bedford 'OB'.

Soon after the cessation of hostilities in 1945, KMB ordered whatever vehicles were available and, through the Military Administration, received an allocation of 30 normal-control (bonneted) Bedford 'OB' chassis which it is believed were fitted with locally built bodies. There were, however, two types of body described below.

As supplied, the Bedford OB was a petrol engined bus but these were later replaced by Perkins P4 diesels.

Those with obviously local bodywork had timber-framework and were of an austere pattern, superficially similar to British wartime bodies on similar Bedford 'OWB' chassis. These bodies had five almost square side windows between the front and rear entrances and the side and skirt panels were flat. One class of passenger accommodation was provided with painted wooden-slatted seats for about twenty-five people. These uncomfortable seats remained a feature of many KMB buses until, in a few cases, the early 1970's. This type of body was almost standard for KMB single-deck buses, merely being adapted to suit differing chassis types. It is not known where these Bedfords were bodied; ie by KMB or by contractor.

An unknown number of Bedford 'OB's' has a much less austere body style which appeared to be metal-framed, and of much more pleasing lines. Instead of being almost square, the side windows were oblong, with only four set between the dual entrances. The lower bodyside panels, were curved, as were the rear skirt and corners, showing distinct 'tumblehome'; features which enhanced their appearance greatly. These bodies could well have been shipped from the UK completely built up and bore a decided similarity to contemporary bodies supplied to the Navy and Army. The doorways could have been local modifications. Another possible explanation for the differing styles arises from the known fact that neighbouring China Motor Bus, which also had similar buses, rebodied many in the early 1950's. There is every reason to suppose that KMB may have done likewise.

On both types of body, sliding metal gates were added about 1950 to protect those standing passengers squashed-in on the steps. The local crowds of refugees were pleased enough to ride at all, let alone in comfort - the bus companies were equally pleased to take their fares!

The registration numbers carried by the Bedford 'OB's' were later transferred to Daimler CVG5's which entered service from 1959-61 - the 'Daimler(b)' type. All these Bedfords were withdrawn during 1959.

VEHICLE SPECIFICATION: BEDFORD OB	
Registration numbers:	42xx / 45xx
Chassis make:	Bedford Model 'OB'.
Engine:	Bedford 6-cyl, petrol, 3519cc, 72bhp; later replaced by Perkins P4, diesel
Gearbox:	Bedford 4-speed, non-synchromesh.
Body Make:	KMB or contractor.
Body layout:	B25FEX,REX,G.
Date introduced:	1946
Length:	25ft 0in
Width:	7ft 6in
Wheelbase:	14ft 6in Turning circle: 59ft.
Unladen weight:	3tons 15cwt (approx.)
Total:	30
Withdrawn:	1959 { April 18 / October 12

BELOW: KMB Bedford OB dual-entrance bus 4249 carries the typically austere style of bodywork standard to KMB in early post-war years. The timber framework gives five window bays between doorways. This view was taken at the Kowloon-side "Star" Ferry circa 1950 *(Fred York*

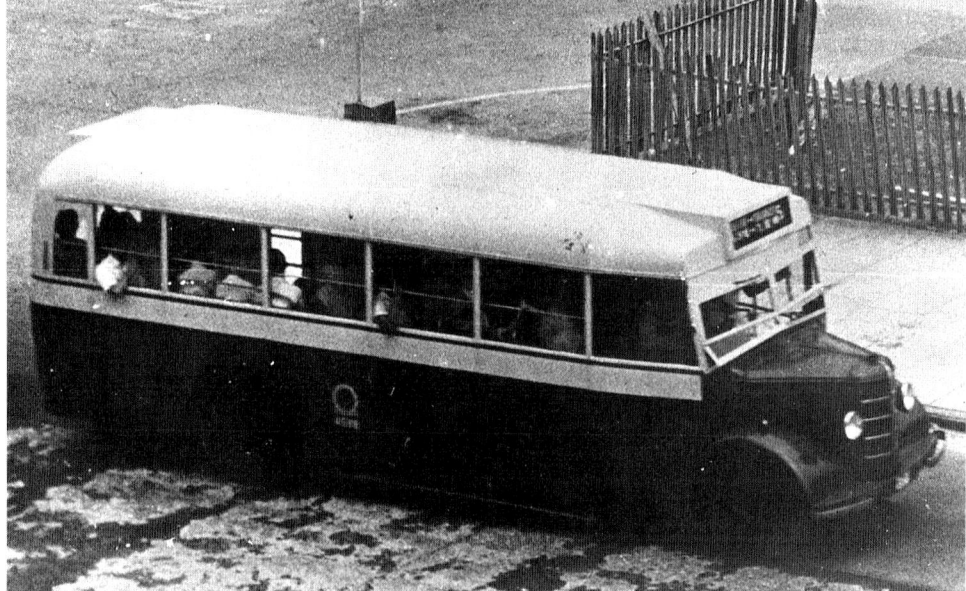

TOP: This 1952 view of Bedford OB No 4244 shows the offside of the early timber-framed body. *(Derek Parsons)*

CENTRE: This offside view of a Bedford OB passing the Kowloon European YMCA, circa 1950, shows off well the curved rear panels of the less austere version of the bodywork. *(Fred York)*

BOTTOM: No 4250 displays a slightly more refined body-style with only four bays between doorways. Evidence might suggest a metal body frame. *(Derek Parsons)*

Tilling-Stevens 'Express', Model K5LA7.

A major purchase of post-war replacement rolling-stock for busy urban routes was a batch of fifty very traditional half-cab single-deckers from Tilling-Stevens Motors (TSM) of Maidstone, deliveries of which commenced during 1947, concurrently with the first of an eventual 108 for China Motor Bus on Hong Kong Island. Tilling-Stevens was never a large volume manufacturer of bus chassis and the 108 for CMB represented the largest fleet of TSM's anywhere, with the 50 for KMB being second.

The chassis was a tandard TSM product, evolved from the 17ft wheelbase pre-war 'Express' model. By 1947 it had a 17ft 3in wheelbase and was better known as Model K5LA7. Tilling-Stevens fitted Gardner 5LW engines of 95bhp and David Brown 5-speed non-synchromesh gearboxes. The controls were very sturdy and basic for hard use under harsh operating conditions. The front dash and bonnet assembly was factory fitted by the manufacturer and the unpainted bonnet-cover remained a feature of these buses throughout their lives.

The bodies for these buses were 27ft 6in long and were locally constructed by, or for, KMB with timber frames sandwiched between a double-skin of sheet metal and, in many respects conformed to past KMB practice in its design for single-deck buses. While mechanically similar to the 27ft 6in TSM's of CMB, it is interesting to compare bodywork differences, some of which become immediately obvious. KMB used sliding panes of frameless glass in side windows, while CMB used full-drop windows with the glass set in teak

VEHICLE SPECIFICATION: TILLING-STEVENS	
Registration numbers:	4600 series - discontinuous.
Chassis make:	Tilling-Stevens Motors Ltd.
Chassis model:	Express Model K5LA7
Engine:	Gardner 5LW, 95bhp
Gearbox:	David Brown 5-speed non-synchromesh.
Body Make:	KMB or contractor.
Body layout:	B38+13FEX,REX,G.
Date introduced:	1947-48
Length:	27ft 6in
Width:	7ft 5in
Height:	9.ft 10in
Wheelbase:	17ft 3in
Unladen weight:	?
Gross vehicle weight:	?

ABOVE: A lini-up of KMB buses at "Star" Ferry in 1949 with Bedford and Tilling-Stevens in evidence. The Tillings in particularlook sparkling new; both they and the Bedfords have hub-caps which were lost in later years. (*Bus & Coach; Oct. 1949*)

LEFT: Tilling-Stevens single-deckers in their heyday, seen here standing at Ngau Chi Wan on layover in the early 1950's. The area today is denseley packed with blocks of multi-storey factories and flats and demands the services of high-density three-axle buses on close-headway schedules. (*Fred York*

ABOVE: KMB Tilling-Stevens, 4685, stands at Kowloon- side "Star-Ferry in 1952. Note the typically unpainted bonnet cover; a feature of TSM's owned by both KMB and CMB. *(The Bishop collection*

frames and had one extra window and body pillar between the front and rear passenger entrances. CMB fitted folding doors as new while KMB opted for tubular metal sliding gates. Both companies originally fitted a single-aperture destination screen with the glass set in a timber frame. These were altered over the years to rubber mounted glass and, in CMB's case only, dual apertures with a separate route number screen. Large flat-lens headlights gave way to more modern pre-focus units, by Lucas, on KMB buses but the originals were retained by CMB to the end.

When new thses buses featured in Tilling-Stevens' advertisements and, from the illustrations therein, it can be observed that in 1948 the fleetname was sign-written in full along the waist-rail, in both English and Chinese. This was superseded circa 1950 by the Company crest incorporating the 'Nine-Dragons' of Kau Loong (Kau = nine; Leung = Dragon), from whence Kowloon takes its name.

Withdrawals of TSM's began in earnest in 1966 and all had left the passenger fleet by the end of 1967, although a few soldiered-on for a few more years as either training buses or service cars.

BELOW: Probably in its last year of service, 4674 passing the Peninsula Hotel on its way to the Kowloon-side "Star" Ferry from So Uk Village, on Route 2, in the summer of 1966. *(Mike Davis*

Commer - small single-deck buses.

During the early post-war period, 1946-50, KMB acquired a considerable fleet of normal-control Commer single-deck buses with some 28 having been identified from KMB records, unofficial sightings, photographs, etc. Of these a number of body types have been identified, the earliest possibly having been a lorry at one time.

4208. This bus was the subject of an early-post-war photograph that shows it as having a somewhat unusual style of body, built largely of timber and having a three piece windscreen, the main section of which was the centre and widest portion and gave the impression of having been adapted from a lorry cab. This bus carried chrome-plated hub-caps and embellishments on the radiator grille. Nothing further is known about this vehicle.

4612 (and perhaos others). Until the discovery, in 1992, of a previously unknown photograph, it had been assumed, from other photographic evidence, that all the Commers prior to the 'Superpoise MkII' model of 1952

VEHICLE SPECIFICATION: EARLY COMMER BUSES

Known registration numbers:	4208, 4612, 4630-3/52/8/9/60/1/2/5/7/8/71 4689/90, 4951/3-7. 4984-7.
Chassis make:	Commer
Chassis model (unconfirmcd):	Commer 'Commando' - 4612 plus three others. Commer 'Superpoise Mk I - remainder.
Engine:	Commer 4-cyl, petrol, later converted to Perkins P4, diesel.
Gearbox:	4-speed, non-synchromesh.
Body Make:	KMB contractor
Body layout:	B24F,G or, (4612 etc) B ? FEX,REX,G
Date introduced:	1948-49 - (4208 possibly earlier.)
Wheelbase:	?

had single front entrances and were of similar size. The 'new' photograph of 4612, however, reveals it to have had a dual, front and rear, doorway arrangement. Commer 'Commando' chassis numbers 17A0703-6, built in 1947, were the only chassis of the type to be sent to Hong Kong and, as these four buses were registered early in 1948, **may** have been 4612, plus three others with low numbers.

The body of 4612 shows that it had five windows between the dual doorways, indicating a longer chassis than those described below which had four-and-a-half windows but only a single entrance. Sliding metal gates were fitted but the seating capacity is unknown.

The remainder of the pre-1952 Commer fleet was later classified by KMB as Commer(a) and photographs show identifiable numbers to have had single entrances. KMB recorded them as having 25-seats.

These buses all had separate headlights on 'storks' protruding from either side of a traditional upright radiator grille, with front wings which were separate from the bonnet assembly. These were probably 'Superpoise' Mk I chassis and were registered in the (discontinuous) series 4630 to 4690 and 4951 to 4957 (others outside this series are possible).

4984-4987. Four additional Commer(a)'s were introduced in August 1949 of generally similar specification to the previous Commer(a)'s except for having the so-called 'new-look' front with curved bonnet, grille and wings, the latter incorporating integral headlights. The body itself was generally similar to the previous Commers but the rear overhang was slightly longer, allowing a fifth full-sized side window. A single entrance was provided.

LEFT: This early post-war photograph shows that Commer 4208 was fitted with a different style body to the majority of KMB's Commer Commandos, having a three-piece windscreen. It also had chrome headlights and hub-caps. *(South China Morning Post*

BELOW: An early style Commer, 4951, built in 1948/9 with vertical radiator grille and separate headlights. *(From a colour transparency by P. Holms for G. W. Morant*

ABOVE: This photograph of 4612 shows just sufficient detail to confirm that the bus had a rear doorway. Can the crew be identified by anyone? *(Kowloon Motor Bus*

LEFT: KMB Commer 4987, one of the last four of the earlier type which was fitted with the then modern bonnet and wings, similar to those on later Commer 'Superpoise' chassis. The bonnet pressing incorporated the front 'wings', with the headlights fitted flush in the space between the wings and grille. Note the absence of a small last side window when compared with later examples. *(From a colour transparency by P. Holms for G. W. Morant*

Commer 'Superpoise' Mk II.

In 1952, KMB took delivery of a further 30 Commers, this time all were of the 'Superpoise' Mk II model, according to the Company's records when examined in 1978.

The Superpoise MkII was, mechanically, a development of the earlier model, fitted from new with a Perkins P4 diesel engine and a heavier-duty 4-speed non-synchromesh gearbox and a slightly longer wheelbase.

The body resembled the last four of the earlier Commers in general appearance, having, like them, the 'new-look' front with rounded bonnet and wings incorporating the headlights. The bodywork was probably supplied by Sparshatts but, if so, it is not known if this was assembled by that firm or by KMB contractors. The longer wheelbase permitted the provision of two additional seats (total 26) and dual, front and rear, entrances which were, protected by sliding gates.

Service vehicles, AA5760, AF7493 and AG8968 (original numbers unknown) were rebuilt with angular 'home-made' bonnets, thereby rendering them unrecogniseable as Commers, even though they remained normal-control.

VEHICLE SPECIFICATION: COMMER SUPERPOISE MK II	
Registration numbers:	HK4016-25, HK4082-89, HK4120-31.
Chassis make:	Commer 'Superpoise' Mk II
Engine:	Perkins P4, diesel.
Gearbox:	4-speed, non-synchromesh.
Body Make:	Probably Sparshatts ckd kits.
Body layout:	B26FEX,REX,G
Date introduced:	1952
Wheelbase:	

BELOW: Commer Superpoise II, HK4123, at Jordan Road Ferry pier in the early 1950's. This is a dual-doorway (front & rear) bus. *P. Holms for G. W. Morant*

Dennis 'Pax'.

With the Tilling-Stevens and Bedford OB's operating on Kowloon urban services, the remaining lorry-buses were sent to the rural New Territories. To replace them in Kowloon, KMB introduced thirty Dennis 'Pax' normal-control buses fitted with the usual austere brand of bodywork. Towards the end of their career, these buses were also drafted to rural routes.

Although only seating 26, they were equipped with both front and rear entrances in order to cope with almost unlimited standing passengers, together with their live ducks and chickens! These bodies were assembled from ckd kits by Sparshatts and had their Dennis chassis numbers prefixed by 'JS' which indi-

VEHICLE SPECIFICATION: DENNIS 'PAX'	
Registration numbers	: HK4090-4114
Chassis make	: Dennis 'Pax' Model D2
Engine	: { Gardner 4LK or Perkins P6, 4730cc, 41bhp @ 2400rpm.
Gearbox	: 4-speed, non-synchromesh.
Body Make	: KMB/Sparshatts
Body layout	: B26FEX,REX,G
Date introduced	: 1952
Length	:
Width	:
Height	:
Wheelbase	:

ABOVE: HK4119, the last Kowloon Dennis 'Pax' to be built, leaving the Jordan Road Ferry terminus and going off-duty in 1966. *(Mike Davis)*

LEFT: KMB sold this Dennis 'Pax', formerly AP7892 in the service fleet, to the Hong Kong Automobile Association who used it as an office under the flyover at the Wanchai approach to the Cross-Harbour Tunnel. Seen here in 1975, it was removed as scrap in 1980. *(Mike Davis)*

cated that they were supplied by 'J. Sparshatts'. The chassis numbers, while known, cannot be related to individual vehicles. The actual chassis was 'Pax, Model D2' and nearly all had Gardner 4LK diesel engines, although a few had Perkins P6, developing 41bhp @ 2400rpm, for hilly routes.

These buses were finally withdrawn from passenger service during 1969, although the first had gone in 1968. Some survived in the 1970's as staff transport.

As part of its limited operation on Lantau Island, KMB kept five Dennis 'Pax' buses on that island and when it relinquished its right to operate there in 1965, 'Pax's', HK4090, HK4104, HK4110, HK4112 and HK4116 were sold to the newly formed 'Lantao Motor Bus Co. Ltd.' and were re-registered AG7942-46. It is not known which bus received which new registration number. All five had the 4730cc Perkins P6 diesel engine. These buses soon fell into a poor condition and were withdrawn during 1969, after failing their annual Government inspection.

ABOVE: Dennis Pax, HK 4092, was passing Kam Tin Clinic and Maternity Home in 1965 when it was on Route 18 from Un Long to Shung Shui via a circuitus rural route. *(Mike Davis)*

LEFT: A view inside Dennis Pax HK 4102, taken in 1953. The pigtail, or queue, worn by the lady on the right was a common sight in those days but is now almost a thing of the past. *(Fred York)*

Bedford 'SBO'.

Thirty Bedford Model 'SBO' single-deckers were introduced in 1956 to help the Tillings on urban single-deck routes.

The 'SBG' Model had a 5.56 litre Perkins P6 diesel engine, located over the front axle, and a Bedford four-speed manual gearbox.

Forty-seat, all-metal bodies were provided but it is unclear as to the body supplier or builder. It is known that they were not Metal Sections bodies but may have been Sparshatts products but this remains unconfirmed. In addition to the seated capacity, 14-standees were officially authorised and two entrances were provided, front and rear, both fitted with the KMB brand of metal sliding gate operated by the conductor.

These Bedford 'SBO's' were supplied with the same bulbous style of front dash, or scuttle, as fitted to contemporary military Bedford petrol-engined 'SB' buses at that time operating in the Colony and world-wide.

VEHICLE SPECIFICATION: BEDFORD 'SBO'	
Registration numbers:	HK4272, HK4283-4311
Chassis make:	Bedford Model 'SBO'
Engine:	Perkins P6, 5.6 litre, 102bhp
Gearbox:	Bedford 4-speed, non-synchromesh.
Body Make:	All-metal - of uncertain origin.
Body layout:	FB40+14FEX,REX,G.
Date introduced:	1956
Length:	27ft 6in
Width:	7ft 6in
Height:	9ft 8in
Wheelbase:	Either 17ft 2in or 18ft.

ABOVE: The typical Bedford 'scuttle' or front panels, readily identify this Bedford SB model as it speeds along Salisbury Road, past the southern part of Nathan Road, on its way to the "Star" Ferry from Kowloon Tong on Route 7, during 1966. *(Mike Davis)*

RIGHT: Despite the degree of enlargement required to reproduce this 1950's photograph of a Bedford SB, the registration number remains illegible. Judging by the condition of the vehicle, it would appear to have been quite new when this picture was taken at the Jordan Road terminus. The chrome hub-caps were still in use at this time. (From a colour transparency by P. Holms for G. W. Morant)

Seddon Mk17/M/3.

This batch of one hundred Seddons was, at the time, the largest order ever placed by KMB for single-deck buses. They were introduced in 1957 (50) and 1958 (50).

The Seddon Mk17 chassis was a very basic vehicle with a somewhat noisy Perkins P6 diesel engine, located at the front with little sound-proofing, and a 4-speed non-synchromesh gearbox which produced a sound of its own. On the plus side, this was a relatively low-floored design when compared with later Albion 'Victor' buses.

Bodywork was all-metal, supplied in ckd form by Metal Sections Ltd., for assembly by KMB or its contractors. Windows had radius corners, similar to those of contemporary bus bodies supplied by the maker to CMB, and the destination screen was flush fitted with rubber gasket mounting for the glass. When new and in good condition, these were quite distinctive looking vehicles for the region, seating 37 passengers plus 13 standees. Dual doorways were protected by tubular metal sliding gates of KMB's usual pattern.

The Seddon Mk17's went to work all over the system and were withdrawn between 1969 and 1971 - one solitary bus having been withdrawn in 1963 as a result of a bad accident.

HK4398 differed from its sister vehicles in having a somewhat odd, square-shaped body which was probably a KMB rebuild following accident damage.

VEHICLE SPECIFICATION: SEDDON MK 17	
Registration numbers:	HK4352-4451
Chassis make:	Seddon Mk 17/M/3
Engine:	Perkins P6 diesel.
Gearbox:	4-speed, non-synchromesh.
Body Make:	KMB assembled Metal Sections ckd kits.
Body layout:	FB37+13FEX,REX,G.
Date introduced:	1957/58
Length:	
Width:	
Height:	
Wheelbase:	

Withdrawn:	
1963	1
April 1969	30
October 1969	9
April 1970	12
July 1970	15
November 1970	17
March 197	16
Total	100

ABOVE: KMB Seddon Mk17, HK4374, leaving the bus stop opposite the European YMCA in Salisbury Road, in 1966 *(Mike Davis*

LEFT: Another enlargement of a background subject, this time showing a Seddon Mk17 at Jordan Road Ferry terminus. *(P. Holms for G. W. Morant*

Trojan Mini-coaches.

During the early 1960's, KMB owned a small fleet of sightseeing mini-coaches built on Trojan van chassis but it has not been possible to determine which model was supplied as Trojan produced two types of chassis; one was a normal-control, bonnetted type while the other was a full-fronted vehicle. Both types were used as buses or coaches both in the UK and the Far East. No photographs have so far been found of this small and elusive class of KMB bus.

KMB records show the Trojans as being 14-seaters used for party-hire and fitted with Perkins diesel engines. They were registered HK4452-59, on 26th February 1960 and were withdrawn six years later on 31st January 1966. Probably because of the more pressing demands of their regular bus operations, the Company allowed the coaching business to die-out. One member of KMB's staff recalled their being used for staff transport but it is unclear whether this was before or after withdrawal as 'Public Omnibuses'.

KNOWN VEHICLE DETAILS: TROJAN.	
Registration numbers:	HK4452-4459
Chassis make:	Trojan, Croydon.
Engine:	Perkins diesel.
Gearbox:	?
Body Make:	Trojan
Body layout:	B14+4F,D.
Date introduced:	1960
Withdrawn:	1966
Total built:	8

THERE ARE NO KNOWN PHOTOGRAPHS OF THIS TYPE.

Ford 'Thames Trader'

For the narrow and difficult roads in the more remote regions of the New Territories, particularly on Route 30, to Rennies Mill (now Route 90), KMB purchased ten small Ford Thames 'Trader 4D' buses with bodies seating 24, plus 10 standing. The actual make of the bodies is uncertain but it has been proved that it was not Metal Sections. There is a possibility that they were constructed from parts supplied by Wallace Harper, the local Ford dealer, who was also an importer of British Aluminium Co. bodies for private sector bus and coach operators.

The distinctive Thames 'Trader' front, or scuttle, of these buses made them easily recognisable but their shortness was their most noticable feature, being of what has more recently become known as 'midi' length. The centre entrance was fitted with KMB's usual sliding tubular metal gate.

These buses were introduced over a period of two years and during that time detail changes were made to the Ford 'scuttle' panel, particularly to the badging but KMB's own modifications included the results of almost regular front-end collisions with other road users on the single-track roads with their many blind bends. This manifested itself in the form of bashed and battered front panels and fog light that frequently pointed skyward.

All ten 'Traders' were withdrawn on 28th February 1969.

VEHICLE SPECIFICATION: FORD THAMES TRADER	
Registration numbers:	AD4713-6/87/8, AD4819-22.
Chassis make:	Ford Thames 'Trader 4D'
Engine:	Perkins P4 diesel.
Gearbox:	Ford 4-speed, nonsynchromesh.
Body make:	Uncertain - possibly KMB or British Aluminium Co. via Harper's.
Body layout:	B24+10CEX,G.
Date introduced:	1961: 4 (AD4713-16) 1962: 2 (AD4787-88) 1963: 4 (AD4819-22)

BELOW: Ford Thames 'Trader', AD4713, at Kowloon City Ferry terminus, early in the summer of 1966. Note the battered frontal appearance! These small buses operated over the very narrow and winding roads to Rennie's Mill. The dented front possibly reflects the number of occasions upon which other vehicles were encountered coming in the opposite direction. *(Mike Davis)*

ABOVE: AD4821 displays detail differences from the earlier AD4713 in having the destination and route number combined on one blind. The lower grille of AD4821 has the name 'Trader' in block capital cut-out letters, in place of the small circular badge of the earlier vehicle. *(Mike Brunning*

BELOW: This near broadside view of a Thames Trader shows well their short length that would, today, classify them as midi-buses. AD4821 sets-down near Kai Tak Airport as it nears the end of its journey from remote Rennis's Mill to Kowloon City Ferry pier. *(Kowloon Motor Bus*

Albion Victor Model VT17AL.

In March 1960, an Albion VT17AL demonstration vehicle arrived in Hong Kong for trials with KMB but it is not known whether it arrived in chassis form or was bodied prior to leaving the UK. When it eventually entered service, following licensing, on 5th February 1961, it carried a British Aluminium Co. (BACo) body and was allocated the registration number HK4502; the chassis number was 79608L. This demonstration bus was obviously a success with KMB, for a further 100 examples were ordered in the late summer of 1960.

Mechanically all 101 of these Victor VT17AL's were similar with Leyland EO.350 diesel engines and Leyland 5-speed constant-mesh gearboxes. The front axle was well forward with only a short front overhang. Following closely behind the Daimler CVG5's, it is worthy of note that, despite the latter's successful preselective gearboxes, the single-deck fleet was to continue with the manual type until 1985.

Bodywork, identical in appearance to HK4502, was prepared in England by BACo and shipped to Kowloon as ckd kits of parts. Two doorways were provided, one each behind front and rear wheel-arches and these were fitted with the usual tubular metal sliding gates. The passenger capacity varied slightly during the life of these buses, partly due to Government restrictions on standing space per

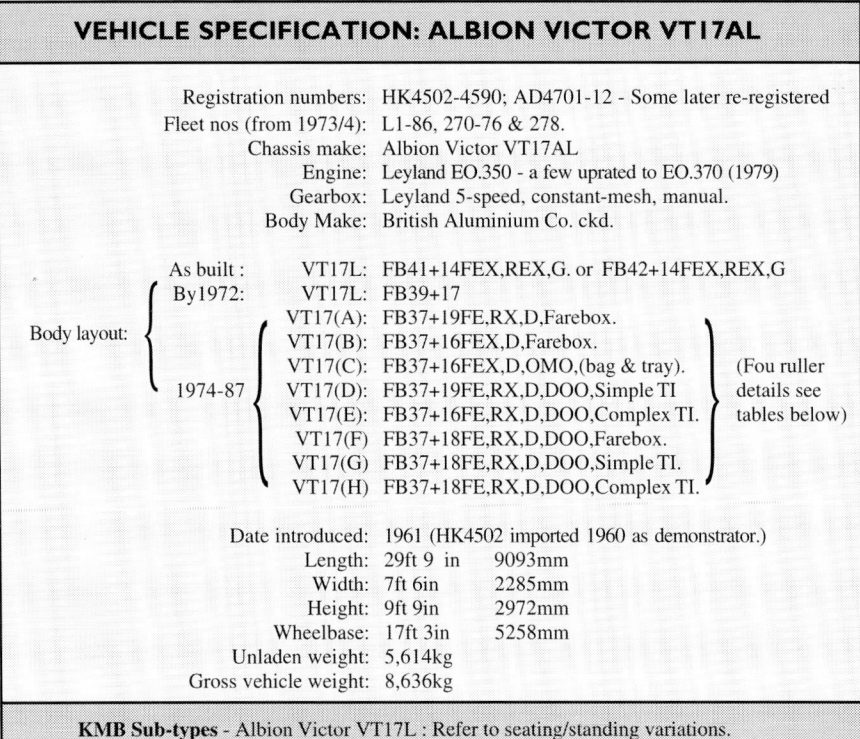

VEHICLE SPECIFICATION: ALBION VICTOR VT17AL

Registration numbers: HK4502-4590; AD4701-12 - Some later re-registered
Fleet nos (from 1973/4): L1-86, 270-76 & 278.
Chassis make: Albion Victor VT17AL
Engine: Leyland EO.350 - a few uprated to EO.370 (1979)
Gearbox: Leyland 5-speed, constant-mesh, manual.
Body Make: British Aluminium Co. ckd.

Body layout:
- As built: VT17L: FB41+14FEX,REX,G. or FB42+14FEX,REX,G
- By 1972: VT17L: FB39+17
- 1974-87:
  - VT17(A): FB37+19FE,RX,D,Farebox.
  - VT17(B): FB37+16FEX,D,Farebox.
  - VT17(C): FB37+16FEX,D,OMO,(bag & tray).
  - VT17(D): FB37+19FE,RX,D,DOO,Simple TI.
  - VT17(E): FB37+16FEX,D,RX,D,DOO,Complex TI.
  - VT17(F): FB37+18FE,RX,D,DOO,Farebox.
  - VT17(G): FB37+18FE,RX,D,DOO,Simple TI.
  - VT17(H): FB37+18FE,RX,D,DOO,Complex TI.

(For fuller details see tables below)

Date introduced: 1961 (HK4502 imported 1960 as demonstrator.)
Length: 29ft 9 in — 9093mm
Width: 7ft 6in — 2285mm
Height: 9ft 9in — 2972mm
Wheelbase: 17ft 3in — 5258mm
Unladen weight: 5,614kg
Gross vehicle weight: 8,636kg

KMB Sub-types - Albion Victor VT17L : Refer to seating/standing variations.

KMB Sub-types: ALBION VICTOR VT17AL. REFERS TO SEATING/STANDING CAPACITY

Sub-type:	KMB designation:	1961-72:	1972-74:	1974-87:
VT17L	Conductor 1961	41 or 42+14	-	-
VT17L	Conductor 1972	39+17	-	-
VT17L(A)	OMO/BP/2-door	-	37+11	37+19
VT17L(B)	OMO/BP/1-door	-	37+11	37+16
VT17L(C)	OMO/Bag&Tray	-	37+11	37+16
VT17L(D)	OMO/BP/SimpleTI/2-door	-	37+11	37+19
VT17L(E)	OMO/BP/ComplexTI/2-door	-	37+11	37+17
VT17L(F)	OMO/BP/2-door	-	37+18	37+18
VT17L(G)	OMO/BP/SimpleTI/2-door	-	37+18	37+18
VT17L(H)	OMO/BP/ComplexTI/2-door	-	37+18	37+18

KMB abbreviations:
- OMO — One Man Operated
- BP — Bell Punch 'Farebox'
- Bag & Tray — Conductor-style moneybag and tickets used by driver: money tray on the driver's left.
- TI — Ticket Issuing machine.
- Simple TI — Issues limited number of ticket values.
- Complex TI — Wide range of ticket values.

ALBION VICTOR VT17AL - FLEET ANALYSIS AT SPECIFIC DATES

	31/12/61	30/9/72	31/5/73	30/6/73	31/8/73	31/11/73	31/5/74	30/6/74	31/7/74	31/12/76	31/8/77	28/2/79	31/12/80	31/12/81	31/12/82	31/12/83	31/12/84	31/12/85	31/12/86	31/12/87
VT17L	-	101	86	62	30	12	11	1	1	-	-	-	-	-	-	-	-	-	-	-
VT17L(A)	-	-	-	24	24	23	23	23	23	24	54	56	54	37	29	27	26	21	6	4	-
VT17L(B)	-	-	-	-	32	35	40	44	43	42	3	-	-	-	-	-	-	-	-	-
VT17L(C)	-	-	-	-	-	16	15	13	12	12	*	-	-	-	-	-	-	-	-	-
VT17L(D)	-	-	-	-	-	-	-	8	8	9	25	24	20	15	10	10	9	5	5	3	-
VT17L(E)	-	-	-	-	-	-	-	-	2	3	9	11	16	14	12	8	7	5	5	3	-
VT17L(F)	-	-	-	-	-	-	-	-	-	-	1	1	**	-	-	-	-	-	-	-
VT17L(G)	-	-	-	-	-	-	-	-	-	-	1	1	**	-	-	-	-	-	-	-
VT17L(H)	-	-	-	-	-	-	-	-	-	-	1	1	3	3	3	3	#	-	-	-
SUB-TOTAL of SERVICE BUSES	101	86	86	86	86	89	89	89	90	94	94	93	69	54	48	42	31	16	10	-
TRAINING BUSES	-	15	15	15	15	12	12	12	12	4	4	7*1	7	27	NO DATA					
UNREGISTERED	-	-	-	-	-	-	-	7*2	3*2	-	-	-	-							
GRAND TOTAL	101	101	101	101	101	101	101	101	101	101	101	101	100	96	-	-	-	-	-	-

NOTES:
- * Last (C) converted to (E) 7/75.
- ** (F) & (G) converted to (H) 1/78 & 11/78 respectively.
- # 3 x (H) withdrawn 10/83.
- *1 3 x trainers returned to training fleet, unused as buses.
- *2 Buses transferred from training fleet to bus fleet.

ABOVE: Albion Victor VT17AL, HK 4528, in almost original condition, including sliding, tubular-metal passenger gates. HK 4528 was departing Jordan Road Ferry terminal when photographed in 1965 *(Mike Davis*

passenger which was altered from 152sq ins per passenger, prior to 1972, to 288sq ins in that year. It was relaxed silghtly in 1974 to 260sq ins.

Driver Only Operation.

The second reason for capacity alterations was the introduction of Driver Only Operation (DOO), then referred to as 'OMO'.

When introduced, the seating capacity was 41 or 42 but it is not on record as to which buses had how many seats, or the reason for the difference. All, however had 14 standees, making a total of 55 or 56 passengers. Sometime between 1961 and 1971, the layout was altered to 39 seats and 17 standing. KMB introduced DOO on this type of bus in May 1973 and converted 24 buses to their sub-type 'VT17AL(A)OMO/BP/2-door' by fitting Bell-Punch "Fareboxes", BP in KMB parlance, but they experienced problems with passengers trying to board by the rear exit and so evade payment. To counter this, the next 32 conversions to DOO entered service in June 1973 with the rear doorway panelled-over and classified as sub-type 'VT17AL(B)/BP/1-door' as a result. In many cases there was no window glass fitted in the former doorway, the panelling merely coming to an abrupt end at waist level but, on others, a window was provided, although in all cases the inside steps were retained.

The third type of DOO conversion involved 16 VT17's which also had the rear doorway panelled-over but, instead of having 'Fareboxes' they were fitted with a money tray on a fixed rail to the left of the driver who was provided with a conductor's money bag and tickets for use on routes with fare-stages. These were known quite simply as type 'VT17(C)OMO/Bag & Tray/1-door'. To relieve the driver from issuing tickets, KMB installed Bell punch 'Autofare' machines in conjunction with the Fareboxes and these came in two varieties. Firstly the so-called 'Simple' type and, secondly, the 'Complex' type. The Simple ticket issuing (TI) machines merely issued tickets of a single value, in multiples if required, while the 'Complex' type issued various values and were on longer routes.

Buses so equipped became either 'VT17(D)OMO/BP/SimpleTI/2-door', or 'VT17(E)OMO/BP/ComplexTI/2-door'

From their introduction until 1974, all the sub-types from (A) to (E) had 37 seats, plus 18 standees but, with the change in regulations for standing passengers, the standee figure was altered as shown in the accompanying table.

At the time of conversion to DOO, most VT17AL's were fitted with power operated folding doors, although a few vehicles retained sliding gates well into the DOO era, especially in the urban New Territories around Tsuen Wan and Kwai Chung.

Rebodied buses.

During 1976, three Albion VT17AL's were completely rebodied by KMB in their Lai Chi Kok workshops and reclassified VT17AL(F), (G) and (H); (F) having Farebox only, (G) having SimpleTI and (H) having ComplexTI. In 1983, types (F) & (G) were converted to (H). All three had 37-seats and 18-standees and the buses involved were (F) HK4516; (G) HK4507 and (H) HK4503. To some eyes, these rebodied VT17's had a 'squared-up' appearance.

To training and back!

Fifteen VT17AL's were transferred to KMB's driving school between 1970 and 1972 and were re-registered in Hong Kong's 'Private Omnibus' series and, as such, were given black letters and numbers on a white plate; at the time 'Public Omnibuses' had white figures on a red background (in the days before white/yellow reflective number plates of the British type). These new numbers were in the series AE, AS, AW, and AX. By mid-1975, many were back in the passenger fleet where they retained their 'new' registration numbers but the plates themselves reverted to 'Public Omnibus' colours of white figures on a red background. During the time that they were with the driver training unit, the KMB bus fleet had been fleet numbered and, due to their absence from bus duties, no fleet numbers had been allocated to them - the series for VT17AL's having finished at L86 - 'L' for Leyland, the parent company of Albion; 'A' having been used for the 'AEC' fleet. The VT17AL's in bus service were fleet numbered L1-86 in registration number order and other Albion/Leyland buses from L87 to L269, so those reinstated from the driving school were eventually allocated fleet numbers from L270 to L276 and L278, in the order that they returned to service (L277 was a later Model VT23L that had also returned to the bus fleet after numbering). The remaining seven VT17AL's were retained in the training fleet until finally scrapped, a date lost from KMB records, although a further three had been set aside for reuse as buses but were returned to the training fleet in 1977 without being numbered.

The latter days.

In order to improve their performance on the steeply graded Lam Kam Valley Road in the New Territories, a few of these VT17AL Victors were re-powered using Leyland EO.370 engines, rated at 106bhp, taken from scrapped Victor VT23L's. It is uncertain which and how many vehicles were involved. The reason for this elaborate engineering exercise was that the VT17AL's were only 7ft 6in wide as compared with the 8ft of the slightly younger VT23L's and thus met Government criteria for vehicles operating this narrow road until it could be widened sufficiently to allow wider vehicles to pass.

By the close of 1982, the VT17AL fleet had been reduced to less than half of its original total, with no less than 20 passing to the driving school, including L273 and L275 which were two of those previously used on training duties form 1971 to 1974, when their registration numbers were changed. By 1982, however, the licensing authorities had simplified the regulations and no further re-registration was necessary; the only change was to repaint the number plates back to 'private' colours.

The VT17AL fleet was pruned by substantial numbers during the early 1980's as routes were either converted to double-deck operation or larger single-deckers were transferred-in to replace them. As this type was the narrowest in the fleet at 7ft 6in wide, a few outlasted the later and wider VT23L single-deckers, the last three, L37, L42 and L 48, being withdrawn on 26th November 1987.

BELOW: Victor VT17, HK4544 approaching Jordan Road Ferry in 1974, having been temporarily modified to have a single-doorway at the front in order to facilitate the introduction of driver-only-operation. The former rear-doorway had been panelled-over but at window level an open, unglazed, space was left. Power-doors replaced the original gates at the same time. *(Mike Davis)*

BELOW: When photographed leaving Kwun Tong Ferry terminus in April 1975, Albion Victor VT17, HK4518, by now fleet-numbered L17, was farebox-operated, with passenger-flow instituted - ie front entrance and rear exit. This view clearly shows the power operated doors. *(Mike Davis)*

ABOVE: AX3283 (ex-HK4589) with white registration plates (at that time, denoting a privately registered vehicle - pre-dating reflective plates) during its time with the KMB driving scghool. Some, but not AX3283, were reinstated as buses, retaining their 'new' registration number. One or two even returned to the driving school for a second time prior to finally being withdrawn for scrap in 1982-4. Seen here returning to the yard at Kwai Chung in 1975. *(Mike Davis)*

BELOW: L270 was was originally one of the Victor VT17's transferred to the driving school in 1970, at which time it received its second registration number, AS6125, vice AD4711. It is seen here in January 1978 at Tai Kok Tsui Ferry having been reconditioned and reinstated to bus duties, retaining its 'new' number which, in the style of the day, was repainted to have white characters on a red plate to denote a Public Omnibus. *(Mike Davis)*

ABOVE: L54 and L57 in July 1983 at Lau Fau Shan, on Routes 57 and 58. Much rebuilt for continued service, these buses, the narrowest single-deckers in the fleet, were retained, at the request of the Government, for use on certain routes which traversed narrow roads. As these routes were also hilly, Leyland 0.370 engines were transferref from scrapped Victor VT23's which had been scrapped. *(Mike Davis)*

BELOW: L20 was reconditioned by KMB on its original British Aluminium Co. framework. It was seen here in January 1978 whilst on lay-over at Tai Kok Tsui Ferry Pier. *(Mike Davis)*

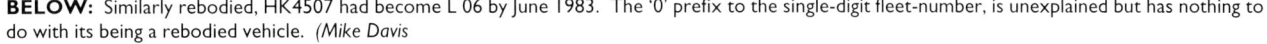

ABOVE: HK4503, by that time L2, had been rebodied and mechanically reconditioned when seen here in November 1976 in the northern part of the New Territories, en route for Shung Shui from Yuen Long. *(Derek Lucas*

BELOW: Similarly rebodied, HK4507 had become L 06 by June 1983. The '0' prefix to the single-digit fleet-number, is unexplained but has nothing to do with its being a rebodied vehicle. *(Mike Davis*

Albion Victor VT23L

The fleet of Albion Victor VT23L's was introduced in 1963 and was, mechanically, a development of the 1961 VT17AL model, with Leyland EO.370 engines of 106bhp and Leyland GB271, 5-speed manual gearboxes. The EO.370 engines gave improved performance, useful on some of the steeper routes in the New Territories, particularly Route 'TWSK', the military built pass over Mount Tai Mo Shan, over 2,500ft up the more than 3,000ft high mountain.

Although of the same overall length as the earlier VT17 model, the VT23 had a shorter wheelbase with set-back front axle. KMB was, in fact, the only customer for the VT23L model.

Instead of BACo bodywork, the VT23's were bodied by KMB using Metal Sections ckd kits which utilized steel frames with aluminium panels. The first batch, AD7000-55, arrived in 1963, while the remainder, AD7056-7999, entered service the following year. The Metsec bodies seated 41-passengers and had dual doorways placed immediately behind the front and rear wheel-arches; they were fitted with tubular metal sliding gates, manually operated by the conductor.

Driver Only Operation.

Upon the introduction of driver-only-operation, power-operated folding doors replaced the gates front and rear, exceptions being the (A) and (B) sub-types which were converted for DOO by removing the rear doorway and panelling the space up to waist-rail level and providing a glazed window above - but not in all cases, when an unglazed space sufficed!

The sub-types are tabulated below but, briefly, they were as follows: (A) sub-type with Farebox and 1-doorway for DOO; type (B) was fitted with a money-tray beside the driver as described under the VT17L heading; (C) with Farebox and Autofare 'Complex' ticket issuing machine; (D) similar but with a 'Simple' ticket machine; (E) was a 2-door bus with Farebox and (F) was represented by a single vehicle for less than a month in January/February 1975. A late re-classification of five (A) type and 22 (E) type to what amounted to a 'sub-sub' type took place in 1981 when these 27 buses were classified '(E)37-seats'.

With the introduction of fleet numbering in 1974, the VT23L's became L87 to L185, one, AD7074, having been prematurely withdrawn due to bad accident damage in June 1973. This was rebuilt and rebodied and returned to service the following year as out of sequence L277 and re-registered BH3001, despite having once been earmarked for conversion to a lorry.

Rebodying

From 1976, a number of these buses were rebodied by KMB themselves, using the same design as that produced for the VT17L's and BH3001; a design not unlike the original in basic layout but of a tidier and more modern appearance. By April 1978, ten Albion VT23L's had received a new type of body; AD7022/46/61/66/70/72/80/89/92 and BH3001.

Withdrawals commenced in 1979 and by December 1982 only 38 members of the class were in passenger service. BH3001 eventually became a goods vehicle in January 1982 and the last VT23L, AD7022, was withdrawn in March 1984.

The decline of the VT23L fleet-strength can be ascertained by reference to the accompanying table. It is believed that one or more of the class were used by the driving school but this could have on the basis of a temporary loan or loans.

VEHICLE SPECIFICATION: ALBION VICTOR VT23L

Registration numbers:	AD7000-99 (BH3001 -ex-AD7074).
Fleet nos (from 1973/4):	L87-185 & L277.
Chassis make:	Albion Victor VT23L
Engine:	Leyland EO.370, 6075cc, 106bhp @ 2200rpm.
Gearbox:	Leyland GB271 - 5-speed, constant-mesh, manual.
Body make:	Metal Sections Ltd. ckd - assembled by KMB
Body layout:	As built: FB41+14FEX,REX,G.
	VT23(A): FB39+14FEX,D,DOO,Farebox.
	VT23(C): FB37+20FE,RX,D,DOO,Complex TI.
	VT23(D): FB37+21FE,RX,D,DOO,Simple TI.
	VT23(E): FB39+19FE,RX,D,DOO,Farebox.
	(see table below for fuller details).
Date introduced:	1963/64.
Rebodying commenced:	1976
Total:	100
Length:	30ft 9144mm
Width:	7ft 10in 2387mm
Height:	9ft 10in 2997mm
Wheelbase:	16ft 1in 5901mm
Unladen weight:	5,639kg
Gross vehicle weight:	8,636kg

KMB Sub-types - Albion Victor VT23L : Refer to seating/standing variations - see separate tables..

KMB Sub-types: ALBION VICTOR VT23L - REFERS TO SEATING/STANDING VARIATIONS

Sub-type:	KMB designation:	1963-72:	1972-74:	1974-87:
VT23L	Conductor 1961	41+14	-	-
VT23L(A)	OMO/BP/1-door	-	39+12	39+14
VT23L(B)	OMO/Tray&Bag/1-door	-	36+12*	-
VT23L(C)	OMO/BP/ComplexTI/2-door	-	37+12	37+20
VT23L(D)	OMO/BP/SimpleTI/2-door	-	37+12	37+21
VT23L(E)	OMO/BP/2-door	-	-	39+19
VT23L(F)	OMO/BP/2-door	-	-	39+11 (Jan/Feb 75 only)
VT23L(E)37-seats	OMO/BP/SimpleTI/2-door	-	37+18	37+18

KMB abbreviations:
- OMO — One Man Operated
- BP — Bell Punch 'Farebox'
- Bag & Tray — Conductor-style money bag and tickets used by driver: money tray on driver's left.
- TI — Ticket Issuing machine
- Simple TI — Issues limited number of ticket values.
- Complex TI — Wide range of values.

ALBION VICTOR VT23L - FLEET ANALYSIS AT SPECIFIC DATES.

	31/12/64	31/12/72	31/6/73	31/7/73	31/5/74	30/6/74	31/8/74	31/1/75	28/2/75	31/7/75	31/12/75	31/12/76	31/12/77	31/12/78	31/12/79	31/12/80	31/12/81	31/12/82	31/12/83	31/12/84
VT23L	100	100	99*	55	22	21	20	18	-	-	-	-	-	-	-	-	-	-	-	-
VT23L(A)	-	-	-	29	61	56	53	52	43	-	-	-	-	-	-	-	-	-	-	-
VT23L(B)	-	-	-	15	10	10	10	7	6	-	-	-	-	-	-	-	-	-	-	-
VT23L(C)	-	-	-	-	6	7	7	13	21	41	41	44	45	45	41	41	37	18	7	-
VT23L(D)	-	-	-	-	-	5	10**	9	10	8	8	7	7	11	11	11	8	5	-	-
VT23L(E)	-	-	-	-	-	-	-	-	2	24	26	28	30	31	32	33	8	1	-	-
VT23L(F)	-	-	-	-	-	-	-	1	-	-	-	-	-	-	-	-	-	-	-	-
VT23L(E37)	-	-	-	-	-	-	-	-	-	-	-	-	-	-	-	-	27	14	4	-
TOTAL	100	100	99	99	99	99	100	100	100	100	100	100	100	100	95	93	83	38	11	-

NOTES:
* AD7074 accident damaged
** AD7074 returned to stock for reconditioning as (D) - re-registered BH3001.

ABOVE: Albion Victor VT23L, AD7049, as built in 1963 with sliding metal gates, open in this official view. Note the original one-piece destination blind with no provision for a separate route number. *(KMB)*

BELOW: AD7087 stopping at Kam Tin Clinic, near the Sek Kong British Army Camp and airfield, in the New territories in 1965. The woman approaching the bus is a Hakka, dressed in the traditional peasant clothes of the time and region. *(Mike Davis)*

ABOVE: Albion Victor VT23L with fog-light and sliding gates - some early versions of this type were not fitted with fog-lights, most necessary on Route TWSK, the route over Tai Mo Shan which, at 2,500ft, attracts mist and low cloud in spring. AD7038 leaves Jordan Rd. in 1965. (*NB:* TWSK denotes Tsuen Wan/Sek Kong, the military term for the mountain pass built by the Royal Engineers, for military purposes, but handed over to the civilian authorities in the early 1960's.) *(Mike Davis)*

BELOW: L104 (AD7017) leaving Jordan Road Ferry in May 1975, after having been fitted with power operated passenger doors. This was an early version never fitted with a foglight. *(Mike Davis)*

ABOVE: In the early days of single-deck OMO (DOO) with farebox operation, many buses had their rear doorways panelled-over to prevent possible fare-dodging by persons boarding at the rear. In this case the window-level space was glazed but on other examples this was left as an open space. L168 (AD7082) still sported a single-doorway when photographed at Tai Kok Tsui on Sunday 8th January 1978, by which time most others had reverted to dual doorways. *(Mike Davis)*

BELOW: L103 (AD7016 in Jordan Road without a foglight in 1975. Note the simplified livery, omitting the cream bands with green lining. This bus was fitted with a farebox and operated as a passenger-flow vehicle with the rear doorway reinstated. *(Mike Davis)*

ABOVE: L183 at Lai Chi Kok Bridge terminus in January 1978. This bus had two functioning doorways and KMB had added their own style of replacement radiator grille. *(Mike Davis)*

BELOW: L178 (AD7092) was one of the Albion Victor VT23L's rebodied by KMB in 1977/78 to a design similar to that used on VT17 Victors and the Chieftains. At Tai Kok Tsui, January 1978. *(Mike Davis)*

ABOVE: L87 (AD7000, was the first of its class and was rebodied by KMB to the appearance shown in this 1980 view, with the distinctively upright front of the rebuilds. *(John Shearman*

CENTRE: Two unidentified VT23 Victors undergoing rebodying at Lai Chi Kok in January 1978. *(Mike Davis*

BELOW: For comparison: A line of Victors at Tsuen Wan Ferry. On the left of the photo is L178, a VT23I while the other two are the older, Baco bodied, VT17AL type. *(Ian Glass*

Albion Chieftain CH13AXL.

Introduced in 1965, these thirty-five little Albion Chieftains replaced the normal-control Dennis, Commer and Thames vehicles on sparsely trafficed or physically difficult routes in the New Territories.

The 24ft 10 in (7558mm) long Albion Chieftain 13AXL was a 13ft 5in (4088mm) wheelbase passenger version of a popular truck chassis which was produced exclusively for KMB. They were fitted with the same Leyland 0.370 engine as the VT23L's, giving them a relatively high power-to-weight ratio, having 106bhp available to pull a maximum gross vehicle weight of about 8tons (8,636kg). The gearbox was the same GB271 non-synchromesh manual type and it is believed that the front and rear axles were similar to those of the Victor. The chassis was nominally based on the Albion Chieftain truck model but with so many Victor fittings that the Chieftain name became almost meaningless. As an aside, KMB also purchased a number of goods vehicles based on the same chassis and, in addition, quite early on, CH13AXL bus AD7202 was withdrawn and converted into a lorry and re-registered AV5269. The cab of the purpose-built trucks was the standard 'LAD' type. (Leyland, Albion, Dodge - all used the same cab fabrication despite the Rootes Group not being part of Leyland, as was Albion.) The converted bus, AD7202, retained its bus-style cab area, as did most of those buses converted to goods vehicles in the twilight of their lives.

The bodywork of the CH13AXL buses was basically a shortened version of that provided for the Victor VT23L's, being one bay shorter within the wheelbase. As built, they had the standard KMB sliding gate arrangement to their dual, forward and rear doorways. All but five were converted for OMO/DOO by the addition of Bell Punch 'Fareboxes', either by themselves or in conjunction with 'Complex' or 'Simple' 'Autofare' ticket-issuing machines and with differing combinations of seating/standing capacity. The differences, and KMB's sub-type code, can most easily be understood by reference to the tables as there were numerous changes during the lives of these vehicles.

There were five exceptions to the standard doorway arrangement that were fitted for 'Bag & Tray' driver-only-operation (OMO to KMB - and most others - in those days) and were provided with Macau-style inward-swinging hinged gates and only one entrance/exit gateway. These five buses also had steeply cutaway rear skirt panels. As 'Bag & Tray' buses, KMB referred to them in-house as 13AXL(OMO/A) when they were unique in operating without a conductor but, when Bell punch 'Fareboxes' were adopted generally, they became '13AXL(A)OMO/BP/Swing-gate'. Used

VEHICLE SPECIFICATION: ALBION CHIEFTAIN CH13AXL.

Registration numbers:	AD7202-36
Fleet nos (from 1974):	L186-219 (AD7202 became goods vehicle AV5269 pre-numbering)
Chassis make:	Albion Chieftain CH13AXL(P) - [P = Passenger.]
Engine:	Leyland EO.370, 106bhp @ 2200rpm.
Gearbox:	Leyland GB271 - 5-speed, constant-mesh, manual.
Body make:	Metal Sections Ltd. ckd assembled ny KMB
Body layout:	As built: FB30+18FEX,REX,G.
	13AXL(A): FB27+25FEX,G. (Swing gate).
	13AXL(B): FB27+17FEX,D,DOO,Farebox.
	13AXL(C): FB27+9FE,RX,D,DOO,Farebox.
	13AXL(D): FB27+14FEX,D,DOO,ComplexTI.
	13AXL(E): FB27+16FE,RX,DOO,Simple TI
	13AXL(F): FB27+14FE,RX,D,DOO,Complex TI.
	13AXL(G): FB27+17FE,RX,D,DOO,Farebox.
Date introduced:	1965.
Rebodying commenced:	1976
Total:	35
Length:	24ft 10^{1}/2in 9144mm
Width:	7ft 6in 2286mm
Height:	9ft 10in 2997mm
Wheelbase:	13ft 4^{5}/8in 4088mm
Unladen weight:	5,055kg
Gross vehicle weight:	8,636kg

NOTE: This CH13AXL(P) chassis was a special specification for KMB and remained unique to them.

KMB Sub-types: ALBION CHIEFTAIN CH13AXL. REFERS TO SEATING/STANDING VARIATIONS

Sub-type:	KMB designation:	1965-72:	1972-74:	1974-87:	KMB abbreviations:	
CH13AXL	Basic type	30+18	-	-	OMO	One Man Operated
CH13AXL(OMO/A)	'Bag&Tray'	?				
CH13AXL(A)	OMO/BP/Swing-gate/1-door	27+25	-	-	BP	Bell Punch 'Farebox'
CH13AXL(B)	OMO/BP/1-door	-	27+9	27+17	Bag & Tray	Conductor-style money bag and
CH13AXL(C)	OMO/BP/2-door	-	27+9	-		tickets used by driver: money
CH13AXL(D)	OMO/BP/ComplexTI/1-door	-	29+9	27+14		tray on driver's left.
CH13AXL(E)	OMO/BP/SimpleTI/2-door	-	27+9	27+16	TI	Ticket Issuing machine
CH13AXL(F)	OMO/BP/ComplexTI/2-door	-	-	27+17	Simple	Issues limited number of
CH13AXL(G)	OMO/BP/2-door	-	-	27+17		ticket values.
					Complex	Wide range of values.

ALBION CHIEFTAIN CH13AXL - FLEET ANALYSIS AT SPECIFIC DATES.

	31/1/65	31/12/71	31/7/73	31/8/73	30/11/73	31/7/74	31/1/75	31/3/75	31/8/75	31/3/76	31/8/76	31/12/79	30/9/81	31/12/82	31/12/83	31/12/86	30/11/87	30/1/88
CH13AXL	30	29*	18	7	-	-	-	-	-	-	-	-	-	-	-	-	-	-
CH13AXL(A)**	5	5	5	5	5	5	5	-	-	-	-	-	-	-	-	-	-	-
CH13AXL(B)	-	-	11	11	13	9	9	13	12	12	11	-	-	-	-	-	-	-
CH13AXL(C)	-	-	-	11	8	-	-	-	-	-	-	-	-	-	-	-	-	-
CH13AXL(D)	-	-	-	-	8	18	16	16	16	-	-	-	-	-	-	-	-	-
CH13AXL(E)	-	-	-	-	-	2	4	5	5	5	5	4	3	3	1	1	-	-
CH13AXL(F)	-	-	-	-	-	-	-	-	1	17	17	16	14	8	5	3	1	-
CH13AXL(G)	-	-	-	-	-	-	-	-	-	-	1	14	12	7	6	4	2	-
TOTAL	35	34	34	34	34	34	34	34	34	34	34	34	29	18	12	8	3	-

NOTES: * AD7202 converted to truck 1971

ABOVE: Comment on the standards of bus parking is beyond the scope of this work. Albion Chieftain L197 (AD7214) was awaiting its next duty as a service extra on trunk Route 70, normally the preserve of second-hand double-deck double-deckers. Such small buses on this service were the result of unreliability problems with the second-hand fleet. *(Mike Davis*

on the lightly used Route 30 (later renumbered 90) from Kowloon City Ferry to the isolated settlement known as Rennie's Mill, on the Sai Keung peninsular, they later received a folding door, devoid of any glazing, and were re-classified as 13AXL(A)OMO/BP.

Fleet numbering.

When the bus fleet was numbered in 1974, the Chieftain's series ran from L186 to L219. As seen above, the premier member of the type, AD7202, had become a goods vehicle in 1971, at the time it was re-registered and thus it was that AD7203 took the lowest fleet-number, L186. (AV5269 remained active into the 1980's when it was joined in the service vehicle fleet by further examples of the type, this time life-expired buses at the end rather than the beginning of their existence.)

Rebodying.

Commencing in 1976, a eleven Chieftains were rebodied to the same basic pattern as used for the VT23L buses similarly treated. Completely new frames were erected at Lai Chi Kok depot and very good, basic, utilitarian body was produced which was plain but functional.

Normally, these buses were allocated to routes where longer types would have been unacceptable but road improvements, or substitution by 'Public Light Bus' services, usually of the green-waistband, regular, licensed service, 'Maxicab' type, reduced the need for such small buses and they were phased-out to the extent that only eight remained by the end of 1986. They had all gone fifteen months later.

RIGHT: The purpose-built goods version of the KMB Chieftain had the so-called 'LAD' cab common to Leyland/Commer/Dodge. *(Mike Davis)*

BELOW: AD7224 was one of the single-doorway buses used on the 90 to Rennie's Mill, along narrow rural roads across some hilly country. Compare the battered panels with the Thames Trader opn page 32. With the demise of the Chieftains, this route has been in the hands of air-conditioned Toyota Coasters. *(John Shearman*

ABOVE: L217 (AD7234). The short-wheelbase version of the Victor was the 24ft 6in (7.46m) Albion Chieftain 13AXL, a passenger version of a popular truck chassis produced exclusively for KMB. L217 was photographed in Yuen Long (Un Long) Main Road in November 1976. *(Derek Lucas*

BELOW: Many Albion buses were rebodied by KMB during 1977/78, including eleven CH13AXL Chieftains. Here an unidentified example takes shape in the old Lai Chi Kok multi storey depot and workshop in January 1978. *(Mike Davis)*

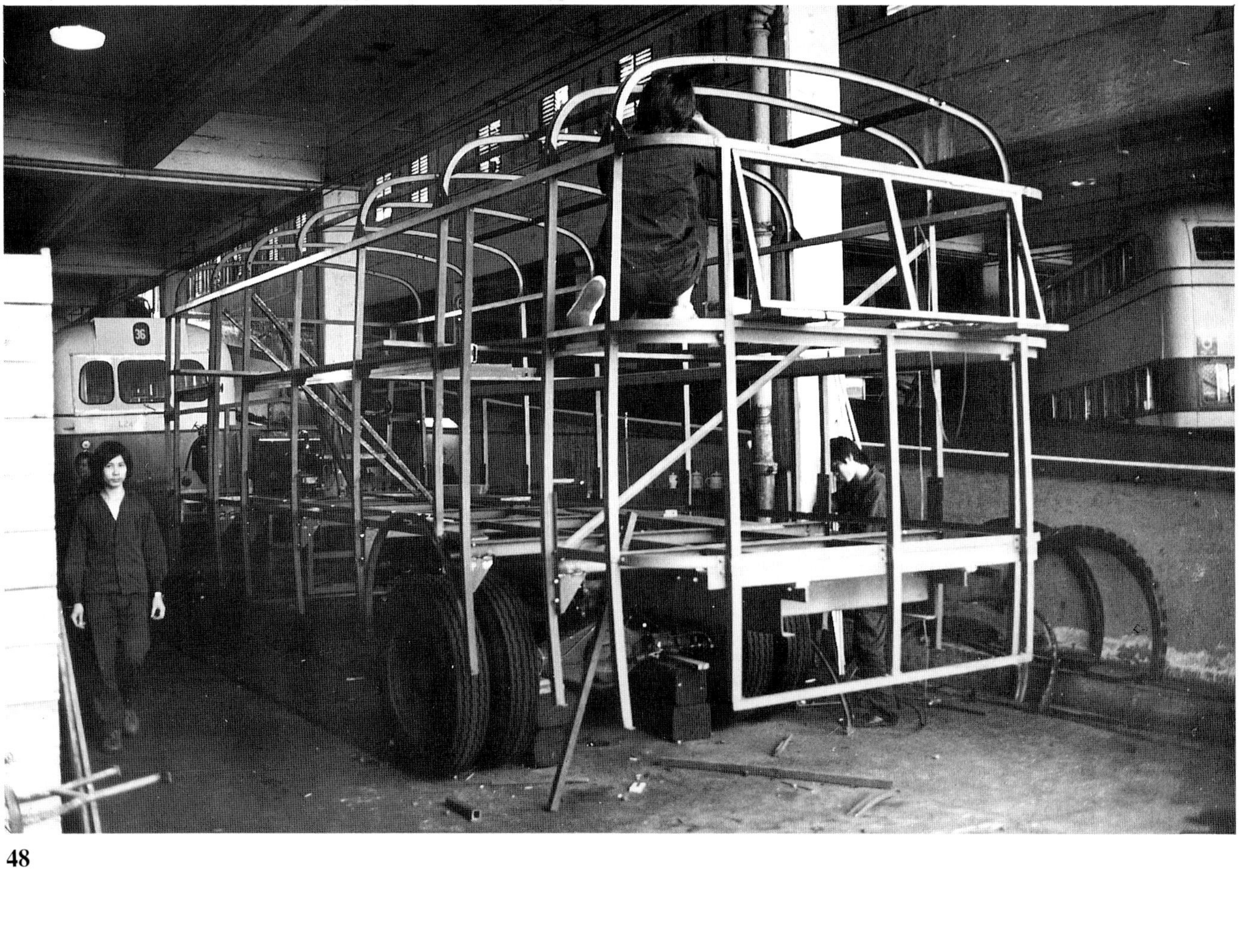

ABOVE & BELOW: Rebodied Chieftain L201 at Luk Keng terminus of route 69K in the northern New Territories in May 1984. The bus displays the then newly introduced KMB insignia. The character preceeding the initial letters 'KMB' is a stylized version of the Chinese character ' ' for 9 (pronounced 'gau'). 'Gau Leung' is the Cantonese for 'Kowloon' and means 'Nine Dragons'. The dragons are represented by the hills of Kowloon so KMB is the Nine Dragons Bus Company, or Gau Bar (Nine Bus) for short The KMB built bodywork was a somewhat angular replacement for the original Metal Sections version and was similar in style to the replacement bodies on the VT17 and VT23 Victors. *(Mike Davis)*

Albion Viking EVK41XL

During October 1970, KMB introduced the first five of what was to become the first 50 of a fleet of Leyland/Albion 'Viking' single-deckers of various marques which would eventually number 184 vehicles, including 100 coaches. The first fifty, however, were to be the longest having a 5.296 metre wheelbase and they were buses. The Viking EVK41XL was a longer, tougher, development of the tried and tested 'Victor' chassis and had a slightly lower frame-height, giving shallower steps and a lower floor. The engine was the Leyland EO.401 and the gearbox a five-speed constant-mesh manual unit, also by Leyland.. As in the Victor, the engine was mounted vertically at the front, beside the driver and ahead of the front-axle. Introduction of the EVK41XL's continued until August 1971 when the final vehicle entered service, a rate much slower than KMB would have liked. They were registered AR7701-50 and later received the fleet numbers L220-269.

The Metal Sections body design was almost 34ft (10.286m) long and followed the established practice of having two doorways, one behind the front wheels and the other behind the rear wheels. These were the first KMB single-deck buses to be built new with folding doors. Although the Viking chassis was suitable for a truly 'front' entrance, beside the driver as opposed to 'forward', the opportunity was missed and quite early in their service lives this became a problem with the conversion to OMO/DOO as the length of the front overhang placed the driver well away from the entrance. When new, these were 46-seaters with an authorised standing capacity for 38; a crush-load indeed, which could only be coped with by the employment of two conductors. This was reduced in 1972 to one seated conductor (sc) which reduced the seating capacity to 36 but increased standees to no less than 42. The lower seating capacity allowed for a circulating area inside the front of the vehicle where passengers could stand prior to paying their fare whilst the vehicle moved off, thus reducing dwell times at busy bus stops. These standing passengers could then pay on the move, pass the seated conductor's desk and either sit or stand in the middle and rear of the bus before leaving by the rear exit.

The transition from seated conductor to driver-only-operation (DOO) was achieved by placing a Farebox remotely to the driver, at the top of the entrance steps but this was never satisfactory.

The front and rear canopies were 'peaked' in the contemporary style but were somewhat shallow. KMB departed from their normal practice by painting the fronts of these buses a pale cream instead of the normal red in order to indicate that it was a seated-conductor bus and that fares could be ready before boarding. Later-on 'moneybox' stickers were added either side of the destination screen to indicate that the approaching bus was now driver-only-operated. The cream front panels made the black wire-mesh radiator grilles appear as a dirty patch.

Unfortunately, British Leyland, as it then was, found itself unable to supply the entire original order for 150 chassis sufficiently quickly and KMB, in urgent need of additional single-deck buses, turned to Seddon who were able to supply the balance of the KMB order with 100 Pennine 4's. That is another story (qv).

KMB devised new sub-type classifications for the Viking EVK41XL series, the main variations being EVK41XL(OCO) as above, followed by 'EVK41XL(OMO)BP' which had 45 seats plus 20 standees, increased to 45+25 in 1975 and which applied to all fifty buses until June/July 1981 when 22 were converted to have 'Autofare' simple ticket issuing machines connected to the Farebox. This left

VEHICLE SPECIFICATION: ALBION VIKING VK41XL	
Registration numbers:	AD7701-50
Fleet nos (from 1973/4):	L220-269
Chassis make:	Albion Viking EVK41XL
Engine:	Leyland EO.401,
Gearbox:	Leyland GB277 - 5-speed, constant-mesh, manual.
Body make:	Metal Sections Ltd. assembled by KMB.
Body layout: As built:	FB46+38FdEX,REX,D - (two conductors).
EVK41XL(OCO):	FB36+42FdE,RX,D(sc).
EVK41XL(OMO)BP:	FB45+20FdE,RX,D,DOO Farebox.
EVK41XL(Simple TI):	FB45+25FdE,RX,D,DOO Simple TI.
EVK41XL(A):	FB41+21FdE,RX,D,DOO Farebox.
Date introduced:	1970/71.
Total:	50
Length:	33ft 7in 10236mm
Width:	7ft 11 in 2413mm
Height:	9ft 8in 2946mm
Wheelbase:	17ft 4 in 5296mm
Unladen weight:	5,893kg
Gross vehicle weight:	10,486kg

BELOW: Albion Viking EVK41XL, AR7729, in almost original condition seen here in 1973 whilst on lay-over at the then remote Lie Yue Mun terminus of Route 14C   *(John Shearman*

ABOVE: Albion Viking (Leyland badge) L228 (AR7709) departing Kwun Tong Ferry Pier on Route 15 in December 1974, having been fitted with a farebox for driver-only-operation.. *(Mike Davis)*

the capacity unaltered but produced the sub-type 'EVK41XL(Simple TI)' In November 1987, an eleventh hour conversion of four buses took place to make them 'EVK41XL(A)' - 41-seats plus 21 standees. Two came from the basic Farebox model (AR7749/50) and two from the 'Simple TI' type (AR7746/48), followed in February 1988 by AR7743 to make a total of five EVK41XL(A)-type. These five buses were the last of their type to remain in service and saw out the decade of the 1980's.

BELOW: On a dull day in Spring 1984, KMB Albion Viking, sub-type EVK41XL(A), L268 (AR7749), one of only four remaining at that time, stands at Sheung Shui bus station awaiting its next duty on Route 73K to Man Kam To, the frontier crossing point for road traffic to and from China. Unauthorised passengers are required to leave the bus prior to its entering the closed-area zone, some distance short of the border proper. Note that the side and rear skirt-panels had been raised to minimise damage following allocation of this type to routes away from urban areas. *(Mike Davis)*

Seddon Pennine 4.

When KMB learned that British Leyland was unable to supply any further Viking EVK41XL's within a reasonable time, the only readily available alternative was the Seddon Pennine 4. As it turned out, this hurried purchase was ill advised but one hundred were duly shipped to Hong kong as complete vehicles to enter service during 1970, '71 and '72. They were purchased 'off the peg' as home market models, with bodywork by Seddon associate 'Pennine'. The only initial concession to the Hong Kong climate was the provision of tropical-style side windows with full-depth opening. The expensive, non-opening, wrap-around windscreen provided no ventilation for the driver and quickly proved unpopular. On the credit side, the Seddons were a most attractive change from KMB's normally austere, kit-built contemporary buses. Their all-glass 'glider' doors and 'peaked' front and rear roof canopies were of similar style to those popular with BET-group bus companies in England and Wales and gave them a clean look which was soon to be lost by KMB's later modifications.

The electrical wiring was the first casualty, being adversely affected by the high humidity in the summer months, when insulation, suitable for relatively dry British conditions, broke-down, resulting in eight buses being completely burned-out, some before the last Seddon was delivered, so that the full one-hundred were never all in service together. Another source of trouble arose from the installation of the very powerful Perkins V8-510 engine, in that the available power encouraged speeding and many more of these buses were withdrawn following the resulting accidents. Soft front springing, unsuitable for KMB's crush loading conditions and Kowloon's roads, combined with spirited driving, failed to prevent grounding of the front overhang, causing the low glass-fibre skirt panels to shatter.

In November 1971, only 18-months after the first 'Pennine 4' entered service, AR7639 was burnt-out and from one cause or another, by late 1973, only thirty-eight Seddon Pennines remained available for traffic and the entire batch was considered for early disposal.

Rebuilding.

This, however, was a time of new bus shortages and the decision was never taken.

VEHICLE SPECIFICATION: SEDDON PENNINE 4

Registration numbers:	AR7601-7700
Fleet numbers (from 1973/4):	S1-93 - **not** in registration number order after S64
Chassis make:	Seddon Pennine 4.
Engine:	Perkins V8-510; one converted to Gardner 6LX.
Gearbox:	5-speed, manual; the 6LX was coupled to an SCG semi-automatic gearbox.
Body Make:	Pennine - a Seddon associated company.

Body layout: (see table below.)
- as introduced: B47+42FEX,CEX,D.
- Pennine(OCO): B40+45FE,CX,D(sc)
- Pennine (OMO)BP: B47+23FE,CX,D,DOO,Farebox - Later 47+25 (see table)
- Pennine(A)(OMO)BP/ComplexTI: B45+25FE,CX,D,DOO,Complex TI
- Pennine (B)(OMO)BP: B52+20FE,CX,D,DOO,Farebox

Date introduced:	1970-72
Length:	32ft - 9750mm (nominal)
Width:	8ft 2½in - 2500mm
Height:	?
Wheelbase:	?
Unladen weight:	?
Gross vehicle weight:	?

KMB sub-types: SEDDON PENNINE 4.

KMB Sub-type:	Capacity: 1970-74	from 1974
Pennine	47+42	-
Pennine (OCO)	40+45	-
Pennine (OMO)BP	45+23	47+25
Pennine(A)(OMO)BP/ComplexTI	-	45+25
Pennine(B)(OMO)BP	-	52+20 (1-only)

BELOW: Seddon Pennine 4, AR7629, in almost original condition, complete with curved windscreens, Seddon scuttle and grille. It is seen here in its conductor operated days, prior to being fitted with Autofare Farebox for DOO. *(Lyndon Rees*

ABOVE: Seddon Pennine 4, AR7655, after fleet-numbering had made it S66. Despite many modifications, including a new KMB-style grille, it retained its curved windscreen in this 1975 view. *(Mike Davis)*

BELOW: S76 (AR7668) on the roof of Lai Chi kok multi-storey depot after being refurbished and immediately before its return to service in March 1975. Note the completely new KMB-style windscreen and grille. *(Mike Davis)*

ABOVE: S51 (AR7680) with small, square, KMB-standard window units replacing the original, larger, Pennine type which had radius corners. Seen here approaching Tai Kok Tsui Ferry on 8th January 1978. *(Mike Davis*

BELOW: The Seddon converted to coach configuration and then reconverted to bus layout without ever entering service as a coach was AR7615, destined to become S8. Seen here in 1976 after it had returned to service - as a bus but retaining some of its coach features, such as the psuedo-AEC grille and twin headlights. *(Martin Weyell*

ABOVE LEFT and CENTRE: Two rear views of Seddon Pennine 4s, showing the original, AR7630 above, and with KMB-style rear windows, AR766 centre. *(Mike Davis)*

BOTTOM LEFT: A front view of the would-be coach, AR7615 (S8) whilst at work on the 35A in 1976. The route number box above the front doorway can be clearly seen. *(Derek Lucas)*

LEFT: Many Seddon Pennine 4's fell victim to fires and were complete write-offs. AR7648, the remains of which are seen here, was scrapped, following fire-damage in 1973. Seen here in the Kwai Chung parking-yard on the site of which the multi-storey depot of the same name was built. *(Mike Davis*

LEFT: All that became of a proposed conversion of a Seddon single-decker into a double-decker was this stripped-down chassis on the roof of Lai Chi Kok depot in January 1978. *(Mike Davis*

Instead, the engineer's staff were given the task of finding ways of making the Seddons suitable for further service. It should be noted that, prior to this instruction, one example, AR7639, the first to be withdrawn, had undergone major rebuilding to become a most unusual looking service car. The driving position was relocated over the set-back front-axle, with the engine remaining ahead of the axle, covered by a short, snout-like bonnet. This vehicle was re-registered BE7470, in the Goods Vehicle series. The curved windscreen was retained in a new set-back position.

One other rebuilding took place but it was not such a drastic operation as with service car BE7470. S8, AR7615, was extensively rebuilt during 1975 when it was reconstructed as a C36FEX,D *coach* with its centre exit removed, high-back seats fitted, twin headlights, an AEC-style radiator grille from a Regent V and front and rear canopy peaks of the shallower type as fitted to the Viking EVK41XL series buses. The purpose of this conversion is unclear as the vehicle was returned to service as a bus, with bus seats and its centre exit reinstated, plus the addition of a route number blind-box over the front entrance doorway. All the additional external trim was retained.

Modifications.

Modifications were made to try to prevent

BELOW: A fitter attends to the nearside mirror of S57 (AR7690), as it lays-over between duties at Jordan Road Ferry in 1975. The original windows were still fitted, but many buses of this type had KMB's small, angular, units fitted as replacements which did little for their looks. *(Mike Davis*

grounding and included heavier-duty front springs and, in order to reduce speeding, the Perkins engines were derated by 10% which cut down accident damage. New front panels of aluminium sheet were fitted in place of the moulded glass-fibre sections and were similar to those of the Viking EVK41XL's, except for the addition of a slatted metal grille over the radiator air-intake. The rear panels were treated similarly, where necessary and most of the curved windscreens were replaced by flat glass with opening ventilation sections. Standing was limited to 25 instead of the originally permitted 42.

Dwindling fleet.

At the time of the 1974 fleet numbering there were 63 Seddon Pennine 4's, ie S1-63, in order of the remaining registration numbers, but with gaps for buses temporarily withdrawn but of uncertain future. As time progressed, the rehabilitation programme continued and others were reinstated to become S64-93, out of registration number order but in order of re-entering service. The highest number allocated was S93. Meanwhile, further casualties had reduced the number in service on 1st January 1978, to 75, showing that there were never even 93 fleet-numbered Seddons in service concurrently.

Following renovation, a few Pennine bodies were returned to service with KMB standard square-cornered, aluminium-framed side windows, considerably shallower than the original rubber mounted units, despoiling the appearance of these one-time very smart buses.

Double-deck proposal.

One Seddon was stripped-down to chassis main-frames for possible reconstruction as a double-decker, a process which would have involved the repositioning of the rear axle to reduce the wheelbase. Preliminary work started but the project was then terminated.

One Seddon Pennine 4 was experimentally fitted with a Gardner 6LX engine and semi-automatic gearbox but it is not known which vehicle was involved or whether it ever ran, so fitted, as a bus in service.

The last few.

Despite the remedial action to prolong their useful lives, the Seddon Pennine 4's were never to achieve the level of reliability required of them by the harsh operating conditions to which they were subjected and withdrawals continued, latterly at some pace. To many, this was disappointing for they were, perhaps, the most suitable for DOO/OMO of any of KMB's contemporary fleet but, during 1979 and 1980, large batches were withdrawn almost monthly, until only four remained in service. These were S69 (AR7601), the first of the type, together with S8 (AR7615) the one-time coach, S90 (AR7684) and S61 (AR7696), all of which ran for the last time on 31st August 1980, ten years and four months from the entry into service of AR7601 in May 1970.

Coach order.

According to PSV Circle publication BB133, KMB had ordered four Seddon Pennine 'Dual-Purpose' coaches with chassis numbers 48417-20 and Pennine body numbers 1067-70 but this order was cancelled and the chassis sold to Cyprus in June 1971. It is also reported (PSV Circle Publication OP2) that the body order specified that the bodies be painted the light brown and cream of KMB's coach fleet.

LEFT: AR7675 was fast approaching its journey's end at Tai Kok Tsui Ferrier in January 1978, closely followed by two Daimlers. *(Mike Davis*

LOWER LEFT: Seddon Pennine 4, AR7639, was converted to become a water-tanker in 1973. The complete driving position was moved back to a point alongside the gearbox in order to avoid the need for a remote gear-lever. The engine, however, was left in its original position ahead of the front-axle and was encased in this strange-looking 'snout', not unlike the arrangement used on many double and single-deck Berlin buses of the immediate pre-war era. It

Bedford Model YRQ Coaches.

A Bedford YRQ demonstrator was sent to the first-ever 'British Motor Show' to be held in Hong Kong, in February 1974. The vehicle shown was a Model YRQ coach with Willowbrook C41FEXD bodywork, finished in a maroon livery and lettered 'KMB' in white block capitals - the first time that the Company's initials had appeared in that style. Three months later a second YRQ arrived, this time with Duple Dominant MkI, C45FEXD coachwork, again painted maroon.

During 1974 these two vehicles were idle in Shing Lok bodyworks alongside the Seddon Pennine 4 which had been converted into a coach. Later that year, the two Bedfords were fitted with Fareboxes and arrangements were almost completed for them to enter service on a new special express route from Lai Chi Kok to Kowloon "Star" Ferry, in competition with an estate resident's association service, also operated by Bedfords but with local bodywork. The idea was abandoned at the eleventh hour when the Hong Kong and Yaumati Ferry Company introduced fast, air-conditioned launches from Lai Chi Kok to 'Central' on Hong Kong Island. As a result, the two coaches remained idle until, in late 1974, KMB purchased them outright, for up to that time they had been demonstrators on loan, and registered them BH8315 and BJ2726.

In common with all KMB coaches, the two Bedfords did not carry their fleet numbers but such numbers *were* allocated for internal administrative purposes and, for simplicity, they were, on paper at least, given the next available number in the 'CA' Albion coach series but being Bedfords, it is believed that became CB67 & 78.

By reason of their non-standard type, both Bedfords were sold in February 1980 and it is believed that they were exported to a nearby operator in China.

VEHICLE SPECIFICATION: BEDFORD YRQ COACHES	
Fleet numbers:	CB67 & CB78 - *not* carried on vehicles.
Chassis make:	Bedford Motors division of General Motors.
Chassis model:	Bedford 'YRQ'.
Engine:	Bedford '466' - vertical mid-underfloor.
Gearbox:	Turner 5-speed, synchromesh.
Body make: CB67:	Willowbrook 'Expressway'.
CB78:	Duple 'Dominant' Mk I.
Body layout: CB67:	C45FEX,D,DOO,Air-conditioned.
CB78:	C41FEX,D,DOO,luggage pen.
Date introduced:	1980.
Length:	31ft 8in 10 metre (nominal)
Width:	8ft 2½in 2500mm
Height:	?
Wheelbase:	16ft 3in 4952mm
Unladen weight:	?
Gross vehicle weight:	?

Willowbrook 'Expressway' bodied Bedford.

In June 1975, some time after the introduction of the Albion coaches (qv) on special, semi-express, urban, services, the Willowbrook bodied Bedford entered service on route 206 which, ironically, was almost the route originally proposed for both Bedfords. For this duty it was licenced for the first time on 16th June 1975 as BH8315, having been painted in what had become the fleet standard coach livery of golden-yellow and brown. Seating in the 10-metre Willowbrook Express body was increased to C45FEXD,DOO prior entering service and a Thermo-King air-conditioning unit was installed in order to cool the interior during the humid summer months. Following front-end damage in 1976, the vehicle was rebuilt by KMB to incorporate a windscreen moulded front roof dome intended for the Seddon Pennine 4 which resulted in the coach having a more bus-like appearance, a look emphasised by the very plain side panels. The large non-opening side-windows remained as they were compatable with air-conditioning which was designed to reduce a humid 33°C in to a dry and comfortable 22°C.

LEFT: A full front view of Bedford YRQ, BH8315, with Willowbrook 'Expressway' coach body, in its original condition as placed into service in 1975. *(KMB*

ABOVE: The Willowbrook bodied Bedford in original condition as seen from the side. *(KMB)*

BELOW: Following accident damage to the front, the Willowbrook body of BH8318 was rebuilt incorporating surplus parts from the Seddon Pennine range, including the winscreen and and canopy. *(Mike Davis)*

Duple 'Dominant' bodied Bedford.

This second Bedford demonstrator was BJ2726, also purchased outright by KMB in 1975 and entered service on Cross Harbour Coach service 200 in September that year. The Duple Dominant MkI body resembled that of the Albion Duple coaches in general but the latter had the later Dominant MkII body with deeper windscreens and other refinements. The destination equipment was located in the front dash panel above the grille instead of in a so-called 'Bristol-dome' (above the windscreen of the Albions). As with the twelve Albion coaches allocated to the Route 200 Airport service, this Bedford was provided with a large luggage holder for the convenience of baggage-laden air-travellers at a penalty of losing four seats, thus reducing the capacity to 41 - no standees were permitted.

The notional fleet number CB78 was allocated but never carried on the coach.

Mechanically the two Bedfords were similar in having the Bedford 466 underfloor engine, mounted vertically, coupled to a Bedford five-speed synchromesh gearbox. It was possibly this underfloor engine configuration that put KMB off the type, as Bedfords have shown themselves capable of long and arduous service in the fleets of private operators in Hong Kong's non-franchised bus sector.

BJ2726 was withdrawn in February 1980, and, as already mentioned above, sold, together with its Willowbrook bodied sister to an operator in China

BELOW: Bedford YRQ, BJ2726, with Duple Dominant bodywork seen shortly before etering service in the summer of 1975. Note the absence of a so-called 'Bristol-dome' and canopy destination screen as incorporated in the similar bodies on the later Albion Viking chassis. *(Mike Davis*

Nash airport coach - 1949.

An earlier version of an Airport coach! This photograph of a Nash coach accompanied a 1949 article in Commercial Motor and was captioned as being an airport coach for transferring Trans World Airlines passengers between Kai Tak and their Kowloon hotel. As the terms of KMB's franchise were exclusive, it is possible that this interesting vrhicle was operated by KMB on contract to TWA.

RIGHT: The Nash coach referred to above. *(Commercial Motor*

Albion Viking EVK55CL coaches.

Following the examination of the Duple 'Dominant' and Willowbrook 'Expressway' bodied Bedford 'YRQ' coaches, KMB placed an order for 50 'Dominant' bodied EVK55CL model Albion 'Vikings', to full express coach specification, for delivery early in 1975. This order was later extended, first to seventy, and eventually to one-hundred vehicles.

These coaches were introduced on new-style urban express services and often running parallel to existing urban crush-loaded double-deck routes but offered passengers a guaranteed seat, standing not being permitted, for a premium fare of HK$1 (10p), more than three times higher than the normal bus fare for the some journey. Longer routes were also operated linking the New Territories with the Kowloon terminus at "Star" Ferry - or more correctly Tsim Sha Tsui Post Office - for the first time, with extra routes on Sundays and Public Holidays, referred to as 'Recreational Routes'. Hitherto all rural routes had terminated at Jordan Road Ferry, Choi Hung or Tai Kok Tsui Ferry terminals. On the longer routes the fare was HK$2 (20p) for up to eighteen miles.

One additional feature of the coach operation was the introduction of scheduled services linking the central tourist areas, on both sides of the harbour, to the International Airport, Kai Tak. There were two such services, one from Kowloon "Star" Ferry and the other from Central District, Hong Kong Island, via the Cross-Harbour Tunnel. This latter service, unlike the bus services was not operated jointly with the China Motor Bus Company which declined, not having suitable coaches. The coaches used on the Airport Service displayed distinctive lettering across the front and each side, otherwise the standard livery of golden-brown and cream was the order of the day.

Until the arrival of these vehicles, the term 'coach', as used in Hong Kong, implied little better than the standard of comfort expected of a service bus in the UK. The Duple Dominant MkII' bodies on the Albions were to largely British standards with the exception that the seats were covered in deeply embossed and perforated PVC, allowing it to breath in hot weather as these coaches were not air-conditioned.

In most cases, without KMB sub-classification, these coaches seated 42 passengers while two which entered service in March 1975, BH5088 & BH5089, were fitted with 46 seats and had the sub-class letter (A). Commencing in August 1975 with three conversions, luggage racks were progressively added to coaches in use on the airport services, Routes 200 & 201. Nine others quickly followed until all twelve coaches allocated to these routes were so fitted but all lost four passenger seats to leave them as only 38-seaters. These were not given sub-class, being known by the suffix (Lug. Comp.), indication 'Luggage Compartment'. During the summer of 1983 a further seven coaches were fitted with luggage compartments, not so much to increase the number so fitted but to allow older vehicles to be scrapped or demoted to loical bus services, such as 5M, and replaced by newer examples which had been reconditioned.

The front entrance was fitted with a one-piece, inward swinging, power door despite the coaches 'Express' function. A 'Farebox' was mounted beside the driver, as the coaches were Driver Only Operated. The large side windows were fitted with tinted glass and each had a central opening section with sliding pane to provide better air circulation. A single-line destination screen was mounted above the windscreen in the 'Bristol-dome' of the front canopy and the route number appeared on a

VEHICLE SPECIFICATION: ALBION VIKING COACHES	
Fleet numbers:	CA1-66, 68-77, 79-102 (*not* carried on vehicles).
Chassis make:	Albion Viking EVK55CL (Leyland nameplate).
Engine:	Leyland EO.401
Gearbox:	Leyland GB277 or 281, 5-speed manual (6th disconnected)
Body Make:	Duple 'Dominant' MkII
Body layout: Standard:	C42FEX,D,DOO. (2x Air-conditioned.)
Luggage pen:	C38FEX,D,DOO + luggage pen - for Airport Service.
Sub-type(A):	C46FEX,D,DOO.
Date introduced:	1975-76.
Length:	31ft 8in - 9650mm
Width:	8ft 2½in - 2500mm
Height:	10ft 6in - 3200mm
Wheelbase:	16ft 1in - 4900mm
Unladen weight:	6,985kg
Gross vehicle weight:	10,674kg.

BELOW: KMB introduced the first of one hundred Albion Viking coaches on limited-stop services during January 1975. They were fitted Duple Dominant MkII body, a later version of that fitted to the Dominant bodied Bedford BJ2726. Here Albion BH3002 was photographed on its first day in service and the day that the coach services commenced in January 1975. The short-lived logo in which the K and the B had a 'pointed' appearance can be seen to good effect in this view. At that stage, the route numbers (all commencing with 200) were displayed on an oval placed behind the nearside windscreen. (Mike Davis

ABOVE: In order to improve public awareness of the Airport Coach Service on Routes 200 & 201, KMB added slogans to the front and bodysides of vehicles operating on these two services (see alos photograph of Bedford BJ2726 on page 60). *(Martin Weyell)*

BELOW: BH3711 passes the familiar Peninsula Hotel in early 1983, en route "Star" Ferry and the nearby Tsim Sha Tsui bus station. Route 207 was one of those to be axed within weeks of this photograph, a casualty of the success of the Mass Transit Railway. *(Mike Davis)*

plate in the nearside windscreen. Below the windscreen, in the space used by some operators for low-level destination screens, there was a large fresh-air ventilator grille, below which was the radiator grille. A metal plate carrying the Company crest was attached to this grille.

When they first entered service, the Albion coaches carried a new KMB logo in which the back of the 'K' was broken to form an arrow-head but this was quickly changed following complaints from local people who claimed that the 'pointed' nature of the 'K' was 'unlucky' - giving the impression of 'arrows of ill-fortune', as shown on the accompanying illustration.

The first of these coaches arrived in December 1974 and the first fourteen were registered in January 1975 and entered service shortly thereafter. All were fully assembled in Duple's factory at Blackpool and shipped completely built up. The livery was one of golden brown and cream which was applied in a complicated pattern of horizontal stripes. By mid-summer, KMB experimented with two Albions and one Bedford, fitting them with air-conditioning, the two Albions being BH8196, introduced in June 1975 and BJ1973 which followed in September. The two units came from different manufacturers, the former being by 'Thermo-King' (USA) and BJ1973 having a Mitsubishi (Japan) unit.

In order to test the body structure and mechanical components and assess their suitability for the heavy use for which they were destined in Kowloon's dense traffic, involving hours of stop-start driving, sometimes over indifferent road surfaces, the prototype was subjected to a 1,000 test at 20-25mph on the pave section at the MIRA proving ground near Nuneaton. Its structure was clearly strong enough to take the punishment which these vehicles were to withstand as a daily occurrence in KMB service.

One problem facing all overseas buyers of completely built-up coaches is that of spare parts. Duple overcame this by despatching to KMB a complete selection of 'Dominant' spares with the finished coaches. Any parts later ordered represented a 'top-up' of the operator's stores.

It is interesting to note that two chassis from the initial batch were diverted to the, 'Auto-Carros "Fok Lei" Ltda' in nearby Portuguese Macau (40-miles to the west of Hong Kong). They had been bodied by Union Auto Body Builders at Kwai Chung prior to being transferred to Macau by barge. Two replacement chassis were later built for KMB.

The engine chosen for the Viking chassis was the reliable Leyland 0.401, driving through an Albion-built five speed constant-mesh gearbox. The chassis was one of Leyland's export range and featured a front mounted vertical engine which made it highly suitable for use in this sub-tropical region.

No fleet numbers were carried by the coaches but, for internal administrative purposes, they were known by KMB as their 'CA-class'. They were allocated CA1-102, the numerical series of which included the two Bedford coaches which took the numbers CB67 & 78.

Due to falling demand on some of the trunk coach routes, forty Albion Dominators were withdrawn from service during 1982, following the opening of the Tsuen Wan extension of the underground 'Mass Transit Railway' that summer. The redundant vehicles met a variety of fates, some being sold to interests in neighbouring China and others were reconstructed as goods vehicles that could be loosely described as a 'Dominant Van'!

An uncertain number of these coaches were sold for further service with the 'Motor Transport Company of Guang Dong and Hong Kong', into whose livery they were painted. Under standard terms of sale imposed by KMB, GDHK were not permitted to use the second-hand coaches within Hong Kong and so they remained on the China side of the Frontier for their remaining useful lives.

Yet another batch was exported via 'Speedybus', a local dealership associated with the Scottish group 'Stage-coach', to Malawi, joining a number of former KMB double-deck Daimlers under the African sun.

Some of those sent to the Chinese hinterland were used by the local operator in the city of Guilin and had off-side rear doorways fitted on the rear overhang.

The Albion coaches were ousted from the Airport services by the Dennis 'Falcon' coaches introduced in 1985/6, also with Duple bodywork.

The last examples were finally withdrawn by KMB in 1990.

KMB donated BH7637 to the Hong Kong Road Safety Association in June 1990 and, interestingly, has 'borrowed' it back for exhibition in a bus rally on at least one occasion.

BELOW: By May 1984, this coach was among a number relegated to urban bus duties on Route 5M, once pioneered by Citybus as a free service, between Tsim Sha Tsui and Tsim Sha Tsui East, a vast new commercial and hotel development. Paper stickers suffice as destination and route number indicators *(Mike Davis*

Albion Viking EVK55CL and EVK41L

EVK55CL L279-308

There were thirty Albion EVK55CL chassis similar ro those of the 100 Albion/Duple coaches. Of these, 20 chassis were probably ordered concurrently with the coaches but were then delayed, possibly by KMB, and the ready-packed chassis were diverted to Zambia. Twenty-two similar chassis were then built as replacements and a further eight were diverted from British Leyland's African subsidiary, 'Leyland-Albion Tanzania', in Dar-es-Salaam, Tanzania, to KMB in November 1975. It was this latter origin that earned the *entire* thirty buses the friendly title, within KMB, of the 'African Diamonds', as the eight rightful holders of the title were, reportedly, obtained at a most attractive price! The remaining 22 were a direct KMB order placed in November 1975.

It is known that Metsec actually prepared a body design for the original twenty chassis which wasd not proceeded with. Upon arrival in Kowloon, the chassis were sent to Union Auto Body Builders in an unusual move by KMB to involve a local outside contractor to supply and construct the bodywork. UABB had recently completed the bodywork on two buses with similar chassis which had been supplied to the Macau operator, Auto-Carros 'Fok Lei', Ltda., and the KMB bodies bore a distinct family resemblance.

Radius corners to the extruded aluminium window-frames gave the design away as being non-standard. The usual dual doorways were provided, but in this case the front entrance was ahead of the front wheels and the exit was almost dead-centre, midway between the front and rear wheels. The attractive body design featured a quartered rectangular radiator grille and twin headlights, complemented by an attractive variation of the standard KMB fleet livery which featured a broad cream band around the body, midway between the lower edge of the windows and the skirt. Seats for 49, plus 12 standees were provided.

When introduced in May/June 1976, the first fourteen, fleet-numbered from new, L279-292, had Bell Punch Farebox and Autofare 'complex' ticket issuing machines and were classified by KMB as EVK55CL(A)OMO/BP/CompTI. These were followed in September by three further (A)-type, L293-5, plus the first seven of type EVK55CL(B)OMO/BP, ie just a Farebox, of which the final six were introduced in November 1976.

The actual eight chassis diverted from Tanzania were those which received the fleet-numbers L285 & L302-308, 51749C, 51727F/J/K, 51728C/B, 51727H & 51728A .

L286 was scrapped after only eight months in service following a serious accident when the vehicle ran away on a steep hill.

During December L305 (BM245) was, as KMB put it - converted to 'new configuration'. The process continued through 1985 and the last to be so treated was L279, the premier member of the type (BK4969). It is not certain as to the extent of alteration also made to the EVK41L's, L309-312, but was not one that warranted a new

VEHICLE SPECIFICATION: ALBION VIKING EVK55CL and EVK41L.	
Fleet numbers :	L269-308 & L309-312.
Chassis:	L279-308 : Albion Viking EVK55CL L309-312 : Albion Viking EVK41L.
Gearbox:	Leyland GB277, 5-speed manual.
Body make:	Union Auto Body Builders, Hong Kong.
Body layout:	B49+12FE,CX,D,DOO.
Date introduced:	1976
Totals:	EVK55CL : 30 EVK41L: 4
Length:	31ft 5in - 9588mm
Width:	8ft 0in - 2440mm
Height:	10ft 2½in - 3120mm
Wheelbase:	16ft 1in - 4900mm
Unladen weight:	6,401kg (55CL) or 6,096kg (41L)
Gross vehicle weight:	10,699kg (55CL

BELOW: L285 (BK4975), an Albion EVK55CL, was bodied in Hong Kong by Union Auto Body builders to a design similar to that produced by them for two similar chassis destined for Macau. This photograph was taken in January 1978, shortly before the type was repainted into the fleet-standard livery layout. *(Mike Davis*

UPPER LEFT: L305 was one of a number of the Viking VK55's to be rebuilt by KMB who retained the general framework but substituted their own shallow side windows. When photographed at Tuen Mun in 1981, L305 had lost the distinctive quartered grille and twin headlights but retained the original front canopy (Mike Davis

CENTRE LEFT: L279 was the first of the type to be completely rebodied and, in this case, the distinct KMB 'style' can be detected, including semi-circular wheel-arches and completly new front panels and roof. L279 was descending Route Twisk in rain when photographed in 1981. (Mike Davis

LOWER LEFT: A rear view of rebodied L279 (BK4969) showing a slightly extanded rear bulkhead. This view was taken near Kam Tin walled village. (John Shearman

body classification, so it is possible that a mechanical alteration took place at this time that brought the EVK41L's into line with the 29 remaining EVK55CL's, their close relatives.

It is interesting to note that Viking chassis were usually supplied to customers in ckd form for local assembly by the operator, but KMB, as a major customer was an exception and having a concurrent order for 100 VK55CL's for Duple to body as coaches, the additional 22 similar chassis for bus bodying was merely a continuation of the production run. The eight chassis originally destined for Dar-es-Salaam were already crated as ckd kits when KMB agreed to take them and, together with the four Cyprus EVK41L's, were assembled by KMB's own staff.

EVK41L L309-312.

These four chassis were originally an order placed by Leyland's Cyprus agent, Demades of Nicosia, and were diverted to KMB on an unknown date, but not before June 1974 when the Demades order was cancelled. The chassis arrived with KMB in January 1976 and were in ckd form, being assembled by KMB's engineering department.

The EVK41L was a slightly lighter version of the EVK55CL, although they shared the same Leyland EO.401 engine and GB277 manual gearbox.

Following completion of chassis assembly, these four Vikings were sent to Union Auto Body Builders to be fitted with bodywork similar to that provided by UABB on the EVK55CL chassis, above. There were no sub-classifications, the type being known to KMB as EVK41L(OMO)BP. This indicated that they were fitted with Bell Punch Fareboxes without ticket issuing machines. They were arranged for 49-seated and 12-standing passengers. They were, of course, DOO with power doors from new.

Both groups were reclassified by KMB in 1985, following mechanical modifications, to become a single class, EVK41L/55CL, the first, BM3032, changing in September, followed in November by BM3028 & 3031 and, finally, in January 1986, BM3029.

EVK55CL/EVK41L; L279-308 & L309-312, less L286.

During 1985/86, KMB made alterations to both the EVK55CL buses and the EVK41L type, bringing them both into a single class of 33 buses, all with the KMB type-code EVK41/55CL, combining the classifications. As the bodies were identical in both types, it is believed that the commonality came about as a result of component rationalization, such as the fitting of heavier-duty rear axles and the provision of exhaust brakes on the EVK41's, but the exact extent of the modifications remains an area of uncertainty. At this time the carrying capacity was altered to 43-seated, plus 27-standing passengers.

Rebodying.

Some of the group were completely rebodied by KMB while others were rebuilt around their original Union Auto frames. In all cases, the original window units were replaced by KMB's standard, rectangular type.

Dennis Falcon 'HC' coaches.   *AF1-19*

Kowloon Motor Bus had, at one time, intended to replace the Duple Dominant bodied Albion coaches used on the Airport Coach Services, then Routes 200 & 201, using double-deck vehicles but the short-sighted design of certain overhead structures at the departure level of the Kai Tak main passenger buildings prevented the implementation of such a measure. A particular obstruction was found to be a footbridge constructed comparatively recently.

The choice of single-deck vehicles being forced upon the company, established KMB vehicle suppliers Hestair-Dennis and Hestair-Duple (Metsec) won the contract to supply twenty Dennis Falcon chassis, fitted with 49-seat bodywork specially designed by Hestair-Duple for local assembly. This was in contrast to the completely-built-up Dominant bodies supplied on the Albion chassis, ten years earlier. At eleven metres in overall length, they were also longer than their predecessors by a full metre.

The Falcon chassis was a conventionally framed structure with rear-mounted Gardner 6HLXB engine driving via a Voith D851 fully-automatic gearbox. Full air-suspension was fitted.

The flat-sided body has been described as having been loosely adapted from the Duple 'Laser' design but the result was more pleasing than its antecedent. The first fifteen Falcons entered service with the intended 49-seats but luggage capacity problems soon emerged and the seating was reduced to 45 from the sixteenth vehicle and luggage pens were provided in place of the first pair of seats on either side. The earlier vehicles were subsequent modified to match.

Full air-conditioning was provided, using Sutrak units centrally mounted in a roof 'pod'.

KMB introduced a new white livery for these coaches, relieved by a narrow red waist band and grey skirt. The surround to the windscreen and side windows was matt black, concealing to some extent the rubber window-mounting gaskets to give an illusion of modernity. It was from this livery that the livery was devised for the then coming generation of air-conditioned double and single-deck buses and minibuses.

VEHICLE SPECIFICATION: DENNIS FALCON

Fleet numbers:	AF1-19 (DH1700 scrapped prior to fleet numbering).
Chassis make:	Hestair-Dennis 'Falcon HC'.
Engine:	Gardner 6HLXB (horizontal underfloor).
Gearbox:	Diwa Voith D851, with retarder.
Body Make:	Hestair-Duple ckd, assembled by KMB
Body layout:	C49F as new - later C45F+luggage pens.
Date introduced:	1985-86.
Length:	11,000mm
Width:	2,500mm
Height:	?
Wheelbase:	?
Unladen weight:	?
Gross Vehicle Weight:	?

Taped announcements over the public address system in English and Cantonese inform passengers of the next stop.

Airbus Services.

Reference to the accompanying promotional information (see page), will help to illustrate the nature of the dedicated services for air passengers. KMB promotes its services as being from major hotels to Hong Kong International Airport, Kai Tak and the services are known as 'Airbus', following the lead of those operated by London Buses between Heathrow and Central London, even using the same 'A' prefix to the route numbers. The A1, A2 and A3 services were replacements for the former 200 and 201 introduced in the days of the Albion Viking coaches, although for a few months the Dennis Falcons ran on the original routes. At the time that the Airbus routes were introduced, the Company added embellishments to the sides of the coaches in the form of blue and white 'swooping' curved arrows and the word 'Airbus' in large blue letters in English only, emphasising the international nature of the intended passengers. Since the publication of the promotional material, overleaf, KMB have added a fourth Airbus service, Route A4 from China Ferry Terminal off Canton Road. All four run every 15 or 20 minutes.

Withdrawals.

The first Dennis Falcon to be withdrawn was DH1700 which suffered serious damage during January 1989 and was never returned to service, being written-off on 29th June same year.

Fleet numbers.

As was the case with the Albion/Duple coaches, the Dennis Falcons were allocated fleet-numbers which they never carried, running from FD1 to 20. When the air-conditioned fleet grew following the introduction of double-deckers, a new series was commenced where all air-conditioned vehicles were allocated fleet-numbers, prefixed with the letter 'A'. Consequently, from May 1991 the 19 remaining Falcon coaches became AF1-19.

Type classifications.

The first 15 which entered service with 49 seats were referred to simply as 'Falcon/Duple(Air-Con). The remaining five entered service with only 45 seats and a luggage pen and were known as Falcon/Duple(A)(Air-Cond), as were the first 14 when altered to the latter specification, DH8180 in February 1986, followed by a further 13 in November 1986 and the fifteenth in the December.

ABOVE: Dennis Falcon coach, DH8180, emerges from the northern exit of the Cross-Harbour Tunnel and approaches the toll plaza whilst running on Airport Route 200, shortly after entering service in November 1985. The 'Airbus' markings had not been introduced and the coach displays the basic livery and its roof-mounted air-conditioning 'pods' to advantage. *(H. C. Steele*

BELOW: Shortly after entering service, KMB applied 'Airbus' logos to the Dennis Falcon coaches used on the two routes, 200 & 201, which served Hong Kong International Airport, Kai Tak. Here DH972 (later fleet-numbered AF3) awaits its departure from the then new airport facility for these services. *(C.Lau*

ABOVE: Falcon coach DH743, the premier member of its class and later to carry the fleet-number AF 1 working on the newly re-numbered Airbus Service A1 as it approached Kowloon-side "Star" Ferry in September 1988. *(Mike Davis)*

BELOW: After the Dennis Lance and Dart Airbus coches were introduced, the Falcons were, one by one, modified for service as single-deck buses with the luggage racks replaced by inward-facing seats. Here unmodified AF6 leaves the Jordan Road terminus in September 1994 on the 203E. *(Mike Davis)*

Toyota Coaster - Long-Wheelbase. AT1-91

1987 saw the first ever Japanese-built Franchised Omnibuses enter service in Hong Kong in the fleet of KMB in the form of 24-seat Toyota Coaster midi-buses with full air-conditioning. The first nine entered service in September 1987 to be followed by subsequent batches of similar vehicles over the next two years. Initially the Coasters were allocated to the remaining coach route 208 from Broadcast Drive in Kowloon Tong to Tsim Sha Tsui East via 'Star' Ferry and thus finally displaced the Albion/Duple coaches from the last of their original routes.

Mechanically KMB's Coasters were the standard long wheelbase Toyota production midi-buses similar to the shorter wheelbase version seen in some numbers in Hong Kong as both Public and Private Light Buses. They have pressed steel bodies by Arakawa Auto Body, an integral part of the Toyota group, and are powered by Toyota 3980cc diesel engines. The air-conditioning equipment is by Mitsubishi and in order to help reduce heat intrusion a sheet-steel 'sun-shade' is provided by as a second skin over the roof. This heat-shield is raised approximately 5cm above the actual roof, the air within the space providing the insulation material; a material that is constantly changed by the movement of the vehicle. Tinted windows also help reduce heat intrusion and glare. Opening windows are provided but remain locked during the hot months but can be opened by the driver should the air-conditioning fail and also in the cooler winter period. A large destination box gives clear route information and the white/grey livery with red lining and lettering distinguishes the KMB Coasters from the hoards of similar Public Light Buses - the major competitor of franchised bus operators in Hong Kong.

VEHICLE SPECIFICATION: TOYOTA COASTER	
Fleet numbers:	AT1-91.
Vehicle make:	Toyota.
Vehicle model:	Coaster LWB.
Engine:	Toyota 3980cc.
Gearbox:	Toyota manual.
Body:	Arakawa Auto Body Co., Japan.
Seating:	24 - no standees.
Toyota(A):	21 + luggage pen. (Airbus service)
Toyota(B):	20 +
Date introduced:	1987
Length:	6900mm
Width:	2300mm
Height:	2700mm
Wheelbase:	?
Unladen weight:	?
Gross vehicle weight:	?

Toyota Coaster (A).

Five Toyota Coasters were specially adapted for use on the Airbus service A4 and, later, A5, having two seats replaced by luggage pens. The vehicles involved, EA5269, EA6127, EJ4740/5883/6276, also carried the special 'Airbus' signwriting as used on the Dennis Falcon coaches. As these small coaches became worn, they were replaced in Airbus service by later Coasters, also called 'Coaster (A)'.

Further conversions took the form of 20-seat Toyota(B) sub-type but the exact identity of these five vehicles remains unclear.

Coasters also work the Rennie's Mill Route 90 along the same narrow winding roads which previously saw Traders and Chieftains, a route for which they are probably most suitable type yet used.

ABOVE LEFT: KMB Toyota Coaster, DV2996 (later AT17) moves off from the traffic lights in Salisbury Road, Kowloon, immediately prior to turning right into Nathan Road on its way to the northern suburb of Kowloon Tong and its terminus in Broadcast Drive where the main television studios are located, near the foot of Lion Rock, in October 1988. *(Mike Davis*

LEFT: Toyota Coaster (A) AT72 seen at Kai Tak Aitport in April 1994 whilst is use on Airbus service A5, a circular route serving Tai Koo Shing on Hong Kong Island. *(Nigel Eadon-Clarke)*

MCW Metrorider. AMR 1 & 2

For comparison with the Japanese-built Toyota 'Coaster', KMB purchased a single example of a wide-body (2.376 metres) MCW 'Metrorider'. As with the Toyota, this was a standard MCW product adapted for KMB's requirements with full air conditioning specified; unlike the Coaster's body, the Metrorider is not provided with any means of opening the tinted side windows. Being of coachbuilt construction and not a van conversion, the roof of the Metrorider offers superior heat insulation and as a result no secondary heatshield is necessary as is on the Toyota. Thirty-three high-backed seats are provided and entrance is gained through a two-leaf folding doorway. Deeper than standard destination screen is provided with three-track route number on the nearside.

The popular and very economic Cummins 6BT engine is provided together with Allison automatic transmission giving a very smart and lively performance in some trying operating conditions.

The white and pale grey livery adopted by KMB for its smaller buses is relieved by red lining and fleet name while the words, in Chinese and English, 'Air-conditioned' appear in pale blue.

DY6050 was registered in August 1988 and was joined by a second example, in November that year, when a demonstrator, built to KMB specification, that had been in Singapore was purchased. This was registered EA4591 and when the fleet of air-conditioned buses was numbered in 1991, the Metroriders were classified AMR 1 & 2.

Due to their small size, the Metrorider failed to find favour in Kowloon and the two AMR's were sold in 1993, AMR1 being purchased by Speedybus Services Ltd. and returned to England for service in England, in conjunction with Ranger, as F737KGJ.

VEHICLE SPECIFICATION: MCW METRORIDER

Fleet numbers:	AMR1 & 2
Vehicle make:	Metro Cammell-Weymann.
Vehicle model:	'MCW Metrorider'.
Engine:	Cummins 6BT, 5.9 litre.
Gearbox:	Allison AT545 automatic.
Body Make:	MCW.
Body layout:	B33+14FEXD.
Date introduced:	Aug. 1988.
Length:	8400mm
Width:	2376mm
Height:	2700mm
Wheelbase:	4750mm
Unladen weight:	5320kgs
Gross vehicle weight:	?

RIGHT: AMR2, the second KMB Metrorider came to Kowloon via a one month period on demonstration in Singapore.

BELOW: Seen here on 29th August 1988, MCW Metrorider DY6050 (later AMR 1) stands at the Broadcast Drive terminus of Route 208, whilst on trial alongside Japanese-built Toyota Coasters. *(Clement Lau*

Fuso-Mitsubishi midi-coaches. AM1-143

In order to provide vehicles of larger capacity than the Toyota Coasters that KMB introduced in initial trial of a 'small coach' type of vehicle, the Company turned to an 'off-the-peg' Fuso-Mitsubishi type which in British terms is smaller in size to the Dennis Dart.

These Fuso coaches were initiakky used on 'Air-conditioned Coach Services' as defined by HK Government Transport Department and may be seen on services radiating from, among other places, Riviera Gardens where a comparative trial was carried out using a single Dennis Dart demonstrator.

After proving a most satisfactory vehicle, they were ordered in some numbers and allocated to a variety of routes ranging from Tsim Sha Tsui in the south to Tuen Mun and Yuen Long in the west and north.

The usual white livery for air-conditioned vehicles was applied with the lettering and lining in red and blue.

VEHICLE SPECIFICATION: FUSO-MITSUBISHI.	
Fleet numbers:	AM1-143
Chassis make:	Fuso-Mitsubishi
Engine:	Mitsubishi 6D16-OA
Gearbox:	Mitsubishi
Body Make:	Mitsubishi
Body layout:	C35+10FEX,A/C
Date introduced:	1990
Length:	9100mm
Width:	2300mm
Height:	3000mm
Wheelbase:	
Unladen weight:	
Gross vehicle weight:	

RIGHT: Fuso-Mitsubishi midi-coach AM136 (FG5027) working on the 208 and seen here passing the end of Nathan Road in September 1994 *(Mike Davis*

BELOW: The first of its class - AM 1 stands outside Tsuen Mun Station on Route 238M between the station and Rivera Gardens, a larhe private high rise estate. *(Nigel Eadon-Clarke)*

Hino - Single-deck Coaches and Buses. AH1-25
On hire from Argos Bus Services.

In 1990, KMB was implementing plans for the introduction of a wider range of air-conditioned bus services as public demand for them increased and, wishing to continue with its expansion of air-conditioned services, fifteen Hino coaches were taken on long term hire from Argos Bus Services Co., a Hong Kong independent operator of both single and double-deck buses.

Chassis for all the 15 Hinos is the Model RK176 with Hino H107C, 6443cc engine rated at 175bhp mounted at the rear. A six-speed manual gearbox is provided.

These vehicles are fitted with Taiwanese built bodies by PDC of Taipei, of two types, both of which look similar and appear to have been modelled on the German products of Neoplan. Six have normal-height coach floors with 58 bus type seats while the remaining nine have 48 coach seats with high backs. Neither type is permitted to carry standees. Tinted window glass is provided throughout and in keeping with KMB practice for air-conditioned vehicles, they are in white livery with the usual blue and red lettering and lining.

Although these Hinos may appear very modern, they do not stand comparison with the modern service buses surrounding them in the streets as they have very hard-riding road-spring suspension where the norm for contemporary buses in Hong Kong is air-bag suspension and has been so for some years. Engine noise is also very harsh and intrusive, particularly towards the rear, near the engine compartment.

AH 1-15 were returned to KMB late in 1992 and EB3247 was exchanged for EB2473, apparently due to its poor condition. The last examples were returned to Argos in 1993.

VEHICLE SPECIFICATION: HINO COACHES AND BUSES

Fleet numbers:	58-seaters: AH1/2/8/10/14/15.
	48-seaters: AH3/4-7/9/11-3/16-25.
Chassis make:	Hino Motors Ltd. (Commercial motors division of Toyota.)
Engine:	Hino H107C, 6443cc, rear-mounted, 175bhp.
Gearbox:	6-speed, manual.
Body Make:	PDC, Taiwan.
Body layout:	low floor: B58FD,DOO.
	high floor: C48FD,DOO.
Date introduced on lease:	1990.
Length:	1100mm
Width:	2500mm
Height:	
Wheelbase:	
Unladen weight:	
Gross vehicle weight:	12,800kg

ABOVE: Hino EB4409 belonging to Argos Bus Services Co. (119) in use by KMB on 14th November 1990 at Jordan Road Ferry. This vehicle was one of the low-deck variety which carried the KMB fleet number AH15. *(Nigel Eadon-Clarke*

BELOW: Two hired Hino vehicles stand side by side at Riveria Gardens between duties on the 238X. Unidentifiable as the flash from the camera has reflected back from the registration numberplate, the coach on the left is a high-floor with high-backed coach-style seats, while that on the right is a low-floor model with bus seats for 58. No standing was permitted on either vehicle. The hired Hinos were returned to Argos commencing in 1992/3 *(Nigel Eadon-Clarke*

Dennis Dart 9-metre Air-conditioned.   AA1 & 2

During the summer of 1990, Dennis Specialist Vehicles sent a pair of their 'Dart' midibuses for evaluation with Hong Kong operators, one bus going to the KCRC and the other to KMB.

The KMB example was registered EP1863 in August 1990 and had a Sutrak air-conditioning unit. It ran for KMB from August to November 1990 when it was transferred to the Kowloon-Canton Railway Bus Division to replace the KCRC demonstrator with Nippon Denso air-conditioning that had failed. Both were delicensed and stored on KCR premises for part of 1991. Both were then returned to the custody of Dennis and they were kept on KMB premises until purchased by the Company in 1992.

KMB's interest lay in comparing the Dart with their similarly sized Fuso-Mitsubishi midis and the demonstrator was placed in service with them on air-conditioned service 234x from, Rivera Gardens and, being airconditioned, was painted in a white livery with blue and red lettering and lining.

The Dart chassis is a semi integral with bodywork of unusual appearance by Carlyle Bus Centre. The KMB demonstration bus was of 9-metre overall length and a wheelbase of 4.3m. The standard Cummins 6BT turbocharged diesel drives the rear wheels via an Allison 4-speed fully-automatic gearbox.

Following their period in store and after negotiations with Dennis, both the Darts were purchased

VEHICLE SPECIFICATION: DENNIS DART	
Fleet numbers:	AA1 & 2.
Chassis make:	Dennis Dart Model 9SDL3002
Engine:	Cummins 6BT
Gearbox:	Allison 4-speed fully automatic.
Body Make:	Carlyle Bus Centre.
Body layout:	B35F,DOO,Farebox, A/C.
Air-con make:	EP1863 - Sutrak
	EP5213 - Nippon-Denso
Date introduced:	August1990 (one on demo)
Purchased by KMB:	October 1992.
Length:	9.5m
Width:	2.4m
Height:	3m
Wheelbase:	4.3m
Unladen weight:	?

by KMB during October 1992 and allocated numbers AA1/2 following which they were fitted with up-rated oil-coolers, increased radiator capacity and lower-ratio rear-axles. They were then allocated to Route 2, where they are regular performers.

The destination displays of both AA1 & 1 were altered to roof-box style.

CENTRE LEFT: AA2, one of the two Dennis Dart demonstrators after purchase by KMB and in service on Route 2. In this view it retained the original and destinctive destination display and windscreens of the Carlyle design. This was the former demonstrator with the Kowloon-Canton Railway which was purchased by KMB from Dennis prior to their ordering twenty 10m Darts. *(Ron Philips collection.*

LOWER LEFT: AA2 again but after replacement of the destination display and still at work on Route 2 from "Star" Ferry to So Uk Estate in mid-September 1994.   *(Mike Davis*

Dennis Dart 10-metre Air-conditioned. AA 3-22

Following its successful trial of the 9m version of the Dennis Dart, in March 1992 KMB signed an order KMB for twenty 10m Darts with Duple Metsec ckd body kits. The first seven chassis arrived in Tuen Mun Works during June that year but the first was not registered until May 1993, by which time the others were under construction.

The air-conditioning plant in AA3-22 was provided by Sutrak.

Urban buses.

It had been the intention, when they were ordered, that all twenty would be fully-seated single-deck buses but it became clear that there was a requirement for additional Airbus vehicles and as a result, only ten, AA3-11 & 21, were fully seated as B43+17FEX.

Airbus Darts; sub-class Dart(A).

The remaining ten, AA12-20 & 22, were converted by Good View Engineering during construction for use on Airbus services to complement the larger Dennis Lances simultaneously being delivered to replace the earlier Dennis Falcon coaches.

These Airbus Darts carry the KMB sub-class Dart10M(A), are fitted with luggage racks for Airbus duties and are DP37+12FEX.

Duple-Metsec/Wadham Stringer bodies.

The purchase of a further twenty 10m Dennis Darts brought a new body make to the Hong Kong franchised bus scene when Wadham Stringer was awarded the contract to assemble in England for the twenty corresponding Metsec bodies due to KMB's main Hong Kong assembly contractor, Goodview Engineering, being fully committed with double-deck bus production.

As with the previous batch, there were two sub-types; Airbus B37+12FEX and urban bus B43+17FEX this latter figure includes three seats over each front wheel-arch where the Airbus version has luggage racks.

The roof-mounted air-conditioning units are by Sutrak.

(As an aside, the first Wadham Stringer bodies in Hong Kong, although not KMB, or even public vehicles, were supplied in 1993 to the Army on Dennis Javelin chassis.)

VEHICLE SPECIFICATION: DART 10m A/C	
Fleet numbers:	AA3-22.
Chassis make:	Dennis Dart Model 98SDL
Engine:	Cummins 6BT, 104kw (145bhp) @2,600rpm
Gearbox:	Allison 4-speed fully automatic.
Body Make:	Duple Metsec AA3-22
	AA23-32 (assembled by Wadham Stringer in UK)
Body layout: Dart 10M:	B43+17FEX,a/c.
Dart10M(A):	DP37+12FEX,a/c
Date introduced:	1993
Length:	9913mm
Width:	2290mm
Height:	3030mm (over air-con.)
Wheelbase:	5115mm
Unladen weight:	approx. 6750kg
Gross vehicle weight:	10,000kg

Stop Press!
Northern Counties bodies.

On the very day that this book went to press, it was announced (*Bus & Coach Buyer No 286*) that Northern Counties had won an order for ten of its 'Paladin' bodies, mounted on Dennis Dart chassis; the bodies to be built in England.

BELOW: An unidentified but nearly completed 10m Dennis Dart with Metsec bodywork stands outside KMB's Tuen Mun body workshop in May 1993. (*Nigel Eadon-Clarke*

ABOVE: AA19, one of the Dennis Darts adapted for Airbus service with high-backed seats and luggage pens over the front wheel-arch. *(Nigel Eadon-Clarke*

BELOW: Wadham Stringer assembled twenty Duple-Metsec bodies on Dennis Dart chassis for KMB in the Summer of 1994. Here the first of the twenty is posed for the official photographer in very English surroundings, prior to shipping. *(Wadham Stringer*

Dennis Lance 11.7-metre Air-conditioned Coaches and Buses. AN-class.

Late 1992 saw photographs taken at Alexander's Falkirk premises of the pilot PS-type body on Dennis Lance chassis for KMB. This represented the first of twenty five similar vehicles, with the balance of 24 bring supplied as kits of body parts for assembly by KMB in Hong Kong.

Twelve of the PS-type bodies are replacements for the Airbus Dennis Falcons and have coach seats for 41 passengers plus luggage racks over the front wheel arches. These are single-entrance vehicles, equipped with roof-mounted air-conditioning pods.

The balance of the order are equipped as maximum-capacity single-deck buses with a multi-standee capacity for 78-passengers. Dual, front and centre doorways are provided, again with roof-mounted air-conditioning units.

VEHICLE SPECIFICATION: DENNIS LANCE.

Fleet numbers:	AN1-24
Chassis make:	Dennis 'Lance'.
Engine:	
Gearbox:	
Body Make:	Walter Alexander & Co (Coachbuilders) Ltd. ckd except AN1.
Body layout:	Coaches: C41FEX,DOO,A/C. (for Airbus service.)
	Buses: B(total capacity 78 incl standees)FE,CX,DOO,Farebox,A/C.
Date introduced:	1993
Length:	11700mm
Width:	2500mm
Height:	
Wheelbase:	5850mm
Unladen weight:	
Gross vehicle weight:	18,000kg

RIGHT: The first Dennis 'Lance' for KMB stands in the parking area at Alexander's factory, Irvine, Scotland, after completion of the pilot bodying. All remaining buses and coaches of this type will be bodied in Hong Kong using Alexander ckd kits to the same 'PX' design. *(David Kat.)*

RIGHT: 11.7 metre Dennis Lance FS7603 emerges from the Cross-Harbour Tunnel toll booths on its way from the Island to Hong Kong International Aitport, Kai Tak on Airbus service A2 on 17th September 1994. At the time, this coach did not carry its fleet number which, by a process of elimination, should have been AN14. *(Mike Davis*

ABOVE: AN23, the penultimate member of the type, in Airbus livery whilst passing through Central District in April 1994 whilst on Route A2. These Airbus Lances have luggage racks and only a single doorway. *(Nigel Eadon-Clarke)*

BELOW: The bus version of the 12m Dennis Lance has two doorways but, unusually for KMB buses they have wide entrance doorways. AN 6 was seen here in Canton Road, Kowloon on the 238X to Rivera Gardens in September 1994. *(Mike Davis)*

Mitsubishi MP618N 'Aero Star' demonstrator. AP 1

This demonstration 'city bus' was taken by KMB as a demonstration vehicle on loan and allocated the fleet number AP 1, in which the letter 'P' denotes the 'MP' of the manufacturer's model designation.

The bus was fitted with a powerful 225bhp diesel engine of 11.149 litres and a dual doorway body with a centre *sliding* door and folding front door; an arrangement popular in Japanese cities. The chassis also featured air-suspension.

While with KMB, the MP618N ran on Route 2, alongside the two Dennis Dart demonstrators, AA1 & AA2.

After its trial period with KMB, no purchase was made and the local Mitsubishi Agent passed the bus to Citybus Limited for another period of demonstration, after which it was passed to the Kowloon Canton Railway (Bus Division) for a further period.

VEHICLE SPECIFICATION: MITSUBISHI.	
DEMONSTRATION BUS NOT OWNED BY KMB.	
Fleet numbers:	AP1
Chassis make:	Fuso-Mitsubishi
Engine:	6D22, 11.149litres; 225bhp
Gearbox:	
Body Make:	Mitsubishi 'Aero Star'.
Body layout:	B46+ ? FED,CXsD (centre sliding door.)
Date introduced:	1992
Length:	10.14m
Width:	
Height:	
Wheelbase:	5.3m
Unladen weight:	
Gross vehicle weight:	

RIGHT: This near-side view of Mitsubishi 'Aero Star', AP 1, shows the combination of folding front entrance-door and sliding centre exit-door. The arrangement, although unfamiliar in Hong Kong, is, apparantly, normal practice in Tokyo.
(Ron Phillips collection)

BELOW: Offside view showing the position of the emergency exit. *(Newton Ng)*

Part Three:
Double-deck Buses Purchased New 1949-1980.

The terms of the 1933 franchise stated that KMB would be permitted to run double-deck buses "subject to the advice of the Director of Public Works as to the suitability of the roads, therefor". In 1939, KMB took delivery of two Daimler COG5DD chassis, suitable for double-deck bodywork. It is believed that it had been the intention to run double-deckers on a trunk route serving Nathan Road, possibly the only route where passenger loadings justified their use. Contemporary newspaper reports state, however, that "the roads are unsuitable due to the overhanging trees". The two Daimlers were thus bodied as single-deckers and only ever ran as such.

After the liberation of Hong Kong in August 1945, the population was found to have been reduced to about 600,000 but by the end of 1947 it was estimated to have risen to 1,800,000, the rate of return reaching a peak of 100,000 in some months. This flood of people was accelerated by the southward sweep of Mao Tse Tung's Communist Armies as refugees joined the returnees. The only way that KMB could cope with the expansion in traffic was by the introduction of double-deck buses into the congested urban area. By this time, the objection to double-deckers on the grounds of overhanging trees could no longer be sustained, for, although there were still a number left, the ravages of the Japanese occupation had seen the majority used for scarce fuel to generate electricity and, so, permission was granted.

ABOVE: In 1949, when KMB introduced double-deck buses to Hong Kong, those trees remaining after the Japanese occupation were trimmed and signboards were raised for clearance *(Kowloon Motor Bus*

KMB's first order for double-eck buses called for twenty vehicles with which they planned to commence their post-war transformation. The first four chassis arrived in Kowloon towards the end of December 1948. They set the precedent for most future deliveries to both KMB and CMB in that the bodies were shipped ckd, or 'completely-knocked-down', kits of parts prepared in this case by Metal Sections Limited, of Oldbury in England and assembled by local Chinese staff. The first four examples actually entered service on Route 1 on 17th April 1949, having been first registered and licensed on 13th April, presumably for testing and driver training..

The practice of shipping bodies in ckd form was not new in principle as bus bodies had previously been shipped in pieces - complete sides, roofs, etc. - but the Metal Sections system involved the supply of the entire frame in component parts, complete with all necessary bolts, screws, rivets and beading as well as the actual pillars and panelling. Metal Sections advertised the method at the time as buying 'A Bus in a Box'. This method was employed principally to save shipping space and thereby substantially to reduce freight charges; an added bonus was use of the inexpensive skilled labour which was readily available in Hong Kong. The system persists today, with the bodies on sophisticated air-conditioned Volvo and Dennis chassis having been assembled from kits of parts shipped in a box, some still prepared by the successors to Metal Sections Ltd., Duple (Metsec) and others by Walter Alexander Ltd. The labour in Kowloon today is very much more skilled but not now quite so inexpensive!

These first double-deckers were very much in the British tradition with 56-seat bodies, open rear platforms and rear staircase. Allowances for the temperatures experienced in the climatic conditions of South China included full-depth sliding windows in all bays on each side, plus half-drop windows at the front, upstairs and the nearside front bulkhead. These were later replaced by sliding units on overhaul. The classic Daimler radiator was fitted; chrome-plated, with the famous fluted top. Only the first 125 of KMB's Daimler double-deckers featured this type of traditional 'exposed' radiator. The dimensions, also, were to British standards with a length of 26ft, height of 14ft 6in and width of 7ft 6ins. In terms of comfort, these first double-deckers, years later to be classified as Daimler(a)'s, were not unlike British wartime utility buses - they even had wooden-slatted seats but these were of a pattern requiring skilled craftsmen to fashion their dove-tailed joints. They were not just slats bolted to a metal frame.

After a few years in service the layout became inadequate for the crowds carried and an additional, narrow, doorway was added at the front of the lower saloon. From the mid 1950's, sliding, tubular-metal gates protected both platforms and, at peak times, three conductors were required to collect the fares and gate-men 'assisted passengers to board' at busier bus stops.

Two further versions of the Daimler CVG5 were added, the first being similar to the original type except for having a pressed-steel bonnet and radiator grille of the 'new-look' - the so-called 'Birmingham' type, while the third version had a lengthened body, 27ft long, built new with the front doorway and gates and having a moulded fibre-glass bonnet and grille assembly of the 'Manchester' type. These were the last 7ft 6in wide double-deckers for KMB and the last with 5-cylinder Gardner engines.

Reflecting changes to vehicle dimensions at home, the next of Kowloon's British-built double-deckers, introduced in 1962, the Daimler CVG6 model, were 30ft long and 8ft wide - metric dimensions were years in the future. These, too, were too small and not for the last time, little Hong Kong overtook current British standards by ordering big double-deckers no less than 34ft in length. These were AEC Regent Mk V's and still had rear staircases and open platforms but (wider) front 'doorways'. These really did become doorways in 1964/65 when the entire double-deck fleet was converted, for safety reasons, to have folding doors in place of the sliding gates hitherto standard. After 210 AEC's had been delivered, KMB turned again to Daimler to supply its future 34ft long bus needs.

The first of these Daimler CVG6-34's (the Daimler(d)'s) also had rear platform bodies but after twenty the layout was altered to front staircases, wide front doorway and narrow centre exit and were classified as Daimler(e) - a shorter, 31ft version was classified Daimler(f). These were the last half-cab buses to be purchased new by KMB but, despite British Leyland's refusal to continue building the type for the Hong Kong market, they were not to be the last front engined 'deckers. Having been given little alternative but to have to purchased 450 of the rear-engined Fleetline, KMB sought front engined buses from other manufacturers and, in deed other countries. CMB had purchased an Indian-built version of the Leyland PD3/5 while KMB took four Guy Victory 'Big-J' buses. The latter were single-deck chassis with a high, straight-frame, upon which the South African bus body builder, Bus Bodies (South Africa) (BUSAF) had constructed 15ft high double-deck bodies. Neither the Indian nor South African options were the real answer and, spurred on by the persistence of the Hong Kong operators, both Leyland and Dennis designed purpose-built, front-engined, front-entrance bus chassis, the Victory 2 and Jubilant respectively, while Alexander and Duple(Metsec) prepared ckd body kits to match. While these buses were a mechanical success they exhibited some unfortunate characteristics, mostly as a result of their short, 15ft 9in, wheelbases. After climbing the steep steps into the saloon, the passenger was soon subjected to a severe bucking motion; the author was forcefully reminded of this when, a few weeks before writing this piece, he rode from Yuen Long to Shung Shui on a Victory, along the 'old' road, beside the new near-motorway, and, before the end of the half-hour run, was feeling decidedly 'queasy'. KMB had been forced by circumstances beyond its control, to purchase what were little more than stop-gap buses.

Later, second-generation rear-engined double-deckers have proved far more suited to Hong Kong conditions by virtue of the attention given by the British manufacturers to adapting their products to the enormous market. In fact, they developed buses specifically for KMB and CMB, thus preventing their customers from turning to MAN or Mercedes-Benz (the latter did receive a small order) or, even, go to Japanese builders.

BELOW: This 1949 view is the earliest available illustration of a KMB double-decker. When new, these early series Daimler CVG5's had open rear platforms with a centre stanchion, upper-deck front windows of a half-drop type and no front doorway. The contemporary fleetname appeared along each side in English above Chinese, applied on the cream band below the side windows. This photograph shows number 4962, one of the original batch of twenty when standing at the Kowloon-side "Star" Ferry, soon after it enteried service in May 1949. The first four (4958-4961) inaugurated double-deck bus operation on 17th April. *(Bus & Coach; October 1949*

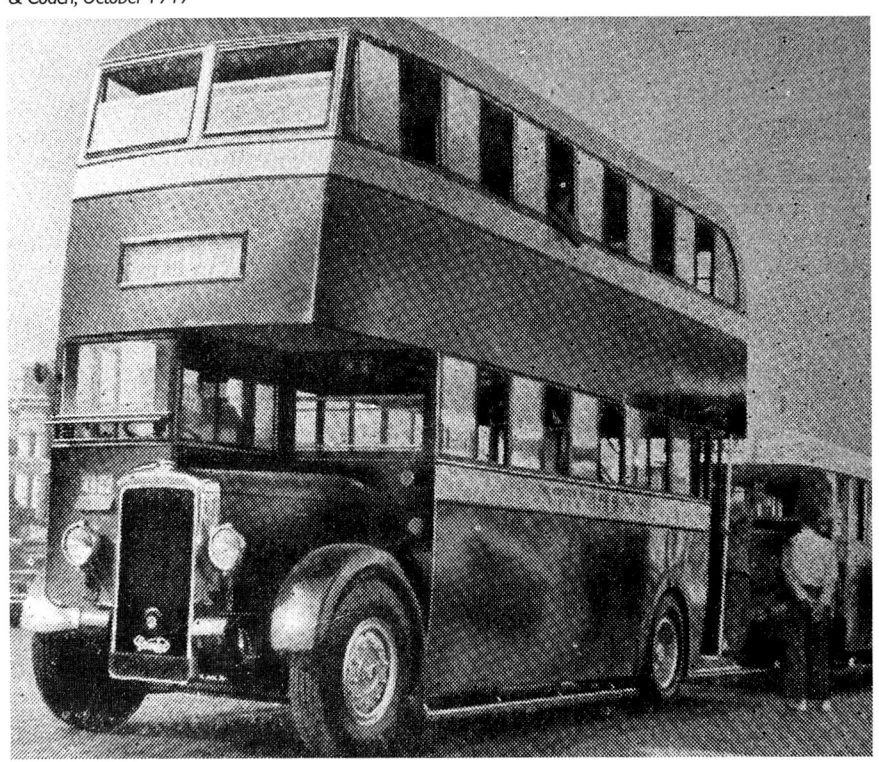

Fleet number prefixes.

2-axle:
A	AEC Regent Mk V - 34ft.
BL	2-axle British Leyland Olympian.
D	Daimler - all double-deck types
DM	Dennis Dominator.
G	Guy (Leyland) Victory.
L	Leyland (Albion) single-deck.
M	MCW Metrobus.
ME	Mercedes-Benz.
N	Dennis Jubilant.
S	Seddon Pennine single-deck.
VMD	Volvo B10MD

2-axle: Air-conditioned.
AA	Dennis Dart.
AF	Dennis Falcon.
AH	Hino.
AM	Mitsubishi 117.
AMR	MCW Metrorider.
AN	Dennis Lance.
AT	Toyota Coaster.
AP	Mitsubishi MP618N demonstrator.

3-axle.
3BL	Leyland Olympian - 12-metre.
3M	MCW Metrobus - 12-metre.
3N	Dennis Dragon - 12-metre.
S3BL	Leyland Olympian - 11-metre.
S3M	MCW Metrobus - 11-metre.
S3N	Dennis dragon - 11-metre.

3-axle: Air-conditioned.
AD	Dennis Dragon - 11 metre.
ADS	Dennis Dragon - 10 metre.
AL	Leyland Olympian - 11-metre.
AS	Scania. N113.
AV	Volvo Olympian 11.3-metre.
3AV	Volvo Olympian 12-metre.

Second-hand.
2A	AEC Regent MkV - 30ft.
2D	Daimler - CCG; CVG & CRG
2L	Leyland - PD3 & PDR1/1

KMB type-classifications.

KMB uses class letters to identify its various types of Daimler and other buses, the main purpose of which is to simplify its vehicle-to-route allocations. The letters refer to passenger carrying capacity and buses can have a separate code to indicate a standing capacity variation of only one passenger. Readers are asked to note the use of upper and lower case letters in certain circumstances. Generally speaking, the traditional half-cab double-deck buses were allocated bracketed lower-case letters, ie Daimler(a), a 56-seat CVG5; Daimler (aa) was a 67-seat modification of a Daimler (a). Later types, such as Daimler Fleetline, were given upper-case letters, *eg* Fleetline (A) or Fleetline(C). In the case of the more modern types the KMB code includes the location of the staircase, eg CS indicated a Centre Staircase, FS a Front Staircase and RS a Rear Staircase. Another refinement to KMB buses was the introduction of cushion seats and this found its way into vehicle codes in the form of (Cush) - likewise OMO and OCO were added to show One Man Operated and One Conductor Operated buses (the author has tried to eliminate the term OMO from the text in favour of the more favoured DOO, or Driver Only Operated, but the OMO was a part of the KMB nomenclature and so it is retained in the codes only when these are used in full). In many cases the basic vehicle has no specific code and here the KMB term is used to describe a type, for instance the basic Fleetline is suffixed merely 'CS', the sub-types being Fleetline (A)/CS/OMO(Cush) which is cumbersome and unnecessary in the text. To overcome this shorthand references are used which are readily understood and the last mentioned Fleetline would be simply referred to as 'Fleetline (A)' as all Fleetlines had soft seat cushions and were OMO (DOO).

Daimler CVG5DD.
Daimler(a)

D1-150
(Fleet numbered from 1974)

KMB's choice for its first double-deck buses, and its fleet for many years to follow, fell to the Daimler CVG5DD chassis, of which 325 were eventually purchased. The first 125 arrived with traditional Daimler fluted-top, chromium-plated radiator shells, while the following 90 had the so-called 'Birmingham' style of pressed steel grille and bonnet, or 'new-look' front, with a semi-circular chromium plated strip across the top, embossed with the famous Daimler flutes. These 215 buses became known by KMB collectively as their 'Daimler(a)' class. Daimler(a)'s with both types of front were 26ft long and 7ft 6in wide.

The actual first vehicle, registered 4958, was sderigistered on 31st December 1970, prior to the May 1974 fleet numbering but 4961, another of the four to enterservice on the very first day of double-deck bus operation, remained in full-time traffic until 1980; by the end of March 1978 it had covered 1,148,805 miles and returned a fuel consumption of 8.9mpg. Following its withdrawal in January 1980, 4961 was transferred to the driver training fleet. Early in 1982 it was taken to Kwun Tong depot where it underwent a major reconstruction and restoration to as near original condition as was possible so as to provide a suitable 'vintage' vehicle for the Company's 1983 Jubilee celebrations, after which it was donated to the Hong Kong Museum of History.

Many other Daimler(a)'s of both types remained in service for near to, or over, thirty years and must be a testimony to the quality of British products then - and still - available to the bus operator in export markets. A programme of complete reconditioning and reconstruction meant that many of this class were available for service until 1983. Four had been totally rebodied.

The specification of the Daimler CVG5DD chassis included the Gardner 5-cylinder '5LW' diesel engine, Wilson-type pre-selector gearbox and Daimler rear axle. The drivers gear control lever on the exposed radiator buses was of the older type and mounted on the right-hand side of the steering column and advanced continuously from first to top. The Birmingham style CVG5's had the newer style of column mounted preselective control, worked by the left hand, the lever being moved through the familiar 'H' pattern. All had fluid flywheels.

Two chassis of this type, Nos 14734-5, were reportedly delivered with Gardner 6LW engines, but these must have been subsequently changed for the standard type as KMB and Government records show 5LW engines when first registered.

Another anomaly that has been impossible to prove or disprove has been the report by the PSV Circle that chassis Nos 17687-96, built in 1950, were single-deck specification CVG5SD's. This was considered by KMB's Chief Engineer as highly unlikely, for some of this batch of chassis were still in service in 1978 when he, together with the author, examined two examples and found them in no way different from the remainder of the class in any chassis measurement; at 16ft 4in, the wheelbase was the standard for the double-deck version.

Following some earlier withdrawals, when the fleet was numbered in May 1974, there was only a combined total of 150 Daimler(a) class of both types and these took the numbers D1 to D150, in registration number order - not age order, so that 1949 buses start at D26, while 1954 buses start at D1. This was due to the reissue of older numbers to later buses.

VEHICLE SPECIFICATION: DAIMLER(a)-type

Registration numbers:	Discontinuous - see fleet list
Fleet numbers - from 1974:	D4, D26-88 - Traditional radiator
	D1-3, D5-25, D89-150 - Birmingham grille
Chassis make:	Daimler CVG5DD
Engine:	Gardner 5LW (a) & (ab), or 6LW (aa).
Gearbox:	Wilson-type; preselective - fluid-flywheel.
Body make: Original:	Metal Sections Ltd.
(aa) & (ab):	re-bodied by Kowloon Motor Bus.
Body layout: 1949-54:	H32/29REXS. - Gates fitted from c1952.
1954-72:	H32/27+20FEX,REXS,G. - Doors from 1964
1972-75:	H32/18+30FE,RXS,D,OCO.
1975-83:	H32/24+17FE,RXS,D,OMO farebox.
Body layout (aa):	H41/28+17FE,CXS,D,OMO farebox
rebodied buses (ab):	H41/28+18FE,CXS,D,OMO farebox
Date introduced:	1949-53 - Traditional radiator
	1954-57 - Birmingham grille
Original totals Traditional:	125 } Total 215
Birmingham:	90
Total at 1974 Traditional:	64 } Total 150
fleet numbering Birmingham:	86
Total 30th June 1983 (a):	4
(aa):	1
(ab):	1 Total 6. Final withdrawal 31st August 1983.

Length:	(a):	26ft 0in	7.92m	Height:	(a):	14ft 7in 4.45m
	(aa):	27ft 0in	8.23m		(ab):	14ft 4in 4.36m
	(ab):	27ft 8in	8.43m		(ab):	14ft 6in 4.42m
Width:	(a):	7ft 6in	2.29	Wheelbase:		16ft 4in 4.98m
	(aa):	7ft 10in	2.39m	Unladen weight:	(a):	8,027kg
	(ab):	7ft 6in	2.29m		(ab):	8,209kg
Gross vehicle weight:		12,700kg			(ab):	8,090kg

Daimler(a)'s with Traditional Radiators.

These Daimler CVG5's were introduced from 1949 to 1953 and in most respects followed contemporary British practice. The steel-framed bodywork was pre-fabricated by Metal Sections Ltd. shipped 'ckd' and assembled by KMB using pre-shaped aluminium panels. The frames of the original sliding windows were of steel, painted black, but later in their lives these were gradually replaced by similarly shaped aluminium frames. As new, all exposed radiator buses were fitted with half-drop upper-deck front windows but, during the mid-1950's, these were changed to full-depth sliding windows. This was at about the time that the first CVG5's with 'Birmingham' fronts were entering service (1954) fitted with sliding windows as new.

The usual specification in Britain for a double-deck bus at that time called for a seating capacity of 56; 30-seats on the upper deck and 26 on the lower saloon. The Kowloon buses, by comparison, had seats for 32 and 29 respectively, the rows of seats being more closely spaced than was possible in the UK by virtue of the shorter stature of the average Hong Kong Chinese in those days. The additional two seats on the upper deck were achieved by adding two single seats at the front - so close were they that had double seats been fitted there would have been insufficient space for the conductor to drop the flap of the destination box in order to change the route details. On the lower deck there were twin seats over each rear wheel-arch (triple seats were then usual) and the front row was reversed to face rearward, providing seats for five.

When introduced, the fleet name was applied in full in both block capitals and in Chinese characters on the cream band, below the lower-deck windows on each side. By 1953 this had been superseded by the Company crest featuring nine Chinese dragons in a circle (Kowloon translates as 'Nine Dragons - 'Kau Leung') together with the Company name in both scripts, plus the initials 'KMB' inside the circle. The registration number appeared below this crest in cream and served as a fleet number until May 1974, when the fleet was numbered.

Early examples featured a traditional 'British' open rear platform, but operating conditions in Kowloon were such that they soon proved to be unsafe and a tubular-metal sliding gate replace the single stanchion in the centre of the platform. At first the bodywork to the rear of the platform remained cut-away, around the nearside rear corner of the platform but, in time, this was enclosed so that the rear edge of the sliding gate completely protected the whole platform in much the same way as was the case where rear platform doors were fitted on buses in Britain. The rear window was replaced with a slightly wider frame that doubled as an emergency exit.

Front entrances was cut into the front bay in 1955 and these were also protected by gates which were power operated and, at first, were provided with an external button for use by

kerbside regulators. Possibly because the power equipment proved unreliable, the gates were later operated manually by the conductor or the on-board gateman whose job was to prevent passengers from forcing their way on or off and to help close the gates against the tide of frustrated humanity when the bus was full. It is a reflection on the low cost of labour that there were, until 1964, two conductors - one for each deck, a driver and a gateman - a crew of four to each double deck bus; all this with a fare on most urban routes as low as 10 cents (HK) or about 0.5p. No buses with traditional exposed radiators were built new with the additional front entrance.

Single aperture, flush fitting, destination boxes were provided front and rear, showing destinations 'to-and-from' in English and Chinese, with a double-ended arrow between the inner and outer destinations. A route number appeared at the extreme nearside, still part of the single blind. Additionally a small aperture immediately above the rear platform showed the route number only. A 'bus-full' sign was fitted over the lower-deck nearside front window, under the canopy, in a position similar to that occupied by the route number on London RT and RM types. Hong Kong regulations require that 'bus-full' signs be displayed clearly, as applicable, and KMB's version was a

ABOVE: HK4007 seen here circa 1952 when passing the Kowloon European YMCA, having just left "Star" Ferry for Sham Shui Po, in north Kowloon whilst operating on Route 2. The 'to & from' style of destination blind can be seen as can the slatted wooden seats. A sliding gate had been provided by this time *(Fred York*

LEFT: Circa 1955-58. The second half of the 1950's saw the addition of a front entrance, also protected by a sliding gate. The upper-deck front windows wad been fitted with sliding panes in replacement of the original half-drop type which were prone to jamming. *(Kowloon Motor Bus*

82

ABOVE: "Star" Ferry line-up circa 1952 with CVG5's HK4004, HK4057 and HK4074, together with a trio of Bedford OB's. Of the CVG5's, only HK4004 survived sufficiently long to be allocated a fleet number in 1974. *(Fred York*

hinged red metal plate that was exposed by the driver who could reach out from his cab to reach the 'flag'. The words 'bus full' appeared in white in English and Chinese. Later in the life of these long-lived buses, the KMB standard destination equipment was fitted and this retained the single glass screen but the destination and number blinds were separated, the route number remaining to the nearside on a single track blind, while the ultimate destination was shown, bilingually, as the main display and was changed at each end of the route. Large, chromium plated headlights complemented the chrome-plated, fluted-top Daimler radiator in a very traditional way.

BELOW: This offside view of HK4009 shows clearly the traditional appearance of the Metal Sections bodied KMB Daimler CVG5 - the Daimler(a)'s. Indeed, they would not have looked out of place in a 1940's British High Street. The cream band extending across the driver-'s cab was not typical of those buses with traditional radiator shells. *(Lyndon Rees*

ABOVE: A 1950's photograph showing a Daimler(a) with dual-entrances, fitted with sliding, tubular metal gates, sliding windows at the upper-deck front and single-display (to & from) destination blinds. *(Mike Brunning*

BELOW: Doors replaced gates during 1964/65, as can be seen in this photograph of HK4001 which was passing the Peninsula Hotel, whilst working on Route 7 in 1966. *(Mike Davis*

ABOVE: Seen here passing the same hotel eight years later, this 1974 photograph of HK4225 shows the yellow bands across the front panels, indicating that the bus was operated by a single conductor who was seated. The stripes were to encourage intending passengers to have their fare ready in good time. *(Mike Davis)*

BELOW: Speeding along Jordan Road to its terminus at the Jordan Road Ferry (Yaumati vehicular ferry), during the summer of 1974, HK4002, by then fleet numbered D41, displays the livery adopted for Driver Only Operation (DOO), following the widespread introduction of farebox fare collection on these buses. The missing bonnet-side, together with the battered front wing and oil covered rear hub, were typical of these old faithfuls at that time. The aluminium passenger doors were usually left unpainted by KMB. *(Mike Davis)*

85

ABOVE: HK4044 stops under an array of advertising signs in Nathan Road, Yau Ma Ti, during the mid-afternoon off-peak. The passengers are forced to walk to the front entrance in order to board and then drop their money into the farebox. Drivers the world over stop with the entrance past the stop! *(Mike Davis)*

BELOW: HK4002 became D41 in 1974 and is seen here leaving Jordan Road Ferry shortly after having its traditional radiator replaced by a moulded cover of the so-called 'Manchester' type as part of KMB's 'face-lift' for elderly buses. It was sent back to Route 4 after the modification. *(Mike Davis)*

ABOVE: An offside view of D29 at the "Star" Ferry late in 1976, after being reconditioned and provided with a moulded glass-fibre 'Manchester' bonnet and grille. Buses thus treated remained classified as Daimler(a). *(Derek Lucas*

BELOW: 1953 saw the whole British Commonwealth and Empire hang-out the flags for the Coronation of Queen Elizabeth on June 2nd. Kowloon Motor Bus followed suit with their loyal tribute, shown here by HK4057, carrying a pair of Union Flags and a Royal Cypher. *(Fred York*

Daimler(a)'s with Birmingham Grilles.

Although most of the foregoing detail about the exposed radiator Daimler CVG5's applied equally to those built with the so-called 'Birmingham' fronts, there were, however, certain exceptions. The upper-deck front windows were always of the sliding type, from number 4201, the first of this sub-type. All were introduced with the KMB crest in place of the fleet-name in full along the side and headlamps were of the smaller pre-focus type set into the side of the grille. The front wheel pressing was less flattened than those fitted to earlier CVG5's with exposed radiators. No 'Birmingham' fronted buses ever ran without the sliding metal gates across the rear platform.

Ninety 'Birmingham' fronted Daimler(a) type were introduced between 1954 and 1957 and the last batch of thirty were of interest in that they were the first to be provided with a separate, single-width, front gateway. Additional front entrance/exits had begun to appear on earlier stock as modifications as early as 1955 but HK4322 to HK4351 were purpose built, the Metal Sections design being modified to suit the extra opening which was immediately behind the front bulkhead. A result of this was that the lower deck seating was reduced by two seats, the 'back-to-the-engine' five seat bench behind the driver being reduced to three; the resulting arrangement being H32/27FEX,REX both entrances being protected by sliding gates.

In other respects the 'Birmingham-fronts' resembled those with exposed radiators.

For comparison -

ABOVE and BELOW: Traditional and 'Birmingham' front designs compared. The two types of Daimler(a) as delivered are shown here in two almost identical poses. Above can be seen the original type with half-drop upper-deck front windows while those of the 'Birmingham' type (below) have laterally sliding units. The earlier type also displays flat front wheels and the cream band does not extend around the cab and engine areas. Both pictures, taken with contemporary single-deckers, date from 1955. *(Fred York*

ABOVE: Birmingham-fronted Daimler(a) 4202 with rear platform gates when fairly new in 1954. *(Fred York*

BELOW: 4201 turning into Jordan Road from Nathan Road in 1966, having then recently been fitted with platform doors. *(Mike Davis*

ABOVE: HK4322 headed for "Star" Ferry in 1973, carrying the yellow stripes indicating that it is One-Conductor-Operated (OCO) with that conductor seated behind a small cash-desk just inside the front entrance. *(Mike Davis)*

BELOW: HK4342 after being fleet-numbered D141 in 1974. It is in the cream over red, half-and-half livery indicating a Driver Only Operated (DOO) bus and was working on Route 2c in 1975 *(Mike Davis)*

LEFT: 'Birmingham' fronted Daimler(a), D16 was one of a number of its sub-type to acquire a 'V' slotted radiator grille shortly before the entire remaining fleet of Daimler(a)'s were rebuilt with 'Manchester-style' fronts, after which, superficially, they resembled the Daimler(b)'s. D16 was leaving Kwun Tong Ferry in 1975. *(Mike Davis)*

A Manila Aside -

At the same time as KMB purchased its Metal Sections bodied, 'Birmingham' fronted Daimler(a)'s, two additional chassis and body-kits were shipped by the Hong Kong Daimler agent, Messrs. Dodwell Ltd., for a client, 'Matorco' in Manila, Philippines, which lies almost due south of Hong Kong. Right-hand drive was retained but the platform was transferred to the right to suit the local rule of the road.

ABOVE: Two views of YH831 photographed in the Matorco depot in 1977, in near original condition. *(John. Shearman)*

BELOW: Only one survived when seen in March 1984, having lost more than just its roof! The engine and gearbox had been replaced by Isuzu units, made in Japan and the registration had become NXH275. The route operated at that time was along the magnificent Roxas Boulevard, from Rizal Park to Paranaque and return, running alongside former US single-deckers with seats on the roof, surrounded by railings. *(Mike Davis)*

RIGHT: D104 had a completely new bonnet and grille to replace its original 'Birmingham' type in 1974, before it was modified to take a more modern 'Manchester' type along with all Daimler(a)'s, regardless of original frontal styling. (Mike Davis

Daimler(a) Class as a Whole.

From this point on, unless specifically mentioned otherwise, all reference is to both exposed-radiator and Birmingham-front buses. From early 1956 new deliveries were built with an additional front entrance/exit and existing buses were modified to this pattern. Seating was maintained at 32/27 by reducing the size of the seats over the wheel arches from three to two each side and adding, in the space made available two normal forward facing seats on each side. At the same time the rearward facing front bulkhead bench seat was reduced from five to three, so allowing free passage through the single width forward entrance. Wooden slatted seats were fitted from new until about 1964 when woven plastic-coated rattan (bamboo strip) was introduced. Unfortunately this was found to harbour bugs, such as lice and fleas, and so was dropped.

1964/5 saw the universal introduction of route indicators showing only the journey destination, again bilingually, which required changing at each terminus. Route numbers were given a separate roller, still on the right, behind a single screen. To accommodate the new gear the destination boxes were raised, proud of the front and rear panels, by about one inch.

During 1964/5, together with all double-deckers, power operated folding doors were fitted in place of the gates and, for safety, an electric bell rang as the doors closed, warning intending passengers to keep clear. These bells did not last long, however, and fell out of use as they failed and were left unrepaired. Gate-men became redundant at this time, reducing the crew to three.

A major change in operating procedure commenced when, in 1971, the Company, in an effort to reduce the labour intensity of its operations, commenced one-conductor operation, with the conductor seated at a small cash desk, just inside the front doorway which became "entrance only". The rear stairs remained and the rear platform doors provided the exit. For this style of working, several rows of seats were removed in the lower saloon to make more space for passengers to stand whilst waiting to pay their fare without delaying the departure of the bus. This reduced the lower deck seating to 18 but 30 standing passengers were allowed. Thus KMB had reduced the wage bill on these relatively low capacity buses by one conductor. At this time urban fares were increased by 10 cents - up to three miles became 20 cents and longer journeys 30 cents with certain 40 cents exceptions. In order to warn passengers that an approaching bus was manned by a seated conductor, broad yellow stripes were painted between decks, across the front of the buses operating this system. As a result of the success of this scheme and with little or no tendency for Chinese passengers to cheat by boarding by the rear doorway the Company in 1974, having established "passenger-flow" embarked upon replacement of the seated conductor by a Bell Punch "Farebox" mounted on the front bulkhead under the eye of the driver.

For one-man/Farebox operation, buses were reseated to 24 in the lower saloon and standing reduced to 17, while outside, the buses were painted in the same livery adopted for the OMO tunnel buses of red lower deck, cream upper deck and green lining out. Yellow-on-red reflective "money-box" signs were applied front and rear and new certificate of fitness issued for three years.

During 1975/6 most of the Daimler(a) class, both exposed radiator and Birmingham, were fitted with "Manchester" style glass fibre bonnet mouldings and grilles making them almost indistinguishable from the B-class which had this front from new. The last two buses to retain their traditional frontal design, D51/71, were withdrawn late in 1977, prior to rebuilding, at which time they were to receive new "Manchester" bonnets, moulded by KMB at their Kwun Tong depot.

At the time seated conductors were introduced rear destination indicators were removed leaving only the route number. The side route number was removed completely.

A new type of moulded glass-fibre seat was also introduced on many buses at about the same time - hard, but a great improvement on wooden slats!

Reconstructed Buses.

Almost unbelievably on 7th July 1976 D149 was relicensed and returned to traffic having been completely reconstructed from chassis up and given a new body of a new design. It thus became obvious that KMB anticipated keeping this bus in service longer than the normal one year recertification given buses of this age. Depending upon condition these Daimler(a) CVG5's were being variously reconditioned, still as Daimler (a) or reclassified as Daimler (aa) or (ab) - KMB were always anxious to impress upon the public the extent to which they have reconstructed the original components for they have been frequently criticised for running old stock. In fact only four examples have been rebodied.

Daimler(a)

These buses were reconditioned mechanically and the body reconstructed to the original MetSec outline, frames being replaced as necessary. Door and staircase positions remained. There was no change in engine, seating or appearance so they remain Daimler (a) class. This includes all those originally fitted with Traditional radiators and all but four of those once with Birmingham fronts that remained in service on 31st November 1978.

Daimler(aa)

D92 and D149 (HK4260 and HK4350) were not only rebodied but also re-engined to have 108bhp Gardner 6LW units in place of their original 5LW units, and were fitted with air brakes in place of engines vacuum. They differed in a number of respects from the later design of the (ab) and similarly treated (ba/bb) class. The roof of the (aa) was flatter and lower; overall height being 14ft 4ins. The rear overhang was extended by one foot to bring the overall length to 27ft. The original chassis width of CVG5's was 7ft 6ins but the two (aa)'s were rebodied as 7ft 10ins wide, so that there was a two inch overhang, almost noticeable at each wheel-arch, made more apparent by the straight side panels and generally box-like appearance. The new bodies had single width front entrances, double width centre exits, and centre staircase. The radiator grille was replaced by using a later pattern Daimler CVG6 type, i.e. wider towards the lower edge. Seating was increased to 41/26+17 giving a total of 84. As already mentioned, D149 was relicensed on 7.7.76 while D92 was not returned to service until 4.10.76.

Daimler(ab) type

D90/1 (HK4259/60), two former Birmingham-fronted buses, were completely rebodied using the design adopted as standard for all later (b) type conversions. This body was 14ft 6.25ins high and 7ft 6ins wide which is normal for a bus of this type. The length however, was extended at the rear overhang by 1ft 8ins to 27ft 8ins overall, giving a passenger capacity of 41/28+18=87. The same front entrance, centre exit and staircase layout as on the (aa) class applied. The engine remained the 85bhp Gardner 5LW. D90/1 were relicensed following reconstruction on 5.10.77.

Although, from the early 1970's, KMB were forced to manufacture quantities of spare parts themselves, the old Daimler(a)'s were running in 1976 at 90% availability - a fact which must have influenced the decision to retain them. In fact, KMB invested heavily in plant and, as a result, almost every item required could be 'home-made', either at Lai Chi Kok headquarters, or in the bodyshop at Kwun Tong or Shing Lok depots. The installation of a gear-cutting machine permitted even major units to be produced. Thus, these 1949-57 Daimlers faced the 1980's completely reconditioned for at least three more hard working years. Other factors, such as public image and official of such elderly vehicles militated against the Daimler(a)'s and three years was to be the very limit as will be seen below. At the time of their final withdrawal in 1982/3, the nearest rivals for bus longevity were only London Transport's 'RT' class, but now overtake by rejuvenated 'Routemasters' operated by the same concern.

Withdrawals.

The first Daimler(a) of all to be withdrawn was, for some unknown reason. No.4205 which went in November 1954, after just 7-months in service. This was followed, in December 1969 by HK4149 which was scrapped. A further example was to be seen after withdrawal in the early 1970's in use as offices at Kai Tak International Airport. On 1st May 1977, HK4059 was donated to the Hong Kong Road Safety Association, who made some minor alterations but continued to operate it as a double-deck vehicle. It thus became the first double-deck ex-bus to operate in Hong Kong while not being the property of either KMB or CMB, if one discounts a short private tour of Hong Kong in May 1975 by Macau LD206, a Bristol 'Lodekka'.

During the influx of Vietnamese refugees in 1979-81, large camps were established by Hong Kong Government at Argyle Street and at the former RAF Station adjacent to Kai Tak civil airport. For the amusement of pre-school

BELOW: D124, was originally 'Birmingham' fronted before being fitted with this 'Manchester' moulding. This bus was amongst the last survivors of the type in passenger service, having been kept for use on Route 54 which, when this photograph was taken in July 1983, still required buses 7ft 6in (2460mm) wide. *(Mike Davis*

refugees, KMB donated a number of old buses, including a Daimler(a), for use as play buses.

Although KMB's driving school received fourteen Daimler(a)'s during 1980, including the oldest survivor, D26 (No.4961), large scale withdrawals in 1981/2 were almost entirely for scrapping, with the exception of D49 (HK4031) which was donated to the Royal Hong Kong Police for use by the Police Driving School and was painted in Police blue and white by staff at the school. D1 and D111 were purchased by dealer Transport Supplies Worldwide (Hong Kong) Ltd. as overhauled 'runners' for further sale to a potential customer. Of those buses which were originally equipped with 'traditional' radiators, the last was withdrawn in August 1982, while seven, formerly with 'Birmingham' grilles survived a further year until August 1983 and included (aa) and (ab) sub-types for use on narrow roads in rural areas.

ABOVE LEFT: Rebodied Daimler(aa), D149, shortly after its return to service in 1976. This front view shows-off well the 7ft 10in wide body on a 7ft 6in wide chassis. *(Both Derek Lucas)*

ABOVE RIGHT: D149 seen shortly before withdrawal in 1983 as it left Shung Shui bus station on Route 78 which ran to, the border town of Sha Tau Kok. *(T. V. Runnacles)*

BELOW: A fuller view of D149 in its early rebuilt days as it stands at Tsuen Wan Ferry terminus circa 1977/8. *(Ian Glass)*

ABOVE: 1957-built Daimler(a), HK4259, rebodied in 1977 as KMB's Daimler(ab)-class No. D91. This bus was 7ft 6in wide and, like the (aa)sub-type (D149) had a centre staircase. Seen here standing outside Star House in January 1977 *(Derek Lucas*

BELOW: The other Daimler(aa) was D90, seen here at work in 1983, on the 14B, between Yau Tong and Ngau Tau Kok. *(Mike Davis*

AFTER WITHDRAWAL some Daimler(a)'s were to be seen in other guises.

TOP RIGHT: The Royal Hong Kong Police took HK4301 for use in the Police Driving School and completely stripped and refurbished it in its own dark blue and white colours. The project was dropped shortly after and the vehicle scrapped. *(Mike Davis*

CENTRE - LEFT and RIGHT: D12 was used as a uniform store and survived to be purchased by Speedybus Services Ltd. who restored it to 'as withdrawn' condition. *(John Shearman (left) and C. Lau (right)*

BOTTOM: The oldest surviving Daimler(a), 4961 from the initial batch of four, was retrieved from the driving school by KMB themselves and was restored to as near original condition as was possible. Here the bus was on exhibition in June 1983, the 50th anniversary of KMB gaining the franchise to operate in Kowloon. *(Mike Davis*

TOP: There are only two recorded occasions when a licensed, KMB owned, CVG5 crossed the harbour to Hong Kong Island. The first occasion was in April 1979 when D26 (4961), by then the only surviving CVG5 from day one (17th April 1949) of double-deck bus operation, undertook a tour, which included the Island. This was to commemorate 30-years of double-deck buses in Hong Kong. The tour commenced from as near as possible to the 1949 terminus at Kowloon City and ran to "Star" Ferry where 4961 stood for a few minutes in the bus station. It then toured the New Territories before crossing to Hong Kong-side via the Cross-Harbour Tunnel. Whilst touring the Island, a visit was even made to the roof of CMB's Chai Wan multi-storey bus depot. 4961 is seen here at Kowloon City at the start of its commemorative run. Kowloon City was the outer terminus of the original double-deck bus Route, appropriately route No 1. (John Shearman)

CENTRE: The second occasion when a Daimler(a) paid a visit to Hong Kong-side was in 1981, when 4222 (D17) took part in the hand-over ceremony of the first MCW 3-axle 'Metrobuses', one each for KMB and CMB. As part of the tableau, 4222 was joined by CMB's restored 1949 Tilling-Stevens, 4943, masquerading as HK104. Thirty-five years of development was on display, even though the KMB Daimler was of 1954 vintage, it represented the type that introduced the double deck bus to Kowloon - and indeed Hong Kong - in 1949, the year not having been forgotten by CMB in friendly rivalry. (John Shearman)

LOWER RIGHT: Also seen here on Hong Kong Island, but not in KMB ownership, is the former KMB Daimler(a) HK4059 which had been donated to the Hong Kong Road Safety Association for use as a mobile classroom. It was photographed on Connaught Road, Central, circa 1980 in its colourful new livery. (John Shearman)

Daimler CVG5DD - Daimler(b)

D151-260
(Fleet numbered from 1974)

In October 1959 the first of the 27ft long x 7ft 6ins wide Daimler(b) class CVG5 double-deck buses entered service and, like the slightly shorter Daimler(a) class were equipped with pre-selective gearboxes, fluid fly-wheels and were, of course, powered by the Gardner 5LW engine, de-rated to 85bhp to obtain smoke-free exhaust in the hot, humid climate. The Metal Sections bodies were again locally assembled having been shipped to Kowloon ckd. Apart from having about one foot longer rear overhang than the Daimler(a) class, they showed very little detail change.

The most obvious evidence of the extra length was in the upper deck rearmost side window which was longer on the (b)'s and in the rear door pillar immediately below it, which was wide enough to house the route number screen. Whereas the 'Birmingham' fronted Daimler(a)'s had pressed steel grilles, these Daimler(b)'s had the very smart moulded glass-fibre bonnet and radiator grille chosen as standard by Manchester City Transport which incorporated the near-side front wing, with its integral headlamp.

Top deck seating (double front seats) was higher than the Daimler(a) at 38, while the lower deck figure was 27 plus 20 standing. The dual entrance/exits, a feature from new, were equipped with sliding tubular metal gates until 1964/5 when KMB fitted all its double-deck buses with power-operated folding doors.

A total of 110 Daimler(b) CVG5's had been delivered when the last batch entered service in May 1961. All these buses were altered to one (seated) conductor (OCO) from 1971/2 and later as OMO/Farebox buses from 1975. For both these operations passenger-flow was introduced with front entrance/rear exit, retaining the rear stairs, as on the Daimler(a)'s.

For use as OCO buses the body layout became H38/19+25FG, RXS D, but later on for OMO they were altered to H38/26+18FE, RX S, D.

It is interesting to note that registration numbers carried by some of this type were originally carried by earlier buses many of

VEHICLE SPECIFICATION: DAIMLER(b)-type.

Registration numbers:	Discontinuous - see fleet list.	
Fleet numbers (from 1974):	D151-260	
Chassis:	Daimler CVG5DD	
Engine:	Gardner 5LW (b) & (ba), or Gardner 6LW (bb).	
Gearbox:	Wilson-type, preselective.	
Body make: original:	Metal Sections Ltd.	
(ba) & (bb):	Re-bodied by Kowloon Motor Bus.	
Body layout (MetSec): Until 1972:	H38/27+20FEX,REXS,G. - D (Doors) from 1964	
1972-1975:	H38/19+25FE,RXS,D,OCO.	
1975-1983:	H38/26+18FE,RXS,D,OMO farebox.	
Body layout (rebodied buses) (ba) & (bb):	H41/28+18FE,CXS,D,OMO farebox	
Date introduced:	1959-61	
Date rebodied:	1976-9	
Original total:	110	
Total at 1974 fleet numbering:	110	
Total 30 June '83 (b):	4	
(ba):	6	= 15
(bb):	5	
Length (b):	27ft 0in	8230mm
(ba):	27ft 8in	8430mm
(bb):	28ft 0in	8530mm
Width:	7ft 6in	2290mm
Height: (b):	14ft 8in	4470mm
(ba) & (bb):	14ft 6in	4420mm
Wheelbase:	16ft 4in	4978mm
Unladen weight (b):	8,230kg	
(ba):	7,976kg	
(bb):	8,205kg	
Gross vehicle weight:	12,700kg	

LEFT: The Daimler(b) class CVG5's were introduced from October 1959 and were fitted with the 'Manchester' style moulded glass-fibre bonnet, front wings and grille as new. At 27ft, they were one foot longer than the previous Daimler(a) type and were fitted with dual entrances and sliding gates as built. *(Kowloon Motor Bus)*

ABOVE: Daimler(b), 4596, passing the Peninsula Hotel in 1966, after being fitted with power doors at both entrances. *(Mike Davis*

BELOW: HK4468 was also passing the 'Penn.' but by this time, 1977, has acquired the yellow bands indicating its status as an OCO bus. *(Mike Davis*

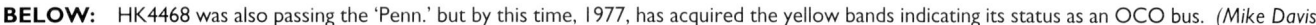

ABOVE: Like the older Daimler(a) sub-type, the (b)'s were also fitted for DOO-farebox operation in 1974/5 and received the half-and-half livery to indicate this. Here D65 (4238) approaches "Star" Ferry on the 7A from Wang Tau Hom in 1974. *(Mike Davis)*

BELOW: Daimler(b) CVG5, 4983 rebodied as a Daimler(ba), retaining the original Gardner 5LW engine. Because each body was individually built, many detail differences existed between supposedly similar vehicles. D214 seen here departing from Jordan Road in January 1978. *(Mike Davis)*

which were Bedford OB's of 1946. The numbers concerned were those without prefix letters.

Prior to the introduction of double-deckers on New Territories trunk routes, 50 and 70, in July 1973, number 4602 had been converted into an open top tree pruning bus. At first the rear dome was retained to act as a guide to the clearance necessary for the safe operation of double-deck buses but later on the rest of the roof was removed and the bus took on the appearance of a derelict open-topper. The remainder of the bus was scrapped during May 1978.

1976 Reconstruction Programme.

To help meet the traffic requirements for the 1980's, when the effects of the Mass Transit Railway could be better assessed, KMB, having invested heavily in plant to manufacture spare-parts for its fleet of ageing Daimler CVG5's, embarked upon a scheme to modernise the Daimler(b) buses. Some were rebuilt to the original MetSec body design, while others were completely reconstructed using a similar KMB designed and built body as those already described under the 'Daimler (ab)-type.

Daimler(b)

These were those buses which retained their original outline but were completely reconstructed, complete with rear platform and staircase. Seating remained as before at H38/28+18FE,RXS,D,DOO.

Daimler(ba)

The (ba)-type buses were those rebodied using a similar body to that designed for the Daimler(ab)-type. These had forward entrances and centre exit and were more suitable for DOO and had a layout of H41/28+18FE,CXS,D,DOO. The overall dimensions of the rebodied (ba) type were 27ft 8ins long, 7ft 6ins wide and 14ft 6.25ins high.

Daimler(bb)

These buses were to a similar design to the (ba) type, except that they were fitted with the more powerful Gardner 6LW engine at the time of their reconstruction. In order to accommodate the longer 6LW unit, a bonnet extension of four inches was necessary, with the result that the grille protruded noticeably in front of the cab. This brought the overall length to 28ft, excluding the bumper-bar at the front, while the width and height were as for the (ba)-type.

Daimler(b) Withdrawals

The first Daimler(b)'s to be withdrawn were HK4482 & HK 4483 which were delicensed on 25th October 1973, before fleet numbering. Interestingly, despite having been withdrawn the previous year, these two Daimler(b)'s were allocated fleet numbers in 1974 at the time of the general numbering. Possibly there had been an intention of rebuilding them as was the case with the (ba) & (bb) sub-types. The next was the first of its class, D151 (4214), delicensed on 25th August 1976, and followed by a handful in 1977. A single example was donated in 1981 to the Vietnamese refugee camp for 'boat people' at Argyle Street, for the use by them as a play school (as opposed to a playbus). Large numbers were scrapped or transferred to the training fleet during 1981 and '82 and those surviving into 1984 were of the (ba) & (bb) types, D178 becoming the very last in service and the only Daimler(b) of any sub-type to be painted into the cream livery or to carry KMB's 50th Anniversary logo and inscription. It was finally delicensed on 31st July 1984 after a spell on Route 74K from Tai Po.

Of the survivors, 33 were taken into stock by the driving school and 4594 was properly converted to a tree-pruning vehicle and re-licensed as 'Goods Vehicle' with a black registration number plate and white numerals.

The withdrawal of D178, a Daimler(bb), brought to an end the era when the Daimler CVG5 graced the streets of Kowloon and the New Territories where D178 had worked its last years on the 74K, uniquely for the CVG5's, in KMB's 1982 mainly-cream livery, plus 50th anniversary slogans..

BELOW: Also in January 1978 and at Jordan Road, D165 with engine conversion to the six-cylinder Gardner 6LW as well as a rebody. It was re-classified as a Daimler(bb). *(Mike Davis*

ABOVE: A rear view of rebodied Daimler(ba), D237 (HK4478) as it turns right into Nathan Road from Salisbury Road, Tsin Sha Tsui, in January 1978. *(Mike Davis*

LEFT: A close-up of the front of D247, a Daimler(bb) with extended bonnet to accommodate the extra length of a six-cylinder Gardner 6LW engine. *(Mike Davis*

BELOW: Daimler(ba), D173, seen here whilst working-out its last summer at Tai Po on Route 74K, before being withdrawn in January 1984. The rear hub-cap is unusual and unexplained. *(Mike Davis*

ABOVE: Daimler(bb), D191, leaving Yuen Long on Route 54 in June 1983. It was scrapped in the following March. *(Mike Davis)*

BELOW: The only example of any type of Daimler CVG5 to be painted into the '1982' livery or to have the 50th anniversary slogans applied, was D178 which was on Route 74K at Tai Po in June 1983. *(Mike Davis)*

ABOVE and BELOW: Both rebodied and unrebodied Daimler(b)'s were latterly used in the driver training school. 4613, **ABOVE**, lays over at Hung Hom Ferry in the company of Citybus D8 (ex-WMPTE) in the summer of 1983, while 4991, **BELOW**, a, rebodied Daimler(ba), stands alongside an original MetSec bodied (b) in May 1984. *(Mike Davis, top; John Shearman, bottom.*

UPPER LEFT: 4602 was converted into a tree pruning vehicle in 1973, prior to the wholesale introduction of double-deck buses in the New Territories, commencing 16th July 1973. It remained licensed as a bus until April 1974 at which time it was 'converted' to a goods vehicle. The rear dome was retained to act as a template, or guide, to indicate the necessary clearance, although this was later removed completely. *(Stan Leeds for Mike Davis*

CENTRE: 4594 was cut down to replace 4602 on tree pruning duties. It was standing at Yuen Long in February 1984. *(John Shearman)*

BOTTOM: For use within its depots, KMB cut-down a number of CVG5's to act as lorries to shuttle parts, etc. around the site. Here 'KB3' is an example of unknown previous identity used in Kwun Tong depot. *(John Shearman)*

Daimler CVG6 - Daimler(c)

D261-330
(Fleet numbered from 1973/4)

The first 30ft long by 8ft wide double-deck buses to be bought by KMB were 70 Daimler CVG6 chassis with Gardner 6LW engines rated at 108bhp, upon which were mounted Metal Section bodies - shipped ckd and assembled by KMB in the usual way. The first of these buses entered service in January 1962 and the last to be completed was licensed at the end of August. The basic design was that of a traditional rear staircase, rear platform double-decker and, as on the Daimler (a) and (b) types, there was an additional front entrance for lower deck passengers only. Originally the platforms were protected by the usual tubular metal sliding gates but during 1964/5 these were replaced by folding power-doors. The original body layout (H42/26+31) was for high capacity crush-loading. Large standing areas were provided on the lower-deck by sacrificing four seats. During the 1965-7 period at least some of the class had even greater standing space provided when lower-deck seating was reduced to 19 in order to provide space for 38 standees - with two roving conductors! This severe loading caused rear axle problems and seating was returned to the more normal H42/30+20FEX, REXS, D in the late 1960's.

AD4727.
One bus of this type, AD4742 had a different style body to the normal MetSec type. The

BELOW: Daimler(c), AD4751, photographed when fairly new and prior to having power operated doors in place of the original sliding gates. The destination blind was, at that time, still one-piece, without separate route numbers. These were the first KMB buses to be built 8ft wide. *(Kowloon Motor Bus*

VEHICLE SPECIFICATION: DAIMLER(c)-TYPES.

Registration numbers:	AD4717-86
Fleet numbers:	D261-330
Chassis:	Daimler CVG6.
Engine *:	Gardner 6LW, 108bhp - all except:- (cc) type: Gardner 6LX, 150bhp
Transmission **:	Fluid-flywheel; .Wilson-type pre-selective gearbox - all except:- (cc) type: semi-automatic
Body make:	Metal Sections - except D286, possibly BACO.
Rebody (cc) type:	KMB
Body layout:	as built: H42/26+31FEX,REXS,G. - D from 1964.
	c1965/7: H42/19+38FEX,REXS,D.
	1969-72: H42/30+20FEX,REXS,D
	OCO: H42/23+42FE,RXS,D,OCO
	(c): H42/32+19FE,RXS,D,OMO farebox
	(ca): H44/31+24FE,CXS,D,OMO farebox
	(cb): H44/31+23FE,CXS,D,OMO farebox
	(cc): H44/31+24FE,CXS,D,OMO farebox
Date introduced:	1962
Rebuilt:	From 1974
Original total:	70
Sub-type totals:	(c): 14
	(ca): 16 before withdrawls
	(cb): 29
	(cc): 11
Length: (c) (ca) (cb):	29ft 10in 9093mm
(cc):	29ft 11½in 9125mm
Width: (c) (ca) (cb):	7ft 11in 2410mm
(cc):	8ft 0in 2440mm
Height:	14ft 6 in 4470mm
Wheelbase:	18ft 6in 5640mm
Unladen weight: (c) (ca) (cb):	9,043kg
(cc):	9,081kg
Gross vehicle weight:	14,931kg

NOTES: * All engines originally 6LW.
  ** All gearboxes originally pre-selective

ABOVE: Daimler CVG6, AD4737, had been fitted with doors when photographed in August 1966. *(Mike Davis*

BELOW: AD4742 had a slightly different body style to the remainder of the Daimler(c) class, including flat 'slab' sides and narrower than usual mouldings beneath the upper and lower deck windows. Investigation showed that this was a one-off body kit produced by British Aluminium Co. (BACo); all other Daimler(c)' had Metal Sections bodies. *(Mike Davis*

ABOVE: In 1972, the Daimler(c) class were converted to OCO and received the yellow stripes across the front to indicate this to the public. D274, seen passing the Peninsula Hotel in 1974. *(Mike Davis*

BELOW: Daimler(c), D294, operating as a Farebox-Driver Only Operated (DOO), passenger-flow, bus in January 1978, when it was seen leaving Jordan Road Ferry on the long trunk Route 50 to Un Long (Yuen Long). *(Mike Davis)*

major external difference was that it had flat lower-side panels between the wheel-arches and was thus very similar to bodies prepared by the British Aluminium Co. (BACo) and fitted to the 1963 batch of AEC Regent Mk V's. Other differences included narrower moulding bands at waist level and a modified front passenger step arrangement. BACo did produce a 'one-off' body for KMB at that time and AD4742's body did have aluminium frames, so this was almost certainly a BACo product. Like the MetSec versions, this 'odd-man-out' was assembled in Kowloon by KMB engineers. It is not known if there were 69 or 70 MetSec bodies.

During 1971/2 the Daimler (c)'s were converted to one seated conductor operation in the same manner as the Daimler(a)'s and (b)'s, for which they were reseated to H42/23+42FE, RXS, D (sc). Following the success of KMB's so-called One-Conductor-Operation (seated conductor) the Daimler (c) type was converted to OMO/Farebox. For this all the lower deck seats were reinstated so that the seating and layout was changed yet again to H42/32+19FE, RXS,D + Farebox. A major rebuilding programme was commenced in 1974 which involved the removal of the rear staircase and doorway to a central position while retaining the original front doorway. This was achieved by three different types of conversion; (ca), (cb) and (cc).

Daimler(ca).

Daimler (ca)-type conversions commenced in mid-1974 when D278 and D317 were taken into workshops and, within the MetSec body shell, had the rear doorway and staircase removed to the central position. D278 was the first.converted bus to return to service. In the (ca) conversions the centre doorway occupied the third bay from the front of the lower-deck with another full bay between it and the rear wheel-arch. The four-piece front doors, to the original pattern remained but were painted red and cream to correspond to the livery and lining-out of the adjacent side-panels. The cream waist-band was not continued across the front of the drivers dash or front bulkhead, while the destination box protruded about one inch from the front panel. The Company crest was applied to the side panel behind the centre doors. Seating and revised layout became H44/31+24FE, CXS, D + Farebox.

Daimler(cb).

The (cb)-type differed from (ca) in a number of ways, although the original body-shell also was used. The principal difference being that the centre exit was not as far forward as on the Daimler (ca), being set back a further two-thirds of a bay, therefore it did not correspond with the body frame, necessitating the use of a short window in front and a small blank panel behind the relocated doorway. The original four sections front door were retained but both they and those of the centre exit were left as unpainted aluminium. The crest appeared on the front panel between the entrance and exit. The cream waist-band crossed the drivers dash but in some cases was omitted from the front nearside bulkhead, but not always.

The front destination box was proud of the front panels as on Daimler (ca). There may have been buses which were exceptions to these guidelines but, if so, they were few indeed. Seating layout was the same as the Daimler (ca) except that the (cb) type was authorised to carry one less standee.

Daimler(cc).

These were similar to the (ca) conversion in general arrangement but were completely new bodies, generally to the MetSec design but incorporating a number of features that help to identify them. The centre door was set in the third bay as on the (ca) type and all passenger doors were painted and lined-out. The front doors were in two parts instead of the original four as on all other variants. The front destination screen was fitted flush in the front panel which was flat. The cream waist-band did not extend across any of the front panels. In many, but not all, cases there was a small window in the panel behind the narrow front door to complete the bay. At H44/31+24FE, CXS, D with Farebox, the layout was the same as the Daimler(ca) but a major difference was that the (cc) type were re-engined; fitted with the more powerful 150bhp, Gardner 6LX engine in place of the original 6LW version. The more powerful engine in turn necessitated a semi-automatic gearbox in place of the pre-selective type fitted from new.

The number of conversions completed by 31st December 1982 totalled 56, including 16 (ca); 29 (cb) and 11 (cc). Due to chassis failure, however, two, (ca) AD4777 and (cb) AD4749 were withdrawn and scrapped in May 1982 and (cb) AD4786 followed similarly in March 1983. No further conversions then took place and various withdrawals took place during 1984, when 16 were de-registered, to be followed in 1985 by a further 11 and 37 in 1986, leaving only two in stock on 1st January 1987, but these soldiered on until June (D284) and, finally, September (D329)

BELOW: Many Daimler(c)'s were rebuilt by KMB to make them more suitable for DOO. They were given a forward entrance and centre exit with centre staircase, retaining the original Gardner 6LW engine. Here D317 had been converted and re-classified as Daimler(ca). It was leaving Jordan Road Ferry in May 1975. *(Mike Davis*

UPPER LEFT: Daimler(c), D315, after being rebuilt to a second style to become a Daimler(cb) with slightly body layout. The centre exit was placed further back and the seating arrangements differed from the (ca) sub-type. Seen here approaching Tai Kok Tsui Ferry terminus on Sunday 8th January 1978. *(Mike Davis*

CENTRE LEFT: the Daimler(c) class buses converted to become (cc) sub-class were more extensively rebuilt than either the (ca) or (cb) varieties and were fitted with more powerful Gardner 6LX engines. The body was largely a KMB structure and the fabricated, angular, rear dome was most noticeable. *(Mike Davis*

LOWER LEFT: D296, an unrebuilt Daimler(c) with rear platform and staircase, seen at Tai Po in July 1983. Buses of this type were destined to be Hong Kong's last rear-platform type. Seen here in 1983 Jubilee livery. *(Mike Davis*

UPPER RIGHT: Without Jubilee insignia in July 1983, Daimler(cb) D285 was destined for an early withdrawal in March 1984, being outlived by many of the type with rear staircase and platform. *(Mike Davis)*

CENTRE and LOWER RIGHT: Body layout differences between Daimler (cb), D290, (top) and Daimler(ca), D320 (lower). Both were photographed at Yuen Long East bus station in July 1983. *(Mike Davis)*

BELOW CENTRE: In 1985 Daimler(c) D296 was converted to a tree pruning vehicle in replacement of previous, Daimler(b) and AEC 2A vehicles. *(John Shearman)*

AEC Regent Mk V.

A1-210
(Fleet numbered from 1973/4)

In 1963, the first batch of 30 AEC Regent V's were purchased in what was a major lead by a Hong Kong bus company towards buses larger than previously acceptable, even in Britain.

In its efforts to cope with the ever increasing demand for its services, KMB, hitherto confined to standard size buses designed primarily for the domestic British market, was granted permission by the Transport Office of the then Hong Kong Police (now Royal Hong Kong Police) to introduce double-deck buses of a size greater than had seen service anywhere else in the world at that time. The specification called for a vehicle capable of safely conveying 110 souls from 'a' to 'b' - no mention was made of comfort in those heady days when newly arrived refugees from China's rapidly escalating political turmoil, to become known as the Cultural Revolution, caused bus queues to increase in length by many thousands per week. Utility was the name of the game.

The problem faced by KMB was that no British manufacturer, to whom they were required, by a clause in their franchise, to turn for new buses, had ever produced a bus of such seemingly gargantuan proportions. The largest product then available from the traditional supplier of KMB double-deck buses, Daimler Commercial Vehicles Ltd., and already in Kowloon service, was 30ft long and 8ft wide - the then standard British domestic dimensions legally permitted and, therefore, in mass production. Daimler felt unable to produce a larger vehicle for what they saw as a limited market and so KMB looked elsewhere.

AEC alone were producing a front engined chassis of 21ft 6in wheelbase. This was designed to be fitted with single-deck bodywork

VEHICLE SPECIFICATION: AEC REGENT MK V - 34ft LONG

Original vehicle numbers	AEC(a)/30:	A1-30 (AD4789-4818)
	AEC(a)/40:	A31-70 (AD4823-4862)
	AEC(b):	A71-110 (AD4863-4900, AD7100-1)
	AEC(c):	A111-210 (AD7102-7201)
1983 numbers AEC (CS) :		A1-210 (in reg. no. order as above)
Chassis:		AEC Regent Mk V, Model 2D2RA.
Engine:		AEC AV690, rated at 135bhp or
		Gardner 6LX, 150bhp (two or three buses only, retrospectively fitted).
Gearbox:		AEC semi-automatic 'Monocontrol' (AEC engines)
		Leyland GB350 with Gardner engines
Body make:	A1-30:	British Aluminium Co. (BACO)
	A31-210:	Metal Sections
	A44 & 48:	Rebodied by KMB.
Body layout as built:		Only total capacity is shown, ie including standees.
		AEC(a)/30: 22 x 110, plus 8 x 120; Layout in all cases
		AEC(a)/40: All 120 H50/?+?FEX,REX,D.
		AEC(b): all 118; AEC(c): all 117 See box below.
Body layout for OCO:		AEC(a)/30 & (a)/40: H50/34+26FE,RXS,D,OCO
		AEC(b) & AEC(c): H50/32+26FE,RXS,D,OCO
Body layout OMO AEC CS:		H51/39+24FE,CXS,D,DOO Farebox.
Date introduced:	AEC(a)/30:	1963
	AEC(a)/40:	1964
	AEC(b):	1965
	AEC(c):	1966
Rebuilt to AEC CS:		1976-82
Total:		210
Length:	34ft 3in	10,440mm
Width:	8ft 0in	2,440mm
Height:	14ft 8in	4,470mm
Wheelbase:	21ft 6in	6,550mm
Unladen weight:	9,246kg	
Gross vehicle weight:	16,257kg	

NOTE: Withdrawls commenced with A32 and A54 early in 1983.

Seating:
The AEC's had a variety of seating capacities, sometimes differing within types. During the 1967 disturbances and strike, perimeter seating was installed in many and, when returned to normal, differing arrangements were installed. The upper-deck figure, however was 50 in all cases

LEFT: An illustration from British Aluminium Co. (BACo) promotional material, showing a prototype AEC Regent MkV of 34ft overall length. Note the lack of either beading or cream relief bands. *(British Aluminium Co.*

ABOVE: The first 34ft long AEC Regent MkV's were fitted with BACo ckd bodies with flat, or 'slab' lower panels and featured sliding metal gates. *(Kowloon Motor Bus)*

BELOW: AD4818, a BACo bodied AEC(a/30), after fitting with platform doors in 1965. Photographed here in July 1966, it will be observed that the wheel trims and AEC grille remained intact at this stage. *(Mike Davis)*

ABOVE: AD4813, an AEC(a/30) in OCO livery in 1974, with yellow bands around the front. Note that, by this time, the AEC grille had been replaced by KMB's own substitute version. AD4813 had been fleet-numbered and was working on Route 5. *(Mike Davis*

BELOW: The second batch of AEC Regent MkV's were provided with Metal Sections bodies and, because there were forty of them, they were classified AEC(a/40); seating arrangements were as for the AECa/30). The curved side panels distinguished Metsec bodies from the BACo type and are displayed here by AD4154, photographed in 1966 *(Mike Davis*

ABOVE: AEC(a/40) with yellow OCO bands, in 1974. Again, a KMB grille replaces the AEC original on AD4861, by this time A69. *(Mike Davis)*

and intended for export, most especially for African markets. This chassis was used to produce long Regal Mk III's and Regal V's model. However, with some degree of redesign, and uprating of such components as springs, axles, steering and wheels, it became suitable to take double-deck bodywork. Thus was developed the unique to Kowloon 34ft long AEC Regent Mk V model, weighin-in at over 16 tons, laden, the heaviest bus type produced by AEC. Although the first examples were designed for 110 passengers, it was not long before this was altered to 120 by increasing the standing area available.

The first order for 30 of these lengthy AEC Regents entered service in 1963 and, like subsequent deliveries, were supplied with AEC AV690, 11.3 litre engines, semi-automatic gearboxes, traditional rear staircases with front and rear entrances. Although the AEC's delivered subsequently became the first KMB double-deck buses to be fitted with power doors from new, the first batch at least entered service with sliding metal gates. So successful were these large buses that KMB re-ordered batch by batch until their total amounted to 210; the last being delivered in 1966. KMB sub-divided these buses by allocating class letters for internal purposes, in a manner similar to that adopted for the fleet of Daimlers, the AEC's becoming known as AEC(a)/30, AEC(a)/40, AEC(b) and AEC(c); all 210 were later rebuilt as centre staircase, front entrance buses for driver only operation and were subsequently all reclassified as AEC(OMO). When the fleet was numbered by chassis make, the AEC's became 'A'-type (A for AEC) and were allocated fleet-numbers A1 to A210, in registration number order.

In all cases KMB's AEC's were shipped to Kowloon in chassis form to be bodied locally using ckd kits of parts provided by either The British Aluminium Co. (BACo), or Metal Sections Ltd., both British specialists in the provision of complete bus bodies for local self assembly in developing nations.

AEC(a)/30 & (a)/40

These first two groups became respectively A1-30 and A31-70 when the fleet was numbered in 1973/4. There were, within the seventy buses two body-kit makers, A1-30 being by the British Aluminium Company (BACo) and the remainder, A31-70, having Metal Sections (MetSec) body-kits.

The first thirty, known as the AEC(a)/30 class (after the number delivered), were built and entered service in 1963, having flat-sided aluminium framed bodies by BACo, similar to, but shorter than, the lone Daimler(c) AD4742, later to be numbered D286.

The second group, forty in number, were delivered in 1964 and became known as class AEC(a)/40, again after the number purchased. This group reverted to KMB's traditional body-kit supplier, MetSec. and thus had that manufacturer's steel framed body with aluminium sheet panelling which, unlike the flat sided BACo product, had lower-body side panels which curved downwards and inwards - the term 'tumble-home' was once used in coachbuilding parlance.

Originally the AEC(a)/30's had a total capacity of 110, except for eight buses, which carried 120, including standees. Both groups were later converted to have similar seating arrangements, the actual body layout, ie window and door spacing being the same by both suppliers. The MetSec bodies of this series could be distinguished from those of later delivered buses with generally similar body arrangements by their having slightly narrower front and rear entrance doors and larger adjacent side windows, particularly noticeable in the case of the first side window, ahead of the forward entrance, which occupies a full bay as opposed to a part bay in later vehicles with wider doorways.

A large area of the forward part of the lower deck area was set aside for multi-standee passengers with stout vertical stanchions and horizontal grab-rails. A rearward-facing bench seat for five had its back to the bulkhead while inward facing seats were provided around the forward doorway area. Single forward facing seats were fitted on the nearside ahead of the rear wheel-arch. Over the wheel-arch the usual inward facing seats were provided.

With a wheelbase of 21ft 6ins, the rear overhang of these 34ft long buses was about 10 feet; this was not all platform, for the inward-facing bench seats over the wheel-arches extended behind the arches by the width of two seats. As delivered these buses were provided with standard tubular metal framed seats 'upholstered' with green plastic covered rattan (bamboo strips closely woven) but after a period of time these were found to harbour bugs and, as with all other KMB buses they were replaced by moulded plastic seats and seat backs which could be easily wiped-clean.

The first 30, ie AEC(a)/30's, A1-30, were the last double-deck buses to be placed in service by KMB which were not fitted with driver-controlled power operated doors and the AEC(a)/40's were the first to be so fitted.

AEC (b) sub-type

The third series of KMB AEC Regent Mk V

were to become the sub-type AEC(b) and were fleet-numbered A71-110 in 1974. These buses were provided with KMB assembled MetSec bodies similar in most respects to those of A31-70, *qv*, with the major difference being the width of the entrance doorways which were wider than those of the AEC(a) types, probably qualifying them as having the biggest rear platform area of any British-style double-deck bus ever built for service anywhere in the world. The steel-framed bodies had curved, or 'tumble-home' lower side panels, a style which was perpetuated, not only with later AEC based buses but which was continued for some years with the later Daimler fleet of similar dimensions, *qv*. Seating was, as built, more conventional than the AEC(a)'s, having 34 lower saloon seats and correspondingly less standing space. During 1967, see below, the interiors of the lower decks on this type were rearranged to have only perimeter seating to permit 'crush-loading' of up to 200 passengers under special conditions.

The AEC(b) type entered service in 1965 and were fitted from new with power operated folding doors which were usually left in unpainted aluminium for the whole of the vehicles life.

AEC(c) sub-type

These were one hundred buses basically similar to the preceding series of AEC(b) type. In fact there was little discernable difference between the body layout of the two sub-classes, except that the lower-deck capacity was only 28-seated plus 30 standing. MetSec prepared the body kits which were shipped ckd during late 1965 and early 1966. In the 1974 fleet numbering scheme the AEC(c) type became A111-210, in registration number order.

Crush-loading 1967

During the politico-industrial disturbances of 1967, when most bus crews were forced to strike, and in an attempt to move the enormous crowds, many AEC's of all sub-types were modified to have perimeter seating on the lower-deck for use as crush-loaders. Legally they were licensed to carry 120 passengers and under normal conditions were frequently overcrowded, but how many they were forced to carry in 1967 is beyond imagination! Two-hundred has been quoted but the author can recall loads which must have been well in excess of this figure. Suffice it to say that it resulted in rear-axle and suspension problems on what were then comparatively new vehicles.

'OCO' modifications.

From 1971-72 all AEC sub-types were modified to suit them for KMB 'OCO' scheme, the use of one seated conductor in place of two roving conductors used hitherto. Yellow bands appeared across the front panels between the upper and lower decks and seating on the lower deck was standardised at 32, except on the AEC(a) type which, with their narrower doors could seat 34 under these conditions. Entry was by way of the forward doorway, to the rear of which sat the conductor. This arrangement allowed for a pool of passengers to board and stand in the seatless front section before filing past the cash desk, thus allowing the bus to proceed as the conductor collected the fare, thus reducing loading time at the kerb-side. Having paid, the passengers passed to seats either in the lower saloon or, via the rear staircase, to the upper deck. Exit was by the rear doorway

Driver Only Operation AEC CS.

Following the success of those Daimler(c) rebuilt to front entrance, centre staircase, centre exit layout, 1976 saw the first example of similar treatment to an AEC Regent V. In the first case the rebuild was almost a rebody as the opportunity was taken to reconstruct the body of A48 following severe accident damage. This was followed by A44, after an almost complete body burn-out. The success of these conversions was such that the entire fleet of AEC's was converted to have a forward, single-width, entrance immediately behind the bulkhead and a double-width centre exit with centre staircase. The conversion programme was completed by the summer of 1982 and as they returned to traffic the buses were reclassified as AEC CS, irrespective of previous sub-type, and the BACo bodies were treated in the same way as the MetSec equivalent, all being H51/39+26FE,CXS,DOO.

Mechanical alterations.

During 1976/77, a few, believed to have been two or three, of the AEC's were modified from AEC engines to have Gardner 6LX engines and Daimler rear axles, following the success of similar treatment afforded to the 30ft long AEC Regent MkV's purchased second-hand by KMB in 1973. In addition one AEC was converted to have fully-automatic gear control during 1977. Due to shortages of AEC axles, a few were latterly fitted with Daimler units.

BELOW: AEC(b)-type buses had wider doorways than the earlier (a) variants and a different seating layout when they were introduced. A98 was OCO fitted when photographed in 1975, leaving Jordan Road; still a good vantage point in 1995!. *(Mike Davis*

ABOVE: AEC(c) AD7130 in original condition when only a few weeks old, having been introduced in January 1966. I was passing the end of Nathan Road, approaching "Star" Ferry, the conductor having changed the blind ready for the return journey to Lai Chi Kok on Route 6. *(Mike Davis)*

BELOW: Two AEC Regent V's at Lai Chi Kok (Mei Foo) terminus, under a flyover, during January 1978. In the foreground is 1966 vintage AEC(c) A189, while behind is A101, an AEC(b) of 1965. *(Mike Davis)*

Withdrawals.

Withdrawal of the AEC Regent V's commenced with the demise of AD4824 & AD4849 on 25th January 1983, followed by AD4846 in July the same year, with a fourth by the year's end. Large scale scrapping took place during 1984, with a year end total of 146 buses, reduced a year later to 118 and to 51 by 31st December 1986. The final 14 AEC's were withdrawn during August 1987.

Of these 210 buses, it is known that at least three are preserved, the first of which, AD4860, was exported to Newcastle, N.S.W., Australia in 1984. In 1987, AD7156 was shipped to the UK by John Shearman for public display in the Oxford BBus Museum. The third, AD4807, was preserved and visually restored by KMB themselves. Interestingly, this latter example was one of the first batch of AEC(a)/30 sub-type, having a British Aluminium Co. body, being AD4807. The restoration to original livery was not accompanied by re-conversion to rear platform and staircase layout and, of course, it retains its centre door/stairs configuration which it received after conversion to AEC CS in 1982.

UPPER: AD7201, the very last 34ft long AEC Regent Mk V in Wh Hu St. on 25th August 1969. This rear view shows off well the rear arrangements of this type which were similar on other double-deckers up until that time. With the advent in the early 1970's of, first, seated conductors and then DOO, the rear destination was reduced universally to show just the route number. *(Julian Osborne*

BELOW: AEC A185 swings into Salisbury Road from Nathan Road and contrasts with the modern lines of the then new Sheraton Hotel in January 1978. *(Mike Davis*

ABOVE: In 1976, KMB extensively rebuilt the body of A48, following a fire and took the opportunity to move the staircase and exit doorway to the centre and make a new single-width front entrance, immediately behind the front wheelarch, in the half-bay once provided with a small window. These alterations made the bus more suitable for driver-only farebox operation. A44 was treated in a similar fashion in 1978 and the remainder followed in a few years. *(Derek Lucas*

BELOW: AEC(c) A124, after rebuilding as part of the main programme to alter all the 34ft long Regents to AEC(CS) configuration. All types were then grouped together as (CS) - Centre Stairs - without recognition of their original type (a), (b) or (c). AD7115 was leaving the "Star" Ferry in 1981 in simplified livery without lining. *(Mike Davis*

ABOVE: An offside view, showing the original rear staircase position of AEC(b), A71, at "Star" Ferry in 1981, prior to its being rebuilt as an AEC(CS) in the October of that year. *(Mike Davis*

BELOW: AEC(CS), originally AEC(a/40), A42, after rebuilding and showing the revised position of the staircase. Seen on the same day, at the same place, as A71, above. *(Mike Davis*

UPPER RIGHT: A204 was turning into Salisbury Road from Chatham Road in may 1984, having received the latest style of livery but retaining the Jubilee insignia into the 51st year. *(Mike Davis*

CENTRE RIGHT: During 1973, AD7152 was fitted with water-filled rubber bumpers at the front and rear in an attempt to reduce panel damage. The bumpers were later removed. *(John Shearman*

BELOW: 34ft (10.36 metres) long AEC Regent Mk V, A44, complete with its 'artificial' front overhang, extending the overall length to 12 metres (39ft 4in) in length to simulate the swept turning circle associated with 12-metre, front-platform, rear-engined double-deckers, awaiting its early morning trial run through Kwun Tong. The metal pennants helped the driver judge the extreme corners from his set-back driving position. *(T. V. Runnacles*

BOTTOM RIGHT: Before setting-off, the water-can man is helped to reach over the artificial front extension. *(T. V. Runnacles*

Daimler CVG6LX - 34   Daimler(d).   D331-350.

(Fleet numbered from 1973/4)

These twenty, 21ft 6in wheelbase, chassis were specially designed for KMB by Daimler in order to regain the market they had lost to AEC. and became known by KMB as the Daimler (d) class. They were the first 20 of an eventual 220, 34ft long, 21ft 6in wheelbase, Daimlers to be bought by the company, and entered service in 1967, the remaining 200 being the Daimler(e)'s, *qv*.

The Gardner 6LX engines of 150bhp, fitted in the Daimler(d)'s, were also the first of their type in the fleet, as were the Daimatic semi-automatic gearboxes. The rear staircase MetSec bodies were identical in most respects to those on the later AEC Regent Mk V's, having seven bays and accommodation for 42 standees in the forward part of the lower deck. They were the last rear-staircase buses to enter KMB service. Since being equipped for seated conductor operation, in 1972, their body layout became H50/32+35FE, RX S, D(OCO); standing was later reduced by new regulations (see note in Vehicle Specification box) to 22 passengers.

Daimler (da).

On 1st June 1979, D333 (AD7239) returned to service having been experimentally converted to DOO, retaining the rear staircase. This alteration had the effect of increasing the lower-deck capacity by six seats and one standee, resulting in a revised layout of H50/38+23FE, RXS, DOO (Farebox). AD7239 was the only bus of this type to be so converted and returned to service. It was temporarily withdrawn from service during February 1980 for further modification to Daimler(d)CS *see below*.

VEHICLE SPECIFICATION: Daimler(d)

Registration numbers: AD7237-56.
Fleet numbers : D331-350 (from 1973/4).
Chassis: Daimler CVG6LX-34DD
Engine: Gardner 6LX, 150bhp.
Gearbox: Daimatic, 4-speed semi-automatic.
Body make: Metal Sections ckd.
Body layout :
  As built: H50/40+42FEX,REXS,D
  OCO: H50/32+35FE,RXS,D,OCO - @ 152sq in per standee
        H50/32+22FE,RXS,D,OCO - @ 260 sq in per standee

OMO(da): H50/38+23FE,RXS,DOO Farebox
OMO(d)CS: H51/39+26FE,CXS,DOO Farebox

Date introduced: 1967
Rebuilt (d)CS: 1979/80
Total: 20
Length: 34ft 4 in   10480mm
Width: 8ft 0in   2440mm
Height: 14ft 8 in   4480mm
Wheelbase: 21ft 6in   6550mm
Unladen weight: 9,449kg
Gross vehicle weight: 16,257kg

AUTHORISED STANDING PASSENGERS
(Applies to all KMB buses)
Prior to 1972, Public Omnibuses in Hong Kong had their standing capacity calculated on the basis of 152 sq ins per standee. This was then revised to 288 sq in, in an attempt to reduce overloading and buses were gradually relicenced at the new figure. Before this was completed, the figure was again altered to 260 sq in during 1974 and so remained in 1991.

BELOW: Daimler(d)-class, D332, one of the first batch of 34ft long Daimler CVG6LX-34DD double-deckers with Metsec ckd bodies. These had rear platforms and staircase, and a front doorway and followed on from the latter style of AEC Regent V, the AEC(c)'s. Seen here in 1974 with OCO bands across the front. *(Mike Davis)*

ABOVE: Daimler(d), D335, rear-staircase bus, carrying a Government Road Safety slogan in January 1978. Compare this photograph with that of A71 (page 120). *(Mike Davis)*

BELOW: Rebuilt to centre staircase, after failure as a Daimler(da), 'Daimler(d)CS', D333 was turning from Nathan Road into Salisbury Road on Route 9 during March 1981. *(Mike Davis)*

D347 was temporarily withdrawn on 24th October 1979 also for rebuilding to Daimler(da) and was largely completed when the decision was taken to rebuild all 20 Daimler(d)}s to centre staircase layout, including D333 & D347. D347 was thus further altered to Daimler(d)CS before the (da) conversion was completed. It returned to service in September 1980

Daimler (d)CS.

Following in-service experience with D333 KMB removed five further Daimler (d)'s from traffic during October 1979 for conversion to centre staircase (CS) - these were D339/44/6/7/50. This method of OMO being more suitable on vehicles of this size. Work proceeded rapidly and the first rebuilt example, D339, returned to service before the end of November 1979. Further rebuilds followed, all becoming Daimler(d)CS - H51/39+26FE, CXS, OMO (farebox) - by November 1980.

Withdrawals.

Two Daimler(d)'s were withdrawn during 1984 and four in 1987, of which one, D342 was to see further service in the nearby Chinese city of Guangzhou (Canton). Fourteen Daimler(d)'s survived until 1988, but all were withdrawn during that year, the last four, D332/5/6/43, finally retiring on 23rd June 1988.

RIGHT: Before the final plan to move the staircase on all 20 of the Daimler(d)}s to the centre, ways were sought to make them suitable for DOO/farebox operation. Ftting a narrow-width front entrance right up to the front bulkhead, whilst leaving the rear staircase and platform was one method and D333 was converted and returned to service thus altered but, before completion work to convert D347 to the same pattern the more radical plan to move the staircase was approved. Here the unfinished work on D347 can be seen stored on the roof of Kwun Tong depot in September 1979. *(John Shearman*

BELOW: D332 leaving the stand at Hung Hom Ferry Pier in 1983 after being repainted into the revised livery. *(Mike Davis*

Daimler CVG6LX-34 - Daimler(e) and Daimler CVG6LX-30 - Daimler(f)

D351-550 (Fleet numbered from 1973/4)

D551-665 (Fleet numbered from 1973/4)

Both the nominally 30ft long and 34ft long Daimlers of KMB types (e) and (f) can be considered as one group because the only differences arise in the wheelbase and overall length. The respective figures for the Daimler (e) are 21ft 6ins and 34ft, while those of the Daimler (f) are 18ft 6ins and 31ft. The additional 1ft in overall length of the 30ft chassis designated length is due to additional body overhang only.

Interestingly, some of the chassis were also assembled by KMB from ckd components; a new step for either KMB or CMB.

As on the previous 34ft long Daimler(d), the Daimler(e) as well as the shorter Daimler(f), have Gardner 6LX engines and semi-automatic gearboxes. One exception to this latter feature is Daimler (f) D600 (AR7535), which in August 1976 was retrospectively fitted with a fully-automatic gearbox unit. Although this measure was successful, the benefits were not judged to be so great as to justify further conversions.

Bodywork on both types were built-up by KMB from MetSec ckd parts, the first thirty of each being to the established standards with the then usual KMB windows, the sliding portions of which were mounted in square cornered aluminium frames which are free to traverse within the fixed outer frames of aluminium channel within which the fixed portion of the window glass is retained. All subsequent bodies on these chassis have radius cornered windows, the frameless glass of which slides in cushioned channels fitted directly as part of the window pan. These later bodies are thus greatly improved in terms of appearance and standard of finish.

The body design represented the first departure by KMB from the traditional rear platform and staircase. All except AD7370/96 (q.v.) had front staircases with double-width forward entrance and a single-width doorway in the centre.

Many of the buses in these two groups operated in three modes but the doorway layout, designed before the advent of OCO and

VEHICLE SPECIFICATION - DAIMLER(e)-CLASS

Registration numbers:	AD7257-7456..
Fleet numbers:	D351-550 (from 1973/4).
Chassis:	Daimler CVG6LX-34.
Engine:	Gardner 6LX, 150bhp.
Gearbox:	Daimatic, 4-speed, semi-automatic.
Body make:	Metal Sections ltd., ckd.
Body layout:	1969-71: H52/40+25FEXS,CEX,D.
	OCO: H52/33+35CE,FXS,D,OCO.
	OMO: H52/40+25FES,CX,D,DOO(Farebox) * #.
	D464/490: H52/33+35FE,CXS,D,DOO(Farebox).
Date introduced:	1969-72.
Total:	200.
Length:	34ft 4½in 10480mm
Width:	8ft 0in 2440mm
Height:	14ft 8½in 4480mm
Wheelbase:	21ft 6in 6550mm
Unladen weight:	9,322kg
Gross vehicle weight:	16,257kg

NOTE: * At 1st January 1978, 86 buses had cushion seats; the remainder had hard fibre-glass resin seats.
     # Some of the earlier DOO buses (Daimler(e)OMO) were permitted only 19-standees but all were later altered to 25.

BELOW: Daimler(e)-class, D351. The first thirty Daimler(e)-class resembled, in many details, the rear-platform Daimler(d)-class but were built new with front staircase and centre exit. These thirty buses can be distinguished from subsequent versions by their angular cornered window-frames. *(Mike Davis)*

VEHICLE SPECIFICATION - DAIMLER(f)-CLASS

Registration numbers:	AD7486-99; AR7500-600.
Fleet numbers :	D551-665 (from 1974).
Chassis:	Daimler CVG6LX-30.
Engine:	Gardner 6LX, 150bhp.
Gearbox:	Daimatic, 4-speed, semi-automatic #.
Body make:	Metal Sections ltd., ckd.
Body layout 1969-71:	H48/34+20FEXS,CEX,D.
OCO:	H48/28+33CE,FXS,D,OCO.
OMO:	H48/34+24FES,CX,D,DOO(Farebox) *
Date introduced:	1969-72.
Total:	115.
Length:	30ft 11½in 9440mm
Width:	8ft 0in 2440mm
Height:	14ft 8½in 4480mm
Wheelbase:	18ft 6in 5640mm
Unladen weight:	8,840kg
Gross vehicle weight:	14,661kg

NOTE: * At 1st January 1978, 23 buses had cushion seats; the remainder had hard fibre-glass resin seats.

DOO, remained unaltered in each case, having a double-width front doorway with front staircase and a narrow, single-width centre doorway.

The earlier deliveries ran at first with the traditional two roving conductors and had body layouts, Daimler (e) of H52/40+25FEX S,CEX,D and Daimler(f) of H48/34+20FEXS, CEX, D. Buses entering service as, or converted to OCO (one seated conductor) were respectively H52/33+35CE, FXS (OCO) and H48/28+33CE, FXS (OCO). The loss of seats was necessary to allow for a 'pool' of passengers inside the bus where they could wait prior to purchasing their ticket thus allowing the bus to proceed without delay. Permitted standees were proportionally increased. In order to provide compatability with CMB buses on jointly operated tunnel routes, KMB introduced one man operation on those routes from the outset in 1972 using the Daimler (e) and (f) types almost exclusively until the advent of the Fleetlines in June 1974. Buses allocated to the tunnel routes were fitted for flat-fare OMO using fareboxes located on the bulkhead under driver supervision. The centre doorway was supposedly the exit in OMO mode but without turnstiles on the wide front doorway this flow system was hard to enforce. Tunnel buses were also fitted with fully upholstered cushion seats. Those buses used on the Kowloon urban routes were fitted with hard, moulded glass-fibre seats and were OCO until the last was converted to OMO in February 1978, having centre entrances and front exits - the opposite of the OMO tunnel buses. These alternative entrance positions were somewhat confusing to intending passengers, for although all buses on one route operated one system, those using the same stops on parallel routes may have had a different entrance position. To aid the unwary the doors were clearly marked bi-lingually in Chinese and English, "ENTRANCE" and "EXIT", near the appropriate doorway. The confusion ended with the cessation of OCO on all KMB routes on which the Daimler (e) & (f) were initially H52/40+19FE S, CX, D, OMO (farebox) and H48/34+16FE S, CX, D, OMO (farebox) but the standing capacity was altered to 25 and 24 respectively with changes in the regulations.

Daimler(e)CS - AD7370, AD7396.

In Hong Kong's circumstances it is more suitable to have single-width entrances and double width exits; this prevents 'crowding-on' while at the same time allowing a free flow of alighting passengers. The Daimler (e) and (f) classes are not therefore particularly suited to this type of operation. As a result, two of the 34ft long Daimler (e)'s, D464 & D490 (AD7370/96) were assembled by KMB from ckd components so that the staircases were in the centre of the bus, although retaining the wide front doorway, with a view to having that layout standard on all future buses of this type. By that time, however, British Leyland had announced that it could no longer manufacture

BELOW: D352, a first series Daimler(e), plainly shows its angular window frames and also the 'Keep Hong Kong Clean' Government slogan. It was leaving Jordan Road Ferry in January 1978 on Route 3C. *(Mike Davis)*

ABOVE: Later Daimler(e)-class buses had radius corners to their side windows but an additional recognition feature was the fixed glazing in the upper-deck first side windows. AD7295 was fitted for OCO working with centre entrance/front exit and moulded plastic seats. *(Mike Davis*

BELOW: D526 entered service as new on Tunnel services and as such had cushion seats and was fitted for DOO with a Farebox. In this mode the entrance was via the front door with centre exit. Seen here on the first section of the newly aligned Queensway, formerly Queens Road East, in May 1975. *(Mike Davis*

ABOVE: Two of the Daimler(e)'s had their body kits assembled with the staircase located at the centre of the bus but retained the wide front doorway and narrow centre exit. D490 (AD7396) was seen at Lai Chi Kok in January 1978 on New Territories Urban Route 44a from Chung Ching to Tai Kok Tsui Ferry. *(Mike Davis)*

BELOW: An offside view of Daimler(e), AD7380, showing-off its great length of these vehicles (as it seemed at the time) and the position of the forward staircase. All-over advertising is for Yellow Pages. *(Mike Davis)*

ABOVE: KMB Daimler(e) deep inside CMB territory as it turns from Queensway into Des Voeux Road Central, on Hong Kong Island, in 1983, having acquired the, then, new livery. *(Mike Davis*

BELOW: 31ft long (9.5m) Daimler(f), AD7486, of the first batch which, like the first Daimler(e)'s had more angular bodies than subsequent versions. Here D551, the premier member of the type, was working on Route 2D, passing a typical block of public housing flats, complete with distinctive washing on bamboo poles. *(Author's collection*

a front engined bus chassis suitable for double-deck bodywork. D469 and D490 thus remained the sole examples of this layout, retaining the wide front doorway and narrow exit at the foot of the stairs and having a layout unchanged over the years of H52/33+35FE, CX S, D, DOO (farebox). They were known by KMB as Daimler (e) CS.

There were no further sub-types of either Daimler (e) or Daimler (f) except that buses with foam-filled seat squabs were referred to as 'Cushion' ie Daimler(e)FE/Cushion. Initially, buses with cushion seats were allocated solely to Cross-Harbour routes in a response to the lead taken by CMB, but later they were allocated to longer routes in Kowloon and the New Territories.

Withdrawals.

Major withdrawals of Daimler (e) type took place between August and October 1986 when 19 were scrapped, followed by 31 in 1987, of which a number were sold for further service in several cities in China, under arrangements made by 'Speedybus Ltd.' - a Hong Kong dealer - and the Scottish 'Stagecoach' organisation. No less than 78 Daimler (e) were withdrawn between by early 1990. Of the 1988 withdrawals, further examples were exported to China, but more interestingly, 50 were purchased, again via 'Speedybus', by 'Stagecoach' for their subsidiary in Malawi, 'United Transport (Malawi) Limited'. The buses were shipped to Durban, then driven overland to Blantyre with remarkably little trouble for vehicles of 16 or 17 years of age. They all left Hong Kong during 1989/90.

The first Daimler (e)'s (D354 and D355) were withdrawn in August 1986 and the last in 1992, after eight buses had been reinstated, following what had been planned as their final withdrawal in 1989/90. D409/17/8/30, 536/46/49/50 were required for about a year on a particular route.

The first Daimler (f), D556, was withdrawn, following accident damage and deregistered on 16th October 1982. The last Daimler(f), D650 (AR7585), was delicensed on 22nd April 1988.

The last half-cabs.

These KMB Daimler CVG6LX-30 & 34 half-cab double-deckers were the last 'traditional' British double-deckers to be built - if Indian-built Askok-Leyland Titans with Rubery-Owen chassis frames are discounted. Later front-engined models can hardly be described as 'traditional'!

TOP: Daimler(f) class, AR7592. Almost identical to the later Daimler(e)'s but having one shorter and with a wheelbase of 18ft 6in. The soon to become D657 was seen here in 1973 on Tunnel Route 103, in Central district, leaving Jackson Road en route from Pokfield Road, on Hong Kong Island, to Wang Tau Hom in New Kowloon. *(Mike Davis)*

LEFT: The unusual sight of an ordinary double-deck bus helping-out on a coach route as Daimler(f) acts as a relief on Route 293 during the Chinese New Year festival in 1977. This was the scene at Sai Keung as the returning crowds boarded for the trip back to Kowloon City Ferry. *(J. Shearman)*

NOTE: Close observation of both these photographs will reveal a small (blue and yellow) disc just inward of the offside side-light. This indicated a bus with lower ratio rear axle for use on routes to Sau Mau Ping, overlooking Kwun Tong. These were short-lived, however, as all Daimler(f)'s were subsequently given lower ratio axles.

ABOVE: Daimler(f) D633 (AR7568), at Yuen Long West bus station in June 1983, still in 1970's half-and-half livery. *(Mike Davis)*

BELOW: D662, one of the last few Daimler(f)'s to be delivered. These buses were new with radius corners to their windows but, in this case, it will be seen that the upper-deck window frames have been replaced using angular corner units, whilst the lower deck retains the original radius corners.. Tsim Sha Tsui temporary bus station in May 1984. *(Mike Davis)*

Daimler Fleetline CRG6LXB or FE33AGR - D666-1115
(Fleet numbered from new)

Between June 1974 and January 1979, KMB placed in service a total of 450 Daimler - later Leyland - 'Fleetlines'. All were 33ft long and powered by the 185bhp Gardner 6LXB diesel engine, mounted transversely and vertically at the rear of the Daimler-built chassis were of Model CRG6LXB-33, or just CRG6 if built after the choice of Gardner 6LX engine was deleted. Later examples, built at Leyland's Lancashire plant, and fleet numbered from D809 upwards, were Model FE33AGR. By 31st December 1982, a total of 235 Fleetlines had either been delivered to KMB with, or had been retrospectively fitted with, fully-automatic gear-control systems, using either Sevcon or CAV511 units. The remainder had CAV semi-automatic transmission.

Following the termination of Daimler CVG6, or any other front-engined bus production in the UK, KMB placed two orders for Fleetlines, each of 150 chassis, complemented by orders for 300 Metal Section body-kits. This was followed by a further tow orders, again of 150 chassis each, but, in the event the final 150 were cancelled in favour of a similar number of the newly available Leyland Victory 2, a front engined double deck bus chassis. The third order, of 150 chassis, was bodied using British Aluminium Co. ckd kits, in preference to MetSec. Of the first (MetSec bodied) 300, two were actually bodied by KMB themselves and one was given a prototype 'BACo' body (D815). A quick check-list of KMB classes for their Fleetlines is as follows:

Fleetline CS MetSec ckd.	Total 297
Fleetline (A) BACo cbu.	Total 1
Fleetline (B) KMB body.	Total 2
	Sub-total 300
Fleetline (C) BACo 'production' ckd	Total 150
	Grand Total 450

VEHICLE SPECIFICATION: Metsec & KMB bodied FLEETLINES

Fleet numbers: Fleetline CS: D666-967, (and, later, D1007 from 4/82) - less 815, 888, 931 & 963/4
Fleetline(B): D888 & D931
Chassis: Daimler Fleetline CRG6LXB-33, CRG6-33 or Leyland FE33AGR Fleetline
Engine: Gardner 6LXB, derated to 165bhp.
Gearbox: SCG 4-speed, semi automatic or with fully automatic* control; either Sevcon or CAV511
Body make: Fleetline CS: Metal Sections ckd
Fleetline(B): KMB with some Metsec parts
Body layout: Fleetline CS: LH61/41+21FE,FdXS,D,DOO(Farebox)
Fleetline(B): H62/44+20FE,FdXS,D,DOO(Farebox)
Date introduced: 1974-77
Total: Fleetline CS: 297 (298 after rebody of D1007 in 1982)
Fleetline(B): 2
Length: 32ft 10in   10,032mm
Width: CS: 8ft 1½in   2450mm
(B): 8ft.0in   2438mm
Height: CS: 13ft 10in   4220mm
(B): 14ft ½in   4270mm
Wheelbase: 18ft 6in   5638mm
Unladen weight: CS: 9,202kg
(B): 9,795kg
Gross vehicle weight: 16,275kg

NOTES: * Fully automatic gear control fitted new to some buses and retro-fitted to others. It has not been possible to identify which buses were so equipped.

D738 converted to Voith fully-automatic gearbox but this was later removed.

This caused there to be three spare Metsec body kits.

A further one hundred Fleetlines were purchased second-hand in the form of ex-London Transport DMS-class buses.and these are described under the 'Buses Purchased Second-hand' section.

Daimler Fleetline CS

The first Metal Sections bodied Fleetlines, D666 et seq, entered service on 26th June 1974, on Cross-Harbour Tunnel Route 102 and many went almost immediately into all-over advertising livery. All these Fleetline CS's were licensed to carry 123 passengers,

RIGHT: D666, KMB's first Fleetline and first rear-engined bus to be bought new in its first few days in service on 1st July 1974, when only five days old. The advertising was applied between the two dates as there reports of D666 in fleet livery on its first day. *(Mike Davis*

ABOVE: KMB's standard Metsec bodied Fleetlines entered service on Cross-Harbour Tunnel Route 102. Here D771 passes a tram in King's Road, North Point, in February 1981 - still on the 102. *(Mike Davis)*

BELOW: Fleetlines soon appeared on Kowloon Urban Routes and D861 is seen here on Route 2 heading for "You Guess Where" in January 1981. KMB had introduced large route numbers by this time, but destination blinds seem to have been a problem! *(Mike Davis)*

having a layout and capacity of LH61/41+21FE,CXS,D,DOO, with single-width entrance to prevent would-be fare dodgers from by-passing the Farebox. The staircase and double-width exit doorway were forward located, immediately behind the front wheel-arches, with a bench seat for three over the offside front wheel-arch, between the driver and the staircase.

It is interesting to note that the high capacity configuration was possible as a result of two factors. Firstly, the overall height of 13ft 10ins, made possible by building the body directly onto the chassis, omitting any intermediate body-bearers, kept the centre of gravity as low as possible. Secondly, because of the somewhat smaller stature of the average Cantonese person relative to Caucasians, it is usual in Hong Kong to have seats arranged 3+2 across the bus. It must be stressed that the

ABOVE: Standard MetSec bodied Dauimler Fleetline, D890 (BM2908) seen approaching the "Star" Ferry terminus in Kowloon during January 1978. *(Mike Davis)*

BELOW: Broadside view of D672 during its first few days in service during July 1974. This bus was sold in 1987 for further service in Guangzhou (Canton), China. *(Mike Davis)*

LEFT: D916, one of a few Fleetlines to which KMB fitted three-track route number blinds, before they were adopted for new buses. *(Mike Davis*

CENTRE: Rear view of a standard MetSec bodied Daimler Fleetline. D735 stands at Lai Chi Kok terminus on lay-over between duties on XHT route 102 in January 1978. *(Mike Davis*

BOTTOM: KMB carried-out a trial installation of a Voith D851, fully-automatic gearbox in Fleetline D738 but the experiment was less than successful without further costly modifications and the bus reverted to standard after about a year. *(Bob Bayman*

same 28 degree tilt test as that applied to double-deck buses in Britain is applied, very strictly, to all Hong Kong double-deck buses prior to type approval and registration.

These were the first new type of KMB bus to be fleet numbered upon entering service and these 297 MetSec bodied Fleetline CS's were in a broken series from D666 to D967, omitting D815 (A)-sub-type; D888 & D931 (B)-subtype and D963 & D964 (C)-sub-type. The omissions were caused by cannibalisation of two MetSec kits for repairs to buses already built and the substitution of a BACo body for evaluation on D815. The last two gaps in the series, D963 & D964 were merely the first two of the BACo production bodied buses entereing service prior to the final three MetSec versions, D965-7.

D666 et seq were to a similar basic design as the China Motor Bus Co's Fleetlines, LF1-30, the main differences lay in the lighting, which was fluorescent in the CMB version and by conventional bulbs in KMB buses; KMB's seats were fully upholstered while CMB only fitted cushioned seat squabs with laminate covered plywood backrests. Other differences were the destination and route number screens which followed each operator's standard practice. KMB's D932 broke with tradition by becoming the first bus in the fleet to have three-track route number blinds, although there were other examples that entered service later.

Rebody.

Following serious accident damage, D695 (BH2596) was the first of the class to be scrapped, in July 1980, but the numbers of the CS-type were restored in April 1982 when the former BACo-bodied D1007 was rebodied using one of the three spare MetSec body kits that should have been fitted to D815, 888 or 931.

Voith gearbox.

An interesting experiment was carried-out when D738 was fitted with a Voith fully automatic gearbox and the power-pack with drive line brought it to similar specification as that of a Dennis Dominator. This was not pursued and the one experimental bus was restored to standard specification after approximately one year in passenger service.

Fleetline (B) - D888 & D931 - with KMB bodies.

Superficially, D888 & D931 resembled the 'improved' MetSec design of China Motor Bus LF106, having similar 'peaked' front domes, but closer inspection revealed that there were substantial differences. The body side panels had KMB's then standard 'mock' beading in the form of pressed 'D' section ribs below each row of side windows which acted as stiffening. Above each of these ribs, the windows themselves differed from standard MetSec type for Fleetlines in being of the earlier type as used on buses up until the early Daimler(e) & (f) types, having square cornered window frames with separate sliding glasses, themselves fitted into square-cornered frames. These older type windows were also shallower than the normal type used on the majority of MetSec Fleetlines. Following the use by KMB of two MetSec body kits to provide spares for accident damaged buses, it was necessary to manufacture many of the parts for D888 & D931 themselves, including body pillars, side panels, roof sections, etc. The resulting 'KMB' body was 2.5 ins higher than the pure MetSec variety and the builders re-arranged the seating so that there were three additional lower-deck seats, one extra on the upper-deck and one less standee, a net increase of three passengers in total, producing a layout of H62/44+20FE, FdXS,D,DOO - a total of 126 passengers. The doorways and staircase were as standard, the latter being forward of centre, immediately behind the front wheel-arch. Of these two buses, D888 had KMB standard route number blinds but D931 has separate glasses for destination and three-track route number.

Mechanically these were standard Daimler Fleetlines and neither was ever fitted with fully automatic gear-change control.

ABOVE: D888 was the first of two Fleetlines which KMB bodied themselves but incorporated various MetSec parts - possibly using up remains of body kits and providing the remaining components, panels, etc. themselves. Seen here turning into Tin Lok Lane, Happy Valley, in December 1978 *(T. V. Runnacles)*

LEFT: The second of the two KMB bodies was assembled on the chassis of D931 and varied in detail; the most noticeable being the three-track route number display. Both buses featured 'peaked' front domes. *(Ian Glass)*

BACo bodied Fleetlines; sub-types (A) & (B)

Daimler Fleetline (A) - D815

Following CMB's move away from Metal Sections products due to corrosion problems with their steel framework, KMB ordered a single body from the British Aluminium Company (BACo). The resulting bus was very different from traditional designs, being, perhaps, the most angular double-deck bus ever built! It was to the same basic design as those supplied for use on just ten of the first twenty Singapore Bus Service AN68 Leyland Atlanteans. The angularity was even extended to the wheel-arches, giving rise to a very austere appearance indeed, as reference to the accompanying photograph will confirm.

Body layout differed from that of the MetSec products in that the staircase was located over the off-side front wheel-arch, as opposed to behind it, the stairs forward ascending to open onto the upper-deck behind the front row of four seats. There was also a penalty in that there was only accommodation for 119 passengers, the actual layout being H58/44+17FE,FdXS,D,DOO and the height, at 14ft 2ins, was four inches higher than the MetSec bodied 'CS' type.

Livery, as delivered, was very much like that adopted for the second-hand fleet, being mainly cream and, there was no front destination aperture, this being provided by KMB to their then standard dimensions. KMB allocated the sub-type Fleetline (A) but this was a reference to capacity; it being mechanically a standard semi-automatic Fleetline until withdrawn. During 1978, D815 emerged from workshops with radius cornered windscreens of the type provided by Alexanders for their CMB Fleetlines.

VEHICLE SPECIFICATION: BACo bodied FLEETLINES

Fleet numbers: Fleetline(A):	D815
Fleetline(C):	D963-4, 968-1115 (less D1007, from April 1982).
Chassis:	Leyland Fleetline FE33AGR
Engine:	Gardner 6LXB derated at 165bhp
Gearbox:	SCG 4-speed eith either semi-automatic or with Sevcon or CAV511 fully automatic control system
Body make:	British Aluminium Co ckd kit (D815 built in Englend)
Body layout: (A):	H58/44+17FE,FdXS,D,DOO(Farebox)
(C):	H57/44+18FE,FdXS,D,DOO(Farebox)
Date introduced: (A):	1976;
(C):	1977-79
Total: (A):	1
(C):	150 (less D1007, from April 1982).
Length:	32ft 11in 10,032mm
Width:	8ft ½in 2438mm
Height:	14ft 2in 4317mm
Wheelbase:	18ft 6in 5638mm
Unladen weight: (A):	9,551kg
(C):	9,919kg
Gross vehicle weight:	16,275kg

Fleetline (C) and (C)(Auto-Trans).

Deliveries of the third KMB order for 150 Fleetline chassis commenced in the early part of 1977 and were fitted with the 'production' version of the BACo prepared body, as on D815, from which it showed some detail differences, most noticeable being the lack of inward body-taper at the front, giving a less unbalanced look.

As on the prototype, D815, the Fleetline (C)'s featured the staircase over the offside front wheel-arch, but, in contrast, it rearward ascended. The standard entrance and exit positions and dimensions were retained, as on the MetSec type.

The first two 'BACo's entered service in August 1977, fleet-numbered D963 and D964, pre-dating the last three Metal Sections Fleetlines, D965-967, by about a month. Known by KMB as their 'Fleetline (C) type, these buses were most unusual in appearance, owing nothing to aesthetics, perpetuating the angularity of D815. The lower edge of upper-deck side windows was so high as to prevent all but the tallest of seated passengers from seeing out of the vehicle without particular effort, and the

BELOW: Prior to its export to KMB, the prototype body built in UK by The British Aluminium Co. was not fitted with any destination or route number gear. The somewhat stark looking vehicle, whose design was similar to BACo bodies that had recently been supplied to Singapore Bus Service, was photographed on a dull day in England. The body style differed from other, subsequent, BACo bodies in having curved-in (tumble-holm) lower body sides. *(Author's collection*

ABOVE: The prototype BACo bodied Fleetline was numbered D815 by KMB who classified it Fleetline(A). This photograph shows the vehicle on the roof of Lai Chi Kok depot after sign-writing and the provision of (in this view) route number blinds. *(Kowloon Motor Bus)*

BELOW: D815 in passenger service on Route 40, late in 1977 and seen at Tsuen Wan Ferry terminus before returning to Kwun Tong Ferry. *(T. V. Runnacles)*

top edge of the same windows extended to within two inches of the edge of the flat roof. Wheel-arches were angular and great gaps showed between the tyre and mud-guard, and the windscreens had square corners and, originally, having traces of the European 'lantern' windscreen about their appearance, brought about by the fitting of narrow vertical strips of glass between the corner of the actual windscreen and the pillar of the doorway, on the nearside and the drivers front side window on the offside. This reduced the 'blind-spot' but many buses emerged from repairs in workshops minus this feature, having plain aluminium in place of the glass.

Probably in an effort to reduce the stark appearance of these vehicles, the livery chosen for these BACo bodies reverted back to the older style, where the red extended to the lower edge of the upper-deck side windows, with cream above, including the roof and also cream lower-deck side-window surrounds. In late 1982 these Fleetline (C)'s were being outshopped in the half-and-half, cream over red, livery, but worse was to come, for with the introduction of the 'Jubilee' livery, of mainly cream with just a skirting band of red and red roof, the bodies looked plain and stark - the red on the roof, not extending over the edge of the roof, could only be seen from above!

Mechanically the Fleetline (C) was a standard Leyland-built FE33AGR Fleetline but, of the 150 vehicles, 117 were eventually converted from semi-to fully automatic gearchange. It has not been possible to determine which buses were converted, but those that were converted were classified as Fleetline (C)Auto-Trans.

D1091-1115, the last 25 BACo bodied Fleetline (C)'s, were modified during assembly in a number of ways in order to reduce the height of the lower edge of the upper-deck side windows which had become the cause of many complaints from passengers of small stature. This involved lowering the window-frames by about three inches and filling the space left above them up to the edge of the roof with plain aluminium sheet and this deeper panel was a most noticeable difference, easily identifying them from the other 125 of their type. The wheel-arches were also modified from the angular shape to a more curved outline, but not semi-circular.

During overhaul, most of the bodies built with 'angular' wheel-arches were modified to have new semi-circular outlines, more closely following the radius of the tyre.

Rebody D1007.

Following what is believed to have been a fire, D1007 was rebodied with a Metsec body, using the spare kit not used for D815, D888 or D931. As a result, it was reclassified Fleetline(CS).

LEFT: D815 as it appeared in a photograph published in a KMB House magazine. *(Kowloon Motor Bus*

BELOW: The standard production BACo bodies had angular wheelarches and very high upper-deck side windows, out of which passengers were only able to see with difficulty. The early livery was revived in order to improve the appearance of these somewhat boxlike buses. In this January 1978 view, D974 was approaching "Star" Ferry on Route 5c. *(Mike Davis)*

ABOVE: D1090 represents the later 'improved' version of the BACo body, where the wheelarches were 'softened' by being partly radiused. The upper-deck windows were lowered by six inches, giving greater depth between the top of the window pan and the flat roof. Note also, the different dash panels and position of the passenger-side windscreen ventilator. D1090 was leaving "Star" Ferry in 1981. *(Mike Davis*

BELOW: D988, after being retrospectively fitted with radius wheelarches, seen parked at Kwun Tong Ferry in March 1981. The thin glazed 'window' between the windscreen and front door pillar was not perpetuated after overhaul. *(Mike Davis*

ABOVE: The 1982 livery did little to soften the harsh lines of the BACo Fleetlines. The reason for painting a completely flat roof red remains unclear. D976 passes the Hong Kong Hilton in 1983. *(Mike Davis)*

BELOW: For comparison, this rear view of a BACo bodied Fleetline, D1011, shows the sloping rear bulkhead and window. Seen between Route 30 duties at Tsuen Wan Ferry in 1976. *(Derek Lucas)*

141

Guy Victory J / BUSAF.

A most interesting purchase was made by KMB early in 1976, when four experimental buses arrived from South Africa, where their bodies had been built by Bus Bodies (SA) Ltd., of Port Elizabeth (BUSAF). The double-deck bodywork was almost 15ft high and 31ft 10ins long - about 14 inches higher than a MetSec Fleetline. The original design had been for 103 seats, but they failed the 'tilt test' and so were down-seated prior to entering service to 100, plus 19 standees, or H57/43+19FE,FdXS,DOO. The high capacity was achieved by the provision of 3+2 seating in both upper and lower saloons. The South African version, from which BUSAF derived these four, did not feature 3+2 seating.

KMB had started their quest for a suitable front-engined double-decker soon after production of all British-built buses of that type had ceased in the early 1970's. Like China Motor Bus, KMB turned to sources other than the traditional British producers - CMB had tried the Indian-built Ashok-Leyland ALPD1/1 - an up-dated Titan PD3/5 - and the Swedish/Scottish Volvo/Ailsa double-decker.

The BUSAF design, based on the Guy Victory-J single-deck bus chassis, was originally developed for the City Tramways Group of Cape Town and Port Elizabeth, which also wanted a modern front-engined double-decker. The design continued to be developed and became of considerable interest in its own right. The suspension of the early buses for South African customers did not require much modification as the tilt test there required only that a bus should remain stable when tilted to an angle of 25 degrees, which was quite easily met with the 2+2 seating fitted. To meet the requirements of the Hong Kong tilt test, which was the same as the British standard 28 degrees and to accommodate 3+2 seating required some considerable redesign of the suspension. The KMB vehicles were modified by the provision of heavier-duty springs incorporating heavy Aeon buffers and a restriction device which limited the extent to which the body may tilt away from the line of the axle. Nevertheless, the ride in actual service compared very favourably with any other double-deck bus - which is the eventual test of any suspension system.

The construction of the Kowloon buses was not simply a 'tidied-up' version of the Port Elizabeth of Cape Town specifications, but represented a thorough redesign to what evolved as a semi-integral vehicle, a description used by the manufacturer. The Guy Victory-J chassis were purchased by BUSAF from Leyland South Africa Ltd., at which point Leyland's responsibility ceased, the chassis then being used to produce what became a Guy/BUSAF product, that is a Guy, modified by BUSAF and bodied by them. The extent that BUSAF was committed to the design became apparent when it was revealed that BUSAF alone carried the cost of development and provided the warranties. The finished product was, therefore, in direct competition with Leyland's own rear-engined bus products in the markets of Africa and Asia where the Atlantean and Fleetline exhibited some limitations. It should be remembered that Guy was also part of the Leyland empire. As a result British Leyland was disinclined to the BUSAF development for obvious reasons but eventually developed a similar product themselves, in the form of the Victory Mk2.

The KMB-BUSAF design utilizes the same basic underframe as the South African vehicles, where the original longitudinal chassis-members were retained but fitted with new cross-members and out-riggers, in line with the body pillars. The out-riggers were bolted through the chassis longitudinals and the flanges on the inner cross-members, using high tensile bolts in reamed holes. Then the cross-member was effectively connected across the top-face by means of a 4ins wide by .25in-deep tension strap, welded to the out-

VEHICLE SPECIFICATION - GUY VICTORY 'J'	
Registration numbers:	BJ9266-9269
Fleet numbers:	G1-4
Chassis:	Guy Victory-J, modified by BUSAF (South Africa).
Engine:	Gardner 6LX, 150bhp, front mounted.
Gearbox:	Leyland (SCG) 4-speed, semi-automatic.
Body make:	Bus Bodies (SA) Ltd, South Africa.
Body layout:	FH57/43+19FE,FdXS,D,DOO.
Date introduced:	1976.
Total:	4
Length:	31ft10in 9702mm
Width:	8ft 1in 2463mm
Height:	14ft 11½in 4559mm
Wheelbase:	15ft 9in 4800mm
Unladen weight:	
Gross vehicle weight:	

BELOW: The hybrid Guy Victory-J with BUSAF body, was produced in South Africa. Here G3 is on Tunnel Route 113 late in 1975, soon after entering service. The high floor-line can clearly be seen in this view, as can the single headlights; a feature G3 shared with G1. *(Stan Leeds for Mike Davis*

ABOVE: The South African built BUSAF body as seen from the offside. These 15ft high bodies were mounted on the straight frames of the nominally single-deck Guy Victory-J chassis. G3 was arriving at Lai Chi Kok on its was to Kwun Tong ferry in January 1978. *(Mike Davis*

riggers and cross-member. This provided a simple but strong and rigid underframe, while the wide tension strap enabled the one-piece plywood floor to be bolted securely to the underframe.

Further development of the concept continued and a design utilising a Japanese chassis was investigated but would have been out of the question in Hong Kong where the terms of both major companies' franchise at that time required the exclusive use of British or Commonwealth produced products.

As a result of the success of these buses and the reappearance of a Dennis bus with front engine, Leyland Truck & Bus felt themselves commercially obliged to offer a version of their own. After considering a Guy Arab MkVI, the decision was made to proceed with the Guy Victory Mk2, which should not be confused with the South African version of the BUSAF-Victory known locally as the MkII. It should be noted that Leyland carefully avoided the use of the Roman numerals to describe their Mk2. The South African version was a totally different concept.

Somewhat surprisingly, the Gardner 6LX engine was chosen for the BUSAF-Victory, rather than the more powerful 6LXB later fitted in later, similar, Leyland-built Victories, *qv*. The choice of Gardner was, of course, in line with standard KMB policy. It was, of course, mounted ahead of the front axle, between the driver and the single-width front entrance, in the style of the ill-fated Guy Wulfrunian of the early 1960's. Since, however, KMB had specified a single-width entrance, it was not necessary to push the engine, and the driver, to the off-side to gain platform space. Both entrance and exit doors were power operated and followed the usual local practice of single-entrance and double exit widths. The exit was located in a forward position, immediately behind the nearside front wheel-arch, while the stairs forward ascended over the offside wheel-box. In view of their having Guy mechanical units, KMB classified these buses as their 'G-class' and they became fleet-numbered G1-4. Some minor differences were apparant between the individual buses when newly delivered. G1 & G3 had single headlights while G2 & G4 had dual headlights.

Following trial use on Cross-Harbour Tunnel routes, G1-4 were transferred to Route 40 where they remained in regular service until they were withdrawn from passenger service.

During the first three months of 1982, all four were modified for use as advanced driving instruction vehicles, having the staircase removed to a more rearward location. An instructor's seat was fitted together with a set of duplicate direction indicators inside the vehicle and extra rear-view mirrors for the instructor's use.

LEFT: G4, as seen here almost head-on, displays its 15ft height as it passes some high rise, high density, public housing en route to Tsuen Wan Ferry. Like G2, G4 had dual headlights. *(T. V. Runnacles*

Dennis Jubilant

Continuing the search for a new source of front-engined double-deck bus chassis, KMB purchased the first four Jubilant chassis from the Hestair-Dennis company (formerly Dennis Brothers Ltd.), following their return to bus manufacturing after almost ten years of concentration on municipal and fire-fighting vehicles. The Jubilant chassis, launched in Jubilee year, was designed specially for KMB, having a straight frame which was dropped just ahead of the front axle to provide for an acceptably low front entrance

Power was supplied by the Gardner 6LXB engine, driving the Eton spiral-bevel rear-axle via a Voith D851 fully automatic, three-speed gearbox. The engine was mounted ahead of the front axle alongside the driver, as on the BUSAF/Guy and Ailsa-Volvo. Power-steering helped the driver cope with the front-end load and a hydra-dynamic retarder aided breaking, the retarder being operated by the footbrake pedal. The most outwardly noticeable feature of the Jubilant was the very short wheelbase of 15ft 9ins on a vehicle of 31ft overall length. This short wheelbase, combined with the weight of the engine canterlevered ahead of the front-axle contributed to the very lively ride associated with this model, especially at higher speeds on main roads. Direct comparison with the BUSAF/Guy Victory shows that the Dennis allowed for a body 6.5ins lower in overall height.

The KMB designed bodywork was built in the Company's Kwun Tong depot and had seats for 102 passengers, plus 14 standing. Seats were arranged 3+2 in both saloons. Body construction was almost entirely of aluminium, but incorporated front and rear mouldings of fibre-glass. The usual narrow front entrance and double width exit behind the front wheel-arch were provided with the staircase rising above the off-side front wheel-arch.

Each of these four bodies was a 'one-off' and variations abounded and could be spotted with relative ease. Following workshop attention associated with annual recertification during the winter of 1982/3, detail changes brought about an altered appearance to the front of these buses, in effect standardising them. New front dash panels were fitted of the type used on the Duple/MetSec bodied version (*qv*) with neat radiator grilles and light units. They were at the same time repainted into the Company's revised livery of almost all cream, relieved with red roof and lower body skirt panels.

KMB chose the fleet number prefix 'N' for the Dennis'- 'D' had already been used for Daimler and the next strongest sound in the name was used. It must be remembered that the Chinese workforce was largely unfamiliar with written English. The first KMB bodied Jubilant, BR4474, was thus numbered N1 and entered service on 3rd November 1977, with N2 following in January 1978; N3 & N4 were in traffic by the spring of that year. KMB's classification was Dennis FS (FS meaning Forward Stairs).

As a result of the success of these four prototypes, KMB placed an order for 50 'production' Jubilants in January 1978, followed by further orders until there was a total of 364

A fifth body to this general design was built to rebody a second-hand Leyland Atlantean, 2L71.

VEHICLE SPECIFICATION - DENNIS JUBILANT DD201

Fleet numbers:	N1-4	
Chassis:	Hestair-Dennis Jubilant, Model DD201.	
Engine:	Gardner 6LXB front mounted.	
Gearbox:	Voith D851, fully-automatic, 3-speed, with retarder.	
Rear axle:	Eaton spiral-bevel.	
Body make:	Kowloon Motor Bus.	
Body layout:	FH60/42+14FE,FdXS,D,DOO(Farebox)	
Date introduced:	1977-78	
Total:	4	
Length:	31ft 0in	9448mm
Width:	8ft 0in	2438mm
Height:	14ft 5in	4397mm
Wheelbase:	15ft 9in	4800mm
Unladen weight:	9,720kg	
Gross vehicle weight:	16,257kg	

BELOW: When the front-engined Dennis Jubilant chassis was announced, KMB purchased four prototypes and bodied them themselves at Kwun Tong to a rather pleasing design. Here N1 approaches Tai Kok Tsui Ferry pier on 8th January 1978. *(Mike Davis*

ABOVE: All four Dennis Jubilant prototypes, N1-4, showed detail differences. Here N1 is seen with Alexander-type windscreens, of the type fitted to contemporary CMB Fleetlines, one of which can be seen following the KMB Dennis. *(T. V. Runnacles)*

BELOW: N2 was the first of the four to be refurbished, late in 1982, and is seen here in Hung Hom, painted in the then new KMB livery with red roof and skirt panels. Note also the use of Duple Metsec type front dash panels and grille. *(Mike Davis)*

145

Leyland Victory Mk2 Series 2. G5-544.

Following the success of the Guy-BUSAF Victory-J double-deck buses for Cape Town, Port Elizabeth and Kowloon and spurred on by the introduction of the Dennis Jubilant front-engined chassis, British Leyland examined the possibility of re-introducing a front engined model in the form of a Guy Arab MkVI for export to these important overseas markets. For a number of reasons the Guy Arab concept was not feasible and as a result the Guy Victory was redesigned and offered in a Mk2 form with all the running units mounted on mainframes which combine features of the Victory-J with the rear configuration of the Leyland World-master, which latter had a 'kick-up' over the rear axle, thus helping to reduce the overall height by 4.5inches when compared with the BUSAF Victory-J which was essentially an unmodified single-deck chassis. This new model was referred to by Leyland as their Victory Mk2, Series 2 - the 'Series 2' indicates a double-deck specification.

KMB's initial order was for two of the original Victory-J chassis, upon which they had intended to build their own design of bodywork, following BUSAF principles, but Leyland pre-empted the idea which had never met with their approval, by offering the Victory Mk2. KMB reacted by cancelling their final order for 150 Daimler Fleetlines, with which type they were at the time somewhat unhappy, and changed the order to a similar number of Victory Mk2's. The order for the two Victory-J's was also converted to the new Victory 2, making a total of 152 chassis, of which one was to be a double-deck coach.

urther orders of 125, 65 & 200 followed and they were fleet numbered in the series started with the Guy Victory-J/BUSAF buses, G1-4; the Victory Mk2's commencing at G5 and continuing to G543 - plus CM3879, the un-numbered coach which later became G544 when converted to a bus - making a total of 542 Victory Mk2 chassis.

Further orders of 125, 65 & 200 followed and they were fleet numbered in the series

VEHICLE SPECIFICATION - LEYLAND VICTORY	
Fleet numbers:	G5 - 544
Chassis:	Leyland Victory 2, Series 2.
Engine:	Gardner 6LXB, forward mounted.
Transmission:	Voith D851, fully automatic, 3-speed gearbox, incorporating a retarder,
Body make:	W. Alexander - ckd, except UK-built pilot vehicles.
Date introduced:	1978 - 1983.
Body layout:	bus : FH60/42+16FE,Fds,CX,DOO coach : FCH43/32+00FE,Fds,CX,DOO (see footnote (b))
Total:	built new as buses : 539 built new as coach : 1 - converted to bus 1983.
Length:	9.75m 32ft 0in
Width:	2.5m 8ft 2½in
Height:	4.47m 14ft 7½in
Wheelbase:	4.8m 15ft 9in
Unladen weight:	10,199kg
Gross vehicle weight:	16,257kg

NOTE: The coach was converted to serve as a bus from May 1983 and was classified as Victory Mk2(D).

Victory Mk2FS -total 151 :	G5-155
Victory Mk2(A) -total 125 :	G156-264,/6-9/72/5-80/3/5/94/8,345.
Victory MK2(B) -total 65 :	G265/70/1/3/4/81/2/84/6-93/95-7/99-344.
Victory Mk2(C) -total 198 :	G346-543
Victory Mk2(D) -total 1 :	G544 (formerly coach)
Victory Mk2(BB) -total 1 :	G544 (formerly G335 (B))

From July 1986 the following body layouts applied:
Victory (A-C) : H60/42+14 from 7/86
Victory Mk2 : H60/42+12 from 8/86.
Victory (BB) : H60/40+17
Victory (D) : H58/37+1

BELOW: The first of the Leyland (Guy) 'Victory Mk2 - Series 2' chassis for KMB was bodied in Scotland by Walter Alexander who eventually supplied 538 similar bodies as ckd kits. Note the very shallow windscreens of the earlier examples. G5 was photographed in Scotland, prior to shipping. *(Walter Alexander*

started with the Guy Victory-J/BUSAF buses, G1-4; the Victory Mk2's commencing at G5 and continuing to G543 - plus CM3879, the un-numbered coach which later became G544 when converted to a bus - making a total of 542 Victory Mk2 chassis.

Generally the specification of the Victory Mk2 is similar to that of its main rival the Dennis Jubilant, having a Gardner 6LXB engine mounted ahead of the front axle, driving via a Voith D851 fully automatic gearbox having three-forward speeds and incorporating a retarder. The chassis introduced for the first time into the KMB fleet the air-released spring operated handbrake with two 'on' positions, one a 'bus-stop' brake and the other a full parking brake.

The body contract was let to Walter Alexander early in 1978 and that company provided the bodywork for the entire class, with all but G5 and the coach (q.v.) being shipped in ckd form for assembly in KMB's own workshops. G5 was exhibited at the October 1978 International Motor Show held at the National Exhibition Centre, Solihull near Birmingham, England. This prototype was the subject of extensive tests and eventually arrived in Hong Kong in May 1979, entering service the same month registered BY8415 and with a body layout of FH60/42+16FE,FdS,CS,D,DOO. From this it will be seen that the desired front entrance necessary for successful Driver-Only-Operation was achieved in combination with a front engine in the manner of the Ailsa Volvo. The staircase is forward ascending over the offside front wheel-arch; a forward rather than a front position. The passenger access from road level to the foot of the staircase has never been satisfactory, commencing as it does with a narrow entrance up three steep steps and along a passage as narrow as 12 inches past the engine. This latter cramped space precludes the fitting of seats over the nearside front wheel-arch, although a single inward-facing seat is located behind the wheel-arch and against the forward of the two centre-exit bulkheads. Standing is also prohibited (in theory) in this forward part of the saloon as there is a real danger of overloading the front axle beyond its legal, and safe, limit. To help compensate for this loss of capacity, extensive use is made of 3+2 seating on both upper and lower decks. KMB bodies differ from those supplied to China Motor Bus in that KMB finishes the interior of its Victory bodies with an application of paint while those of CMB are faced with plastic laminate, reducing to a marked degree the cost of reconditioning.

G5-155 - Victory Mk2.

A feature of the first series of Victory Mk2 bodywork as assembled by KMB is the externally sliding driver's cab door which initially caused the licensing authority to refuse to accept the additional width which caused the body to exceed the 2.5m (8ft 2 in) maximum as permitted by the, then, 'Construction and Use Regulations'. Faced with 151 buses, each over-width (plus a number of similar Dennis Jubilants), as a fait accompli, a concession was finally agreed upon and 'Excess Width Permits' were issued on the understanding that future orders specified internally sliding cab doors. It is also worthy of note that the actual opening of the doorway on G5, etc., is located forward of the driver, whereas the later, amended, version has it located beside the driver. (CMB did not incur this problem as their Victories had hinged cab doors

On this first type Victory Mk2 the windscreen is very shallow and the dash panel above the radiator grille is very high. This is easily identified by reference to photographs and making comparison with the later types (A), (B), & (C). It will also be noted that the opening vents on both sides of the windscreen are very shallow. The coach, which was ordered as part of the initial 152 Victory Mk2's, has a type (A) windscreen - see G156 etc. for details.

G19, 131 - 150 were fitted with air-conditioning units in the driver's cab in an effort to reduce the degree of discomfort associated with the sitting alongside the internally located engine compartment, especially during the hot and humid summer months. G19 was retrospectively fitted, while the remainder had the air-conditioning units fitted during body assembly of the body. The success of the air-curtain type of air-conditioning units was very limited and they fell out of use, being physically removed during 1982. No further attempt to cool the driver's working environment has been made.

Victory Mk2 - Urban Coach CM3897.

CM3897 was built as a fully air-conditioned coach for use on the trunk route 206, between Lai Chi Kok and the Kowloon-side "Star" Ferry Pier. The Alexander body was completely-built-up in Scotland by Alexander. Drawing on the experience gained in fitting air-conditioning in the New York Atlanteans, Leyland and Alexander joined forces to pro-

BELOW: G141 from the first batch of Victories with shallow windscreen, heads north out of the Cross-Harbour Tunnel in the 1982 livery and carrying an early version of the still (1994) current KMB logo. Seen here in June 1984. *(Mike Davis*

ABOVE: One of the first 152 Victory Mk2's was built to urban coach specification and was fully air-conditioned - a feature that quickly proved unreliable. Note the shallow windows over the rear wheel-arches to allow for a raised saloon floor and forward-facing seats over the rear wheel-arches. *(Mike Davis*

BELOW: Victory Mk2 of the first production batch for KMB with the shallow windscreen and high dash panel which incorporated a hinged flap over the radiator filler cap. Here, G131 was in King's Road, North Point, in February 1981. *(Mike Davis*

ABOVE: The second type body for the Victory still featured a high dash panel but the windscreen was two inches taller. The radiator filler has no flap to conceal its opening. G244 was photographed in North Point at Chinese New Year 1981. *(Mike Davis)*

BELOW: G383, of the ultimate sub-type (C), approaching the Cross-Harbour Tunnel in 1982 livery and 'large logo'. *(Mike Davis)*

vide the same facility on the Kowloon coaches (one of the Dennis Jubilants *(q.v.)* was similarly treated) but, despite protracted experimentation, met with no better result than that achieved on the ill fated New York Leyland Atlanteans. There was a separate diesel engine located behind the lower-deck rear seats, in the position occupied by the main power plant in a conventional rear-engined bus. This secondary and smaller engine was specifically required to power the air-conditioning plant which would otherwise drain the main engine to an unrealistic degree. Thus the coaches had the distinction of having an engine at both ends! Seating and layout was FCH43/32FE,Fds,CX,D,DOO. No standing passengers were permitted and seating was to full coach standards. As became the fate of most of the single-deck urban coaches, CM3897 became redundant upon the Mass Transit Railway's Tsuen Wan Extension in the summer of 1982 and, with it being too new to warrant disposal, had its air-conditioning plant removed prior to it being regarded as a bus, when it became G544 but retaining coach seats. KMB classified it as Victory (D). As a coach it had no fleet number. The cab-door was of the hinged type and the windscreen as for type (A).

G156 etc. Victory Mk2(A)

In keeping with KMB's more recent type classification, the second version of the Victory Mk2, comprising 125 units, was deemed to be "Victory Mk2(A)" - the initial version having the distinction of only "FS", which indicates "Front Staircase". (All Victories have a front staircase.)

The major obvious differences between the first (G5) and second (G156) series are, as noted, the design and location of the driver's cab door and depth of windscreen which, on the type(A) is slightly deeper by approximately 5cm at the top. The high dash panel - the lower margin of the windscreen - of the first series is retained. This second type windscreen also has narrow (shallow) opening vents on both sides. In addition the radiator water filler cap is, or was originally, concealed behind a top-hinged flap on the earlier type but from G156 to 345 access is provided by an unconcealed opening in the front dash panel. The flap on earlier models was later removed to aid the task of the "water-can-man" at each terminus.

The highest fleet numbers of the type (A)s are blurred into the earliest of the type (B)s; both types being delivered in January 1981, but the very last type (A), G345, was not licensed until 5th June 1981, almost five months after the penultimate bus of its type, G298! G298 was itself unfortunately burned beyond repair and was scrapped in December 1982.

Victory Mk2 (B)

The first of the 65 type (B) Victories were introduced during January 1981, coinciding with the last of the type (A)s and, as a result, fleet numbers allocated are intermixed between the two types but the lowest numbered type (B) carries the number G265.

The third type windscreens as fitted to Victory (B)s are similar in overall dimensions to those of the Victory (A) but differ in having a slightly deeper passenger side opening vent. No other obvious differences in appearance have been recorded and the carrying capacities are identical.

Victory Mk2(BB)

For some unknown reason the passenger capacity of G335 was reduced during August 1983 by a figure of one, to 117 persons - two less lower-deck seats and one additional standee, making the layout 60/40+17FE,Fds,CX,D,DOO, instead of 60/42+16=117. The bus was reclassified by KMB to type (BB).

Introduced in May 1981, G335 was shipped back to Leyland for investigations into mechanical problems, in November that year. It returned in July 1982.

Victory Mk2 (C) - G346-543

KMB's final order for Victory Mk2's was for 200 chassis and an equal number of Alexander body kits, similar in most respects, including capacity, to the previously set standard for the type. One major difference to the appearance, and a great aesthetic improvement, is a result of relocating the water radiator water filler-cap on the interior dashboard inside the bus immediately behind the windscreen. The desired objective was to provide the driver with a better field of vision towards the road directly in front of the bus. The top edges of the fourth sized Victory windscreens remain at a similar level to those of the types (A) & (B) but the bottom edge is approximately 10cm lower. It is understood that a person was killed when a stationary bus moved-off, the driver being unable to see the hapless victim who was out of his line of vision, too close to the radiator. Buses with the high dash panels were provided with an additional convex mirror directed to show the front of the bus more clearly.

Although KMB ordered 200 buses of this (C) sub-type, only 198 entered service due to

BELOW: KMB sub-class 'Victory(B)', G307 was passing under the then newly constructed elevated approach road to the Aberdeen Tunnel at the junction of Tin Lok Lane and Morrison Hill Road when photographed here in late 1981. No obvious external differences have been determined to distinguish the sub-type (A) from sub-type (B). *(Mike Davis*

ABOVE: G431 represents the final batch of 200 Alexander/Victory 2 buses ordered by KMB and this view shows the most major change to the design, where the radiator filler-cap has been relocated behind the windscreen, so permitting a lower border to the windscreen as requested by Government Vehicle Examiners. Seen here in Canal Road East, September 1982. *(Mike Davis*

LEFT: Following the withdrawal of most coach services, the two double-deck coaches were converted to buses but the external appearance remained the same, with shallow lower-deck side windows above the rear wheel-arches. The air-conditioning unit was removed. Here, now numbered as bus G544, the former coach is seen in Nathan Road, heading for the "Star" Ferry on Route 6 in March 1984*(John Shearman*

Still at work in 1994; **ABOVE:** **G128** from the first batch passes to the side of Tai Wai bus and railway station in the New Territories whilst on Route 87B to San Tin Wai on a dull September afternoon. **BELOW:** G215 seen arriving at the bus station behind the railway station at Tai Wai during September 1994. *(Both Mike Davis*

the sale of two newly assembled but unregistered buses by KMB to New Lantau Bus Company.

G544 - Victory Mk2 (D)

This vehicle originally entered service as double-deck coach CM3879, on route 206 but, as already noted, was too late to stem the defection of passengers to the Mass Transit Railway following the opening of the Tsuen Wan Extension. Together with the similar Dennis, CM3879 was downgraded to bus duties, losing the air-conditioning plant in the process but retaining the 2+2 coach seats, and was additionally licensed to carry 15 standing passengers, giving the vehicle a layout of FH43/32+15FE,FdS,CX,D,DOO. A fleet number G544 was then allocated to it, G544, and the livery changed to the then new bus style of red roof and skirt panels with the main body panels cream. G544 was up-seated to H58/37+12FE,FdXS,D,DOO in January 1985.

Victories scrapped

A number of Victories have been the victims of serious fires. In some cases complete or partial rebodying has been possible, but G94, G298 and 440 were deemed to have been beyond repair and were written-off and the chassis cut-up.

G95 was a severe fire victim in December 1980 but was, nevertheless, recoonstructed and rebodied, returning to service in December 1983, a period longer than the two years permitted by HK Government regulations for the registration of a delicensed vehicle to be retained. It was thus that upon its return to service G95 was re-registered with the out of sequence number, CY4529. The original fleet number was kept. There is no knowledge available to show the source of the new body.

Victories - general

The Victories were soon found to have weaknesses which developed quite early on and affected both body and chassis although the question of which came first is a very delicate matter between the two manufacturers involved! Perhaps the most noticeable problem for the general public was that of squealing (screeching?) brakes which could be heard for miles as the penetrating sound carried up hillsides to residential areas. After two years the brake squeal was cured but other demons beset the Victories of both KMB and CMB and in order to examine a defective vehicle under the eye of senior engineers, a KMB example (G335) was returned to UK for thorough diagnosis. Fortunately the problems were identified and the vehicle returned to Kowloon where the necessary modifications were carried out on the rest of the KMB fleet.

Re-classifications 1986

During September 1984 KMB grouped Victory (A), (B), (C) together as one Victory (A-C) group, which also included the first un-suffixed buses G5-155, making a total (A-C) group of 532 buses with a body layout H60/42+16. Further alterations were made under new regulations in July 1986 when 149 buses reverted to un-suffixed Victory Mk2 retaining 16 standees. The Victory (A-C) then became 383 buses with 14 standees. To further complicate the issue the 149 Victory Mk2's had their standing capacity reduced in August 1986 from 16 to 12.

Victories sold

In March 1983 KMB sold two buses, G526 & 527, to the New Lantau Bus Co (1973) Ltd. followed in October 1987 by G498 & G505. A further two were, G542 & G543 were sold to NLBC in January 1991.

Driver's doors.

LEFT: The first series, G5, etc., had externally sliding driver's doors which, strictly speaking, made the over-width. *(Roger Bailey*

LOWER LEFT: In order to regularize the door situation, the second and subsequent batches had an internally sliding door, located further to the rear than the first type. *(Mike Davis*

Dennis Jubilant.

N5-364

Following on from successful trials with the four prototype Dennis Jubilants, N1-4, KMB placed orders for 50 chassis, progressively increased in groups of 75, 35, 50 and 150 to a total of 360 units.

Apart from minor alterations, the 'production' chassis resembled those of N1-4 in general specification, being similarly powered by a Gardner 6LXB engine, driving the rear-axle via a Voith D851, 3-speed fully-automatic gearbox incorporating a retarder; the engine being located ahead of the front axle, alongside the driver.

Alexander bodies

Alexander bodies of almost identical design to those provided for the Leyland Victory were ordered for the first 210 Jubilants, known by KMB as their 'Dennis(A), (B), (C), & (D) types' corresponding to the batches of 50, 75, 35 & 50 respectively. These were numbered:-

- Dennis(A) N5-53 omitting coach CF4180
- Dennis(B) N54-126,128/9
- Dennis(C) N127, 130-163
- Dennis(D) N164-213.

Unlike the similar prototype Alexander bodies on the first two Leyland Victory Mk2's (ie G5 and the coach), the prototype Alexander bodies on the corresponding Jubilants (ie N5 and coach) were not assembled in Scotland. (source PSV Circle publication BB199/27.)

The first 'production' versions of the Dennis Jubilant entered KMB service in May 1979 and, of the order for 50, only 49 were delivered as buses, the remaining example being bodied as an experimental double-deck coach to be

VEHICLE SPECIFICATION: Dennis Jubilant.

Fleet numbers:	N5-364	
Chassis make:	Dennis Specialist Vehicles Ltd.	
Model:	'Jubilant'	
Engine:	Gardner 6LXB, 165bhp, forward mounted.	
Gearbox:	Voith D851, fully automatic, 3-speed.	
Body Make: (A) to (D):	Walter Alexander - ckd	
(E):	Duple (Metsec)	
Body layout: (A) to (D):	FH60/42+18FE,FdS,CX,DOO.	
Coach:	FCH43/32+00FE,FdS,DOO,A/C.	
(E):	FH60/40+16FE,FdS,CX,DOO.	
Date introduced:	1979-82	
Length: (A) to (D):	9728mm	31ft 11in
(E):	9640mm	31ft 7½in
Width: (A) to (D):	2500mm	8ft 2½in
(E):	2450mm	8ft 0½in
Height: (A) to (D):	4470mm	14ft 7½in
(E):	4500mm	14ft 9in
Wheelbase:	4800mm	15ft 9in
Unladen weight: (A) to (D):	9,741kg	
(E):	?	
Gross vehicle weight:	16,275kg.	

Totals:	
(A):	49
(B):	75
(C):	35
(D):	50
(E):	150
(F)-(ex-Coach):	1
Total:	360

NOTE: The coach was converted to serve as a bus in May 1983 and was classified as Dennis(F). form.

BELOW: An early version of the KMB Dennis Jubilant, N8, followed closely by two later examples, all with shallow windscreens, similar to those fitted to the early G-class Victories. The radiator water filler-cap was located inside on all Jubilants as built new. *(Mike Davis)*

known only by its registration number, CF4180.

All the bodies of the Alexander sub-types have a similar layout, being arranged FH60/42+18FE,Fds,CX,D,DOO. It will be seen that two additional standees are authorised over the equivalent Alexander bodied Victory Mk2's. The single-width entrance doorway gives access to three steep steps past the farebox and engine cover, alongside which runs a narrow passage to the foot of the staircase which forward ascends over the offside front wheel-arch. Immediately behind the near-side front wheel arch is a single, inward facing and adjacent to the forward bulk-head of the exit doorway, which itself is opposite the foot of the staircase.

As is the case on the Victory Mk2's the density of seating in a 31ft long vehicle is achieved by extensive use of 3+2 seating on both decks.

A feature that helps to distinguish Jubilants from Victories is that on all Jubilants the windscreen is shallow and there is no external radiator filler. On Victories, those buses with shallow windscreens have external fillers, while those with internal fillers have deeper windscreens.

ABOVE: N17, passing the shored-up Supreme Court, undermined by excavations for the Mass Transit Railway, was one of very few Jubilants to be fitted with spot-lamps. Seen here in January 1978. *(Mike Davis*

BELOW: N103 of the second batch of Jubilants, the (B) sub-type, seen here leaving the "Star" Ferry for Tse Wan Shan in 1981. *(Mike Davis*

155

ABOVE: N60 of the (B) sub-type with windscreens slightly higher at the top than the (A) sub-type. The difference is quite small, amounting to no more than two inches (5cm) between the two types. Seen here in 'new' livery in Tsim Sha Tsui, Kowloon during 1984 *(Mike Davis*

BELOW: The Dennis equivalent of the Victory coach was CF4180 which, like its Leyland counterpart was relegated to bus work, with the air-conditioning plant removed, after the opening of the MTR brought about the termination of many urban coach services. *(Mike Davis*

ABOVE: N15 shows its offside and the externally sliding driver's door which made these buses slightly over-width. Only the earlier deliveries of Jubilant - and Victory - wore this livery; they also featured the KMB crest on the grille. *(Mike Davis*

LEFT: N140 has a deeper driver's cab door and is typical of the majority of the type. *(Mike Davis*

Chassis designations:
KMB Dennis Jubilant chassis are as follows:-
N 1-4: DD201
N 5-53 & 364: DD204
N 54-162 & 178: DD206
N163-177 & 179-213: DD208
N 214-363: DD207

Reference to photographs may help the reader to understand more easily.

The provision of an external sliding cab door proved equally as troublesome on the Dennis(A) as it had on the Victory Mk2 'FS'. From observation, it has been recorded that sliding doors of varying depths are fitted but for what reason is unclear. Cab doors on the subsequent Dennis(B), (C), & (D), ie G54 to 213 are internally sliding, within the overall body-width of 8ft 2 ins.

CF4180.

As already commented upon, unlike its Victory counterpart, this Jubilant coach was bodied by KMB in Kowloon and not in Scotland and, as a fully air-conditioned double-deck coach, it was visually similar in most respects to the Leyland Victory coach. An additional small diesel engine was mounted at the rear specifically to provide power for the Thermo king air-conditioning plant. All seats were to high-backed coach pattern and fitted 2+2 throughout, including over the rear wheel-arches where a raised plinth was provided. The adjacent lower-deck side windows were of an accordingly reduced depth. Body layout was as for the Victory coach, being FH43/32FE,Fds,CX,D,DOO.

With the curtailment of Urban Coach Route 206, CF4180 became redundant and was withdrawn for conversion for bus work when it was allocated fleet number N364. For its role as a bus little was changed except that standees are permitted and the air-conditioning equipment was removed.

Duple-Metsec Bodywork

Duple-Metsec Bodywork was specified for the final 150 Dennis Jubilant chassis, N214-363, which became KMB's Dennis(E) type. This new design from the Duple-Metsec stable is - or was when new - a very modern style, somewhat 'squared-up' but not unattractive.

As a product it showed sufficient improvement over earlier Metsec designs to lay the ghost of their previous reputation and put the company in the forefront as a supplier of quality ckd bodywork. General arrangement is as described for the Alexander type with regard to entrance, exit and staircase positions. The seating arrangements are slightly different with two less lower-deck seats and two less standees, their layout being FH60/40+16FE,Fds,CX,D,DOO.

A distinct difference exists between the earliest of the Duple-Metsec bodies and subsequent batches which can be seen in the shape of the central pillar dividing the upper-deck front windows. Reference to illustrations is the most certain way to appreciate this difference which is that on earlier types the pillar is wider at the top than at the bottom giving the appearance of a distinct 'v' shape. On later bodies the pillar is of equal width throughout its length, between radiused corners.

ABOVE: The earlier version of the Duple(Metsec) bodies for the Dennis Jubilants, from N214 onwards, can be easily distinguished from later versions by the distinctive pillar between the two upper-deck front-windows which is of a 'V' shape. N225 was leaving Nathan Road, leading a convoy of Victories and Jubilants in 1982. *(Mike Davis*

BELOW: The later, and majority, style featured a parallel pillar, as shown here by N302, leaving the Cross-Harbour Tunnel toll plaza for Tsz Wan Shan, on Route 116 in 1983. *(Mike Davis*

ABOVE: N316 was wearing the KMB 50th anniversary slogan beneath the company's initials when photographed leading a CMB Fleetline, XF101 (ex-London DMS882), in Central District during 1983. It had not, by then, acquired reflective number plates. *(Mike Davis)*

BELOW: Former coach, now bus N364 at Kowloon Station, showed the shallow windows as it worked on Route 1K in March 1984, before receiving its fleet number. *(John Shearman* **INSET:** The rear of the ex-coach was unaltered despite the removal of the air-conditioning plant. *(John Shearman)*

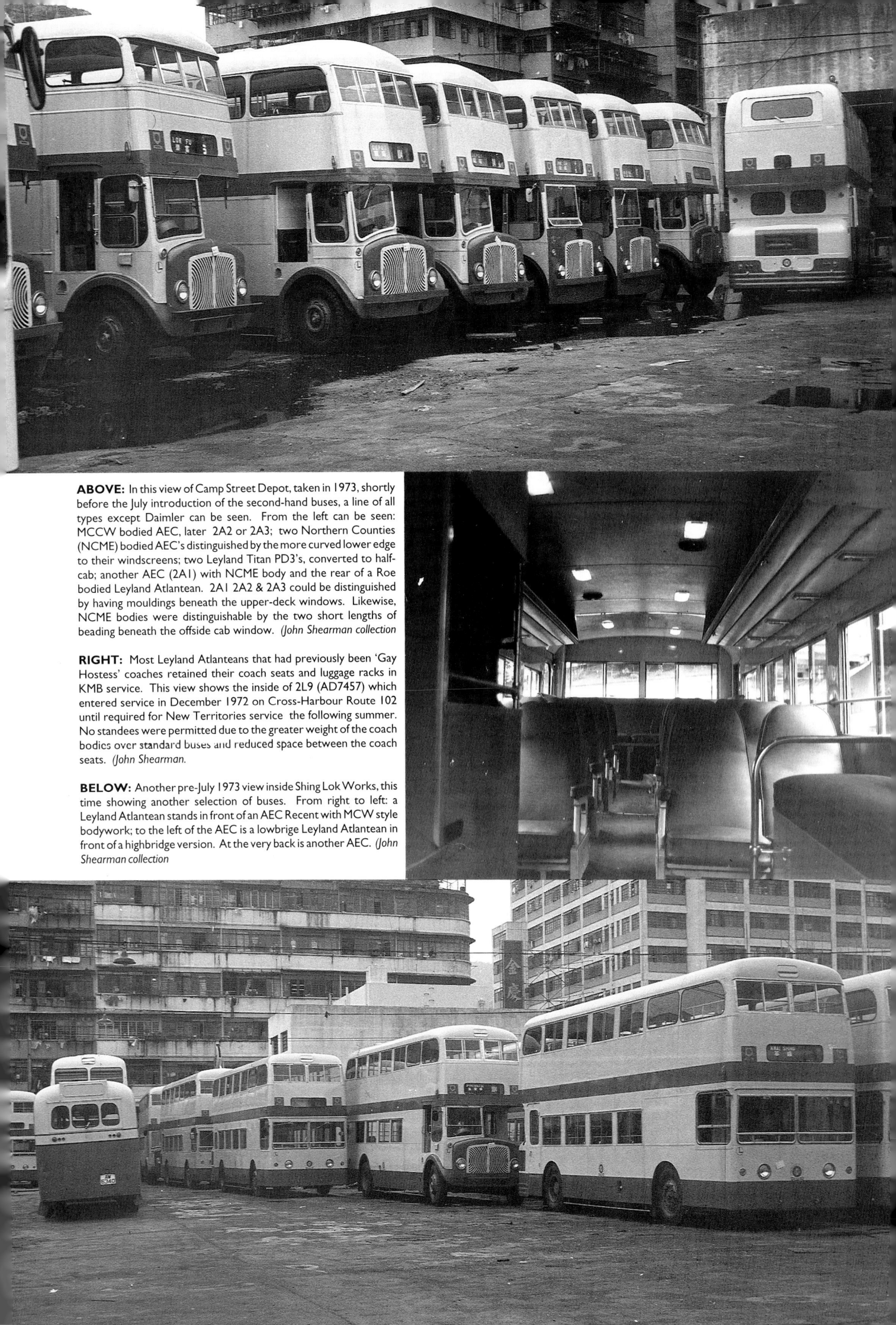

ABOVE: In this view of Camp Street Depot, taken in 1973, shortly before the July introduction of the second-hand buses, a line of all types except Daimler can be seen. From the left can be seen: MCCW bodied AEC, later 2A2 or 2A3; two Northern Counties (NCME) bodied AEC's distinguished by the more curved lower edge to their windscreens; two Leyland Titan PD3's, converted to half-cab; another AEC (2A1) with NCME body and the rear of a Roe bodied Leyland Atlantean. 2A1 2A2 & 2A3 could be distinguished by having mouldings beneath the upper-deck windows. Likewise, NCME bodies were distinguishable by the two short lengths of beading beneath the offside cab window. (John Shearman collection)

RIGHT: Most Leyland Atlanteans that had previously been 'Gay Hostess' coaches retained their coach seats and luggage racks in KMB service. This view shows the inside of 2L9 (AD7457) which entered service in December 1972 on Cross-Harbour Route 102 until required for New Territories service the following summer. No standees were permitted due to the greater weight of the coach bodies over standard buses and reduced space between the coach seats. (John Shearman.)

BELOW: Another pre-July 1973 view inside Shing Lok Works, this time showing another selection of buses. From right to left: a Leyland Atlantean stands in front of an AEC Recent with MCW style bodywork; to the left of the AEC is a lowbrige Leyland Atlantean in front of a highbridge version. At the very back is another AEC. (John Shearman collection)

Part Four:
Buses Purchased Second-hand.

The standard of service provided by KMB in the rural areas of the New Territories had, by 1970, failed to keep pace with ever-growing demands. The Government had plans to develop certain small urban areas into full-scale 'New Towns', with massive provision of high-rise accommodation and it was obvious that the exclusively single-deck bus service network would become even more inadequate by the mid-1970's than it already was.

The only practical way to increase capacity within the Company's staffing capability was by progressive conversion to double-deck operation but, at that time, there was a desperate shortage of new double-deck rolling-stock. There were several reasons for this situation, of which the principal one was the pressure on UK double-deck bus manufacturing capacity following the acceptance by British trades unions of driver-only-operated (DOO) double-deck buses with the result that operators undertook wholesale replacement of single-deckers. This was further encouraged by the introduction of generous British Government financial assistance in the form of 'Bus Grant'. The shortage was aggravated by a number of industrial disputes which affected key component suppliers, making even the purchase of spare parts a difficult process. On top of all these UK originated problems, nearer to home, the Hong Kong to Kowloon 'Cross-Harbour Tunnel', a fixed-link road facility, was due to open in the second half of 1972, four months early and KMB knew that it would require large numbers of double-deckers to carry the additional traffic that was expected to be generated.

Given the circumstances, KMB opted to purchase a fleet of second-hand double-deck buses from British dealers - a policy already adopted by CMB to speed its expansion programme. As a result, KMB obtained one hundred and fifteen buses from three dealers, North's, Sykes and Martin's, a fleet that had originated with a variety of operators from all over England and South Wales. The original intention had been to obtain as many as possible of types known to KMB but its first choice, Daimlers, were in short supply and only seven were obtained. Already satisfied with its fleet of 210 AEC Regent Mk V's, built to 34ft length, twenty-eight AEC's were purchased but with a nominal overall-length of 30ft - the UK norm at the time.

Failing to locate sufficient numbers of its preferred front-engined bus types drove KMB to take Leyland products, untested by KMB, all of which, apart from the first eight, were rear-engined 'Atlantean' PDR1/1 models, a type which had not proved particularly reliable even under relatively calm British operating conditions. The eight exceptions were Leyland 'Titan' PD3/5 models, similar mechanically to some later obtained by CMB.

It will be seen from the foregoing that the entire assemblage of second-hand buses was composed of the following very non-standard fleet:

   3 x Daimler CCG6, 27ft 6in long;
   4 x Daimler CVG6, 30ft long;
  28 x AEC Regent MkV, 30 ft long;
   8 x Leyland Titan PD3/5 and
  72 x Leyland Atlantean PDR1/1.

Of the Atlanteans, seven were of the more unsuitable lowbridge type, 14 were coaches and one never entered service.

The majority, but not all, of this motley fleet had bodies of the MCW's 'Orion' type, on which the lower-deck side windows were deeper than those of the upper-deck and the front dome and window surrounds were made as one glass-fibre moulding. Upon delivery, KMB modified them all to have their own style of tropical opening window units which had square cornered aluminium frames, ill fitted for installation in the radius-cornered 'Orion' openings. The new windows were fitted in all locations except above the staircase and, on some buses, the first window on each side of the upper-deck. All the second-hand buses required extensive rebuilding and reshaping of the bodywork in areas surrounding the windows, especially on the lower deck with the larger openings, in order to take the shallower KMB-standard window pans.

Some former 'Maidstone & District' Atlanteans, quite apart from being amongst the more reliable of the rear-engined second-hand buses, had the distinction of having the only MCW-style bodies to retain their original upper-deck front windows, rubber mounted and complete with 'push-out' vents. Most buses had upper-deck emergency exitsdoors which were provided with a single opening window unit of standard 'side' pattern, mounted between two sheet-aluminium blanking pieces. The 'M & D' Atlanteans remained the most original-looking of the immigrant buses except, perhaps, for the four ex-Bolton Daimler CVG6's with East Lancs bodies.

The most drastically rebuilt of the second-hand buses were the eight ex-Ribble Motor Services Leyland Titan PD3/5's which had their full-width fronts altered to half-cab layout, retaining the original interior bonnet cover in an external role, combined with a KMB designed front grille as used for the AEC's. Buses which originally had sliding doors had been altered to have folding doors by late 1975. All buses were re-upholstered and largely repanelled by KMB's Shing Lok body works before entering service.

A feature of the KMB second-hand fleet from the first day of operation was its unreliability. This, combined with non-availability of spare parts for Leyland and AEC engines, resulted in all the AEC Regents and Leyland Titans being re-engined using Gardner 6LW or 6LX engines, thus bringing at least the front-engined second-hand buses into line with KMB's own double-deck fleet. Most were also subsequently rebodied and these re-engined buses thus became almost standard with the rest of the fleet.

There were also two attempts at fitting Gardner 6LX engines in rear-engined Leyland Atlanteans but this was a much more difficult task and no other conversions were undertaken. Since the Gardner engine was longer than the original Leyland 0.600 unit, it only just fitted under the rear engine cover and, in fact, caused the off-side mounted gearbox to protrude through the side of the cover to sit flush with the bodyside.

Some of the AEC Regent Mk V's were from the same batches as those purchased by CMB but those for the latter were purchased for their bodies only and the chassis were discarded even though many chassis parts would have been useful to KMB as spares. Ironically, CMB required bodies and not chassis and KMB required chassis and not bodies!

For its second-hand fleet, KMB introduced a new livery of cream, relieved only by two broad bands of red, a scheme inspired by the livery of the former 'Gay Hostess' Atlanteans of Ribble/Standerwick, in whose livery some were delivered. All KMB buses were fleet numbered in 1975 when the second-hand buses received numbers prefixed by the figure '2', for 2nd-hand. Type letters were to indicate chassis maker and thus the 2nd-hand AEC's became '2A1 to 2A28', the Daimlers '2D1 to 2D7' and the Leylands '2L1-79' etc, regardless of Titan or Atlantean.

Second-generation second-hand.

A second generation of second-hand bus appeared in 1981 in the form of ex-London Transport 'DMS' type Daimler 'Fleetlines'. These buses, which eventually numbered one hundred, were given '2D' prefix fleet numbers to follow on from the CVG/CCG types, 2D1-7. From the start these buses were painted in the standard fleet livery, which was soon to change to a scheme not unlike that adopted for the 1972 second-hand buses.

Being similar to their own Fleetlines in most respects, once KMB had removed London Transport's idiosyncratic gadgetry, these buses performed considerably better than their Atlantean predecessors. A weak point was to become the Leyland 0.680 engine which failed to meet Hong Kong's demanding expectations with the result that many examples were replaced by Gardner 6LX units as the Leyland enginesfailed.

From Singapore.

One last vehicle in the second-hand category was a DAF-engined Dennis Dominator that had been a Dennis demonstration vehicle with Singapore Bus Service. This bus was never placed into KMB passenger service, becoming instead an advanced driver training vehicle.

Driver only operation.

It should not go unrecorded that the second-hand fleet all entered service as driver-only-operated (one-man) buses and carried fluorescent yellow-on-red coin-in-the-box stickers on front and rear but, by the time the ex-London 'DMS' type arrived the entire fleet was DOO and the use of the stickers had been discontinued.

Leyland Titan PD3/5: New to RIBBLE Motor Services. 2L1-8
(Fleet numbered from 1973/4)

Although sequentially numbered ahead of the former 'Gay Hostess' coaches, these eight Leyland Titan PD3/5 double-deckers with MCW bodies did not enter service until May and July 1973, with the last one appearing in traffic in December 1973. Of all KMB's buses purchased second-hand at that time the eight Titans received the most radical change to their external appearance. Built as full-fronted buses, KMB converted them to half-cab arrangement in order to reduce heat gain in the cab generated by the presence of the engine. The original interior bonnet cover was retained in use as the external cover, the cab nearside corner pillar, nearside windscreen and the nearside cab structure were removed down to the level of the top of the bonnet. The nearside front wheel-arch remained as a part of the original side panelling. The restricted opening of the radiator grille was enlarged and covered by the type of KMB made grille used on the AEC-classes. The 'Orion' bodywork was further disfigured by the fitting of the KMB standard square cornered window fittings. The aluminium frames of these were mounted directly to the bodysides without the use of pans.

The windows across the front of the upper-deck were particularly narrow and were an unhappy fit in the MCW dome. KMB indicator screens were fitted, thus ensuring that these buses lost their identity completely. Traces of the original survived in the handful of windows not replaced - the first two windows each side upstairs and the emergency door behind the rear offside wheel-arch. The upper-deck emergency window was unaltered and, being bottom hinged and frequently left open to improve ventilation, it invariably hung down, banging continually against the rear of the bus. Nobody seemed to mind!

It is interesting to note that CMB, who operated over a hundred ex-Southdown and Ribble Titan PD3's, did not immediately deem it necessary to alter their full width cabs to half-cab until 1983/4, after eleven years service in Hong Kong.

After some months in service, KMB removed the Ribble-style sliding passenger doors and replaced them with power operated folding doors as the originals were found to be cumbersome in KMB's hurried operating conditions.

Mechanically the Titans were little altered for their Kowloon service but after a time the engines gave trouble and KMB took the opportunity of replacing them with their fleet-standard Gardner. First 2L6 was fitted with a 6LX unit and then the others followed with 6LW units. In order to keep the running units compatible, the rear-axles of 2L1/ & 5/6/7 were changed to Daimler CVG type.

VEHICLE SPECIFICATION: LEYLAND TITAN PD3/5

Fleet numbers:	2L1-2L8
Registration numbers:	AD7470-77
Chassis:	Leyland Titan PD3/5
Engine: originally:	Leyland 0.600
as rebuilt:	Gardner 6LW & 6LX, later all 6LX.
Gearbox: originally:	Leyland 'pneumocyclic', semi-automatic.
as rebuilt:	SCG GB340 semi-automatic.
Body make: originally:	MCCW 'Orion'
as rebodied:	KMB (2L1, 2, 5, 6 & 8 only)
Body layout: MCCW body:	H41/31+12FEXS,DOO*
KMB body:	H41/31+17FEXS,DOO (2L1, 2, 5, 6 & 8 only)
Date built:	1961
Date introduced by KMB:	1973
Date rebodied:	1976 (2L6) or 1977 (2L1/2/5/7)
Total:	8
Length:	30ft 1in 9.168m
Width:	8ft 0in 2.44m
Height:	14ft 6in 4.42m
Wheelbase:	18ft 6in 4.95m
Unladen weight:	9136kg
Gross vehicle weight:	14961kg

NOTE*: Standing capacity revised to 17 in MCCW body from Nov 74. Also AD7475 converted to 'Leyland PD3 (A)' from 11.8.76 until 30.9.76 with standing limited to only 10.

During 1976/77, 2L1, 2, 5, 6 & 7 were rebodied by KMB to the same design as that used for their second-hand AEC Regent Mk V's, to the extent that even the AEC Regent bonnet, curved wings with grille were incorporated, making AD74xx registration numbers the only quick means of spotting the Leylands from the AEC's. Because of the change of rear-axle, not even the rear hub caps could offer any clue!

Of the eight Titan PD3/5's, 2L4 (Ribble PCK389) was scrapped on 29th September 1976 and 2L3 & 8 (PCK362/392) followed in February 1979, after lying derelict for almost three years. At the time of rebodying, those buses earlier given Gardner 6LW engines were further re-engined using 6LX units and Daimler semi-automatic gearboxes, thus they virtually became Daimlers with Titan chassis frames.

LEFT: Former Ribble Motor Services Leyland 'Titan' PD3/5, PCK326 which became KMB AD7472, later fleet numbered 2L5 (the 2 indicating a 2nd hand bus), seen here in 1973 while still retaining its original sliding passenger door. The half-cab conversion was unusual but functional (Lyndon Rees

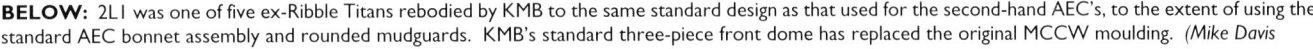

ABOVE: AD7472 was seen here in May 1975 after fleet numbering had made it 2L3. By this time the door had been changed from air-operated sliding to electric powered, folding. *(Mike Davis*

BELOW: 2L1 was one of five ex-Ribble Titans rebodied by KMB to the same standard design as that used for the second-hand AEC's, to the extent of using the standard AEC bonnet assembly and rounded mudguards. KMB's standard three-piece front dome has replaced the original MCCW moulding. *(Mike Davis*

Leyland Atlantean PDR1/1:
New to RIBBLE/STANDERWICK as 'Gay Hostess' coaches.

2L9-21 & 61
(Fleet numbered from 1973/4)

The first second-hand buses to arrive were, in fact, coaches! A small fleet of 13 former Ribble Motor Services/Standerwick 'Gay Hostess' motorway coaches were shipped during 1972 (AD7457-74), followed some months later by a fourteenth example (AD7485) that had latterly been in the service of a small operator in the UK, namely Kirby Coaches of Kirby, Lancashire. This last vehicle and ten of the others retained their coach seats, although bus-type seating was fitted in the space formerly occupied by the toilet and kitchen. These eleven vehicles were classified by KMB as the Atlantean G/H. The three remaining buses had been with the London independent, City Coach Lines, London NW1, complete with dramatic 'City Sightseeing' over-all Union Jack livery and included the star of the motion picture 'Mutiny on the Buses'. This trio arrived in Kowloon with few, or no, seats and were equipped by KMB with bus seats throughout. With the lesser weight of the bus-type seats, 13 standees were permitted and the buses were dubbed KMB's Atlantean G/H(B)-type. There is no record of there ever having been an (A) sub-type, but perhaps this had been an option not taken-up.

Mechanically, the former Gay Hostesses differed from the rest of KMB's aquisitions in having the more powerful Leyland 0.680 engine and high ratio rear axle. The pneumocyclic gearbox also differed in having five forward speeds as in the Leyland Leopard coach.

Standard KMB windows were fitted into the MCW window-openings which, on the lower-deck, were much deeper than their replacements, thereby necessitating considerable rebuilding and repanelling above the window-frames. All these replacement window-frames were attractively set in wood. Exceptions were the first and last side window on the upper-deck, the window at the head of the stairs and the last, smaller, bay on the lower-deck. In time, even these were replaced with sliders on some examples. Replacement windscreens were fitted, having opening sections at their lower borders but generally retaining an outline similar to the original. The exception was AD7485, ex-Kirby Coaches, which acquired windscreens of the type fitted to the Albion Victor single-deckers where the lower edge slopes downwards and outwards from the centre pillar.

Other external changes made by KMB included removal of all the distinctive trim strips from the Gay Hostess bodywork, except for the strip on the roof cant panel, above the upper-deck windows, leading to the belief that reports of complete repanelling were correct.

On buses retaining the bulky coach seats (KMB sub-type: Atlantean G/H) with accordingly less gangway space, plus their heavy coach bodywork, no standees were permitted and their layout was H34/30+00FEXS,DOO. Painted notices were carried by the door "On trial, Long distance coach, no standing, please co-operate." Attempts to prevent standing, however, are often futile in Hong Kong!

AD registration numbers distinguished the Gay Hostesses from all other KMB Atlanteans, AD7457 being the first second-hand bus to enter service during December 1972, on the Tunnel Route 102 and, during the Christmas period seasonal music was played over the retained cassette player and the interior was decorated festively. During January 1973 it was joined by AD7458, which was painted in an all-over slogan for the Government's 'Fight Crime' campaign. All fourteen Gay Hostesses had their overhead luggage racks retained in situ, ie, including the bus-seated versions.

No other Atlantean G/H's were registered until May and in June 1973 AD7457 and AD7458 were withdrawn from Tunnel service and prepared for service in the New Territories, for service on trunk Route 50, commencing on 16th July.

Withdrawals.

The unreliability of the KMB Atlanteans as a whole became legend and the Gay

VEHICLE SPECIFICATION: LEYLAND ATLANTEAN PDR1/1 COACHES

Fleet numbers:	2L9-21 & 2L61
Registration numbers:	AD7457--69 & AD7485
Chassis:	Leyland PDR1/1 'Atlantean' special
Suspension:	front: air-bag; rear: conventional leaf-springs
Engine: Original:	Leyland 0.680
2L15 as rebuilt:	Gardner 6LX
Body make: original:	MCW
2L15 as rebuilt:	KMB
Body layout: G/H:	H34/30+00FEXS,DOO.
G/H(B):	H34/30+13FEXS,DOO - 16 standees from 1974/5
G/H(C):	H34/30+9FEXS,DOO
Date introduced by KMB:	1972-73
Date built:	1960/61
Total:	14
Length:	30ft 00in 9.143m
Width:	8ft 00in 2.44m
Height:	14ft 6in 4.42m
Wheelbase:	16ft 3in 4.95m
Unladen weight:	10233kg
Gross vehicle weight:	14000kg

LEYLAND ATLANTEAN PDR1/1 FORMER GAY HOSTESS SUB-TYPE BREAKDOWN AT DATES SHOWN.

DATE:	12/72	3/74	5/75	6/75	8/75	10/75	1/76	3/76	4/76	5/76	7/76	12/76	6/77	1/78	7/79
SUB-TYPE: G/H	11	10	9*¹	8	7	6	3	3	2	2	1	1	*²	-	-
G/H(B)	3	3	3	3	3	3	3	3	3	2	*³				
G/H(C)	-	-	-	-	-	-	1*⁴	1	1	1	1	1	1	1	#

NOTES: *¹ AD7463 accident - later rebodied as G/H(C) 3/76   *² Last G/H - AD7465 de-licensed 'temporarily' 26/6/77
*³ Last G/H(B) - AD7565 & AD7468 withdrawn 26/5/76   *⁴ AD7463 reintroduced as G/H(C) and # withdrawn 19/7/79

LEFT: A rear view of AD7457 (later 2L8) as it pulls away from a stop in North Point in January 1973. The additional mesh-covered cooling grille was cut in the rear engine cover by KMB and is quite clear in this view, as are the large rear bulkhead windows that were a feature of KMB's lower numbered forme Gay Hostess coaches. *(Lyndon Rees)*

ABOVE: The first former ex-Ribble/Standerwick 'Gay Hostess' to enter service with KMB was AD7457 (NRN613) which made its Far East debut in December 1972. Complete with coach seats and luggage racks, it was allocated to Cross-Harbour Route 102, until required for New Territories service in the following July. Seen here in May 1973, this bus subsequently became 2L9. Note the moulded 'beading' intended for the application of the then standard lined livery that was never applied. The 'Gay Hostess' mouldings were removed, with the exception of those on the roof above the upper-deck side windows. *(John Shearman.*

BELOW: The second former 'Gay Hostess', AD7458 (SFV418) (later 2L10), entered service 2nd Feb 1973 in time for Chinese New Year; it was, however painted in a Government sponsored 'Fight Crime' livery - not festive decorations!. It can be seen from this photograph that the mouldings were omitted from this second conversion, and all subsequent second-hand buses except 2A1 & 2. The cab area was left in fleet livery which looked somewhat strange. *(John Shearman.*

ABOVE: 2L14 leaving Jordan Road Ferry for the 26-mile long journey to Yuen Long (then the Colony's longest route) in May 1975. Like all the ex-Gay Hostess coaches, the destination screens were slightly higher than on ex-bus Leyland Atlanteans. *(Mike Davis)*

Hostesses were no exception, failing in service from the first day. The problem was made worse by a lack of readily available spare parts. Gradually, the failures resulted in the buses being 'temporarily' de-licensed, awaiting spare parts or, in a few cases, pending a decision as to whether or not to rebuild them. AD7476, fleet numbered 2L19, was the first to be de-licensed 'awaiting spare parts', during March 1974, only seven months after entering service.

AD7463 (2L15) was involved in an accident when it overturned on the Castle Peak Road in 1975 and was set aside for rebodying, with others, but, in the event, it was the only former Gay Hostess to be rebodied. Withdrawals on a so-called 'temporary' basis were, generally, a death knell for the vehicle which was usually later scrapped without ever returning to traffic. A glance at the table on Page 164 will show the steady drop in the number of Atlanteans G/H's available for traffic. The last unrebuilt example, AD7461, was 'temporarily delicensed' on 26 June 1977.

The second-hand Atlanteans were given fleet numbers in 1974, prefixed by the figure '2' for 2nd-hand, and 'L' for Leyland chassis, and thus were all '2L' class, with the former Gay Hostesses being 2L9-21 and 2L61, the late arrival.

LEFT: AD7485 (SFV416 ex-Kirby Coaches) was fitted with windscreens of the type used on Albion 'Victor' single-deckers. Seen here at work on Route 50 shortly after its introduction in 1973. *(John Shearman.)*

2L15 (AD7463) rebuilt: Atlantean G/H(C) sub-type.

As noted above, AD7463 (2L15) was involved in an overturning accident in 1975, suffering extensive damage, both from the accident and during the awkward recovery operation. Despite this, much of the structure was undamaged or easily repairable and complete rebodying was unnecessary. Instead a completely new front was grafted-on and any remaining non-KMB standard side windows were replaced with standard units, the most noticeable being the 'D'-shaped last upper-deck side window, fitted awkwardly into the MCW dome, the rear of which was somewhat more vertical than suited the contours of the glass. The body was repanelled, using the original pillars and pillar-spacing and included the fillets between the window frames and pillars. The original closely spaced headlights were retained but the side-lights were relocated higher than before. Seating was altered from high-backed coach seats to low-backed bus seats of KMB's standard pattern with chrome plated hand-rails across the upper edge and cushioned seats and seat-backs but the actual number of seats remained as previously; that is with 34 on the upper-deck and 30 in the lower-deck but the use of lighter and less bulky bus seating permitted authorisation of nine standees where none were previously allowed.

At the same time, the mechanical engineers were busy fitting a Gardner 6LX engine, a difficult task that resulted in the end of the gearbox being flush with the body-side and exposed through an opening in the side of the new and cumbersome engine cover, at the maximum vehicle width.

Despite the extensive renovations carried out on it, 2L15 was withdrawn from service on 19th July 1979.

ABOVE: The only ex-Gay Hostess to be rebuilt by KMB was 2L15 (AD7463) which had been NRN615 in its Ribble days. This vehicle was also re-engined to Gardner 6LX at the same time and fitted with bus seats in place of the high-backed coach type. *(Mike Davis*

LEFT: A rear view of 2L15, the rebuilt Gay Hostess, showing the untidy arrangement of the 'home-made' engine cover provided by KMB to conceal the longer Gardner 6LX engine. As seen here, on the offside, the end-plate of the gearbox protruded through the side-cover to lie flush with the body-side. AD7463 was photographed at Tsuen Wan Ferry terminus in 1978. *(Mike Davis*

Leyland Atlantean PDR1/1
New to various UK operators.

2L22-60, 62-79

In General.

Following the former 'Gay Hostess' Atlantean coaches, KMB were successful in securing a further 58 PDR1/1 'Atlantean' buses; 51 highbridge and 7 lowbridge

All these PDR1/1 'Atlanteans', both the highbridge and the lowbridge buses, but excluding the 'Gay Hostesses', were fitted with similar mechanical equipment in that they were powered by the Leyland 0.600 engine, driving through a fluid-flywheel and Leyland's four-speed semi-automatic gearbox. Gear-control was Leyland's Pneumocyclic system with a pedestal-mounted miniature gear-lever. Power was transmitted to the rear-axle by way of an angle-drive from the off-side mounted gearbox. For operation in Hong Kong's hot and humid summers, the radiators were up-rated and additional fan-blades fitted to improve cooling but, in addition, problems were experienced with the gland-seal in the fluid coupling and also the fluid-coupling, both of which were apparently unsuited to Hong Kong's harsh driving conditions.

Probably due to unsatisfactory cooling, the Leyland engine was not found to be reliable and buses so powered became Cinderellas in a fleet with Gardner-familiar engineers and fitters. Attempts were made to fit Gardner 6LX units in a small experimental batch but, in fact, only three were completed as problems were encountered due to the combined overall width of the gearbox and Gardner engine being fractionally greater than the eight foot overall width of the bus. It was found necessary to allow the gearbox to protrude beyond the off-side bodywork by about 20mm.

The window arrangements of all the bodies, whatever their make, were altered to take the KMB standard-type full-depth sliding glass in square-cornered aluminium frames, mounted directly into the body frame and thus the deep MCW-type lower-saloon windows were replaced by the shallow KMB type, except for an incongruous few which were left as original. Some ex-Ribble and Maidstone & District buses retained their original upper-deck front windows where UK-style shallow push-out vents. were fitted. Windscreens were replaced by KMB's standard type, with a 150mm deep opening portion at the lower edge. Some of these windscreens were of the type usually fitted to Albion Victor single-deckers, where the lower edge formed an inverted 'V' shape.

All electrics were replaced and air-operated doors replaced the humidity-prone electric type. Most external panels were removed and any necessary repairs were carried-out to pillars and framework.

All the Atlanteans had their destination and route number equipment changed to KMB's standard type. Most of this work was carried-out at the Company's Shing Lok body works, now long since closed.

VEHICLE SPECIFICATIONS: LEYLAND ATLANTEAN PDR1/1 BUSES.

Fleet Numbers:	2L22-60, 62-79 plus 604EUP.
Chassis:	Leyland Atlantean PDR1/1.
Engine:	Leyland 0.600 - 2L52 & 64 converted to Gardner 6LX.
Gearbox:	Leyland/SCG 4-speed semi-automatic with pneumocyclic control.
Body make:	MCW, Roe, Alexander, Weymann - see table below.
Body Layouts: L/B	L39/34+13FEXS,DOO
H/B	H44/34+13FEXS,DOO
H/B(B)	H44/34+11FEXS,DOO
H/B(C)	H44/34+6FEXS,DOO
H/B(D)	H44/34+8FEXS,DOO
H/B(E)	H44/34+9FEXS,DOO
As rebodied: H/B(A)CS	H45/32+10FE,FdXS,DOO
H/B(F)CS	H47/34+00FE,FdXS,DOO - **no** standees
H/B(OMO)FS	H44/34+9FEXS,DOO
Date built:	1959-61
Date introduced by KMB:	1973-74
Dates rebodied:	1976 (2L52/67); 1978 (2L71)
Length:	30ft 1in 9168mm
Width:	8Ft 0in 2240mm
Height: lowbridge:	13ft 6in 4110mm
highbridge:	14ft 6in 4420mm
Wheelbase:	16ft 3in 4950mm
Unladen weight:	9318kg (H/B)
Gross vehicle weight:	14000kg

See table for 1975 revisions

The Lowbridge Buses.

Of these seven vehicles, five came from East Midland Motor Services and two from Ribble Motor Services. They were of the unusual, but standard, early Atlantean layout where the upper-deck was effectively lowheight for the forward two-thirds of its length and lowbridge for the remaining rearward one third. This resulted in the front portion having standard 2+2 seating, with a central gangway, while the rearward rows of seats were located on a raised section of the floor and were reached by a sunken side gangway. In the lower saloon, the raised rear upper-deck floor gave adequate headroom where the lower-deck floor was built-up over the rear-axle, which was of the straight variety. (Later, true low*height* examples were fitted with a drop-centre rear-axle to reduce the overall height whilst retaining the level floor and 2+2 seating upstairs. Five of these, secondhand from Nottingham, went to Hong Kong in 1982 for service with Cascade Tourist Service, with whom they were used as works buses at a power station construction site - but that is another story.)

New to EAST MIDLAND Motor Services Ltd.
KMB class 'Ley.At.PDR1/1 L/B' Nos: 2L23, 33-36.
(Fleet numbered from 1973/4)

Five lowbridge buses, new to East Midland Motor Services in 1959, were purchased by KMB from North's, the UK dealer, in October 1972, having been D138, 137, 134, 139 & 131 in the East Midland fleet. Body make was Weymann and the layout in Kowloon was L39/34FEXS,DOO.

New to RIBBLE Motor Services Ltd.
KMB class 'Ley.At.PDR1/1 L/B' 2L45 & 2L60.
(Fleet numbered from 1973/4)

These two buses were new to Ribble Motor Services in 1961 (PCK336 & 337) and, like the East Midland buses above, were obtained from North's, having been Ribble 1703 and 1704 respectively. KMB body layout was L39/34FEXS,DOO and has been variously described as being either MCCW or MCW.

NOTE: Both these two groups were classified by KMB as Atlantean 'L/B'. It should also be noted that lowbridge buses were purchased only because there were insufficient full-height models available at the time.

BELOW: 2L35 was new to East Midland Motor Services in 1959 as their D139 (139BRR) and was one of five similar buses with Weymann lowbridge bodies purchased by KMB in 1973. Lowbridge buses were purchased only because they were easily available, KMB not having any routes with restricted headroom to make them either necessary or desirable. *(Mike Davis)*

BELOW: 2L45 was also a lowbridge bus; one of the two that had been new to Ribble Motor Services in 1961, starting life as 1703 (PCK336). It was travelling along Jordan Road towards the ferry when seen here in 1975. *(Mike Davis)*

The Highbridge Buses.

The 51 Highbridge Atlanteans were all sourced for KMB by North's but in some cases were initially obtained from other dealers, North acting as middle-man and shipper. The buses obtained were new to a number of operators; Trent Motor Services, Ribble Motor Services, Maidstone & District Motor Services, the Northern General Group and Sheffield Corporation. The variety of bodywork included the products of Roe, Alexander, Weymann and MCCW.

LEYLAND ATLANTEAN PDR1/1 - *SUB-TYPES REFER TO SEATING/STANDING VARIATIONS*

Sub-type & KMB description			Period	Totals		Capacity 1973-75	Capacity 1975-79
Ley.At. PDR1/1 H/B	(OMO/Cushion)FS		1973-77	41 (incl 604EUP)		44/34+13	44/34+13
Ley.At. PDR1/1 L/B	(OMO/Cushion)FS		1973-76	7		39/34+13	39/34+13
Ley.At. PDR1/1 H/B(B)	(OMO/Cushion)FS		1973-75	2	58:	44/34+11	44/13+11
Ley.At. PDR1/1 H/B(C)	(OMO/Cushion)FS		1973-76	3	of	44/13+ 6	44/34+ 5
Ley.At. PDR1/1 H/B(D)	(OMO/Cushion)FS		1973-77	4	which	44/34+ 8	44/34+ 8
Ley.At. PDR1/1 H/B(E)	(OMO/Cushion)FS		1973-75	1	3 were	44/34+ 9	44/34+10
Ley.At. PDR1/1 H/B(F)	(OMO/Cushion)CS		1977-79	1 Rebody 2L67	rebodied	- -	47/33+00
Ley.At. PDR1/1 H/B	(OMO)FS		1977-79	1 Rebody 2L52		- -	44/34+ 9
Ley.At. PDR1/1 H/B(A)	(OMO/Cushion)FS		1978-79	1 Rebody 2L71		- -	45/32+10

FS + Front Stairs.   Cushion = cushion seats
CS + Centre Stairs. 2L52 = fibre-glass moulded seats after rebody; capacity unaltered from H/B sub-type..

New to RIBBLE Motor Services Ltd.

KMB class: 'Ley.At.PDR1/1 H/B'; 2L22, 25, 30, 46, 48, 49, 50, 54, 55, 56, 74 (Fleet numbered from 1973/4).

Thirteen buses new to Ribble Motor Services were obtained from North (dealer) between October 1972 and May 1973, having been 1618, 1608, 1622, 1634, 1609, 1638, 1607, 1619 1636, 1637, & 1642 respectively in the Ribble fleet, and were new in 1959 and 1960. KMB classified them 'Ley.At.PDR1/1,H/B' and they had a layout of H44/34+13FEXS,DOO throughout their entire time with KMB. The last unmodified example was withdrawn on 24th November 1977 but one, BE4166, which had been rebodied in December 1976, remained in service until August 1979, albeit with hard, fibre-glass seating in place of the cushions.

BELOW: 2L49, an ex-Ribble highbridge version of the Atlantean, formerly Ribble 1638 (NCK627), with MCW (or MCCW) bodywork. The retained original sliding window can be seen in the second upper-deck side position and generally this, together with that at the head of the staircase, was retained, as were the side windows in the domes and the emergency door and the nearside equivalent. 2L49 was rushing towards the Jordan Road Ferry terminus, along the thoroughfare of that name one week-end in 1974. *(Mike Davis)*

LEYLAND ATLANTEAN PDR1/1 HB OMO - SUB-TYPE ANALYSIS AT DATES FLEET INCREASED/DECREASED.																																							
DATE:	6/73	7/73	8/73	9/73	10/73	11/73	1/74	2/74	3/74	4/74	7/74	11/74	12/74	4/75	5/75	6/75	8/75	9/75	12/75	1/76	2/76	3/76	4/76	5/76	6/76	8/76	9/76	1/77	2/77	3/77	4/77	11/77	8/78	7/79	8/79
BUSES IN:	2	13	2	4	5	6	11	5	1*1	1	-	-	1*2	1*3	-	-	-	-	-	-	-	-	-	-	-	-	-	2r5	-	-	-	1r7	-	-	
BUSES OUT:	-	-	-	-	-	-	-	-	11*1	-	2	1	-	2	1	1	8	1	1	3	1*4	1	2	2	2	1	3	2	2	3	1	16	1	1	2
TOTAL IN SERVICE	2	15	17	21	26	32	43	48	38	39	37	36	37	35	34	33	25	24	23	20	20	19	17	15	13	12	11	8	8	7	4	3	2	3	2	2

NOTES:
- r Rebodied vehicle - 5 2L52, 5 2L67 & 7 2L71.
- *1 3/74 Includes BG655 which was introduced and withdrawn within the same month.
- *2 12/74 BE7947 (2L64) returned to service after conversion to Gardner 6LX.
- *3 4/75 BE2419 (2L48) returned to service in standard form after repairs.
- *4 4/76 604EUP scrapped without ever being registered in Hong Kong.
- *5 1/77 2L52 (BE4166) and 2L67 (BE8656) - rebodied and returned to service.
- *6 11/77 Last unrebuilt H/B, BE7947 - 2L64, withdrawn 24/11/77.
- *7 8/78 2L71 rebodied and returned to service.

Other than the three rebodied buses and BE2419, no other of the HB Atlanteans were ever returned to passenger service having once been withdrawn "Awaiting spare parts" - in KMB's vernacular.

New to SHEFFIELD Corporation Transport.

All KMB class 'Ley.At.PDR1/1 H/B' Nos: 2L41, 52, 68 (Fleet numbered from 1973/4).

Originally in the fleet of Sheffield Corporation, 2L41, 52, 68 were obtained from North's in March 1973, having been 921, 924 & 917 respectively in the Sheffield fleet where they carried distinctive single index letter registration numbers, 5921W, 5924W and 5917W. The first two entered service for KMB in July and October 1973 but 2L68 (5917W) was a late entrant in January 1974, amongst the former Maidstone & District Atlanteans. KMB classified these three buses in their 'H/B' series, their having a layout of H44/34+13FEXS,DOO (44, 33 in Sheffield). They were built new in 1960.

BELOW: The only photograph so far located of a former Sheffield 'Atlantean' is this of an unidentified example in Shing Lok body shop early in 1973, prior to being reconstructed by KMB. All Sheffield insignia had been painted-out prior to sale. *(John Shearman.)*

Formerly TRENT Motor Services Ltd.

KMB classes:
   Ley.At.PDR1/1 H/B: 2L26, 27, 31, 37, 47 - with MCCW bodies.
       H/B(B): 2L58.
       H/B(C): 2L44
       H/B(D): 2L51, 59, 62.   with Roe bodies.   (All fleet numbered from 197/34)
       H/B(E): 2L53.

Of the eleven former Trent Atlanteans, there were two body makes, 2L26/7/31/37/47 were MCCW (Metropolitan Cammell Carriage & Wagon), while 2L44/51/3,/8/9/62 were by Charles Roe.

The MCCW bodied vehicles were sold by Trent to Cowley (dealer) in September 1972 and were then passed to North's for sale to KMB. They were new to Trent in 1959/60 and carried the fleet numbers 426, 432, 431, 430, 429 respectively.

The Roe bodied vehicles were new in 1959 and were sold by Trent, in a similar manner as the MCCW buses, in 1973, having carried the Trent fleet numbers 439/55/38/49/37/54. The standing authorisation on the six Roe bodied buses differed from standard and KMB issued sub-type letters to denote these differences, of which there were four, and the reader is directed to the accompanying table for clarification. The Roe bodies were to 'standard' specification for the basic 'H/B' type.

LEFT: Hong Kong's traffic is a nightmare for the bus photographer! Here a typical Kowloon taxi of the period out-accelerates former Trent 438 (RRC 72) in Jordan Road during 1974. The flatter roof-line distinguishes these buses from their MCCW bodied former fleet-mates. *(Mike Davis.)*

BELOW: The former Trent MCCW bodied Atlanteans looked very similar to those from Ribble, Maidstone & District and Northern General. Here former Trent 432 (ORC666), by now Kowloon Motor Bus 2L26 (BD5106), whilst departing Jordan Road Ferry for Un Long; the spelling of the town name by KMB, more usually seen spelt 'Yuen Long'. *(Mike Davis)*

ABOVE: 2L53, again. This time leaving Jordan Road Ferry terminus for Yuen Long, despite the destination being set to the contrary! This photograph was taken about half an hour after the view opposite. This bus was classified by KMB as sub-type H/B(F), uniquely, being authorised to carry 9 standees. *(Mike Davis*

BELOW: 2L59 was previously 437 (RRC 71) in the Trent fleet and although almost identical to other Roe bodied Atlanteans, was classified H/B(D), together with 2L51 and & 62, all of which carried (theoretically) 8 standees.. *(Mike Davis*

New to MAIDSTONE & DISTRICT Motor Services Ltd.

All KMB class Ley.At.PDR1/1 H/B: 2L63-67, 69-73, 75-78. *(Fleet numbered from 1973/4)*

Maidstone & District sold these fourteen buses to Macclesfield dealer 'T.P.E.' in May 1973 who then passed them on to North for sale to KMB. All were new in 1960 when they carried the fleet numbers DH535/63/30/41/27/40/42/49/45/29/43/54/52/57. Of all the second-hand Atlanteans, KMB regarded these as having been in the best condition, although this did not help them survive any better mechanically than similar buses form other operators. KMB did not replace the upper-deck front windows on the majority of the ex-M&D buses, probably because the opening ventilators with which they were fitted were felt to be satisfactory. The front domes of the MCW body had a double-skin which may have made the fitting of KMB windows more complicated. Standard seating, standing arrangements were authorised and, at H44/34+13FEXS,DOO they were classified 'H/B' by KMB.

Two of the ex-Maidstone & District Atlanteans, 2L67 & 2L71, were rebodied by KMB and in this form lasted in service until July and August 1979 respectively.

BELOW: Atlanteans from Maidstone & District Motor Services retained the original upper-deck front windows which were fitted with useful push-out hopper vents and the resulting buses were arguably the most British looking of all KMB conversions. The diversity of form taken by many other windows did, however, alter the appearance of the vehicles when seen from other than the front. Ex-M&D DH543 (543HKJ) became KMB's 2L75 in 1973 and was seen at the photographer's favourite vantage point near Jordan Road Ferry Terminus in September 1974. *(Mike Davis*

Fleet numbers (all 2L)	UK operator	Body make	Height
26/7/31/7/47	Trent Motor Traction	MCCW	Highbridge
44/51/3/8/9/62	Trent Motor Traction	Roe	Highbridge
22/5/30/45/6/9-50/4-6/74	Ribble Motor Services	MCCW	Highbridge
45 & 60	Ribble Motor Services	MCCW	Lowbridge
63-7/9-73/5-8	Maidstone & District	MCCW	Highbridge
24/38/40/43+604EUP	Northern General Group	Alexander	Highbridge
28/32/57/79	Northern General Group	MCCW	Highbridge
29/39/42	Northern General Group	Roe	Highbridge
41/52/68	Sheffield Corporation	MCCW	Highbridge
23/33/4/5/6	East Midlands Motor Services	Weymann	Lowbridge

TABLE: Leyland Atlantean bus fleet as purchased second-hand in 1973-4. Ex-Northern General Group (Sunderland) 604EUP was never prepared for service, its parts being cannibalised for spares. Seven other Atlanteans were due for shipping but the order was curtailed in December 1973.

KMB classifications for Atlanteans.

KMB's type letters for the Atlanteans were initially to indicate either lowbridge - L/B - or highbridge -H/B. After this the usual bracketed sub-type letters were added to the H/B type but there were none for the L/B's. The letters H/B(B), (C), (D) & (E) indicated minor differences in standing capacity, often of only one person. H/B(A) & (F) were added for rebodied buses as noted above.

ABOVE: Although retaining its upper-deck front windows, former M&D DH552 (552LKP) was fitted with windscreens of the type more usual on the Albion 'Victor' single-deckers. The rear of the bus in the background illustrates the rear window conversion used for many, but not all, Atlanteans, with a single side window-frame in the emergency door. *(Mike Davis*

BELOW: Exceptions often prove the rule. Here 2L63, ex-M&D DH535 (535HKJ) had not only gained Albion Victor-style windscreens, it had, in addition, lost its MCCW-style upper-deck front windows in favour of KMB sliders. The location is probably by now familiar to the reader as 2L63 heads away for Yuen Long - or Un Long in KMB-style - in May 1975. *(Mike Davis.*

New to NORTHERN GENERAL Group.

KMB class:
'Ley.At.PDR1/1 H/B': 2L24, 28, 29, 32, 40, 43, 57, 79 plus 604EUP.
H/B(B): 2L38.
H/B(C): 2L39, 42.

(Fleet numbered from 1973/4)

Again, the reader is directed to the accompanying table for details of the variations in standees authorised in the sub-types. All were new 1959/60.

While the MCCW & Roe bodies were similar to those from other operators, the presence of Alexander bodies resulted in some dramatic changes in appearance, although the deep front dome was a feature that remained unmistakable. The fronts of these once shapely buses were rebuilt almost completely flat and even more upright than the MCCW version.

These vehicles were obtained from a different dealer, namely Martins of Middlewich, in December 1972 and had been owned by Northern General Group constituent companies as follows:

TYNEMOUTH & DISTRICT with MCCW body: later fleet numbered 2L28.

KMB classified tex-Tynemouth & District No 243 (CFT643) as a basic H/B type with 13 standees. Externally this 1960 vintage vehicle resembled the other Atlanteans with MCCW bodywork. This bus was registered BD5108 and remained on the active list for less than a year and it is believed that it was never photographed but was allocated the fleet number 2L28.

GATESHEAD & DISTRICT with Alexander bodies: later fleet numbered 2L24 & 40.

Of these two buses, 2L24 (BD3906) was classified H/B, with 13 standees, while 2L40 (BD6203) was H/B(D) with only 8 standees. Both were new in 1959 when they were registered KCN183 &. KCN187 (Gateshead & District 83 & 87).

BELOW: 2L24 came to KMB from Gateshead & District (Northern General Group) and had a body by Walter Alexander. It was seen here on the first day of double-deck services in the New Territories. *(Lyndon Rees).*

SUNDERLAND DISTRICT with Alexander bodies: later fleet numbered 2L38 & 43.
Also 604EUP, which never entered KMB service.

2L38, previously Sunderland District 305 (605EUP), was classed H/B(B) with 11-standees, while 2L43, Sunderland 302 (602EUP), was a standard H/B type with 13 standees. Both buses were withdrawn in 1974.

The third bus, the former Sunderland 304 (604EUP), was cannibalized for spares and this became the only KMB second-hand bus never to take to the road although the former SBS Dennis Dominator *q.v.* never entered passenger service.

UPPER RIGHT: An unidentified Alexander bodied ex-Northern General Group Atlantean stands with others prior to registration in summer 1973. *(John Shearman.*

LOWER RIGHT: 2L38, Ex-Sunderland District (Northern General) 605EUP, also with Alexander body, speeding through a more urban environment than that in the left-hand illustration. *(Lyndon Rees.*

General Note
Many of the above Leyland 'Atlantean' buses were registered in Hong Kong using the chassis number actually carried on the chassis and this differed from that originally issued by Leyland, due to the chassis number-plates being fixed to a removable part of the rear chassis extension. They had inadvertently been swapped around during major component changes, prior to export.

NORTHERN GENERAL TRANSPORT: later fleet numbered 2L32/29/39/42/57/79.

2L29/39/42 with Chas. Roe bodies.

These buses were previously NGT Nos 1927/23/24 (927/3/4GPT) and, of these, one, 2L29, was KMB type H/B with the standard 13 standees, while 2L39 & 42 had only 6 standees and were H/B(C). All three were new in 1960.

2L32/57/79 with MCW bodies.

Formerly NGT Nos 1895//93/94 (895/3/4) respectively, all three were of the standard H/B type and all three had also been new in 1960.

TOP and CENTRE: Two views of ex-Northern General MCCW bodied Atlantean, 896EUP, which became 2L32 in the KMB fleet. Seen here soon after the introduction of KMB double-deck routes in the New Territories during the summer of 1973. *(Lyndon Rees*

BELOW: The very last Atlantean to enter service with KMB was ex-Northern General 1894 (894EUP), which became 2L79 when it entered service concurrently with the introduction of fleet numbering in April 1974. It was withdrawn two years later but not before it was photographed in 1974 leaving Jordan Road Ferry for Un Long. *(Mike Davis*

Second-hand 'Atlanteans' rebodied by KMB.
2L52; 2L67; 2L71.

The unfortunate unreliability of the Leyland PDR1/1's whilst in KMB service resulted in one premature withdrawal after another until, by the close of 1976, the 'cripple-yard' near Lai Chi Kok depot was almost full with withdrawn Atlanteans, most of which were brought to scrap value by the accountants on 31st January 1977. Actual breaking-up took place slowly over the following twelve months. Four buses escaped, however, of which two, 2L52 and 2L67, were rebodied during 1976. Of the remaining two, 2L71 was rebodied during 1978 and 2L64 remained in service, having been fitted with a Gardner 6LX engine, until November 1977, when it was set aside pending a decision to rebody it but it was eventually broken-up in its original state during February 1979.

2L52. (BE4166): KMB body.

KMB rebodied 2L52 during 1976 and it returned to service on 30th December of that year. The design followed quite closely that of the Metal Sections body built for the KMB 'Fleetlines', D666 et seq. and possibly incorporated many MetSec parts held as spares - or KMB fabricated the parts themselves, using MetSec originls as patterns. The resulting body was shorter than that on the 33ft long Fleetline chassis and incorporated only one doorway. The window frames were of KMB's old standard square-cornered aluminium variety, possibly salvaged from the original body. The frontal appearance differed from the MetSec type, having very thick upper-deck corner pillars. The staircase was over the offside front wheel-arch and the body layout was H44/34+9FEXS,DOO. but the original cushion seats were replaced by hard moulded fibre-glass units as found on many contemporary KMB urban buses. When returned to service, 2L52 was painted in an all-over advertising livery for "Supreem Fruit Juice" (sic.). Mechanically, the specification was altered from Leyland 0.600 engine to Gardner 6LX; the engine being covered with a 'Fleetline' style 'bonnet'.

ABOVE: 2L52, ex-Sheffield 924 (5924W), rebodied by KMB to a similar pattern as the MetSec Fleetline body but with only one doorway. It was seen here parked in Yuen Long (Un Long) Depot in 1977. *(John Shearman collection)*

BELOW: There have been only a few photographs discovered of 2L52 after rebodying an of these none are of it in service. This view shows the rear engine cover of Fleetline pattern and all-over advertising livery for "Supreem" (sic.) fruit juice. *(John Shearman)*

2L67 (BE8656) - Atlantean HB(F): ICL body.

2L67 was new to Maidstone & District in 1960 as their DH527, being first registered in Hong Kong in January 1974. During the latter part of 1976, 2L67 was rebodied by International Containers Limited (ICL) at their Kwai Chung premises. In degrees of angularity, its external appearance was rivalled only by the products of BACo, then in service in some numbers in both Kowloon and Singapore. The ICL body was, perhaps, marginally more conventional, being somewhat like a squared-up London DMS-class bus (see 2D8-107). The ICL roofline was particularly angular and box-like. 2L67 differed from 2L52 in more fundamental ways in having a standard KMB doorway layout, narrow front entrance with centre exit and staircase. Body layout was H47/33FE,CXS,DOO - no standees; cushion seats were fitted. The bus was painted in standard half-and-half, cream over red, 'urban' livery. This unique vehicle was classified by KMB as Atlantean H/B(F). The Leyland 0.600 engine was retained in 2L67 as the cost and complication of a Gardner conversion was not considered necessary. It is believed that the body of 2L67 was the only bus body ever built by ICL.

ABOVE: A close-up of 2L67 very kindly supplied by the operator. *(KMB*

BELOW: Another photographically shy bus was 2L67, rebodied in a most angular fashion by International Containers Ltd. (ICL) to KMB's design. Apologies are offered for the quality of this illustration but no alternative can be found. *(John Shearman.*

2L71 (BE8983) - Atlantean H/B(A): KMB body.

This bus was also new in 1960 to Maidstone & District, with whom it was DH549. It received its new body at Kwun Tong depot during the summer of 1978, to a design similar in general appearance to those of N 1-4, the prototype Dennis 'Jubilants'. The body design incorporated separate entrance and exit in standard KMB fashion, with single-width front and wide centre doorways. The forward ascending staircase was located over the offside front wheel-arch.

Seating, body layout and arrangements were H45/32+10FE,FdS,CX,DOO and the rebodied vehicle was classified 'Atlantean H/B(A)'. (The H/B(A) sub-type never having been used when this class was originally introduced.) Again, the original Leyland 0.600 engine was retained, as in 2L67.

2L71 re-entered service in August 1978 and received a four-year certificate of fitness. It was thus remarkable that the vehicle was withdrawn from service almost exactly one year later, in August 1979. In fact, it actually ran in service for considerably less time than this.

In its rebodied form it is believed that it was the only Atlantean to have operated from Kwun Tung Depot and, as such, the only one to have operated on urban Kowloon routes, probably 1 and 1a.

ABOVE: Rebodied 2L71 inside Kwun Tong depot in 1979, having had its engine removed in readiness for scrapping the bus, complete with its relatively new body *(T. V. Runnacles*

2L64 (BE7947).

Partly because it had been expensively converted to have a Gardner 6LX engine, 2L64 was set aside for rebodying, having had its licence suspended in November 1977. During February 1979 the decision was taken not to rebody any more Atlanteans and 2L64 was broken-up with its original MCW body which, like the chassis, was built in 1960 for Maidstone & District, with whom the bus had been fleet numbered DH563.

BELOW: Possibly the only KMB Atlantean to work on Kowloon urban services, 2L71 was rebodied to a design similar to that of the four prototype Dennis 'Jubilant' front-engined buses. In this case, however, the body was modified to suit the rear engine and different proportions. Like 2L67 it also had dual doorways but despite its modernity 2L71 was destined to survive for only a short time, being finally withdrawn in August 1979. *(Bob Bayman*

AEC Regent Mk V
New to various UK operators.

During 1973, KMB acquired a varied assortment of AEC 'Regent' Mk V's which were, with minor exceptions, all fairly standard UK models and, like the Atlantean buses, had started life with several English and Welsh operators.

All the AEC's arrived in Kowloon fitted with AEC AV590 diesel engines, standard for the type in Britain but which, unfortunately, were soon to prove under-powered for the hilly Tai Po Road. This, together with overheating problems, and a poor spare parts situation, brought about a decision to re-engine all the second-hand AEC's, using Gardner 6LW units, although the first two buses to be re-engined, 2A25 & 2A26 (BG657 & 658), formerly 350/HWE, were, prior to entering service, fitted with the more powerful Gardner 6LX engine during March 1974. These were chosen as the Gardner matched-up better with the AEC semi-automatic gearbox fitted in those two buses than they would have with the synchromesh gearboxes fitted in all other second-hand AEC's, except 2A27 (BG4139). It was BG4139 that became the pilot for conversion from AEC to Gardner 6LW power, a task commenced at To Kwa Wan engine shop in March 1974, the vehicle not entering service until 3rd June.

At the same time, the first two of the 25 synchromesh gearbox AEC's entered the To Kwa Wan plant. All were to have both engine and gearbox transplants; the semi-automatic gearboxes coming from new and used stocks held by KMB, supplemented by reconditioned units from UK dealers. These two buses were 2A3 (BD3901) and 2A16 (BD5123), MCW bodied examples from Rhondda and Yorkshire Woollen District respectively. By the end of 1974, many AEC's were on the road again, not only with better power characteristics but improved fuel consumption. In order to accommodate the longer Gardner engine and a larger capacity radiator, the bonnet cowl was extended forward by about four inches, a measure that required a short offside bonnet side panel of that length.

As the AEC's were standard British specification, with limited opening windows, KMB re-styled them to meet its tropical ventilation requirements by fitting its standard aluminium framed windows with squared corners. As was the case with the Atlanteans, the replacement of neat rubber mounted glazing by the standard unit altered the appearance of these buses considerably.

Unlike CMB, who adapted the existing destination equipment to its own needs, KMB fitted completely new units to its standard specification, with the separate destination and route number displays arranged side by side behind a single glass screen. An interesting comparison can be made between KMB and CMB approaches to tropicalisation when similar MCW family bodies from the same Yorkshire Woollen District and Rhondda batches are examined (CMB placed bodies from these sources on Guy Arab chassis, discarding the AEC chassis).

VEHICLE SPECIFICATIONS: AEC REGENT Mk V; Purchased second-hand

Fleet numbers:	2A1-28
Chassis:	AEC 'Regent' Mk V
Engine: As purchased:	AEC AV590
Re-engined 1974:	Gardner 6LW - except 2A25/6; Gardner 6LX.
at rebody 1975-8:	Gardner 6LX derated at 150bhp.
Gearbox 2A1-24,28:	AEC synchromesh
converted to:	SCG GB340 (or 350?)1974
Gearbox 2A25-27:	AEC semi-automatic, Monocontrol.
Body layout As purchased: AEC(N)	H39/31+11FEXS,DOO
AEC(M)	H39/31+13FEXS,DOO
AEC(S)	H40/32+12FEXS,DOO
1975 Recalculated standing capacity: AEC(N)	H39/31+19FEXS,DOO
AEC(M)	H39/31+18FEXS,DOO
AEC(S)	H40/32+18FEXS,DOO
Date new:	1959-64
Date introduced by KMB:	1973-74
Total:	28
Length:	30ft 1in 9115mm
Width:	8ft 0in 2424mm
Height:	14ft 6in 4757mm
Wheelbase:	18ft 6in 5639mm
Unladen weight:	9673kg
Maximum permitted weight:	14961kg

UK Dealers were:
AEC(N) North's   2A2, 3, 6, 7, 8, 9, 10, 11, 12, 13, 14, 15, 16, 17, 18, 19, 22, 25, 26.
AEC(M) Martin's 2A1, 4, 5, 27, 28.
AEC(S) Sykes   2A20, 21, 23, 24.
AEC(2A)   For one month only (8/76 to 9/76), 2 unknown AEC(N) licensed for 9 standees.

Fleet nos; all 2A:	Original operator:	Body:
1, 4, 5, 28	Yorkshire Woollen District	Northern Counties
2, 3, 7-15, 17-19, 22	Rhondda	MCCW
6, 16	Yorkshire Woollen District	MCCW
20, 21, 23, 24	Hebble/Halifax-area operators	MCCW
25, 26	Sheffield JOC via YWD	Park Royal Vehicles
27	Huddersfield	East Lancashire Coachworks.

NOTE: All except 2A26 & 27 rebodied by KMB; all converted to Gardner 6LX at time of rebodying.

RIGHT: New to Hebble Motor Services as their 312, this bus was to emerge from Shing Lok body works as KMB's BD8782 (later 2A20). It is seen here in 1973, prior to commencement of reconstruction work. *(John Shearman)*

Degrees of rebuilding varied from bus to bus; 2A1, 2A2 and 2A3, like 2L9 *(qv)*, were repanelled using the then standard KMB practice, where double moulding strips were formed to represent 'beading' below both upper and lower deck windows, a practice designed to ease the application of the hitherto standard KMB 'urban' fleet-livery where cream bands appeared below the windows, separated from the surrounding red by green beading, usually a 'D' shaped pressing in later years. A policy decision was reached, prior to 2A1-3 (and 2L9), being painted, to adopt a new, largely cream livery, for all the vehicles in the second-hand fleet where the only relief was a band of red around the skirt, extended to include the front wings and bonnet of front engined buses, and a red band around the bus at upper-deck floor level. No part of this colour scheme coincided with the traditional mouldings which were omitted from all other second-hand buses. These were repanelled using plain sheets of aluminium, often reused from the original body.

After these buses had been rewired, the interiors were repanelled using a formica laminate sheeting, which resisted the attentions of the idle passenger who, the world over, likes to pass his journey quietly scraping away the paint with edge of a coin.

As was the case with many buses purchased second-hand by both CMB and KMB, the effect of salted roads from their British service days had taken its toll, particularly of such underfloor items as chassis outriggers, which required replacement.

Rebodying.

Unlike the troublesome Atlanteans, the AEC's could be, and were, successfully modified to become reliable enough for KMB to embark upon a programme to rebody its second-hand AEC fleet.

The bodies were stripped off and the chassis completely re-engineered, including a second engine replacement programme to fit the more powerful, but similarly dimensioned Gardner 6LX engine as even the thrifty 6LW units, installed in place of the original AEC AV590's during 1974/5, had not proved to be sufficiently powerful on the steep hills of Route 50 from Jordan Road Ferry to Sheung Shui.

The pilot rebody/re-engine was carried-out on two buses, 2A11 & 2A12 early in 1975, the vehicles returning to traffic in March 1975. Seven rebodied buses returned to service in 1976, five in 1977 and ten in 1978.

2A26, 2A27 and 2A28 were exceptions to the general rebodying scheme; 2A26 was converted into a tree pruning 'open-top' service car in 1978 and 2A28, a late-comer from YWD, was scrapped without rebodying in 1979. 2A27 (BG4139) had already received extensive repairs to serious accident damage late in 1975 following which it was rebuilt and largely repanelled, but not rebodied in so far as the original pillar spacing and original body outline were retained. It received the moulded side panels with 'mock' beading, described above, but retained the cream livery.

In addition, some question exists as to whether 2A25 was actually rebodied but, if it was, it was done in mid-1978, the bus returning to traffic in the August of that year after being off the road for a considerable length of time for some reason.

The new bodies were similar in design to that used for the Daimler(cc) but modified to have a single, wide, front entrance/exit and front staircase. While the design was based on the Metal Sections style, the bodies were all made from scratch by KMB's own staff.

KMB sub-type suffix letters were allocated to the second-hand AEC's in a somewhat unexpected way, in that they were classified by the initial letter of the dealer's name that supplied them. Thus, those from North's became AEC(N); those from Martin's AEC(M) and those from the Paul Sykes Organisation, AEC(S). In the latter case, the four buses involved had a different seating/standing layout with one extra seat on the upper-deck and one on the lower-deck, when compared with all the other 25 buses of this type. These four AEC(S) type, 2A20, 21, 23 24, initially had a different authorised standing capacity of 12. This was altered in November 1974 to 18, in line with the AEC(M) sub-type, when the official calculation was altered from 288 sq in per standee to 260 sq in. As the chassis of all 28 buses were similar, it has remained a mystery why, in 1975-78 when they received new bodies, all of a similar pattern, the capacity differences persisted.

Of the second-hand purchases, these AEC's were, after the four Daimler CVG6's (qv), the most successful from the operator's point of view.

BELOW: A Welsh expatriate from Rhondda, 464KTG, later to become 2A10, was seen here shortly after having been unloaded from the ship that had brought it to Hong Kong, standing at Kwai Chung Container Port in 1973. *(Lyndon Rees*

ABOVE: A line of un-registered second-hand buses at Shing Lok in 1973, following reconstruction and awaiting licencing. The two buses on the left are AEC Regents while those to the right are Leyland Titan PD3/5s which have had their full-fronted cabs converted to half-cabs. The left-hand AEC is an unidentified bus with MCW body while that to its right has a more pronounced curve to the lower edge of its windscreen, identifying it as having an Northern Counties body; furthermore, the beading below the windows positively identifies it as the only bus with thay make body to have this feature, BD925, later 2A1 and ex-Yorkshire Woollen District FHD121. Only 2A1 and 2A2 (below) were panelled with mouldings for standard KMB lined livery. *(John Shearman.*

BELOW: Another line-up. Here the bus nearest the camera is MCW bodied 2A2 which, was new to Rhondda, the second of the two AEC's with full beading. the bus to its right is another (unidentified) MCW bodied AEC and the third vehicle is a Northern Counties bodied The labels tucked in the radiator grilles were inspection notices placed there by the Government Motor Vehicle Examiner. *(John Shearman*

AEC - Grouped by former UK operator.

Formerly RHONDDA Transport Company, via WESTERN WELSH Omnibus Co.
KMB-class: AEC(N): 2A2,3,7,8,9,10,11,12,13,14,15,17,18,19,22 *(Fleet numbered from 1973/4).*

The most numerous of the AEC's were these fifteen, ex-Rhondda, via Western Welsh, MCCW bodied and new in 1961. Initially, fourteen were supplied by North's of Sherburn-in-Elmet, the purchase being made in October 1972. The late-comer was the former 458KTG which was to become BE4682 and take the fleet number 2A22 when the fleet was numbered, in 1973/4.

These vehicles were distinctive at the rear in that the lower-deck window fitted by KMB was offset to the nearside, it being the standard rear unit fitted by KMB to buses with rear staircases and platforms. This new window became the emergency exit, as on the ex-Bolton Daimlers *qv*. Two buses, 2A2 & 2A3 were repanelled using aluminium sheets embossed with 'mock beading' as hitherto used on buses built new by KMB. This was dropped on othe rebuilds when the special 'second-hand' livery was introduced, prior to their being painted.

During 1974, KMB set about replacing the AEC engines and synchromesh gearboxes in the AEC's, using Gardner 6LW engines and GB340 semi-automatic gearboxes.

All fifteen buses were rebodied in 1975-78, their original MCCW H39/31F bodies being replaced using a body type similar to the Daimler(cc) type, except that a single forward doorway was provided. Seating remained the same as on the original body.

ABOVE LEFT: 2A2 (BD3900) AEC Regent V from Rhondda, via Western Welsh. The KMB frontal treatment and deep mouldings gave former 450GTX a somewhat antique appearance. *(Mike Davis*

ABOVE RIGHT: 2A10, the former 464KTG (also seen on page 183 when newly arrived in Hong Kong) seen here in the summer of 1974 on the Tai Po Road, near the Chinese University, prior to fleet numbering. *(Mike Davis*

LEFT: This 1977 view of 2A3 (BD3901) shows the modifications made to the former Rhondda 462GTX shortly before it was withdrawn for re-bodying. (It returned to service with its new body in February 1988.) The upper-deck nearside corner pillar shows signs of a makeshift repair - careful study reveals that the original corner and offside front window remain as original. Another unusual feature of this bus is that it has regained an AEC radiator grill in replacement of the KMB 'circles' pattern. *(John Shearman*

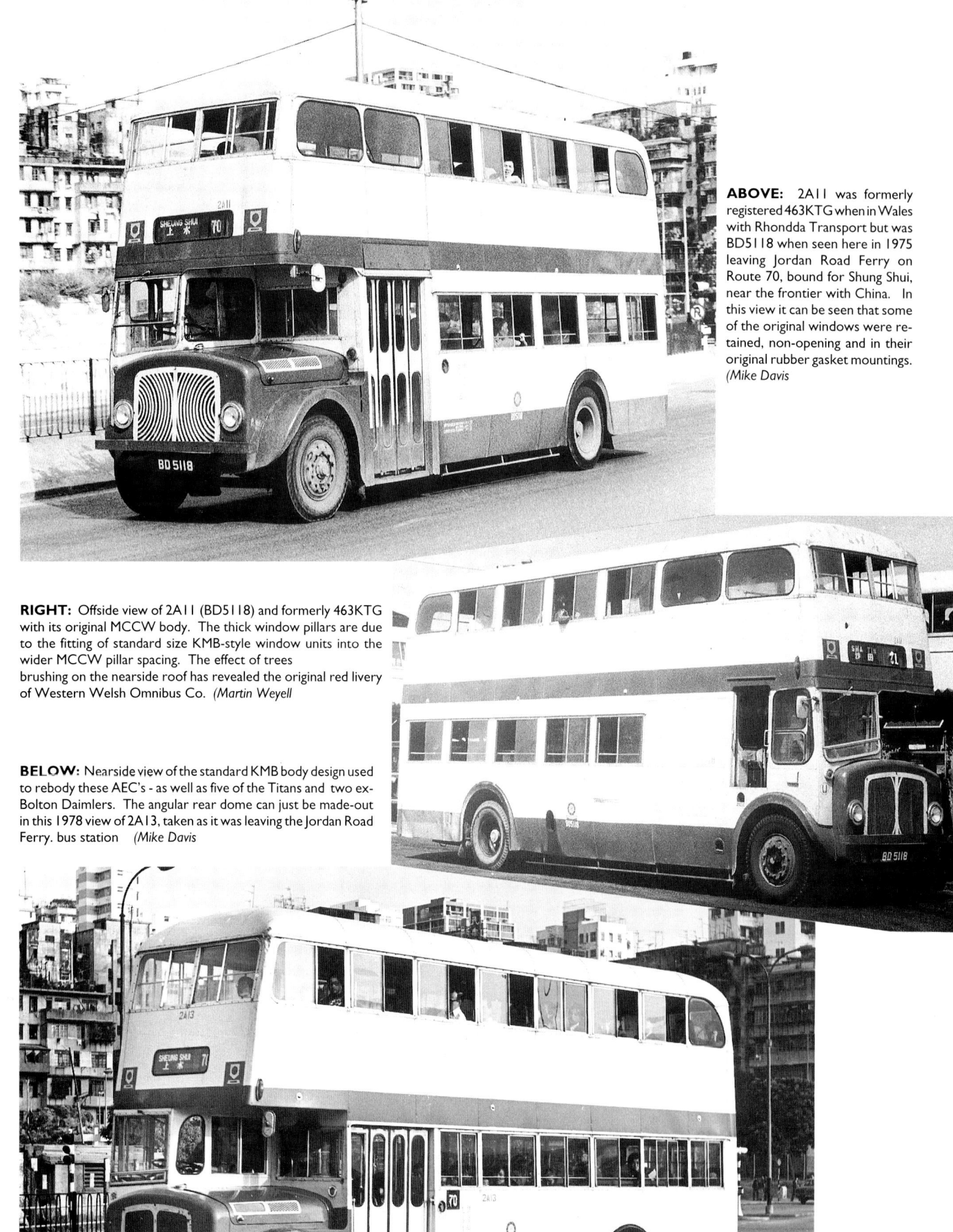

ABOVE: 2A11 was formerly registered 463KTG when in Wales with Rhondda Transport but was BD5118 when seen here in 1975 leaving Jordan Road Ferry on Route 70, bound for Shung Shui, near the frontier with China. In this view it can be seen that some of the original windows were retained, non-opening and in their original rubber gasket mountings. *(Mike Davis)*

RIGHT: Offside view of 2A11 (BD5118) and formerly 463KTG with its original MCCW body. The thick window pillars are due to the fitting of standard size KMB-style window units into the wider MCCW pillar spacing. The effect of trees brushing on the nearside roof has revealed the original red livery of Western Welsh Omnibus Co. *(Martin Weyell)*

BELOW: Nearside view of the standard KMB body design used to rebody these AEC's - as well as five of the Titans and two ex-Bolton Daimlers. The angular rear dome can just be made-out in this 1978 view of 2A13, taken as it was leaving the Jordan Road Ferry. bus station *(Mike Davis)*

New to YORKSHIRE WOOLLEN DISTRICT Transport.
KMB-class: AEC(N): 2A6, 16 - MCCW bodies.
AEC(M): 2A1, 4, 5, 28 - Northern Counties bodies *(Fleet numbered from 1973/4)*

There were six ex-Yorkshire Woollen District AEC Regent Mk V's with KMB (excluding those two with bodies by Park Royal Vehicles which originated with Sheffield *qv*); two with MCCW bodies, supplied by North's [AEC (N)] and four NCME (Northern Counties) examples supplied by another dealer, Martins of Middlewich [AEC (M)]. These six buses were from the same batches as those supplied to CMB, although CMB only required the bodies. They were built in the period January 1959 to January 1961.

The bus that became BD3908 had been acquired via an independent operator and it was this vehicle that was unique in having a rear bumper fitted by KMB. Both at the front and rear, massive American-made hydraulic bumpers with large black rubber water containers were fitted. It is thought that these were probably those that had been fitted to 34ft long AEC Regent, AD7152.

One of the NCME buses, BD3907 - 2A4 - (there may have been others), had a lower-deck emergency window as on the ex-Rhondda Regents (*qv*), although the original offside emergency door was retained.

It is of passing note that these buses from YWD with NCME bodies were the first and last Regents to enter KMB service; ex-FHD121 which became 2A1 (BD925) being the first and ex-FHD119, later 2A28 (BG8544) was the last. 2A28 was held-back until last while it was undergoing an engine and gearbox conversion, prior to having its bodywork modified for the local climatic conditions. Unusually, this conversion was undertaken in the To Kwa Wan engineering workshop and not in the Shing Lok bodyworks, where all the other 113 second-hand buses were prepared (one Atlantean not prepared for service).

The carrying capacity of the two body groups, AEC(N) (MCCW) and AEC(M) (NCME), differed by no more than two standees. The MCCWs were licensed to carry 11 standing (later 19) and the NCME bodies were licensed for 13 (later 18). The AEC(N)'s thus entered KMB service as H39/31+11FEXS,DOO and AEC(M)'s were H39/31+13FEXS,DOO.

BELOW: 2A6 (BD3909) was originally DHD181 in the Yorkshire Woollen District fleet and represents the appearance given to many buses as a result of KMB's 'customising'. Many MCCW bodies of this type, to the 'Orion' design, retained some original glazing on the upper deck but, here, the last of the deeper lower-deck side windows was retained, giving an unequal look. Seen here at Jordan Road Ferry in May 1975. *(Mike Davis)*

Formerly HALIFAX area buses new to Halifax Cpn., Halifax J.O.C. and Hebble M. S.
KMB-class: AEC(S): 2A20, 21, 23, 24. *(Fleet numbered from 1973/4)*

The Paul Sykes Organisation (dealer) supplied the four ex-Halifax MCCW bodied AEC's and they wer the only KMB second-hand buses from that source (others went to Cascade). They entered service a little later than the majority of the ex-Rhondda and ex-YWD Regents and were the only ones with the higher capacity of H40/32FEXS,DOO; one additional seat on each deck when compared with the other 25 second-hand AEC buses. All four were new in 1960; 2A20 in the December and 2A21/3/4 in February, March and January respectively. By the time these buses were purchased by KMB in March 1973, they had all passed to the ownership of the Calderdale Joint Omnibus Committee (JOC) before sale to Sykes.

BD8782 (2A20), had been new to Hebble Motor Services of Halifax as their No 312 (NCP475), passed to Halifax Corporation in April 1971 as No 75 and to Calderdale JOC in March 1972 as No 375.

BE725 & BE6793 (2A21 & 24) were new to Halifax Corporation as Nos 11 & 12 (LJX11 & 12), passing to Halifax JOC in March 1970 without fleet number change but being renumbered 311/312 in June 1970; they passed to Calderdale JOC in September 1971 as 311/312.

BE4683 (2A23) was new to Halifax Joint Omnibus Committee in March 1960 as No 212 (LJX212), passing to Calderdale JOC in September 1971.

All four buses were rebodied by KMB in 1976-78, using their standard body design similar to the Daimler(cc)'s, complete with the somewhat angular rear dome but with single forward entrance/exit and staircase. Gardner 6LW engines and GB340 gearboxes replaced the original AEC AV590 engine and synchromesh gearbox soon after entry into service in 1973, while Gardner 6LX engines were fitted at the time of rebodying.

LEFT: 2A24 was new to Halifax Corporation as No 12 (LJX 12), passing to Halifax Joint Omnibus Committee in March 1970 and was later (6/70) renumbered 312. It passed to Calderdale J.O.C. in September 1971 and to KMB in early 1973. *(Mike Davis)*

LOWER LEFT: There is little to show that this is the same ex-Hebble bus illustrated at the foot of page 182. 2A20 (BD8782) had not been in service long when photographed in 1973. *(John Shearman)*

Formerly SHEFFIELD; to KMB via HEBBLE and YORKSHIRE Woollen District.
KMB-class: AEC(N): 2A25 & 26. *(Fleet numbered from 1973/4)*

In the Autumn of 1973, somewhat later than the entry into service of many of the previously mentioned AEC's, North's supplied two more AEC Regent Mk V's; the first for KMB with semi-automatic gearboxes. They were also the first buses for KMB to have Park Royal bodies.

New to Sheffield in January 1964, they were also the newest second-hand buses thus far acquired, having started life owned by British Railways as part of the Sheffield Joint Omnibus Committee fleet in which they were numbered C1150/51. They passed to Hebble Motor Services, Halifax, in January 1970, becoming their Nos 317 & 318. They were later renumbered 622 & 623 before passing to Yorkshire Woollen District in December 1971 where they took fleet numbers 144/45 before being renumbered, yet again, as YWD Nos 531 & 532. They did not enter KMB service until March 1974 and did so on Route 70, registered BG657 & 658 and were, along with one Atlantean, the first KMB buses to receive BG-numbers. The distinctive triangular panel behind the staircase, and the corresponding triangle of glass, was retained, as was the H39/31F layout.

The reason for the long delay before entry into service was a result of their being taken into To Kwa Wan works for the replacement of their AEC AV590 engines by Gardner 6LX units, a process that required the bonnets to be lengthened. These two buses were, as already mentioned, chosen as the first AEC's to be fitted with Gardner 6LX engines as that type was compatible with the semi-automatic gearbox and Mono-control. They were, however, to remain jointly unique only until the main body of AEC's, previously converted to Gardner 6LW, were rebodied and given the more powerful 6LX engine as fitted to 2A25 & 26.

While it is known for certain that 2A26 was not rebodied, it being converted to an open-top tree-pruning goods vehicle in May 1978, there remains a very small doubt that 2A25 may not have been rebodied, although some internal KMB records suggest that it *was* included in the programme.

LEFT: 2A25 and 2A26 came to KMB from YWD but had been new to British Railways as part of their contribution to the Sheffield Joint Operating Committee, in which fleet they were numbered C1150/51. They had been transferred to YWD in 1971, finally carrying the fleet numbers 531/2, before shipping to Kowloon to become BG657 & 658. BG657 shows-off the distinctive triangular window behind the staircase. Seen in the New Territories in 1974, shortly after entering service. *(Lyndon Rees*

BELOW: 2A25 (BG657) again, this time in Jordan Road, hotly pursued by one of KMB's MetSec bodied, 34ft long AEC's, in late 1974. *(Mike Davis*

New to HUDDERSFIELD Corporation.
KMB-class: AEC(M): 2A27 *(Fleet numbered from 1973/4).*

The last second-hand bus to be delivered to KMB, but not the last to enter service, was also an AEC Regent Mk V with Mono-control, semi automatic gearbox and was built in 1961, having been new to Huddersfield Joint Omnibus Committee, later passing to Huddersfield Corporation, where it retained its fleet number, 194. It was unique with KMB as the only East Lancashire Coachbuilders bodied AEC and it came to KMB via Martin's who had procured it from another dealer, Hartwood Exports. 2A27 also had a triangular panel and glass arrangement behind the staircase, a feature not found on the East Lancs bodied ex-Bolton Daimler CVG6's (qv). Like the two Park Royal bodied Regents, also with semi-automatic gearboxes, 2A27 was delayed from entry into service by being set aside for conver-

ABOVE: Formerly with Huddersfield Corporation, BD4139 was new to the Huddersfield Joint Omnibus Committee as 194 (SCX194). It was seen here leaving Jordan Road Ferry early in 1975, before receiving its fleet number, 2A27. It was unusual in Kowloon terms in having an East Lancs. body. *(Mike Davis)*

RIGHT: 2A27's East Lancs. body was extensively rebuilt after an accident in which it overturned on the Tai Po Road sustaining serious damage. The extent of the rebuild did not involve the replacement of the body frame, as on other AEC's and the result was a much less austere vehicle, which retained its original front and rear domes. The mouldings used on 'urban' buses were incorporated in the rebuilt body. *(Mike Davis)*

LEFT: An offside view of BG4139 before it became 2A27, still showing the majority of its original East Lancs body features, including a vestige of beading below the driver's cab window. In this view it is apparant that the original radiator was still in place as it the bonnet has not been extended. *(Mike Davis*

sion to Gardner 6LW (ie *not* 6LX) power. When it did enter service in June 1974, it was registered BG4139 and retained its UK seating layout of H39/31F but, like all other Regents supplied by North's, it was authorised to carry 11 standees; later revised to 19 when the regulations were relaxed to reflect reality. Following a serious overturning accident in 1975, 2A27 was extensively rebuilt, as opposed to rebodied, using much of the original structure and pillar spacing. It was completely repanelled and acquired the mock mouldings below both upper and lower deck windows, as used on the urban bus fleet but it retained the mainly cream 'second-hand' livery. It was finally withdrawn in April 1980.

BELOW: The former UK AEC's lasted long enough to be repainted into standard fleet livery after the special paint scheme for the second-hand fleet was dropped.   2A10 (see photos of 464KTG on a earlier pages) in its final, rebodied, condition, early in 1982, in fleet-standard livery, without lining. *(Derek Lucas.*

Daimler CCG6: 2D1-3 *(from 1973/4).* New to CHESTERFIELD Corporation.

With Daimler front engined chassis having been KMB's choice since the very first double-deckers were introduced in 1949 a major disappointment was that so few were available. It was, at the time, difficult to locate suitable front entrance examples of 30ft overall length which was the KMB requirement.

The purchase of three 27ft long Daimler CCG6's was curious, especially as they were fitted with constant-mesh gearboxes to which KMB's drivers found it difficult to adapt, despite their having previously driven Albion single-deckers with crash gearboxes. KMB allocated them the sub-type 'Daimler CCG6/27' - they were of similar size as the Company's 26/27ft long Daimler(a) & (b) CVG5's.

The three Daimler CCG6 models - the second 'C' indicating the constant mesh gearbox - were new to Chesterfield Corporation in 1963 with fleet numbers 258, 256 & 253, taking, respectively, the Hong Kong registration numbers AD7478-80 and KMB fleet numbers 2D1-3, when the fleet was numbered. All three were purchased from North's in October 1972 being received in Hong Kong during January 1973. They were prepared for service by KMB in a similar manner to that applied to the 28 AEC Regent's, being fitted with tropical opening windows in all main bays and lining the interiors with formica laminate panels. They were also completely rewired. They retained their H37/28F Weymann bodies which had been fitted with folding doors from new. A solid bar front bumper was fitted as on all KMB front-engined buses. The Gardner 6LW engines were retained as were the Guy constant-mesh gearboxes, the latter until late in 1974, when they were replaced by SCG GB340 semi-automatic epicyclic gearboxes.

2D1 - Rebodied as KMB sub-type Daimler CVG6(2A)/27.

2D1 was rebodied during the thirteen months between its withdrawal in October 1975 and its re-entry into service on 18th November 1976 when it re-emerged with a revised seating layout as H41/26+17FE,CXS,DOO. At this time, the engine was changed to Gardner 6LX, still with the semi-automatic gearbox fitted earlier. KMB took the opportunity to re-classify the type from CCG6/27 to CVG6(2A)/27, the '(2A)' would appear to indicate a similarity to the rebodied Daimler(aa) type - ie two a's. The external appearance was very similar to the Daimler(aa), except that the front of the body was near vertical, as on the MetSec Fleetlines, D666, etc. and the chassis was to the same 7ft 11in width as the body - on the (aa)-type the chassis was 7ft 6in, leaving 2 ins between the body and tyres on each side. Although no moulded panels were fitted, simulating beading, the body was out-shopped in the standard fleet livery of half-and-half, cream over red, without green lining-out. The destination screen was narrower and only accommodated the destination and one track for the route number, where KMB standard displays had two number tracks, the second rarely being used.

When the rebodied bus was returned to service, it was regularly on the 30 between Jordan Road and Tsuen Wan, and other NT routes. In mid-1979, however, 2D1 became one of only two first-generation, second-hand double-deckers (the DMSs being second-

VEHICLE SPECIFICATION:	DAIMLER CCG6	
Fleet numbers:	2D1-3	
Registration numbers:	AD7478-AD7480	
Chassis: as purchased:	Daimler CCG6LW	
as rebuilt:	Daimler CVG6LX (originally CCG6).	
Engine: as purchased:	Gardner 6LW	
as rebuilt:	Gardner 6LX	
Gearbox: as purchased:	Guy constant mesh, manual.	
as rebuilt:	Leyland/Daimatic epicyclic, semi-automatic.	
Body make: as purchased:	Weymann	
as rebodied:	KMB - 2D1 only.	
Body layout: as purchased:	H37/28+8FEXS,DOO (7 standees from 6/75	
as rebodied:	H41/26+17FE,CXS,DOO (2D1 only)	
Date built:	1963	
Introduced by KMB:	1973	
Date rebodied:	1977 (2D1 only)	
Length:	27ft	8181mm
Width:	7ft 11in	2399mm
Height as purchased:	14ft 7in	4440mm
Height as rebodied 2D1:	14ft 4in	4370mm
Wheelbase:	16ft 4in	4978mm
Unladen weight:	8,107kg;	2D1 as rebuilt: 8,098kg
Gross vehicle weight:	12,218kg;	2D1 as rebuilt: 12,727kg

BELOW: Ex-Chesterfield Corporation Daimler CCG6, 3258NU, after becoming KMB's AD7478 but before being fleet numbered 2D1. It was photographed in 1973 at Shung Shui, near the frontier with China. *(Lyndon Rees)*

ABOVE: AD7480, shortly after becoming 2D3, and formerly Chesterfield 3253NU. These Daimler CCG6's were of similar size to the early KMB Daimler(b) CVG5's, although somewhat more modern. Seen here in 1975 at Jordan Road Ferry terminus *(Mike Davis)*

generation) ever to be allocated to Kwun Tong depot when it was transferred there from New Territories operations to work Route 19, the last urban route to be worked by single-deckers, which, after the steep climb to Ngok Yue Shan, terminated in a small cul-de-sac with a tight turning circle; a combination requiring both the power of a 6LX engine and a short wheelbase.

The only KMB double-decker possessing both attributes was 2D1. The second bus allocated to Route 19 had to remain a short-wheelbase, single-deck Chieftain because, unfortunately, 2D2 & 2D3, the other two Daimler CCG6/27's had been scrapped as recently as February 1979, without having been rebodied and after lying derelict for some time in Tuen Mun depot.

2D1 was withdrawn from passenger service in the early 1980's but was kept at Kowloon Railway Station, still licensed but used as a staff rest room until January 1984.

These three buses were sister vehicles to the bus that became "Fok Lei", Macau, number CC203, whose chassis number was 20019; KMB's trio were 20017/5/2.

LEFT: 2D1 after being rebodied by KMB to their own design, not dissimilar to the two rebodied Daimler CVG5's of the Daimler(aa) class, except for the more uprghit front panels; KMB then reclassified it CVG6(2A)27. It is seen here at Ngok Yue Shan during the test run on Route 19, despite 52 (and 71) still being displayed. *(Mike Davis)*

Daimler CVG6-30:
New to BOLTON Corporation then to SELNEC.[1]

2D4-7
(Fleet numbered from 1973/4).

The only four of the 115 first generation second-hand buses purchased which were exactly of the type required by KMB (ie 30ft long, front engined Gardner Daimlers with semi-automatic transmission) were these four ex-Bolton Daimler CVG6-30's which were supplied by North's. In Kowloon they became AD7481-84 and were later fleet numbered 2D4-7.

As delivered, they had attractive East Lancs H41/32F bodywork; a layout that was retained, although the sliding passenger doors on three were replaced, quite soon after entry into service, by folding doors. The fourth, unidentified, bus retained its sliding door much longer, possibly into 1975.

Their lower-deck emergency back window was offset towards the nearside, as those on the ex-Rhondda AEC's, described earlier. The push-out ventilators which formed the upper quarter of the upper-deck front windows were retained in the same way as those on the ex-Maidstone & District Atlanteans. Solid bumper-bars were added by KMB as were ventilation grilles in the bonnet cover - a feature of all KMB front engined buses, new and second-hand.

These buses which were built in 1960 were fitted from new with semi-automatic gearboxes and were thus almost standard KMB specification. Before reaching North's, they had been absorbed into the SELNEC[1] PTE fleet - later Greater Manchester PTE - and were purchased by KMB in December 1972. They arrived in Kowloon in February 1973 carrying SELNEC fleet numbers 6644/6647/9/50 (PBN662/5/7/8).

The usual modifications were undertaken by KMB, including re-wiring, re-panelling inside and out and the fitting of full-depth sliding windows in all main side bays, together with standard KMB destination indicators. AD7481 overturned early in 1974 but was only superficially damaged and was soon returned to service.

The chassis of these four buses was similar to the Daimler(cc) sub-type, as rebuilt with semi-automatic gearboxes. KMB, however, christened the ex-Bolton buses merely Daimler CVG6/30.

Two (2D6/7) were rebodied during 1976 & 77, to the same basic design as the rebodied AEC Regent V's and two (2D4/5) to the same design, with dual doorways, as the rebodied Daimler(cc) type. All four were given Gardner 6LX engines.

Daimler CVG 6/30(cc): 2D4 & 2D5.

These two buses were reconstructed and rebodied during 1977 and returned to service in August that year. Reflecting the Daimler(cc) style of bodywork, with dual entrance/exit doorways and centre staircases, the sub-type letters (cc) were added to the original designation to become CVG6/30(cc). Unlike their two sisters, both buses were returned to service in standard livery with green lining, but did not have the raised mouldings applied to earlier second-hand buses that never carried a livery that corresponded to the mouldings! Narrower destination screens with only a single-track route number replaced the standard type provided prior to rebodying. Seating and layout as rebodied became: H44/31+24FE,CXS,DOO.

[1.] SELNEC PTE; South-East Lancashire and North-East Cheshire Passenger Transport Executive.

VEHICLE SPECIFICATION: DAIMLER CVG6LW-30

Fleet numbers:	2D4-7
Registration numbers:	AD7481-AD7484
Chassis:	Daimler CVG6-30
Engine: as built:	Gardner 6LW
as rebuilt:	Gardner 6LX
Gearbox:	4-speed semi-automatic with ep control.
Body make: as purchased:	East Lancashire Coachbuilders.
as rebodied:	KMB.
Body layout: as purchased:	H41/32+12FEXS,DOO - Sliding door; later folding door.
rebuilt(cc):	H44/31+24FE,CXS,DOO
rebuilt 1-door:	H41/32+19FEXS,DOO
Date new:	1960
Introduced by KMB:	1973
Date rebodied:	1976 & 1977
Length:	30ft 1in 9115mm
Width:	7ft 11in 2399mm
Height as purchased:	14ft 6in 4757mm
as rebodied:	14ft 8in 4811mm
Wheelbase:	18ft 6in 5639mm
Unladen weight: CVG6/30:	9100kg; or,
CVG6/30(cc):	8896kg
Gross vehicle weight:	14961kg

LEFT: AD7482, ex-PBN665, later to become 2D5, stands at Jordan Road Ferry awaiting departure on Route 70 on 1st January 1974. The sliding door was retained for a short period before being replaced by the power-operated folding type with four leaves. *(John Shearman*

ABOVE: The four ex-Bolton Corporation Daimler CVG6-30's were numbered 2D4-7 by KMB. Here 2D7 in original condition leaves Jordan Road in 1975. It was originally registered PBN668 when in the United Kingdom. *(Mike Davis)*

Daimler CVG6/30 type: 2D6 & 2D7.

The higher numbered pair, 2D6 & 7 were rebodied and returned to service eighteen months earlier than the two that had become CVG6/30(cc)'s (2D4 & 5) and were to the then standard rebody design for front-engined double-deckers in having a single front doorway with front staircase, almost identical to those fitted to the Leyland Titan PD3/5's and AEC Regent Mk V's. The original KMB type code, Daimler CVG6/30, despite being based on the chassis type, was retained as the passenger capacity was unaltered at H41/32+19FEXS.

These buses came from the same batch as CV201 & 202 of "Fok Lei", Macau (40 miles west), which had chassis numbers 30075/8, while KMB 2D4-7 were 30076/9/81/2.

BELOW: Of 2D4-7, 2D4 & 2D5 were rebodied as Daimler(cc) class and received standard fleet livery, although continuing to work on routes to or in the New Territories. Here 2D5 with 2-door, centre-staircase layout was leaving Jordan Road Ferry in January *(Mike Davis)*

ABOVE: Ex-Bolton 2D4 (PBN662), showing the offside with centre staircase panels as rebodied by KMB in August 1977 and seen here in 1978. *(Mike Davis)*

BELOW: 2D6 and 2D7 were rebodied as single-doorway buses with front staircases, to KMB's standard outline, as per the AEC Regent V's and Leyland Titan PD3's, 2A & 2L classes. 2D6 heads away from Jordan Road Ferry at the commencement of journey on Route 71 in January 1978. *(Mike Davis)*

ABOVE: One of the four Daimler CVG6's new to Bolton Corporation. Positive identification is not possible but it *is not* the bus which became AD7482, judging by the position of the vents on the bonnet cover. Seen in early 1973 at To Kwa Wan undergoing reconstruction. The sliding door was initially retained. *(John Shearman.*

CENTRE: Another unidentified ex-Bolton partly rebuilt at Camp Street. *(John Shearman*

BELOW: Two Boltons and a Chesterfield. Daimlers from those two municipalities stand awaiting the attentions of the Motor Vehicle Inspector (MVI) prior to being registered and licensed for the first time in Hong Kong. The two ex-Bolton buses are those with the original upper-deck front windows with UK style push-out vents. The ex-Chesterfield CCG6 in the centre has a sign hung on the radiator filler-cap which reads 'M. V. I. Inspected'. *(John Shearman*

Daimler Fleetline CRL6: 2D8-107.
New to LONDON TRANSPORT.

Following on from the successful operation of former London Transport DMS-type Fleetlines by neighbouring China Motor Bus, KMB ordered one hundred for their own fleet; delivery commencing early 1981. Unlike CMB, however, KMB took buses with Leyland 0.680 engines. KMB's delivery included the first Metro-Cammell Weymann bodied versions to be exported to Hong Kong, although the actual first into service was Park Royal bodied ex-DMS725, numbered 2D8 by KMB. 2D12 (DMS1554) was the first MCW bodied DMS to enter KMB service.

The first batch arrived in Kowloon in late February 1981 and the first two were licensed on 1st April. An initial batch of nine entered service on 8th April on Route 2C, from "Star" Ferry to Tai Hang Tung, in replacement of Daimler CVG5's.

KMB's modifications to the DMS type (some had actually been DM and D types in London days but the term 'DMS' is used here as a generic term except when referring to specific former D or DM type vehicles) followed a similar pattern to those carried out by CMB but were more extensive. The front entrance doors were restricted by allowing only the front pair of leaves to open, the rear pair being firmly secured in the closed position by fixing a steel bar across the inside, half way up the doors, the object being to channel boarding passengers past the Bell Punch 'Autofare' Farebox, to which no ticket issuing machine was ever attached during KMB service. An exception to the doorway arrangement was 2D13 (DMS698) which, like CMB XF34, had a complete conversion to single-width front doorway. The centre exit doors were retained unaltered in all cases. London Transport's

VEHICLE SPECIFICATION: ex-LONDON TRANSPORT DMS-class.	
Fleet numbers:	2D8 -107
Chassis:	Daimler or Leyland 'Fleetline} CRL6 (Note 1)
Engine:	Leyland 0.680 or Gardner 6LX (see note 2)
Transmission:	Semi-automatic, four-speed gearbox
Body make:	Park Royal or MCW.
Body layout:	H45/28+21FE,CXS,DOO.
Date built:	1973-74
Date introduced by KMB:	1981-83
Length:	9300mm
Width:	2500mm
Height:	4420mm
Wheelbase:	4953mm
Unladen weight:	10,059kg
Gross vehicle weight:	16,257kg

NOTE 1: The last Daimler (Coventry-built) DMS was chassis number 67261, DMS660, later CMB XF56. After production had been moved to Leyland, Lancashire, construction resumed at 67262 (DMS1675) and thus those chassis with this number and higher are correctly termed 'Leyland Fleetline CRL6 or CRG6', according to engine make. The CRL designation was retained for vehicles ordered from Coventry but actually built at Leyland. None of the DMS-type Fleetlines purchased by KMB waerethe Leyland FE30ALR type.
NOTE 2: Certain buses in this group which suffered failed Leyland engines were fitted with Gardner 6LX units; ie not 6LXB as fitted to DMS Fleetlines as original equipment.

BELOW: 2D28, MCW bodied ex-DMS1599, showing the standard KMB treatment of the destination gear as applied to the majority of the type, with low placed aperture and screen. Seen here on Route 7 at Kowloon Tong in 1981. *(Mike Davis)*

UPPER LEFT: 2D12 arriving at Kowloon "Star" Ferry in 1981, showing the MCW-style of offside emergency door to good advantage. Note how the top of the door is higher than the upper margin of the lower deck windows. *(Mike Davis*

LEFT: 2D14 awaits its departure time outside Star House, Tsim Sha Tsui, during August 1981. This view allows comparison with 2D12, above, and identifies 2D14 as a Park Royal bodied bus; ex-DMS758 in fact. The original destination screen is retained but the lowest screen (the ultimate destination in London terms) is masked. *(Mike Davis*

automatic fare collection (afc) machinery was removed by Ensign, the supplying dealer, and replaced by inward-facing seats for five passengers. On former DM-class vehicles, which were LT's crew-operated version of the driver-only-operated 'DMS', no afc equipment was installed.

Hong Kong regulations in force at the time of the DMS's arrival allowed for six passengers to be carried on the rear bench seat where the body width was 2.5m (8ft 2 in), the UK norm being five due to the greater bulk of the average European. As a result, the capacity and arrangement in KMB service became H45/28+21FE,CXS,DOO, regardless of the vehicle being originally DM or DMS. KMB renewed all seat upholstery which was originally leather-trimmed moquette and unsuitable in a humid climate, and replaced it with their standard very thin cushioning and, as a result, the seats were considerably lower, ie closer to the floor as the original seat frames were retained. No attempt was made to install 3+2 seating, even on a trial basis; in fact, the buses carried bodyside notices in large characters asking passengers not to overload the bus - possibly KMB remembering the problems with its Atlanteans in the 1970's.

KMB standard destinations were fitted on the initial conversions, up to 2D13. These were located approximately at the level of the LT ultimate destination screen, thus enabling the display to be changed from within the driver's position. 2D14 & 15 and 19 & 20 were as per CMB, with the destination in the LT via box and the route number in the LT position. A third version appeared on 2D21 where standard KMB layout was fitted but with the screen over the LT via and route number rollers, an easily observed difference as the screen was considerably higher than that of 2D8 etc. The three variations subsequently occurred indiscriminately through the series to 2D107. During their period of KMB service, the Company replaced both types of display where the LT apertures were retained, with or without KMB screens, installing their own standard type as 2D8, it not having been possible to change destinations on either type from the driver's cab; unacceptable on a DOO bus. Until replaced, the problem of blind-changing was overcome by the provision of slip-boards placed in the windscreen to indicate the outer destination while the inward destination, typically '**STAR FERRY**', was left on the blind display more or less permanently. The rear LT-style route number box was retained and used by KMB but the side route number and destination screens were either panelled-over or painted-out and then the Company went to the trouble of providing a new route number display unit inside the first nearside window, the standard position on KMB's Fleetlines (D666-1150) purchased new.

KMB tropicalised the windows throughout using 'Planet' aluminium framed full-depth sliding window units, including the upper-deck front and first side windows. Windscreens were only altered as replacement became necessary and were of non-standard pattern. 2D8 re-appeared after accident damage with very angular front panels and windscreens apparently 'grafted-on' and of a pattern very similar to the BACo bodied Fleetlines, possibly utilising spares held for that type. 2D28 was followed by 2D82, and others, similarly modified. 2D85 appeared in February 1982 sporting flat windscreens but with radiused

ABOVE: 2D26, like a number of other former London Transport DMS-class Fleetlines, retains its original destination equipment but, because it could only be turned from the upper saloon, it was often left unused and recourse made to the slip-board so popular with bus crews in Hong Kong. *(Mike Davis)*

LEFT: KMB-style destination boxes mounted in the high position, approximating to the London 'via' box. This arrangement was also unacceptable as the winding handles were upstairs. *(Mike Davis)*

outside corners. Yet others were given front panels and windscreens of MetSec pattern, with slightly recessed headlights.

Brief mention has been made that the DMS fleet was purchased through the Ensign Bus Company (Ensignbus) of Purfleet, who undertook much of the preparation and tropicalisation work prior to shipping. Following the example of the way it classified the second-hand AEC Regent fleet, KMB identified the DMS type from other Fleetlines by allocating the supplier's name as the sub-type so that they became 'Fleetline/Ensign', using the whole name as opposed to just the initial letter, as was done with the AEC's.

All the DMS's were shipped by Ensign in their somewhat faded London Transport livery, all painting into fleet livery being carried-out by KMB at Tuen Mun. Ensign also removed all engine shrouds while most retained their twin fog/spot lights. After some months in service most had succumbed to damage and their locations panelled over and solid bumper-bars were fitted to afford protection as on CMB's fleet of similar vehicles. In line with other operators of the type in Hong Kong, KMB removed the troublesome fully-automatic gear-control option and all became semi-automatic.

It was not long before KMB began to regret their choice of Leyland 0.680 engine and commenced a gradual, but never completed, programme of replacement by Gardner

ABOVE: 2D13 was the first KMB ex-DMS to have its front entrance doorway reduced to single-width, rather than having the rear portion locked-up in the closed position. *(Mike Davis)*

BELOW: Perhaps the most radical rebuild of a KMB ex-DMS was applied to 2D34 which, following fire-damage, was rebuilt with inset window pans, similar to those of the BACO bodied 33ft Fleetlines. New front panels and windscreens were also provided. *(Mike Davis)*

UPPER RIGHT: KMB relaced accident damaged front dash panels and windscreens by drawing on stocks of spare parts intended for their fleet of Metal Sections and BACO bodied Fleetlines. 2D28 (ex-DMS1559) shortly after receiving a BACO front and about six-weeks after being photographed for the picture on a previous page. *(Mike Davis*

CENTRE: The Metal Sections style of front as applied to un un-nmbered 2D18 CM8999 and ex-DMS1551). *Mike Davis*

LOWER RIGHT: 2D17 (ex-DMS1587), with new windscreen of yet another pattern, turning into Salisbury Road from Nathan Road, early in 1984. The sign writing along the sides of the ex London Fleetlines reads 'please don't overload the bus'. This was said to underline KMB's distrust of second-hand buses following the experience of the PDR1/1 Atlanteans in the mid-1970's. *(Mike Davis*

6LX units. This is not a straightforward substitution and requires replacement of many other parts so that changes were only carried-out on an 'as necessary' basis, Gardners being substituted when a Leyland unit failed and was found to be-yond economic repair. It has not been positively identified which buses were converted or how many but those that were converted for certain included: 2D14/7/9/24/ 31/33/46-7/64/9/90 & 106.

2D34 was damaged by fire and was rebuilt using BACo-style, deeply inset, window pans and the front windows were also replaced using non-standard parts.

Withdrawals.

An early victim of accident damage was 2D15 which was withdrawn after only weeks in service. Other odd vehicles were scrapped or transferred to the training fleet, where they again replaced Daimler CVG5's. but it was not until 1985 that the numbers were significantly culled; when 2D102 was withdrawn, no less than ten were sent to become training buses. This process continued until 2D26, 62 & 69 became the last to be withdrawn, in 1990.

Numerous examples were sold to Hong Kong dealer 'Speedybus' for further service in Chinese cities, some a long way from Hong Kong, in cold northern climates.

ABOVE: As with the remainder of the double-deck fleet, the '1982' livery was applied to the former DMS-class buses and to good effect. 2D86 turns into Salisbury Road in 1984, after the introduction of reflective registration number plates. *(Mike Davis)*

BELOW: In their latter days with, many of the former DMS-class buses were converted for use as trainers and received the special Traininig School livery. Here CM8752 (formerly KMB 2D21 and London DMS765) was being manoeuvred by a novice driver, reversing between the cones in April 1992. *(Nigel Eadon-Clarke)*

Dennis Dominator - *Demonstrator via SINGAPORE Bus Service.*

Along with the manufacturers of similar competing products, Hestair Dennis were fortunate enough to be permitted to provide a demonstration double-decker for trial operation in Singapore for evaluatory service with Singapore Bus Service (SBS). Accordingly, a 10.3metre Dennis Dominator was despatched, complete with East Lancashire Coachbuilders (East Lancs.) bodywork and, surprisingly ,a DAF DK1160 engine, made in The Netherlands. Prior to leaving the UK, Dennis registered the bus for demonstration purposes and it ran for a time with the number ACM408X.

Unfortunately, no SBS order was forthcoming and, in accordance with the laws of Singapore, the vehicle had to be removed following the completion of its authorised period of demonstration. It is understood that it was felt more prudent to send the Dominator the relatively short distance to Hong Kong, with the possibility of a sale locally to one of the operators already using Dominators. CMB's DD1 was similar in many ways in both body and chassis, with the exception, of course, of the DAF engine.

VEHICLE SPECIFICATION: DENNIS DOMINATOR TRAINING BUS	
Registration number:	Not allocated
Chassis:	Dennis Dominator DDA144/343
Engine:	DAF DK1160, 180kw
Transmission:	Voith fully-automatic, three-speed, D831 gearbox
Body make:	East Lancashire Coachbuilders Ltd.
Body layout:	H00/14FE,FdS,CX,Trg.
Date built:	1982
Date arrived in HK:	May 1984
Length:	10,300mm 33ft
Width:	2500mm 8ft 2½in
Height:	4419mm 14ft 6in
Wheelbase:	5639mm !8ft 6in
Unladen weight:	?
Gross vehicle weight:	16,257kg
NOTE:	Originally registered ACM408X in the United Kingdom. Subsequently registered in Singapore as SBS7003Y.

LEFT: Former Singapore SBS7003Y seen shortly after its arrival in Hong Kong, parked on the roof of KMB's 'Tuen Mun 80' engineering facility in the north-west New Territories town of that name. Mechanically, this bus differed from other KMB and CMB Dennis Dominators in having a DAF engine. *(George Luke)*

Still carrying its Singapore registration number, SBS7003Y, the bus arrived in Hong Kong early in May 1984 and arrangements were made to stable the vehicle with KMB at that Company's Tuen Mun facility. Despite a demonstration run with Citybus Ltd., and its close proximity to KMB's large fleet of Dominators, the one time SBS7003Y remained unsold until the summer of 1985 when KMB acquired it with the intention of converting it into an advanced driver training vehicle. The conversion included the provision of water containers in place of the upper deck seats in order to simulate varying load factors, something which most novice bus drivers only experience the hard way on their first day in passenger service.

It is perhaps unfortunate that this bus was scrapped circa 1991, having never carried a fare-paying passenger in Hong Kong.

BELOW: A 1989 view of DL9739 in the training bus livery after being fitted-out for advanced driver training. On the roof of Lai Chi Kok depot. *(Mike Fenton)*

Part Five:
KMB Double-deck Buses 1980 to 1994.
Section A:
The Development of 12-metre, 3-axle Buses for Hong Kong.

A Larger population, larger buses and a larger number of axles.

During the early 1980's, KMB purchased relatively small numbers of the second-generation British, two-axle double-deck bus in the form of Leyland Olympians (the BL-class), Dennis Dominators (the DM-class) and MCW Metrobuses (the M-class). They also purchased 41 Mercedes-Benz O.305 double-deckers, based on the German-style space-frame chassis, and a single, ill-fated Volvo B10M.

ABOVE: KMB's 1949 method of meeting increased demand was the introduction of the Daimler CVG5 double-deck bus to Hong Kong. *(Author's collection)*

ABOVE RIGHT: Prototype Leyland Olympian with ECW body seen here at the bodybuilder's factory in a livery that was only modified by cream lower window surrounds. *(Brian Ollington)*

Before progressing further, the reader is reminded that, during 1978 and 1979, MCW had built for KMB's neighbour, China Motor Bus, a lengthened version of their standard Metrobus: a vehicle some 11.45-metres in length and weighing-in, fully laden at 18.3 tonnes - on two axles. While these 146-passenger vehicles proved to be some of the most reliable buses ever to have been purchased by CMB, their great weight did require them to have 'overweight' permits in order to legally operate in Hong Kong. CMB made it known that it had intentions of ordering further examples of the type but the highway engineers of the Hong Kong Government came down heavily against further vehicles of 18 or more tonnes on two axles. From this it will be appreciated that there was no prospect of further dispensation being forthcoming from the authorities.

During 1979 legislation in the Colony was revised to bring it into line with that in the UK and the then EEC, permitting buses on three-axles to be built up to 12-metres in length and 24-tonnes gross vehicle weight and this was the direction that the Hong Kong Government Transport Department (HKGTD) encouraged both CMB and KMB to take. CMB has shown interest in even larger buses as early as 1978.

The Initial Concept and Government Cooperation.

In view of the concern being expressed by both lay and informed commentators, over the possibility of the design tyre loadings being exceeded on the CMB 11.45 metre, MB-class, Metrobuses, particularly on the front wheels, the HKGTD adopted an active role in the design of future buses which, it was envisaged, would have a carrying capacity larger than any previous designs but within contemporary technology and acceptable axle-loadings.

Fodens Ltd.

During 1979/80, a number of chassis and body manufacturers were approached and, as a result, agreement to proceed came initially from an unexpected quarter when Fodens, the world-famous multi-axle truck manufacturers, indicated their willingness to construct a three-axle bus chassis which also incorporated the much expressed preference of Hong Kong bus operators, a front engine. Even at this early stage, Kowloon Motor Bus was anxious to advance the concept and placed an order for six prototype chassis 'off the drawing-board'; of course, these were to have been from Fodens. Thus, CMB, despite having initiated the concept, were not the leaders when it came to placing orders but it was, nevertheless, a joint effort by CMB, KMB, HKGTD and Fodens that led to a feasible design being presented during 1980.

Fodens were anxious in 1979 to develop their bus manufacturing interests and

Had the Foden Twin-Steer bus project reached production, this montage shows what it might have looked like with Alexander bodywork

Hong Kong offered an unusual market place for large double-deck buses and, as it was known that both KMB and CMB were keen to lessen dependency on a limited number of manufacturers, Fodens found favour with all parties.

Having already prepared a front engined design, similar in some respects to that of the Leyland Victory and Dennis Jubilant, Fodens found themselves too late to develop that particular market and decided to adapt what they had dubbed their 'BG' model into a 12-metre, 3-axle design as demanded by Hong Kong. Fodens were thus nailing their colours to the mast of a front-engined design, to be powered by either the then new Gardner 6LXC or Rolls-Royce Eagle 220, and, in view of the great combined front-end weight, of the engine

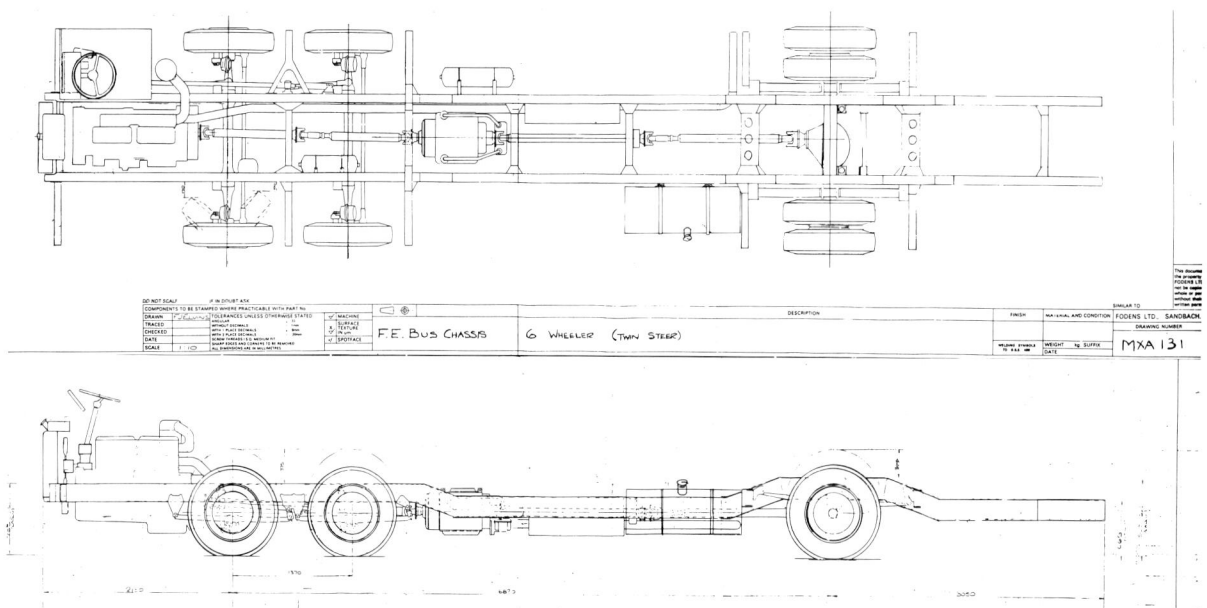

ABOVE: Drawing reproduced from the original Foden works drawing of their proposed twin-steer, three-axle bus chassis with front engine. (*Fodens Ltd.*)

and high passenger loadings, the Foden proposal was to have had twin-steering front axles in order to present the lightest possible front-tyre stresses and loadings. Fodens achievements in the field of multi-axle steering systems need no amplification here so, perhaps they would have been the ideal manufacturer for such a design.

It is believed that Fodens had approached body manufacturer Walter Alexander with a view to cooperation form the earliest possible stage.

As a preliminary, and necessary, precaution, KMB, in conjunction with the Police, HKGTD and others, undertook clearance tests on some very tight corners and at bus stations, using a specially modified 34ft long AEC Regent Mk V double-deck bus which had been fitted with a skeletal extension ahead of the front axle in order to simulate a 12-metre long vehicle. After the driver had accustomed himself to both the great length of the vehicle and the artificial driving position, the test was deemed to have been a success and very shortly afterwards came the already mentioned announcement that KMB was negotiating for six front-engined, twin steering, three-axle, 12-metre prototype double-deck buses.

And that was where the matter might have finished!

Fodens went into receivership and their eventual purchaser, Paccar Cpn, of the USA, declined to continue bus building operations and withdrew from the Hong Kong negotiations.

The Alternatives.

In the vacuum which followed Foden's withdrawal, other manufacturers who had been approached in the earliest days, showed renewed interest but all, except Dennis, with rear-engined, single front axle concepts.

Hestair-Dennis.

By early 1980, Hestair-Dennis, recently established in the Hong Kong bus market with their front-engined Jubilant, were also prepared to cooperate in the design of 3-axle, 12-metre double-deck buses, expressing willingness to produce both front and/or rear engined models, the former a lengthened Jubilant with Foden-like twin steering. At about this time, CMB and KMB had become aware to some of the disadvantages related to the then current generation of front-engined models, the Jubilant, Victory and Ailsa, mainly as a result of high entrance steps and saloon floor levels, together with wasted space alongside the engine casing. Drivers who had become accustomed to the relative coolness of their working environment in the rear-engined types, were less than enthusiastic about having the heat of the engine reintroduced alongside them. Additionally the front engine contributed to less than acceptable steering characteristics. Nevertheless, over a thousand were sold to KMB and CMB. The intrusion of the wheelboxes into the saloon of a twin-steering bus would have exacerbated the loss of passenger space alongside the engine where no seats or standing was permitted on the Jubilant/Victory/Ailsa. It was thus that Dennis settled upon the successful Dennis rear-engined 'Dominator' model, to be known by KMB as the 'Dragon' and by CMB as the 'Condor'. Dennis refer to the type generically as the 'Dragon'

The Dragon chassis featured a second, or trailing, axle immediately ahead of the driven rear axle, although this is not correctly described as an axle as the wheels were attached to two independently sprung stub-axles which had no steering capacity as on some other similar designs. This centremost 'axle' was located 1.6 metres ahead of the rearmost axle and supported sufficient additional weight so that, with 12.00x20 tyres, front tyre loadings were reduced to an acceptable level. The driven, or rearmost, axle was provided with double-wheels and tyres in the conventional manner, while the centremost trailing 'axle' had only single tyres.

Orders were placed for Dennis Dragon/Condor prototypes during 1980, three for KMB and two for CMB. The three for

BELOW: The first Dennis Dragon 3-axle, 12-metre bus for KMB, 3N1, after being adorned in an advertising livery. (*Mike Davis*

BELOW: Dennis Dragon badge carried by all three-axle derivatives of the Dennis Dominator except those for the China Motor Bus Company which are badged 'Condor' although essentially similar. (*Nigel Eadon-Clarke*

KMB and one for CMB were specified to have Alexander 'R-type' bodywork, while the remaining CMB chassis was to have Hestair-Duple(MetSec)bodywork, similar in many ways to that fitted on later KMB Jubilant chassis and on all subsequent KMB Dragon chassis up to 1991.

Ailsa-Volvo.

The products of this Scottish/Swedish manufacturer were already represented in Hong Kong, albeit in small numbers, by the Ailsas of CMB, their AV1-8, and, being satisfied with them in most respects CMB alone responded and ordered two

Metro-Cammell Weymann

Having initially shown its intention to purchase six Foden three-axle buses, KMB had only ordered one alternative bus type, ie the Dennis Dragons. Hitherto, Metro-Cammell Weymann (MCW), which had pioneered the large bus concept in conjunction with CMB, to produce the 11.45 metre MB-class Metrobus, had not ventured into yet another field of extended buses. However reluctant they may have felt in committing themselves to this new generation, commit themselves they did, to their eventual profit. In order to keep CMB's market and to break into KMB's, 12-metres it had

generation buses. The typical asymmetrical windscreen layout of earlier Metrobuses was omitted.

While CMB opted for a maximum capacity configuration, KMB chose a more fully seated layout with less standees, giving a total capacity of 160, of which only 36 were not provided with a seat.

Choice of engine was a point of difference between KMB and CMB, where the

ABOVE: The Ailsa/Volvo, front-engined, 12-metre bus that KMB did not take. This one is seen in service in Jakarta. *(John Shearman)*

former specified Rolls-Royce Eagle 230L's and the latter the new Gasrdner 6LXCT. KMB later replaced its Rolls-Royce engines with Gardner units.

For the first time an MCW Metrobus was supplied in ckd kit form for local assembly, in this case by KMB engineering Ltd., at their Tuen Mun workshops. This was to become the last into service of KMB's order for

ABOVE: The MCW Metrobus at the official handover. *(Mike Davis*

to be. Once embarked on the project, MCW pursued their goal relentlessly, from behind Dennis and Volvo by more than four months. KMB ordered three and CMB two. This completed KMB's original requirement for six 3-axle, 12-metre buses. MCW's late start notwithstanding, the first two 'Super' Metrobuses arrived in May 1981, ahead of their competitors by seven months.

The MCW product was, like the Dennis, a rear-engined design with the rearmost axle being driven and equipped with double tyres, while the centremost axle was undriven and fitted with single wheels and tyres but, unlike the Dennis, the trailing axle was a solid beam and did not feature any steering facility.

MCW fitted their own make of bodywork, in all five cases to its then new 'Apollo' design, which in many ways superficially resembled the Alexander 'R-type', a feature which helped to promote a similar 'type-image' to the fleet of new

three and was fleet numbered M3 - later changed, together with the other two, M1 & 2,, to 3M3, with the prefix '3' indicating a 3-axle bus, a feature added to all KMB' prototype 12-metre buses retrospectively.

It was to MCW that the first production order for new 12-metre buses went when CMB placed an order for thirty to be delivered in late 1982, early 1983, but that is not part of the story of the development of the type.

Leyland Vehicles Limited.

The story so far shows that three manufacturers had provided a combined (KMB/CMB) total of twelve prototype 3-axle chassis, namely MCW, British-Volvo and Hestair-Dennis, in order of delivery, and these were provided variously with bodies by MCW, Alexander and Duple-MetSec.

Notable by their absence from the preceding account are British Leyland, or, more correctly Leyland Vehicles Ltd, the bus and truck manufacturing arm of the BL empire. At the time that the initial offers to participate in the 12-metre project were made, Leyland was hoping to develop a market in Hong Kong for its new Olympian and, indeed, secured orders for five prototypes. BL were also reluctant to develop another new concept, following the less than expected demand for the Titan and early uncertainty over the success of the Olympian itself, which in the event, proved groundless. As a result, Leyland felt unable to participate at the very time that orders for prototypes were being placed but, almost a

BELOW: KMB's Leyland Olympian, 3BL1, with ECW bodywork, was the only one of its kind to enter sevice as a bus in Hong Kong. It is seen here at ECW's Lowestoft works prior to shipping. It was repainted cream with red relief in Hong Kong. *(Brian Ollington*

year behind, took the bold step of speculatively developing a three-axle version of the Olympian and shipping it unsold to Hong Kong as a demonstrator. CMB rejected it on the grounds that it had a Leyland engine, a type that they were having particular difficulty with at the time, and KMB took the vehicle on three months trial from 8th April to 8th July 1982, at which time KMB purchased the bus. Mechanical layout was similar to the other two rear-engined, 3-axle prototypes, in that the rear axle was that which was driven and had twin rear wheels, while the centre axle had single tyres. In the case of the Olympian, the centremost axle had a self-tracking device, similar to that on the Ailsa, where the wheels turn on king-pins to follow the curve that the bus was following. This effectively reduced tyre-scrub on routes with tight corners. The engine fitted when delivered was the Leyland TL11, driving a Leyland 4-speed Hydracyclic gearbox fitted with LVA45 control system.

Bodywork was by BL's Eastern Coachworks (ECW) factory at Lowestoft and was of similar design to that fitted to the prototype two-axle Olympians already operating in Hong Kong and based on the manufacturer's standard design. At 4.16m (13ft 8in) high, it was the only lowheight version of the 12-metre genre and, as such, it great length was accentuated, adding to the impression of enormity when most buses were of only 10 or 10.4m (33 or 34ft) in length.

Three doorways were provided, as on CMB's second Metrobus, ML2, but the centre and rear doorways were somewhat wider than on the Metrobus. The forward ascending staircase had its foot opposite the centre of the three doorways. Seats were provided for 104 passengers, plus 53 standees.

The Outcome.

As has been seen, CMB placed an order early in 1982 for a production run of thirty MCW 'Super' Metrobuses and followed this with further examples, later turning to Dennis 'Condors' for the 1990's. KMB, on the other hand, opted for the Dennis (Dragon) option first, followed by Leyland Olympians with Alexander bodies, and by the early 1990's had many hundreds of three-axle buses of both 12-metre length and the slightly shorter 11-metre later found more acceptable on many routes where clearances were too tight for the longer versions. The 1990's also saw the widespread introduction of air-conditioning to double-deck buses in Hong Kong The three-axle bus had become established and increasing numbers would offer air-conditioned comfort into the next century.

The progress of the air-conditioned double-decker is described later in Section Three of this chapter.

Historical Note.

The development of the three-axle bus as outlined here does not mean that the type is new on the China coast - far form it.

Prior to the second World War, there was a fleet of three-axle Tilling-Stevens and AEC 'Renown' double-deck buses operating in the fleet of the 'China General Omnibus Company' of Shanghai, prior to the destruction wrought upon the region by the forces of Imperial Japan. The ultimate fate of these buses can only be imagined, but it is almost certain that none reappeared in 1945.

LEFT: At the time of writing (November 1994) the air-conditioned Olympian and Dragon chassis represented the ultimate in urban bus operation. Here 12m Leyland Olympian 3AV41 in 1994 *(Mike Davis*

LEFT: After thirty years, the Hong Kong Mercedes-Benz dealers, Zung Fu Motors, finally succeeded in obtaining a toe-hold in 1983 but not in three-axle form. *(Mike Davis*

LEFT: The China General Omnibus Company operated 68-seater double-deck Tilling-Stevens TS15A's, with aluminium bodies like this, in Shanghai nearly 50-years before the three-axle bus was to make its debut in Hong Kong. They were later joined by 3-axle AEC Renown double-deckers also with all-aluminim bodies. The fresco painted along the between-decks panels presents an illustration of the buildings along the famous waterfront thoroughfare known as the 'The Bund' *(John Shearman collection, taken from the June 1937 issue of 'Transportation', an Australian magazine.*

Part Five:
Double-deck buses from 1980.
Section B:
Prototype buses for evaluation.

Chassis type:	Class:	Page:
Leyland Olympian B45 9.5m - 2-axle	BL1-3	210
MCW Metrobus 12m - 3-axle	3M1-3	213
Dennis Dragon 12m - 3-axle	3N1-3	215
Leyland Olympian 12 - 3-axle	3BL1	218
Mercedes-Benz 0.305 11m - 2-axle	ME1	220
Volvo B10MD (Citybus*) 10m -2-axle	VMD1	222

*Citybus name not used by KMB.

BELOW: The first of the prototype MCW Metrobuses for KMB, later 3M1, is swung ashore at Kwai Chung Container Port, from the container-ship that carried it from Europe. Many, but not all completely built bused are transported on roll-on roll-off car-carriers which ply between Japan and Europe with new cars and often return lightly loaded. On other occasions, buses are carried as deck or hold cargo, as in this case. *(John Shearman*

Leyland Olympian Model B45 or ON.   BL1-3

During the closing months of 1980, KMB completed arrangements to purchase, for evaluation purposes, three examples of the then new Leyland double-decker, named 'Olympian' by the time the vehicles were delivered. The first two had prototype Model B45/TL11/2R chassis while the third was a similar ONTL11/2R, the third in the ON series, chassis number ON3, which was actually registered first and become BR1.

Unlike CMB, who specified one 9.5 metre and one 10.5 metre version, all three for Kowloon Motor Bus were of the longer type, with 5.6 metre wheelbase. Mechanical specification of the KMB vehicles also differed from those for China Motor Bus in that the standard Leyland turbocharged TL11 engine of 11.1 litres was installed, together with Leyland's own GB350 fully-automatic gearbox.

Standard outline Eastern Coachworks (ECW) low-height (4.16m - 13ft 8in) bodywork was adapted for local climatic conditions by the provision of full-depth sliding windows along each side and in the upper-deck front. (CMB specified three-quarter depth sliders each side and push-out hopper vents at the front.) Windscreens were of flat glass arranged as two panes in a very shallow 'V' shape in place of the curved ECW standard type in order to allow for opening vents at the bottom on the driver's side and at the top on the platform side.

The usual narrow entrance was provided, as was a double-width forward-of-centre exit, spaced about 330mm behind the nearside front wheel-arch, while the staircase was of the forward ascending type over the offside front wheel-arch, its foot being opposite the exit doorway.

Seating was arranged in the usual Hong Kong fashion with 3+2 seats wherever possible and bench seats over the wheel-arches. When introduced, the actual arrangement was LH61/38+14FE,FdXS,D,DOO. In June 1987, BL3 returned to service from overhaul with its standing capacity reduced to 7 but without the allocation of a new sub-type. BL1 & BL2 were also nominally changed on paper as they remained out of service for overhaul until October 1987 when they returned to service with their standing capacity differently reduced, in this case to only four. As BL3 was still running with seven standees, BL1 & BL2 were allocated the (A) suffix to their type-code, to be followed by BL3 in March 1988 when it too was retrospectively altered to a nominal four standees. How this was to be enforced was not apparent!

When introduced, the rear panels of these B45 Olympians were protected by 'Eon' hollow rubber springs to a heavy-duty specification.

During the period of overhaul mentioned above, all three buses were re-powered by substitution of their Leyland engines by Gardner 6LXB units which were then the fleet standard engine.

VEHICLE SPECIFICATION:	Leyland Olympian 10.3m - B45.
Fleet numbers:	BL1-3
Chassis make:	British Leyland Motor Corporation
Chassis Model:	Olympian B45/TL11/R2 (BR2 & 3) or ONTL11/2R (BR1)
Engine: as built:	Leyland TL11, turbocharged, 11.1 litre.
from 1987/8:	Gardner 6LXB, 10450cc
Transmission:	Leyland Hydracyclic, incorporating a retarder.
Body layout: Body make:	Eastern Coachworks (ECW).
as built:	LH61/38+14FE,FdS,DOO (BL3 had 7 standees 10/87-3/88
from 1987/88:	LH61/38+4 from 10/87 (BL1&2) or 3/88 (BL3).
Date introduced:	1981
Length:	10280mm
Width:	2452mm
Height:	4170mm
Wheelbase:	5400mm
Unladen weight:	10,460kg
Gross vehicle weight:	16,257kg
Total:	3.

BELOW: KMB's first three Leyland Olympians were the 10.3-metre version with classic Eastern Coachworks bodywork, modified for hot weather conditions by the addition of full-depth sliding windows. In this view, BL 1 turns under the Canal Road flyover in late 1982. *(Mike Davis)*

ABOVE: BL3 turns into King's Road, North Point, surrounded by a phalanx of CMB buses. The only outward difference between BL3 and its two sisters, BL1 & BL2, would appear to be the location of the Leyland nameplate. *(Mike Davis*

BELOW: BL1 from the offside, showing the forward location of the staircase as it heads along Causeway Road heading for Causeway Bay and the Cross-Harbour Tunnel whilst on Route 112 in December 1982. *(Mike Davis)*

ABOVE: During 1983/4, BL1-3 were repainted into the mainly cream '1982' livery that had been introduced at about the time that the trio were new. *(John May)*

BELOW: In their latter years, the BL-class were fitted with replacement front dash panels and windscreens similar to early Alexander bodied Olympians. Here BL2 stands at Kowloon Station in the company of Metrobus M2 whilst working the 69X to Tin Shui Estate near Yuen Long. *(Dennis Dao)*

MCW Metrobus; 12-metre - 3-axle   3M1-3

KMB had originally intended to purchase six of an ill-fated Foden twin-steer, 12-metre bus chassis for which body designs had been prepared by both Alexander and Metal Sections. Following Foden's fall into the hands of the receiver and subsequent sale to Paccar, other manufacturers expressed their readiness to build a bus to the newly permitted size but, of these, Metro-Cammell Weymann were the last to conclude an order for prototypes. In the event, however, they were the first to deliver by over six-months.

So it was that, in May 1981, KMB, along with CMB, received the first of its order for three 12-metre, three-axle MCW Metrobuses, or 'Super Metrobuses' as they were known in some quarters at the time. Official handing-over took place at Queen's Pier at a public ceremony when Directors of MCW were present as well as the Chairmen, some Directors and senior management of both customers.

KMB initially allocated its first 12-metre bus, which was originally numbered M1, to trunk cross-harbour Route 112. During 1982, following the arrival of the 12-metre Dennis Dragons (qv) all KMB's 3-axle buses were given the additional prefix '3' to indicate 3-axle, so that M1 became 3M1 (likewise M2 & 3).

The MCW chassis featured the style of three-axle concept where the driven axle has traditional double-wheels, while those on the idler, or 'tag' axle are single. The driven axle, in the case of the Metrobus remains that which is rearmost and is located closest to the rear-mounted engine. The Trailing 'tag' axle is a solid beam with no capacity for steering as on some other makers products.

The engine chosen to power KMB's three trial Metrobuses was initially the 230bhp Rolls-Royce 'Eagle 230L', a turbocharged diesel mounted transversely and vertically at the rear. The Hong Kong preference for Gardner engines eventually prevailed and, when the Rolls-Royce unit fitted in 3M1 failed it was replaced by the standard engine then fitted in KMB's other 3-axle buses, namely the Gardner 6LXCT, then rated at 210bhp. Transmission to the GKN hub-reduction rear axle with either engine was by way of a Voith/DIWA D831, three-speed, fully-automatic gearbox incorporating torque converter and integral retarder.

Bodywork was also by MCW to their MkII or 'Apollo' type which departed from earlier MCW body designs in a number of ways, the most noticeable being the absence of the asymmetrical windscreens and the upper-deck front windows which extend upwards to within a few centimetres of the roof line. Entry is in the usual Hong Kong manner, through a narrow front doorway beside the driver and farebox, while egress is by way of a wide, centrally placed doorway opposite the foot of the centrally located, forward ascending, staircase.

The second, similar, 3-axle Metrobus, like the first, was shipped fully built up and arrived almost three months after M1, taking the fleet number M2 and being registered on 11th August 1981. Like its forerunner, this bus was powered by a Rolls-Royce Eagle on arrival but the eventual failure of the unit saw it also replaced by a Gardner 6LXCT.

The third vehicle of the type, M3, was to be the most interesting in that it was sent out to Hong Kong with its body as a ckd kit for local assembly; the first MCW excursion into the ckd field. It was assembled by KMB engineers under the supervision of a specialist team from MCW. Mechanically M3 was delivered to a similar specification to M1 & M2.

Body layout on all these three 12-metre Metrobuses was fully seated H77/47+36FE,CXS,D,DOO, whereas CMB opted for large standee areas. The upper-deck capacity was reduced to 76 in August 1986.

VEHICLE SPECIFICATION: MCW Metrobus 12m trial buses

Fleet numbers :	3M1-3 by mid-1982 (M1-3 on introduction)
Chassis:	MCW Metrobus 12MDR116/1
Engine:	Rolls-Royce Eagle 230L, 230bhp; replaced by Gardner 6LXCT in 1983/4.
Gearbox:	Voith/DIWA D851, fully-automatic, 3-speed.
Rear axle:	GKN hub reduction.
Body make:	MCW cbu, except 3M3 which was assembled by KMB.
Body layout:	H77/47+36FE,CXS,D,DOO reduced to H76/47+36 in 8/86.
Date introduced:	1981
Total:	3
Length:	11990mm 39ft 4½in
Width:	2500mm 8ft 2½in
Height:	4430mm 14ft 6in
Wheelbase:	5650mm+1600mm 18ft6in + 5ft 5½in
Unladen weight:	13,152kg
Gross vehicle weight:	24,000kg, later 21,800kg

BELOW: Sporting an all-over advertising livery for a brand of tea, KMB's first Metrobus, 3M1 (originally M1) in Canal Road East in September 1982. By this time it displayed a 'Gardner Turbo' plate on the front, replacing the original 'Rolls Royce Diesel' plate. *(Mike Davis)*

ABOVE: The body of 3M3, the third KMB Metrobus was built-up from a ckd kit in KMB's workshops, under the eye of an MCW engineer. The bus retained its Rolls Royce 'Eagle' engine in September 1982 when photographed at the same place as 3M1, above. *(Mike Davis*

BELOW: 3M3 in 1994 at Mei Foo running on Route 6, a route on which it is not usually deployed. *(Newton Ng*

Dennis Dragon 12-metre, 3-axle.   3N1-3.

KMB's first order for 12-metre buses, following the demise of the Foden project, was for three Dennis Dragon chassis which were a development of the two-axle Dominator. The reason for the specialised name for the three-axle variant was that MCW, Dennis' arch rival in the 12-metre bus stakes, had launched its 12-metre Metrobus with a Rolls-Royce 'Eagle' engine and adopted the slogan 'the Eagle has landed' for the launch. It was then decided that an ornithological theme should be pursued and the 3-axle Dennis for CMB was named the 'Condor'. Being fierce competitors, Dennis felt that to supply KMB with a chassis specifically named for CMB would result in considerable loss of oriental 'face' and so it was that the KMB version was renamed 'Dragon', picking-up the second character of KMB's Chinese name which means Nine Dragons or 'Kau Leung' (Kowloon, when transliterated into the Hong Kong English form). Although ordered before the MCW product, the Dennis chassis did not enter service until after the former by almost ten months.

The Dennis chassis, like that of the MCW, featured a non-steering rear-axle arrangement where the drive axle, nearest the power pack, was the rearmost. The trailing, or centremost, wheels had independently sprung stub-axles mounted on swing arms. By comparison with the MCW Metrobus, the Dennis chassis had a slightly shorter wheelbase, resulting in a longer rear overhang with consequent wide rear out-swing.

Power was supplied by a non-turbocharged Gardner 6LXC engine of 205bhp, driving a Voith D854, four-speed, fully-automatic gearbox incorporating a retarder. The rear axle was of the GKN hub reduction type.

Bodywork was the Alexander RX-type, derived from the R-type and featuring a single-width entrance, double-width centre-exit and tropical-style sliding windows which had three-quarter depth sliding sections with a fixed glass above. Because the upper front windows extended up to almost roof level the fixed portion was slightly deeper than along the sides. The staircase and exit door were set back to a position just ahead of the rear wheelbox, while the front wheels were closer to the front of the vehicle than was usual with rear-engined buses in Hong Kong and, in this respect, the type resembled the Dennis Jubilant front engined bus.

When introduced, the 4360mm (14ft 4in) high body on 3N2 & 3N3 had a layout of H73/55+24FE,CXS,D with farebox for DOO, while 3N1 had one less upper-deck seat and, as a result, was classified by KMB as 'Dragon(A)' - the other two being merely 'Dragon 12-metre'. The extra seat on the upper-deck of 3N2/3 was the result of fitting a staircase with its front to back dimensions reduced from the R-type standard. It will be seen that KMB chose conventional seating for its three-axle Dennis while, as with the Metrobus described earlier, CMB opted for a multi-standee arrangement on the lower-deck.

During August 1986, 3N1 had its seating layout revised to H72/52+27 to bring it into line with new Government regulations. Alteration of 3N2/3 followed in October 1986, when they became H73/52+27. As the one upper-deck seat differential was retained, 3N1 continued as Dragon(A) sub-type.

VEHICLE SPECIFICATION: Dennis Dragon trial buses.

Fleet numbers:	3N1-3N3.
Chassis:	Dennis Dragon 12-metre.
Engine:	Gardner 6LXC, 205bhp
Gearbox:	Voith D854, four-speed, fully automatic gearbox.
Body make:	Walter Alexander, built in Scotland.
Body layout: 3N1 as new:	H72/55+24FE,CXS,D,DOO Dragon(A)
3N1 8/86:	H72/52+27FE,CXS,D,DOO
3N2-3 as new:	H73/55+24FE,CXS,D,DOO Dragon 12M
3N2-3 10/86:	H73/52=27FE,CXS,D,DOO
Date introduced:	1982
Total:	3
Length:	11990mm 39ft 4½in
Width:	2500mm 8ft 2½in
Height:	4360mm 14ft 4in
Wheelbase:	5410mm+1600mm 17ft 9in+5ft 2½in
Unladen weight:	12,498kg
Gross vehicle weight:	22,908kg

BELOW: KMB Dennis 'Dragon', 3N1, advertises sunglasses as it descends from Canal Road Flyover in September sunshine in 1982. KMB's three Dragon prototypes each have Alexander two doorway bodywork. *(Mike Davis)*

ABOVE: The second Dragon, 3M2, also seen on Route 112, turning into Morrison Hill Road from Canal Road East in 1982, was delivered in KMB's mainly cream livery but the KMB initials were in a non-standard place, being further forward that usual and they were smaller. *(Mike Davis)*

LEFT: The tail of a Dragon! 3N2 leaves Marble Road, terminus of Route 112, headed for Lei Chen Uk, in Kowloon. *(Mike Davis)*

ABOVE: 3N3 entered service on Kowloon Urban Route 5 and remained there, being the only prototype three-axle bus not to regularly work on Route 112. Here 3N3 leaves Choi Hung for "Star" Ferry in March 1983. *(John Shearman*

BELOW: A more recent photo of 3N1, taken in 1994 at Kowloon City Ferry Pier whilst working on the 6C between there and Mei Foo. The replacement radiator grille is clearly seen in this view. The other two early Dennis Dragons, 3N2 & 3 were not similarly treated. *(Wan-Shing Mak*

Leyland Olympian 12-metre, 3-axle — 3BL1

Leyland were not in a position to offer their Olympian chassis in 3-axle, 12-metre form at the time that orders were placed for evaluatory prototypes. When circumstances changed, however, the Bristol factory built a single 3-axle chassis, sent it to Eastern Coachworks (ECW) for bodying and shipped the complete vehicle as a speculative demonstrator to Hong Kong where KMB agreed to operate it on loan for three-months from 8th April 1982, after which time they purchased it. It received the fleet number 3BL1 and registration CV184 while still on loan.

The chassis was virtually a stretched version of the standard 2-axle Olympian and was powered by the then new Leyland TL11, turbocharged engine, uprated to 210bhp and driving the rearmost axle by way of a Leyland (SCG) 5-speed, fully-automatic, hydra-cyclic gearbox, controlled by the LVA45 control system. In most respects, the chassis had standard Olympian proportions, with a long front overhang and standard engine-to-rear-axle dimensions, the extra length being between the front and rearmost axles. The centre axle was the first in Hong Kong to have a self tracking, or steering, facility, allowing the centremost axle of the rear wheels to follow the line of curvature being taken by the bus. The method of operation was said to follow the principle of the castor on a tea-trolley and was set to lock-up in a straight line when a speed of about 12km/h had been reached and when reverse gear was selected. The purpose of this steering function was to help reduce the effects of tyre-scrub when turning on a tight lock. The engine was cooled by the front mounted radiator, assisted by a thermostatically activated electric fan. Very heavy-duty front tyres, size 13.00x20, were fitted but those on the six rear wheels were 10.00x20, a combination felt best able to withstand the stresses of the high loading of such a large vehicle with a Gross Vehicle Weight of up to 24 tonnes. The fact that the centremost wheels followed the curve, effectively extends the wheelbase to 7030mm (23ft 8in), a length requiring great driving skill and careful route planning.

The ECW bodywork (body number EX19 in the experimental series) was unique amongst the prototype 12-metre buses in being, like KMB's 2-axle Olympians, low-height at 4.16m (13ft 8in). The front entrance was single width, as per standard local practice, but there were two separate, and slightly wider, exit doorways within the wheelbase. The foot of the forward ascending staircase was opposite the foremost of the two exits which were, like the front entrance, fitted with two-leaf, fully glazed, 'glider' type doors. This plethora of doorways made for some very non-standard pillar spacing and, thus, window sizes which was most noticeable on the upper-deck where the third, fourth and fifth bays were shorter than standard and the first was also shorter but of a different size. The second sixth and seventh windows were of ECW standard dimensions. The body layout of this prototype ECW Olympian was LH73/31+53FE,FdS,2CX. A standard Bell Punch (Control Systems) Farebox without ticket issuing machine was fitted.

It will be seen that this bus alone amongst the KMB three-axle prototypes had a multi-

VEHICLE SPECIFICATION:	3-axle Olympian Demonstrator.	
Registration number:	CV 184	
Fleet numbers:	3BL1	
Chassis:	Leyland Olympian ONTL11/3R	
Engine:	Leyland TL11 - turbocharged.	
Transmission:	Leyland 5-speed, fully-automatic Hydracyclic gearbox incorporating a retarder	
Body make:	Eastern Coachworks	
Body layout:	LH73/31+53FE,FdS,2CX,D,DOO,Farebox	
Date introduced:	1982	
Total:	1	
Length:	12,000mm	39ft 4½in
Width:	2500mm	8ft 2½in
Height:	4166mm	13ft 8in
Wheelbase:	5430+1600mm	17ft 9½in+5ft 2½in
Unladen weight:	12155kg	
Gross vehicle weight:	23,000kg	

BELOW: Eastern Coachworks applied their standard body style the three-axle version of the Leyland Olympian demonstration vehicle which KMB eventually purchased after trials. The pillar spacing means that no two adjacent windows are of the same width. 3BL1 was the first bus delivered new in the '1982' cream livery. *(Mike Davis*

ABOVE: By 1994, 3BL1 had acquired a replacement dash and windscreens of a pattern similar to some Alexander bodies and like its two-axle sisters, BL1-3.. *(Dennis Dao*

LEFT: The 'steering' middle axle helps 3BL1 around the sharp 90-degree turn out of Marble Road, North Point in 1982. *(Mike Davis*

BELOW: Rear view of 3BL1 seen in a queue of buses waiting to turn right. *(John Shearman*

standee lower-deck arrangement, reflecting that it had been hoped to demonstrate the bus initially to China Motor Bus who, although they favoured the standee concept, turned the Olympian down, reportedly, on the grounds of its having a Leyland TL11 engine as opposed to that Company's standard Gardner. In fact, the vehicle followed the specification of CMB's two-axle Olympians, BR1 & 2, in many respects particularly as regards trim and finish.

A second prototype (body EX20) was partly constructed at ECW and had a Gardner 6LXCT engine but this bus was also declined by CMB and the vehicle was reconstructed as a double-deck coach for Citybus Limited for its cross-border service from Hong Kong into China.

Mercedes-Benz 0.305 - Demonstrator.   ME1

When it arrived in Hong Kong in June 1983, this Mercedes-Benz demonstration bus caused quite a controversy in view of the franchise terms applying at that time to both bus companies which included a 'Commonwealth Preference' clause. To enable KMB to register any bus not of British or British Commonwealth manufacture, special sanction was demanded and, in this case, the sanction was hard-won at very high level indeed given that the terms apparently required special authority from The Governor-in-Council.

Having reached the shores of Hong Kong, the 'Benz', in local vernacular, was demonstrated to both franchised, and some non-franchised, bus operators before being handed-over to KMB on 1st August, after which it appeared on Route 101, alongside CMB's 3-axle Metrobuses.

First impression was of a very big bus to be on only two axles but it was actually, at 11m, about 450mm shorter than CMB's MB-class Metrobuses which were also on two axles. It was, like the MB's, in excess of the 16 tonnes limit set for a two-axle bus when a maximum seating arrangement was adopted, so a very spacious layout was adopted being H58/46+27FE,CXS. The lower-deck ceiling height was increased above the rear axle in order to retain the minimum 1.75m (5ft 9in) headroom, resulting in a corresponding height penalty at the rear of the upper-saloon where a strange 'blanking-off' structure replaced the last row of seats. The lower-deck seating was subsequently reduced by two and the standing capacity raised by the same number to compensate so that post-July 1986 the layout became H58/44+29.

VEHICLE SPECIFICATION:	Mercedes-Benz 0.305 Demonstrator
Registration number:	CZ6686
Fleet numbers :	ME 1
Chassis:	Mercedes-Benz 0.305
Engine:	MB OM407H - horizontal rear-underfloor mounted.
Gearbox:	MB fully-automatic gearbox with hydrocyclic retarder
Body make:	Walter Alexander - built in Scotland.
Body layout as new:	H58/46+27FE,CXS,D,DOO
from 7/86:	H58/44+29FE,CXS,D,DOO
Date introduced:	1983
Total:	1
Length:	11,000mm
Width:	2490mm
Height:	4478mm
Wheelbase:	5600mm
Unladen weight:	9,240kg
Gross vehicle weight:	16,257kg

The chassis was the well-proven Mercedes-Benz 0.305 which, with an MB OM407H, 6-cylinder, horizontal, rear-underfloor engine was, when introduced, unique in Hong Kong, although buses of similar specification were currently being supplied to Singapore and South Africa. Power developed by the derated engine was approximately 240bhp (179KW), transmission being via the MB fully-automatic gearbox which incorporated a brake-pedal actuated hydrodynamic retarder. The rear axle was a planetary-gear, hub-reduction type, allowing for a small differential size which, in turn, reduced the problem of saloon floor intrusion. In addition to the transmission retarder, three independent brake systems were installed; the service brake, the spring parking brake, incorporating the partial application 'bus-stop brake' which, acting on the rear-axle only and, thirdly, an exhaust brake which could be disconnected in slippery conditions, giving fuller control. Suspension was by airbags and was self-levelling, thus providing a very smooth ride.

The bodywork, by Walter Alexander was an adaptation of their 'R-type', similar to KMB's first three Dennis Dragons, 3N1-3. A major difference was the windscreen which was not recessed and fitted almost flush with the front panels and had narrow corner fillets between the

BELOW: During 1983, KMB took one demonstration Mercedes-Benz 0.305 double-deck bus for evaluation. Alexander R-type bodywork was fitted, similar in most respects to those supplied to Singapore on similar chassis. A striking livery of red flashes on the standard cream base was applied which served to highlight the length of this eleven metre, **two**-axle bus. *(Mike Davis*

ABOVE: This offside view serves to emphasise the length of the Mercedes-Benz 0.305 in 11-metre double-deck form. Seen here on demonstration to China Motor Bus, Citybus and TSW (HK) Ltd. in July 1983. *(Mike Davis)*

lateral extremities of the screens and the body-pillar, the original frame having been designed for UK-style curved windscreens, not flat glasses. The single headlights were mounted very low on the front dash panel, either side of the radiator grille, on which was featured the traditional Mercedes-Benz three-pointed star. The light units incorporated the side-lights and direction indicators, resulting in a very plain, but uncluttered, front aspect.

All side windows were fitted with sliding glass in the upper two-thirds only, thus removing the necessity for an internal safety rail. Passenger entry was by a narrow entrance with a two-leaf door and two shallow steps. The one exit had double sets of doors and also two shallow steps.

A striking livery was devised by a leading member of KMB's management team, which incorporated dramatic red flashes rising from the skirt to the roof and around the rear of the vehicle. The Singapore origins of the body design were evident in the provision of roof marker lights front and rear, a feature not required in Hong Kong.

On entering, service the Mercedes was allocated the registration CZ6686 and fleet numbered ME1 and, having proved successful, it was purchased by KMB and a further batch of forty were ordered.

ME1 suffered frontal accident damage during 1984 and when it returned to service it featured standard fittings such as round headlights and separate side and indicator light clusters. The three-pointed star was relocated above the grille.

LEFT: Following unfortunate frontal damage, ME 1 was rebuilt with with the result that it was rather plain-looking, with round headlights in place of the original recrangular type. Seen here leaving Statue Square in Central on Route 106 bound for Lai Chi Kok. *(Nigel Eadon-Clarke)*

Volvo B10MD VMD 1

One example of this underfloor mid-engined chassis was accepted by KMB on loan for evaluatory purposes in July 1984. Known elsewhere as the Volvo 'Citybus', the name was dropped in Hong Kong at the request of KMB to whom the private bus company known as Citybus Limited was an arch rival, at that time challenging the franchised operator's right to exclusivity on a certain route! As the chassis was derived from the Volvo B10M single-deck chassis the Hong Kong double-decker had a 'D' added to become B10MD - not to be confused with a later European single-deck B10MD!

Although sharing the drive-line and accessories of the basic B10M, the Citybus- for that was what it was - had many features of the Ailsa double-decker. Gone was the front mounted engine but the successful perimeter frame was retained; a ladder frame, more suitable for mounting the mid-underfloor engine, replaced the single spine of the Ailsa.

The chassis was powered by the turbocharged 9.6 litre Volvo THD100EC diesel engine derated from 245bhp to 210bhp, driving the rear-axle via a Voith/DIWA D851, three-speed, fully-automatic gearbox/retarder.

The body was, again, by Alexander to their R-type specification and was built up in Scotland, having a layout of H51/40+30FE,FdSX from which it will be seen that the exit was immediately behind the front wheel-arch while the staircase forward ascended over the off-side front wheel-arch. Full-depth sliding windows were provided along each side and the upper-deck front windows had their lower three-quarters sliding, with the upper part fixed. The flat driver's windscreens were in a shallow V and were recessed, unlike the Mercedes-Benz demonstrator where an awkward looking arrangement was adopted. Opening ventilators were provided in both sides of the windscreen with that on the driver's side being at the bottom while on the platform side it was at the top. Fleet-numbered VMD 1, the demonstration B10MD entered service on loan in July 1984 and was painted in a distinctive style, although somewhat less flamboyant than that of the Mercedes-Benz, ME 1.

KMB subsequently purchased the Volvo which suffered a fire whilst in the Cross-Harbour Tunnel and was written-off on 23rd February 1988. The almost intact chassis was damaged when the perimeter frame was cut away with the bodywork and the remains were sold for scrap to a New Territories dealer from whom it was subsequently purchased by Ranger Roadways Limited, a South London coach operator.

The chassis of VMD1 was shipped to the UK during October 1988, the intention being to rebody and operate it as a single-deck urban coach. Unfortunately this was not to be and the remains were sold to Barnsley Engineering and Transmissions for final disposal and scrap.

VEHICLE SPECIFICATION: Volvo B10MD (Citybus)	
Registration numbers:	DC4146
Fleet numbers :	VMD 1
Chassis:	Volvo B10MD ('Citybus').
Engine:	Volvo THD100EC, derated to 210bhp
Transmission:	Voith DIWA D851, 3-speed, fully-automatic.
Body make:	Walter Alexander, cbu
Body layout:	H51/40+30FE,FdS,DOO.
Date introduced:	1984
Total:	1
Length:	10,000mm (nominal)
Width:	2500mm
Height:	?
Wheelbase:	4,950mm
Unladen weight:	?
Gross vehicle weight:	?

BELOW: For comparative trials, KMB ran Volvo B10MD 'Citybus' demonstrator, DC4146, from July 1984. Seen here sweeping across the tram tracks in Wanchai, surrounded by CMB Metrobuses while on Tunnel Route 101 *(Roger Bailey)*

Part Five
Section C
Production Double-deckers buses from 1980.

1. 2-axle buses: | Class: | Page:
   - MCW Metrobus 9.7m — M-class — 224
   - Mercedes-Benz 0.305 11m — ME-class — 226
   - Leyland Olympian 9.5m — BL-class — 228
   - Dennis Dominator 9.5m — DM-class — 230

2. 3-axle buses:
   - Leyland Olympian 12m — 3BL-class — 232
   - Leyland Olympian 11m — S3BL-class — 236
   - MCW Metrobus 11m — S3M-class — 240
   - Dennis Dragon 12m — 3N-class — 242
   - Dennis Dragon 11m — S3N-class — 245

BELOW: After large orders for 12-metre, 3-axle double-deckers, KMB turned its attention to even bigger orders for 11-metre vehicles, assing a prefix letter 'S' in front of the fleet number. Here S3BL291 (DY9758) was arriving at Tai Wai bus station on Saturday 10th September 1994. *(Mike Davis)*

MCW Metrobus 9.7m

M-class

KMB's first major order for MCW Metrobuses was not, as might have been expected, for the 12-metre, tri-axle version of which three had been in service since 1981. Choice for 1983 delivery fell upon the short, 9.7 metre, two-axle, Model DR102 which was similar in overall size and mechanical specification to the twelve coach-seated versions delivered to China Motor Bus in 1978 and known by them as the MC-class. Similar in appearance and function to the MC's they were not. KMB's 1983 delivery, numbered M1-20, were the first two-axle Metrobuses to have the redesigned MCW 'Apollo' bodystyle, omitting the asymmetrical windscreens and carrying the upper-deck front windows up, almost to the roof line.

Mechanically the 9.7m Metrobuses were powered by the Gardner 6LXB engine, naturally aspirated and rated at 185bhp, driving a GKN hub-reduction, drop-centre rear axle via a Voith/DIWA D851, three-speed, fully-automatic gearbox incorporating a torque converter retarder. Exceptions included M58-60 and M81-88 which had Cummins L10 engines.

VEHICLE SPECIFICATION: 9.7m MCW Metrobus	
Fleet numbers :	M 1-88
Chassis:	MCW Metrobus 9.7 metre, Model DR102 or (M58-60) DR132
Engine:	Gardner 6LXB, 185bhp (M1-57, 61-80. or;
(B):	Cummins L10, 245bhp (M58-60)
(D):	Cummins L10, 280bhp (M81-88)
Transmission:	Voith /DIWA D851, fully-automatic gearbox/retarder
Body make:	MCW ckd - metrobus Mk II.
Body layout:	H58/34+18FE,FdXS,DOO or;
(A):	H58/34+20, or;
(B):	H58/34+19, or;
(C):	H58/34+9, or;
(D):	H58/34+3
Date introduced:	1983-5; (D): 1989.
Total:	88
Length:	9.7m
Width:	2.5m
Height:	4.37m
Wheelbase:	4.95m
Unladen weight:	10,211kg (Gardner)
Gross vehicle weight:	16,257kg

1982 order:	20 x Model DR102/34
1983 order:	10 x Model DR102/41
1984 order:	27 x Model DR102/46
plus:	3 x Model DR132/7 (B)
1985 order:	20 x Model DR102/49
1988 order:	8 x Model DR102/013

Sub-classes.

In line with KMB's usual classification system, the first delivery, M1-20, had no sub-classification, being called merely Metrobus 9.7m. Classifications reflect the passenger carrying capacity which in this case was H58/34+18FE,FdXS when new. The staircase was situated above the offside front wheel-arch and forward ascended with the foot opposite the forward-of-centre exit doorway.

Metrobus 9.7(A) - M21-57, 61-80.

A further thirty similar 2-axle Metrobuses were ordered in May 1984 for entry into service in 1985, followed by another twenty for introduction in October of the same year. This second batch took the classification Metrobus 9.7(A) as they were authorised to carry an additional two standees. Of these fifty buses, three, were fitted with Cummins engines *(see type (B) below)*.

Metrobus 9.7(B) - M58-60.

The three chassis missing from the (A) type were ordered to have Cummins L10 engines rated at 245bhp for service evaluation, to the benefit of both KMB and the engine manufacturer. Once again Hong Kong's demanding operating environment were seen as a testing and unusual proving ground for new products. Transmission remained as before. Only 19 standees were permitted - one less than the (A) type.

BELOW: KMB's first short (9.7m) Metrobus, M 1, in Morrison Hill Road, Happy Valley, a very few days after entering service in June 1983. *(Mike Davis*

Metrobus 9.7(C) - M21.

During July 1987, M21 was modified to sub-type (C) when its standing capacity was reduced to only 9.

Metrobus (D) M81-88.

These otherwise standard 9.7m, 2-axle, MCW Metrobuses were specially ordered with the 280bhp version of the Cummins L10 engine, as fitted to the larger 3-axle buses. The purpose of this was to allow M81-88 to work KMB's most difficult route, hitherto single-deck operated, namely the 51 which passes over the ridge of hills that separate the Tsuen Wan area from the northern part of the rural New Territories and which is surmounted by Mount Tai Mo Shan, a peak of some 915 metres (over 3,000ft). Although Route 51 does not actually reach the peak, it does climb over the shoulder of the mountain via a pass not far short of 760 metres (2,500ft) above sea level. Known now as Route Twisk[1], this steep and sinuous former military road was built by The Royal Engineers to connect the town of Tsuen Wan with the British Army encampment and RAF station at Sek Kong. Until the introduction of these Metrobus double-deckers, the route had been worked in the author's memory, firstly by Albion Victor VT17L's, followed by new VT23L's in 1963. These were replaced in 1976 by Union Auto bodied Viking VK55L's themselves remaining the mainstay of the service until the arrival of the Metrobuses - truly thirteen years hard labour.

M81-88 were known as Metrobus (D) and the body layout on introduction was H58/34+3FE,CX,FdS,DOO, reflecting the limited standing considered appropriate on a route with such a long climb and descent over TW/SK (or Twisk). The allocation is for eight buses so these are dedicated to the 51.

Additional vehicles were provided for the 51 by upgrading earlier M-class buses.

NOTE: [1.] This route was known to the military as 'TW/SK' (the initials of Tsuen Wan to Sek Kong) but was universally spoken as a word, and pronounced *Twisk*. Since the hand-over of the road, the written word has become spelt Twisk; the 'i' in the middle being a recent civilian adoption now seen on sign posts.

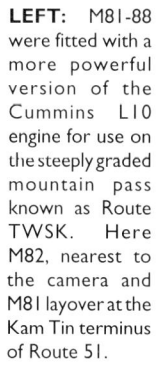

LEFT: M81-88 were fitted with a more powerful version of the Cummins L10 engine for use on the steeply graded mountain pass known as Route TWSK. Here M82, nearest to the camera and M81 layover at the Kam Tin terminus of Route 51.

BELOW: Another member of the class, M 9, in Des Voeux Road Central later in 1983. These vehicles entered service from new with the 'Jubilee' insignia. *(Mike Davis)*

Mercedes-Benz 0.305 11m ME2-41

Following the lifting of the 'Imperial Preference' restrictions, KMB ordered forty Mercedes-Benz Model 0.305 double-deckers from the West German manufacturer, Daimler-Benz of Mannheim.

The space-frame type chassis is powered by the Mercedes 11.4 litre, 6-cylinder OM407H engine, rated at 177kW and mounted longitudinally beneath the floor at the rear. An MB fully automatic gearbox drives the rear axle and incorporates a hydracyclic retarder. The performance characteristics were described as being generally similar to the first 0.305, ME 1, used for evaluation and subsequently taken into stock.

Having chosen a non-British chassis manufacturer, KMB specified Alexander R-type bodies for local assembly at its Tuen Mun assembly works. The doorway layout is similar to the trial bus, ME 1, as is the seating but the permitted number of standees is reduced from 29 to 21, giving a layout of H58/44+21FE,CXS,DOO. Detail differences can be observed to distinguish ME 1 from the main batch of 40; most noticeable being the deeper V-shaped windscreen, shallower dash and larger destination screens - the KMB standard display size, however remains as small as usual.

The striking red flourishes of the prototype livery have, surprisingly, been perpetuated, even through repaints. Large black fenders are fitted and the grille/headlight area is also finished in black. Round headlights were fitted from new.

In view of their power and comfort, the ME-class is used extensively on the Tuen Mun Road (incorrectly referred to locally as Tuen Mun Highway) on Route 68 between the MTR railhead at Tsuen Wan and the new town of Tuen Mun.

In line with their policy regarding double-deck 0.305's, Mercedes have not required either the Singapore or Kowloon buses to retain their distinctive front scuttle.

VEHICLE SPECIFICATION: Mercedes-Benz 0.305.

Fleet numbers:	ME2-41.
Chassis:	Mercedes-Benz 0.305, space-frame.
Engine:	MB OM407H - horizontal rear underfloor.
Transmission:	MB fully-automatic gearbox with hydracyclic retarder
Body make:	Alexander, ckd
Body layout:	H58/44+21FE,CXS,DOO.
Date introduced:	1985.
Total:	40.
Length:	11m
Width:	2.5m
Height:	4.8m
Wheelbase:	6.5m
Unladen weight:	9,240kg.
Gross vehicle weight:	16,257kg.

BELOW: Production Mercedes-Benz, ME26, stands at Yuen Long (East) bus and tram terminus on 26th September 1988, prior to returning to Jordan Road Ferry and Canton Road in Kowloon, on Route 68, a journey of 41.8km (26.1 miles). The larger destination screen and deeper windscreen distinguish these buses, ME2-41, from their progenitor, ME1. *(Mike Davis)*

ABOVE: By coincidence, the author photographed ME26 again in 1994, this time leaving Jordan Road terminus, now known as Jordan Road (Canton Road) as the erstwhile ferry has ceased to operate and the site is now, due to reclaimation, inland. By 1994 the strikingly distinctive livery, only ever applied to the Mercedes ME-class, had been replaced either by the standard style or advertising livery. *(Mike Davis)*

BELOW: By September 1994, many of the the once prestigious ME-class had been painted in advertising liveries of one kind or another and few which suited them. Here ME33 advertised a product known as 'Barbeque Paradise' with somewhat faded green lower panels representing grass. *(Mike Davis)*

Leyland Olympian 9.5m BL4-123

BL4-53.

As is usual with KMB bus types, the initial batch receive no sub-type classification, even after the delivery of subsequent variants. Although, as far as KMB are concerned, BL4-53 are all similar, there are two distinct body sub-types; BL4-23 and BL24-53.

As part of their 1982 order for two-axle buses, KMB introduced, in 1983, twenty short - 9.5m - Leyland Olympians. The choice of chassis length was understandable in view of the three trial buses, BL1-3, being 10.5m in length, with a total capacity, including standees, only two more than on these 9.5m buses.

Mechanically, BL4 onwards differ from the three B45's, BL1-3, in having Gardner engines from new, with Voith/DIWA D851 three-speed, fully-automatic gearboxes.

Bodywork is by Walter Alexander and was supplied in ckd form for local assembly, being of that manufacturer's R-type, incorporating a front entrance and forward exit, with the foot of the forward ascending stairs opposite the exit doorway. The staircase rises over the offside front wheel-arch. Seating and layout was originally H57/34+20FE,FdSX,DOO but this was altered to H57/32+22 in April 1986 in accordance with Government requirements.

BL4-23 had a very angular black grille covering the slots in the front dash that serve as air intakes for the front-mounted radiator.

The side windows on both decks have only about $^2/_3$ of their depth as opening segments, with a lower fixed portion to obviate the necessity of providing internal safety rails. The reduced air-flow was unpopular with passengers.

BL24-53 differ from BL4-23 in body detail only, having full-depth sliding side windows throughout. The front dash is very much tidier, having a shallower grille with radius corners and is wider than on the earlier type. Side-lights and indicators are moved above the grille and look much neater.

Leyland Olympian(A) - BL54-93.

The 1984 order for short Olympians was for 40 chassis and a similar number of Alexander ckd R-type body kits, all of which entered service during February and March 1985. Similar in most respects to the previous series, except that the seating arrangements differed, body layout being H60/35+16FE,FdXS,DOO when new but this was altered to H60/33+18 in January and February 1986 so that all forty buses are now part of the enlarged Olympian(B) type (see below).

Leyland Olympian(B) - BL93-120 (plus BL54-93 in 1986).

A further thirty 9.5m Olympians were ordered for entry into service from August 1985. Again these are similar to previous batches with Alexander ckd bodies. Of these 30 buses, 27 became classified Olympian(B) while the final three, BL121-123, became Olympian(C) type (below).

Although the *total* passenger capacity remained as before at 111, the layout differed, having the arrangement H60/33+18FE,FdXS,DOO. In January/February 1986 the ranks of the (B) type were swollen by the re-seating of the (A) type to (B) type specification, all forty buses being added to the (B)-type inventory.

Leyland Olympian(C) - BL121-3.

These two buses differ in having two additional standees, being arranged H60/33+20FE,FdSX,DOO. KMB describes these buses as having 'aircraft roofs', referring to the honeycomb structure with which the upper-deck floor and roof was constructed, using materials developed by the aircraft industry.

VEHICLE SPECIFICATION:	9.5m Leyland Olympian.
Fleet numbers: Olympian 9.5m:	BL4-53
Oly 9.5M (A):	BL54-93
Oly 9.5M(B):	BL94-120 (& BL54-93 from 2/86)
Oly 9.5M(C):	BL121-123
Chassis:	Leyland Olympian ONLXB/1R
Engine:	Gardner 6LXB, 185bhp.
Transmission:	Voith/DIWA D851.2 fully-automatic.
Body make:	Alexander R-type, ckd.
Body layout: BL4-53: as new:	H57/34+20FE,FdSX,DOO
4/86:	H57/32+22FE,FdXS,DOO
Oly(A):	H60/35+16FE,FdXS,DOO
Oly(B):	H60/33+18FE,FdXS,DOO
Oly(C):	H60/33+20FE,FdXS,DOO
Date introduced:	1983-85
Total:	120
Length:	9.5m
Width:	8.5m
Height:	4.37m
Wheelbase:	4.95m
Unladen weight:	?
Gross vehicle weight:	16,257kg

ABOVE: Following the success of the three Olympians purchased for trial and evaluation, KMB purchased a further fifty in two groups, one each of twenty and thirty. Alexander bodywork was specified, rather than ECW. BL 8 was in Des Voeux Road Central in October 1983. *(Mike Davis)*

ABOVE: The second batch (of thirty) two-axle Alexander R-type bodied Olympians display detail differences to their bodywork, including improved front scuttles, grille and headlight assemblies and also full-depth sliding side-windows, as opposed to, roughly, $^2/_3$-depth on BL4-23. This photograph illustrates BL34 on its first weekend in service in May 1984 and still without fleet numbers. *(Mike Davis)*

BELOW: BL38 on its first day in service, 9th May 1984, standing at "Star" Ferry, awaiting its next duty on Route 6, the first allocation of new vehicles to non-Tunnel routes for some time. *(Mike Davis)*

Dennis Dominator 9.5m. DM1-40

Concurrently with its tri-axle cousins, the Dennis Dragon(B)'s, the first of KMB's two-axle Dennis Dominators, DM 1, entered service during December 1983, with DM2-20 following in January 1984.

Strangely, for a company struggling with overcrowding on so many routes and with such a sizeable fleet of low capacity Victories and Jubilants, the chosen chassis are of the short, 9.5m length. The power and transmission train is the familiar Gardner 6LXB engine driving via a Voith/DIWA D851 fully-automatic gearbox/retarder.

Bodywork is by Hestair subsidiary Duple (Metsec) Ltd. and was supplied in ckd form for local assembly under the supervision of the manufacturer's representative at KMB's 'Tuen Mun 80' installation. Externally the general appearance is similar to the Duple (Metsec) bodies on Dennis Jubilant chassis and are a shorter version of those on three-axle Dennis Dragon chassis. The usual front entrance, centre (forward) exit arrangement is combined with a forward staircase which rearward ascends over the offside front wheel-arch. Layout is H57/35+11FE,FdXS,DOO. A distinctive feature of this type of body on whichever chassis is the protruding front destination box.

A second batch of Dominators was ordered in 1983 for entry into service in 1984 and these follow on in fleet number sequence from DM21-40 with DM 21 entering service on 1st August 1984 and the last, DM40, on 3rd September.

VEHICLE SPECIFICATION:	9.5m Dennis Dominator.
Fleet numbers:	DM1-40
Chassis:	Hestair-Dennis Dominator.
Engine:	Gardner 6LXB, 185bhp at 1850rpm.
Transmission:	Voith?DIWA D851, three-speed, fully-automatic gearbox/retarder
Body make:	Hestair-Duple (Metsec), ckd.
Body layout:	H57/35+11FE,FdXS,DOO.
Date introduced:	1983-85.
Total:	40 (1982 order 20; 1983 order 20)
Length:	9.5m
Width:	2.5m
Height:	4.42m
Wheelbase:	4.953m
Unladen weight:	?
Gross vehicle weight:	? (All dimensions nominal)

BELOW: KMB's 9.5-metre Dennis Dominators share a basically similar body style with the longer Dragons. DM16 was leaving the Cross-Harbour Tunnel when photographed in June 1984. *(Mike Davis)*

ABOVE: The same Tunnel exit location saw un-numbered DM3 in advertising livery on the same Sunday in June 1984. The mainly blue livery suited the bus well, giving it a fresh air. *(Mike Davis)*

BELOW: DM1, the premier member of its class was adorned in advertising livery early on in its career. As with the other two photos of this class, it was photographed in May 1984 by the Toll Plaza at the north portal of the Cross-Harbour Tunnel. *(Mike Davis)*

231

Leyland Olympian - 12m - 3-axle.   3BL2-163

Olympian(A)12M.

3BL2-21 comprised KMB's first order for 'production' versions of the 3-axle, 12m, Leyland Olympian, designated ON6LXCT/3R by the manufacturer. Powered by the successful Gardner 6LXCT turbocharged engine, rated at 230bhp and provided with Voith/DIWA D851 fully-automatic gearbox, these buses feature the self-steering, or tracking, rear-axle mechanism on the forward wheels of the twin rear-axles which are non-driven. This differedrs from the prototype in having actuators on both sides, 3BL1 only having had one actuator. From this, it will be understood that none of the major mechanical units has been perpetuated from the all-Leyland prototype.

Also, unlike the all-Leyland/ECW prototype, 3BL2 *et seq* have Alexander R-type ckd bodywork, assembled at KMB's Tuen Mun engineering facility and similar in general appearance to the first Dennis Dragons, 3N1-3, except that the Leylands have two exits within the wheelbase, the forward of the two being opposite the foot of the forward ascending staircase. Seating is provided for 101, plus 48 standees, giving a licensed total of 149, a figure that appears to be exceeded during the morning and evening peak periods. The actual body layout is H71/30+48FE,FdXS,CX,DOO. The windows on 3BL2-21 have a ²/₃-depth sliding portion, with the lower part fixed, as on the contemporary 2-axle version, BL4-23.

The first production series, ordered in 1982, was introduced to traffic between September and November 1983.

VEHICLE SPECIFICATION:	12m Leyland Olympian
Fleet numbers:	Oly(A)12M: 3BL2-41 Oly(B)12M: 3BL42-61 Oly(C)12M: 3BL62-163
Chassis:	Leyland Olympian Model ONLXCT/3R
Engine:	Gardner 6LXCT, 10450cc, 230bhp
Transmission:	Voith/DIWA D851, 3-speed gearbox/retarder.
Body make:	Alexander R-type.
Body layout:	3BL2-41: H71/30+48FE,FdXS,CX,DOO. 3BL42-61: H74/31+58FE,FdXS,CX,DOO. 3BL62-163: H74/30+60FE,FdXS,CX,DOO.
Date introduced:	1983-86.
Total:	162.
Length:	12m
Width:	2.49m
Height:	4.37m
Wheelbase:	5.3+1.6m
Unladen weight:	?
Gross vehicle weight:	23,000kg

3BL22-41 represent a second order, placed in 1983, for a further twenty 12m Leyland Olympians, again with Alexander bodies and basically similar mechanically to 3BL2-21, hence the same KMB classification, 'Olympian(A)12M', as their predecessors. Visually, 3BL22-41 present a tidied-up look when compared with the previous twenty, having a redesigned front dash panel, headlight and grille arrangement. The sliding side windows are of the full-depth sliding type and so internal safety rails are fitted. Seating is the same as for the previous batch.

This series was introduced in April 1984 when they appeared on urban trunk route 5 from Choi Hung to Kowloon "Star" Ferry Pier; the Cross-Harbour Tunnel route for which they were intended required some last minute civil engineering work to ease a tight turn in Causeway Bay which was quickly at-

BELOW: The 'production' version of the three-axle, 12m Olympian was provided with similar-style bodywork to that of ts two-axle BL-class counterpart but had a high-capacity, multi-standee, lower-deck configuration. Two exits are provided and, here, 3BL12 carries a full load - including (illegal) upper-deck standees on the first Sunday in June 1984. *(Mike Davis)*

232

ABOVE: KMB converted Route 101 to 3-axle operation using Leyland Olympians and, here, 3BL27 heads back to Kwun Tong when nearly new on a Sunday morning in May 1984. Again, following the example of the 2-axle Olympians, this second batch of 12m buses showed detail differences from earlier versions, including improvements to the frontal arrangements and side windows. *(Mike Davis)*

BELOW: 3BL58 serves to illustrate the detail differences introduced on the third Alexander bodied Olympian body type. The front grille is obvious but less quickly noticeable are the larger screens for the front and side destination/route number displays. Close examination reveals a dot-matrix, electronically controlled, front route number - the numbers of which seem remarkably small. *(John May*

ABOVE: 3BL139 [Olympian(C)], passes the Sheraton Hotel in Salisbury road on its way from Tsim Sha Tsui East to Shun Tin in the south-eastern New Territories, a relatively short 12.8km (8 miles). The later grille arrangement is readily seen in this October 1988 view. *(Mike Davis*

tended to and the 5 reverted to its usual Fleetline allocation.

Leyland Olympian(B)12m - 3BL42-61.

The twenty buses of the 1984 order for Alexander (ckd) bodied 3-axle Olympians were introduced between the October and December of the year they were ordered. The major difference between this group and their immediate predecessors is that they have a different interior layout, permitting a higher passenger carrying capacity with a layout of H74/31+58FE,FdXS,CX,DOO - a total of 163 as opposed to 149 on the (A) type.

It is a matter for reflection that the standing capacity alone exceeds the total seating capacity of the not long withdrawn Daimler(a)'s in their latter DOO days; buses of similar size to the London Transport RT-class!

An unknown number of this type were equipped with dot-matrix route number displays, front, nearside and rear.

Leyland Olympian(C)12M - 3BL62-163.

Obviously satisfied with its growing fleet of 12m Olympians, KMB placed orders for an additional 102 units for introduction in late 1985 and in 1986. Once again Alexander R-type ckd bodies were specified and assembled in KMB's Tuen Mun workshops. Lower-deck seating is, however, reduced by one to 30, but standing, at 60, was two up on the Olympian(B)12M-type. Upper-deck seating remains at 74.

Delivery of these last 3BL-class buses brought the acquisition of 12-metre buses to a halt as the physical restrictions of the remaining high density routes in need of large buses were such that shorter vehicles were the order of the day. The next order for Olympians was for 11-metre, 3-axle buses. Other factors made this move more desirable, particularly the fact that new legislation required more space per passenger while retaining the existing gross vehicle weight restrictions, making the maximum weight capacity limit for an 11m bus the same as for a 12m version. With no capacity advantage in the extra metre, it was more expedient to standardise on the 11m version.

LEFT: A well loaded 3BL28 leans heavily as its driver turns into the bus stop area at the north end of the Cross-Harbour Tunnel in May 1984. *(Mike Davis*

ABOVE: The rear of 3BL33 as it turns from Chatham Road into Salisbury Road, en route "Star" Ferry, only a few hundred metres distant. Kowloon urban routes were often used for running-in new vehicles and saw types not regularly seen thereon and this was true on this occasion in March 1984. *(Mike Davis)*

BELOW: 3BL104 (DH7439) pursuing a tram in Queensway during September 1994, en route for Kennedy Town, to the west of Hong Kong Island, working on Cross-Harbour Tunnel Route 104. *(Mike Davis)*

Leyland Olympian - 11m short 3-axle   S3BL1-420

In line with KMB's policy of buying 11-metre, three-axle, double-deck buses rather than 12-metre, all further Leyland Olympians following on from the last longer bus, 3BL163, were, until 1994, to the shorter, 11m length. This helps to reduce the space available into which passengers can crush-load themselves and to keep the loaded vehicle within the 21,800kg limit, even during the peak loading periods.

When the 12m, 3-axle concept was first introduced, the gross permitted weight was 24 tonnes but, following new legislation and consequent gross weight reduction to 21.8 tonnes, it was deemed more suitable to reduce the floor area of the bus by restricting the overall length, so limiting the capacity by physical constraint. Under Hong Kong regulations, the gross permitted vehicle weight is the same for both 11m or 12m buses. The maximum passenger load can thus be carried in the shorter, more acceptable, length.

The mechanical specification of the first 351 buses of the Short 3BL, or S3BL, class was similar to that of the preceding 12m version, having a combination of Gardner 6LXCT engine, rated at 220bhp, and Voith/DIWA D851 three-speed, fully-automatic gearbox/retarder. Change came with the advent of the Cummins L10 engine and ZF gearbox, often, but not exclusively, matched as some Gardner/ZF and Cummins/Voith examples were evaluated. A single air-conditioned example was purchased, followed by large ongoing orders. *(See section on air-conditioned double-deckers).*

Alexander bodywork is fitted to all S3BL buses and these were built-up by KMB from

VEHICLE SPECIFICATION: 11m Olympian 3-axle

Fleet numbers :	S3BL1-420
Chassis:	Leyland Olympian 11 metre, 3-axle **Note 1.**
Engine:	Gardner 6LXCT **Note 1**
	Cummins L10
Transmission:	Voith D851, 3-speed or **Note 2**
	ZF4HP500, 4-speed.
Body make:	W. Alexander - ckd
Body layout: S3BL1-105:	H69/30+51FE,FdXS,DOO KMB class: Olympian 11M
S3BL105-420:	H69/40+34FE,FdXS,DOO
Date introduced:	1986-1991
Total:	420
Length:	11m
Width:	2.5m
Height:	4.37m
Wheelbase:	4.9m+1.6m
Unladen weight:	?
Gross vehicle weight:	21,800kg

NOTE 1

All	ONLXCT/3RV	All	S3BL1-47/50/4/6
Gardner	ON6XCT/5RV	Voith	S3BL48/9/51/2/3/5/7-99/105
6LXCT	ON6LXCT/5RV	gearbox	S3BL106-263
engines	ONLXCT/5RV		S3BL273-370
Cummins L10	ONCL10/A5RV	Voith gearbox	S3BL100-104
engines	ONCL10/5RZ	ZF gearbox	S3BL264-272
	ONCL10/5RV	Voith gearbox	S3BL371-420

NOTE 2:

Voith gearbox Model D851	3-speed fully automatic	
ZF gearbox Model 4HP500	4-speed fully automatic	

BELOW: A 1986 view of 11-metre 'Short' 3BL, or S3BL66, whilst working on the 6B when only a few months old. The length was reduced both at the rear overhang and within the wheelbase when compared with the 12-metre 3BL-class *(Nigel Eadon-Clarke*

236

ABOVE: S3BL175, Olympian 11M(A), demonstrates how the side-light clusters and surrounding mouldings differed from both S3BL66 (opposite) and S3BL181 (below). Seen here on Route 1, leaving Nathan Road for "Star" Ferry in October 1988. *(Mike Davis*

BELOW: S3BL181, also on Route 1, turns into Salisbury Road, Tsim Sha Tsui, from Nathan Road minutes after S3BL175 (above). *(Mike Davis*

the conventional ckd kit. Standardised layout of the body-shell is employed with the usual narrow front entrance but, unlike the 12m version, only one exit is provided, this being a double-width doorway, forward of centre, just behind the nearside front wheel-arch and opposite the foot of the staircase which forward ascends.

Ventilation arrangements similar to those incorporated in the later 12m type are provided, the most easily visible evidence of this being a full-length valance above the upper-deck side windows and a row of louvres above the lower-deck side windows.

Seating and layout for S3BL1-105, the KMB class 'Olympian 11M', was, when delivered, H69/30+51FE,FdXS,DOO, while later buses, 'Olympian(A)11M' provided ten more lower-deck seats, being H69/40+34. The air-conditioned prototype was registered late in May 1988 just before S3BL257.

These 11m Olympians were ordered in batches of sometimes small numbers and detail differences are many but chiefly centre around front grille and side-light types from an external recognition point of view.

S3BL1-99/105 (Total 100).
KMB Class: Olympian 11M.
This group comprised the first and second KMB orders, the first for 50 chassis and the second for 55 chassis - of the latter, the five chassis that became S3BL100-4 were fitted with Cummins engines - *see below* - effectively making two orders for 50 similar buses each, of which all had Gardner engines. The Leyland designation for the first 50 was ONLXCT/3R, having Gardner 6LXCT engines, three axles and Voith gearboxes (R=Right hand drive). The second batch were mechanically similar to the first but Leyland, possibly seeking to differentiate between Gardner 6LXCT and 8LXCT in certain other models, changed the engine code by dropping the L of LXCT and replacing it with the number of cylinders, in this case 6, to become ON6XCT. At the same time the length/axle code was altered from 3 to 5, possibly to differentiate between the 11m and 12m models.
The last few of the second batch entered service before the last of the first batch and so fleet numbers between S3BL47 and 57 are intermingled between batches.

S3BL100-104 (Total 5).
KMB-class: Olympian 11M.
This small group were specified to have Cummins L10 engines for evaluation purposes but retained the Voith gearbox as used with the Gardner engined main batch, S3BL1 etc. Leyland type code for these chassis is ONCL10/A5RV; the A indicating an alteration to accommodate the Cummins engine, not regularly offered by Leyland at that time.

S3BL106-263
S3BL264-370 (Total 256)
KMB-class: Olympian 11M(A)
This large group was introduced between 1987 and 1991 and represent orders in batches of 45, 50, 61 and 110 but, of the 110, ten were to a different specification - *see below*. They were generally similar mechanically to the first fifty but, again, Leyland changed its nomenclature to include the entire Gardner engine code, becoming ON6LXCT/5RV from which it will be recognized that the Voith gearbox was again fitted. From S3BL273 onwards, this code was changed yet again, this time dropping the figure indicating the number of cylinders so that these chassis are ONLXCT/5RV.

11-metre Olympians with Cummins engines.
S3BL264-272 (Total 9).
KMB-class Olympian11M(A).
Introduced in 1988, these are nine of the balance of ten chassis from the order for 110 - *above* - of which 100 have Gardner engines; the tenth bus is the air-conditioned prototype DX2473 that became AL 1. All ten buses with Cummins L10 engines are fitted with ZF 4-speed gearboxes and the chassis designation is ONCL10/5RZ.

S3BL371-420 (Total 50).
KMB class: Olympian 11M(A).
Ordered in two batches, one of 30 and one of 20 and introduced in 1990 with another change of specification, Cummins L10 engine but this time coupled to the Voith gearbox. By this time the Cummins/Voith combination was a standard option and so the chassis code omitted the 'A' of earlier examples of this combination, S3BL100-4.

S3BL421-470.
KMB-class: Olympian 11M(B)
An additional fifty buses introduced 1993-94 with similar Alexander R-type bodies but with revised carrying capacity; H69/40+33FEX,FdXS - ie one less standee than the Oly.11M(A). A further thirty were under construction in August 1994, during which month a new order was placed, again for 30.

BELOW: S3BL278 sets-down - via the front entrance - at Tuen Mun Town Centre in October 1988, while it was working on New Territories Express Route 60X, from Town Centre to Jordan Road Ferry Pier, 31km distant. *(Mike Davis)*

ABOVE: Cummins engined S3BL381 prepares to turn right into Nathan Road from Salisbury Road en route from the railway station to Yiu On on Route 81C in September 1994. *(Mike Davis)*

BELOW: S3BL450, seen here in April 1994 clearly displays the newly introduced route number display where the number is printed on a rectangle of transluscent plastic and slotted behind the glass. The destination is also provided in a similar manner but is arranged so that the display can be changed either end of a route; the new destination dropping down while the old one is raised. The change is effected by means of a conventional looking winding handle in the cab. This is a simple and low-cost soloution where buses remain allocated to the same routes for long periods of time. This batch of S3BLs also featured deeper destination screens. *(Nigel Eadon-Clarke)*

MCW Metrobus 11m - 3-axle. S3M1-254

MCW failed to win any orders for three-axle buses following the evaluation trials of the 12 metre buses, 3M1-3, until 1985, when 11m buses on three-axles were ordered in line with the developing policy of having 11m instead of 12m buses on routes requiring a high capacity vehicle. Having decided to prefix the fleet numbers of three-axle buses with the figure '3', the shorter versions received an additional prefix 'S', for short, hence Short 3-axle Metrobus No1 became S3M1 when it entered service on 26th May 1988 on Route 6A. The type soon appeared on other Kowloon urban routes, ousting the former London Transport DMS Fleetlines, Guy Victories and Dennis Jubilants, as well as the few remaining AEC Regents still operating in the Kowloon area.

Mechanically the S3M-class are standard MCW products with Gardner 6LXCT engines driving the rearmost axle via a Voith/DIWA D851 fully-automatic gearbox. The wheelbase is reduced in proportion to the overall length.

The MCW bodywork is similar in general appearance to the three prototype 3M-class buses but is one bay shorter. A narrow front entrance is provided, fitted with a single, two-leaf, folding door, while the exit is located some 400mm to the rear of the front nearside wheel-arch and opposite the foot of the staircase which forward ascends through 180°. A single inward facing seat is fitted above the wheel-arch, between the driver and the staircase. Seating and standing arrangements vary, causing there to be four sub-types, known by KMB as MCW11M; MCW11M(A); MCW11M(Air-con) and MCW11M(B).

VEHICLE SPECIFICATION: 11 Metre MCW METROBUS

Fleet numbers:	S3M1-254.
Chassis:	MCW Metrobus Models DR137 & 138.
Engine:	Gardner 6LXCT.
Transmission:	Voith/DIWA D851, 3-speed, fully-automatic.
Body make:	MCW ckd - except DP1932 (S3M145) cbu)
Body layout: MCW 11M:	H69/33+48FE,FdXS,DOO
MCW 11M(A):	H69/41+32FE,FdXS,DOO
MCW 11M(Air con):	H65/38+32FE,FdXS,DOO,A/C
MCW 11M(B):	H69/42+32FE,FdXS,DOO
Date introduced:	1986-90
Total:	254
Length:	11m
Width:	2.5m
Height:	4.43m
Wheelbase:	4650mm+1600mm
Unladen weight:	12550kg
Gross vehicle weight:	21330kg

Orders:
- 9/85 : 20
- 1/86 : 25
- 6/86 : 28 (A)
- 9.86 : 1 (Air-con)
- 11/86 : 50 (A)
- 5/87 : 40 (A)
- 9/88 : 90 (B)

MCW 11M. S3M1-45.

This group consisted of two orders, one of 20, placed 9/85, and one of 25 placed in 1/86. These 45 buses entered service with a layout of H69/33+48FE,FdXS,DOO. Fleet numbered S3M1-45, the first bus entered service in May 1986 but the second example was not ready until 28th August, the majority following in September. Stragglers continued to be introduced until the last appeared on 1st December.

MCW 11M(A)0. S3M46-180 (with gaps).

The lower saloon seating/standing arrangements were altered on buses of this sub-type to provide more seats and less standing space on longer distance, lower density services. The resulting arrangements were H69/41+32FE,FdXS,DOO. Orders were placed in three batches in 6/86 (28 buses), 11/86 (50 buses) and 5/87 (40 buses). Fleet numbers start cleanly at S3M46 but finish-up with S3M155-180 being mixed-in with early examples of the (B) sub-type. S3M180 was the last (A)-type.

MCW 11M(Air con). DP1932 (later S3M145).

Anxious to introduce air-conditioned double-deck buses, KMB placed an order in Sep-

BELOW: S3M23 turns into Salisbury Road en route "Star" Ferry whilst working on Route 7 in October 1986. The 7 was once the stamping-ground of Tilling-Stevens and Bedford single-deckers, then Daimler CVG5 double-deckers. The bus was barely a month old when photographed. *(Nigel Eadon-Clarke)*

tember 1986 for one prototype air-conditioned 11-metre Metrobus. Mechanically, this was similar to the main batch of this type but with the addition of an engine-driven Sutrak 'air-con' unit mounted at the rear, above the engine.

This bus was shipped to Hong Kong completely built-up and it entered service in April 1987 but was withdrawn a year later and the air-conditioning equipment was removed. After standing idle for another year it was returned to service as a conventional vehicle in April 1989, the problems with the air-con. units apparently arising from the compressor and associated drive arrangements. A direct drive was taken from the engine to a single large compressor but this was not totally satisfactory and unacceptable vibrations developed. In the later Olympian and Dragon applications, three smaller compressors are provided, with the Dragon also having direct engine-drive, while the compressors on the Leyland are driven by a shaft from the geargear-box.

As built, the air-conditioned MCW was known only by its registration number, DP1932, and had slightly fewer seats (H65/38+32) as the plant intruded slightly into the passenger space and increased the weight. Following its return as a non-air-con. bus, it was allocated the fleet number S3M145 and was re-seated and included in the MCW 11M(A) sub-class.

The front door was unusual on DP1932 in that it was a one-piece 'glider' type, where the main ckd body type has two-piece 'jack-knife doors'.

See also: **Airconditioned buses, Page 252**.

MCW 11M(B). S3M156-254 (with gaps).

KMB's final order for MCW Metrobuses (MCW ceased its bus-building activities in 1988/89) was placed in September 1988 and was for 90 chassis and 90 MCW body kits. Layout was again changed to H69/42+32 and, as noted above, fleet numbers of earlier examples were intermixed with the last of the (A)-sub-type, the lowest (B) being S3M156.

ABOVE: The experimentally air-conditioned MCW Metrobus, DP1932 (later to become S3M145 after removal of the air-con. plant) showing the distinctive livery applied to this bus. *(KMB press release*

BELOW: S3M180 in 1994, near Tai Wai KCR Station, Shatin. Note the improved ventilation introduced from 1987 - a length-wise vent above the upper-deck side windows and louvres above the lower-deck windows. *(Mike Davis*

Dennis Dragon 12m - 3-axle.   3N4-191

During November 1983, the first two of KMB's first production batch of twenty 12m, 3-axle Dennis Dragons entered service on cross harbour duties. Basically a three-axle version of the Dennis Dominator, the production buses are similar in general specification to 3N1-3, with the major exception that the engine is the turbocharged Gardner 6LXCT - the earlier buses had naturally aspirated 6LXC units. The turbocharged 6LXCT produces 230bhp. Drive to the rearmost axle is via a Voith D851.2, three-speed, fully-automatic gearbox fitted with an integral retarder. As was the case with the earlier vehicles, the 5.4m (17ft 9in) wheelbase, combined with the short front overhang produces a very long rear overhang with considerable out-swing on tight turns. The unpowered centre axle is in fact two independently sprung stub-axles and no steering facility is incorporated to reduce tyre 'scrub' on tight turns. Full air-suspension is provided on all axles and ZF power assisted steering is incorporated. Braking is fully air-operated. The chassis is of pressed steel channel section with heavy duty axles which enable the vehicle to run at a gross vehicle weight of 21,800kg.

Bodywork by Hestair-Duple (Metsec) in the form of a ckd kit was assembled at KMB's Tuen Mun facility and in general outline resembles CMB's DL1, similarly bodied by Duple (Metsec), except that the latter has a two-door layout where the KMB buses have three, of which two are exits, each slightly narrower than KMB's standard for single exit doorways. Both exits are set within the wheelbase and all three doors are two-leafed, hinged at the front. There are also similarities in general appearance to the later Dennis Jubilants and the DM-class Dominators.

VEHICLE SPECIFICATION: 12-metre Dragon (B), (C) & (D).	
Fleet numbers:	Dragon (B): 3N4-93 Dragon (C): 3N94-154 & 158-191 Dragon (D): 3N155-157
Chassis:	Dennis Dragon: Model DDA605 - Dragon (B) Model DDA607 - 3N19-140, 155-7. Model DDA1701 - 3N141-54, 158-91
Transmission:	Voith/DIWA D851, fully-automatic gearbox
Body make:	Duple MetSec
Body layout:	Dragon (B): H75/33+58FE,FdXS,CX,DOO Dragon (C): H75/33+33FE,FdXS,CX,DOO Dragon (D): H75/33+50FE,FdXS,CX,DOO
Date introduced:	1983-86
Length:	11.993m
Width:	2.5m
Height:	4.369m
Wheelbase:	5.410m+1.6m
Unladen Weight:	?
Gross Vehicle Weight:	21,800kg.

Dennis Dragon(B)12M. 3N4-93.

The KMB classification Dragon(B) was given to the first production batch of 90 buses, fleet numbered 3N4-93, introduced between November 1983 and January 1985. As with all 12m Dragons, the seated capacity was 108 but the permitted number of standees varied. The service layout of the Dragon(B) is H75/33+58FE,FdXS,CX,DOO. Large areas of the lower saloon are set aside for standees.

Dennis Dragon(C)12M. 3N94-154 & 158-191.

The order placed in February 1985 specified a further 50 similar buses, to be followed in July of the same year by a further 48, of which all but three were to be classified by KMB as their Dragon(C) class. These were to be similar in most respects to the previous (B) type except that the standee capacity was reduced to 53, giving a layout of H75/33+53FE,FdXS,CX. Only 47 of the first fifty were to be classified Dragon(C), the final three (3N155-7) becoming Dragon(D). All 47 (C)'s of the first group were designated Model DDA607 by Dennis, while a second group of 48 were DDA1701.

Dennis Dragon(D)12M 3N155-57.

The three chassis not included with the Dragon(C)-class but which had DDA607 chassis were to enter service three weeks later than the remainder of their group, on 18th June 1986 and were permitted only 50 standees, making their layout H75/33+50FE,FdXS,CX.

BELOW: The Hestair-Dennis contribution to KMB's massive purchase of three-axle buses came in the form of the Dragon, derived from the Dominator and bodied by Hestair Duple (Metsec). Three doorways are provided as with all other 12m types purchased concurrently. 3N4, a Dragon(A) and the premier member of the production series, leaves the toll-plaza of the Cross-Harbour Tunnel in June 1984 *(Mike Davis*

ABOVE: 3N23, another Dragon(A), was on Hong Kong Island, passing Admiralty Commercial Centre in May 1984. *(Mike Davis*

BELOW: A wider destination screen identifies 3N47 as a third series Dennis Dragon with Duple (Metsec) body but still a Dragon(A). Seen here with a full load as it leaves Canal Road East en route Tse Wan Shan on the lower slopes of the Kowloon hills. *(John May*

ABOVE: 12-metre 3N131 [Dragon(C)] displays clearly the increased size of the ventilator in the front dome and the louvred vents above the lower-deck side windows. *(Mike Davis)*

BELOW: One of only three Dragon(D)s, 3N155, in advertising livery, approaches the "Star" Ferry terminus in October 1988 with the Hong Kong Sheraton Hotel as a backdrop. Again, the larger ventilator louvres can be seen. *(Mike Davis)*

Dennis Dragon 11m - 3-axle.   S3N4-270

Dragon 11M & (A) to (C)

In addition to, and following-on from, the 12-metre Dennis, 3-axle derivative of the 2-axle Dennis Dominator, KMB ordered a large number of 11-metre, 3-axle Dragon double-deck buses with bodies from Duple MetSec, the Hestair in-house partner with Dennis.

These chassis differ from the 12-metre variety only in the spacing between the front and centremost axles. Power is, in most cases, the tried and tested Gardner 6LXCT, driving the rearmost axle via a Voith/DIWA D851, 3-speed, fully-automatic gearbox. S3N51-55 were specified to have Cummins L10 engines for evaluation and comparison with the Gardner used as standard.

The ckd bodywork as supplied by Duple MetSec had two doorways; a single width front entrance and a double forward exit, the latter being immediately behind the nearside front wheel, opposite the foot of the staircase which forward ascends over the offside front wheel-arch. Improved ventilation was incorporated, the most obvious outward signs of which were large louvres in the front dome and a line of louvres along each side above the lower saloon windows. After some rationalisation, three types of seating arrangement became standard in 1987; these were the basic 'Dragon 11M' and the sub-types (C) & (D), the details of which are shown in the accompanying table.

Differences in licensed standing capacity

VEHICLE SPECIFICATION: 11-metre Dennis Dragon, Dragon(C)11M; Dragon(D)11M; Dragon(E)11M.
(For details of (A) & (B) sub-types see table on page 514.)

Fleet numbers:	Dragon 11m: S3N1-55 includes former (A) series Dragon(C)11M: S3N56-140 includes former (B) series Dragon(D)11M: S3N141-221 .
Chassis make/model:	Dennis Dragon:- Model DDA1801 - S3N 1-50 Model DDA1802 - S3N51 Model DDA1803 - S3N56-140 Model DDA1804 - S3N141-186 Model DDA1805 - S3N187-221
Engine:	Gardner 6LXCT, or Cummins L10 - S3N51-55
Body make:	Duple MetSec
Body layout:	Dragon 11M: H67/29+55FE,FdXS,DOO Dragon(C)11M: H67/39+31FE,FdXS,DOO see table below. Dragon(D)11M: H67/41+33FE,FdXS,DOO
Date introduced:	1986-91
Length:	11000mm
Width:	2500mm
Height:	4370mm
Wheelbase:	4900mm+1600mm
Front overhang:	1600.m
Rear overhang:	3.02m
Unladen Weight:	12450kg
Gross Vehicle Weight:	21,800kg

```
Dragon 11m      H67/29+55   S3N 1-28
  28 buses new 12/86                         in 28
  27 buses ex-Dragon(A)11M                   in 27
                                         Total 55

Dragon(A)11M    H67/29+48   S3N29-55
  27 buses new 1/87                          in 27
  27 buses transferred to basic'Dragon 11M' 5/87-7/87
                                            out 27
                                         Total nil

Dragon(B)11M    H67/39+31   S3N56-90
  35 buses new 4/87 & 5/87                   in 35
  35 buses transferred to 'Dragon(C)' 6/87 & 7/87
                                            out 35
                                         Total nil

Dragon(C)11M    H67/39+38   S3N91-140
  50 buses new 4/87 & 5/87                   in 50
  35 buses ex-Dragon(B) 6/87 & 7/87          in 35
                                         Total 85
```
TABLE ABOVE: Breakdown of Dragon 11m sub-types S3N1-140

BELOW: S3N2, built new as KMB's basic type 'Dragon 11m' in December 1986 was photographed as seen here almost eight years later in September 1994 whilst trapped in an enormous traffic jam caused by the collapse of a building under demolition on Nathan Road. The original small dome ventilator and lack of side louvres identify the bus as an early member of its class. *(Mike Davis)*

resulted in the first 140 eleven-metre Dragons being divided into four, rather than two, sub-types. It was the intention to have the basic types with greater or lesser standing capacity on the lower deck but each type emerged with two different standing figures, each a difference of only seven passengers.

Within six months of the first delivery the situation had been resolved when each type was approved for its respective higher figure, as illustrated in the table on the previous page.

The Dragon (D) sub-type, see below, was introduced subsequent to the rationalisation and has an additional differing layout.

Dennis Dragon (D) 12M.

This sub-type was the last to have the original body-style and comprised 81 vehicles; S3N141-221. The Dragon(D)s were delivered between September 1987 and November 1988. The odd total of 81 was reduced to eighty in August 1994 when EA2263 (S3N221) was prematurely withdrawn.

RIGHT: S3N47 is a 'Dragon 11M', having started life in 1986 as one of the 27 'Dragon(A)11M's' but was converted to the basic type in 1987 when the authorised standee capacity was increased from 48 to 55. The bus was photographed leaving the Jordan Road Ferry terminus in September 1994.

BELOW: S3N126, a Dragon (C)11M from new, advertises for international petroleum group Caltex and shows that the 11m S3N-class has only two doorways rather than the three of the 12m version. The enhanced ventilation seen on the 3N-class is also evident on the S3N's. Seen passing the Peninsula Hotel in October 1988. *(Mike Davis)*

ABOVE: S3N144, a Dragon(D)11M with two more seats than the (C)-type, exposes its flank to the camera to show the reduced length and better proportions of the 11m type when compared with the 12m version with similar Duple (Metsec) bodywork. Seen in October 1988. *(Mike Davis)*

BELOW: S3N163, also a Dragon(D)11M but in advertising livery, in 1988 running on the 6A, which had been Metrobus S3M operated in 1986. *(Mike Davis)*

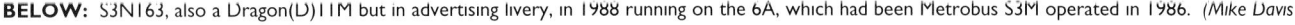

247

Dennis Dragon(E) & (G).   S3N222-270 & 320-369

In addition to the Duple(Metsec) bodied Dragons with air-conditioning (*qv*), KMB ordered similar buses but without any provision for air-conditioning at all, the (E) & (G) sub-types, together with fifty that had all the ducting, etc. to enable them to be retrospectively air-conditioned. (The latter are the (F) sub-type which are described separately but whose details are presented in the same specification box as the (E & (G)s.)

The Dragon 11m's are numbered from S3N222-270 (E) and 320 to 369 (G), and are mechanically similar to earlier buses in most respects

The bodies were assembled in Tuen Mun by Goodview Engineering Ltd., on behalf of KMB, and feature the new Duple(Metsec) body shell introduced for the air-conditioned Dragons, the AD-class (*qv*), having wider spaced, and fewer, window bays, a more domed roof and generally 'tidied-up' appearance. They differ from those of the AD-class, however, in respect of glazing and ventilation. Windows are not bonded, as on the AD-class, but fitted by means of rubber gaskets and are of the full-depth sliding type. An exception is the window over the nearside front wheel-arch which has only a half-depth vertical opening section. Additional ventilation is provided by means of full-length vents along the side of the roof on each side although the extensive louvres in the front dome of early-style Dragon bodies are omitted. Windscreens have opening sections to provide the driver with fresh but, perhaps, polluted air.

Large destination boxes have provision for the new KMB destination blind display with the destination and route number being printed on rectangular white translucent plastic panels, slotted into holders which, in the case of the destination, can be changed from the driver's position; the new destination drops while the old one rises. This arrangement makes for a very clear display, stays clean and does not tear. A route number box is provided at the rear and in the small side window immediately behind the front entrance doorway.

Window-frames, previously polished aluminium, are now black.

The original order for Dragon (E)s had been for 50 chassis and 50 bodies. During November 1989, KMB notified Government Transport Department that it had cancelled one chassis due to its having been damaged beyond repair in the UK prior to shipping. No mention was made of cancelling the corresponding body kit from Walter Alexanders.

Passenger capacity is H69/41+33FE,FdXS,DOO. Standard cream livery with red roof and skirt was applied from new.

VEHICLE SPECIFICATION: Dennis Dragon 11m(E) (F) & (G).

Fleet numbers:	(E) S3N222-270; total 49 (one chassis destroyed in UK prior to shipping). (F) S3N271-319 & 370; total 50 (G) S3N320-369; total 50
Chassis:	Dennis Dragon DDA1808 (E) Dennis Dragon DDA180? (F) Dennis Dragon DDA ?? (G)
Engine:	(E) Gardner 6LXCT (F) Gardner LG1200 (G) Cummins LT10
Transmission:	(F) Voith D863 (G) ZF 4HP500
Body make:	Duple Metsec (improved model.)
Body layout:	(E): H69/41+33FE,CXS,DOO. - introduced 1991 (F): H65/37+41FE,CXS,DOO. - introduced 1993 (G): H67/41+35FE,CXS,DOO. - introduced 1994
Totals:	(E) 49; (F) 50; (G) 50.
Length:	11100mm
Width:	2500mm
Height:	4369mm
Wheelbase:	4900mm+1600mm
Unladen weight:	1,3330kg
Gross vehicle weight:	21,800kg

NOTE: Of the last batch of fifty, commencing S3N222, one chassis was destroyed in transit and was not replaced making the last number S3N270 instead of 271.

BELOW: KMB sub-class 'Dragon (E)11M, No S3N268 prepares to turn right into Nathan Road from Salisbury Road en route Yiu On, when on the 81C in Seprember 1994. *(Mike Davis*

Dennis Dragon 11m(G).

These buses, introduced in 1994 numbered S3N320-369, are mechanically similar to the (E)'s but there are minor body detail differences; while internal handrails on the (E) type are bright stainless steel, those on the (G)-type are covered in high visibility green plastic.

The passenger capacity of the two types is similar but the (G)-type has two fewer upper-deck seats and two more lower-deck standees at H67/41+35FE,FdXS,DOO.

RIGHT: S3N266 turns into Salisbury Road from Nathan Road when heading for Kowloon KCR Station in September 1994. *(Mike Davis)*

CENTRE: Rear view of an unregistered Dragon (G), parked outside the Tuen Mun assembly works awaiting completion. Unlike the Dennis (F) type, qv, these were not fitted with ducting for future air-conditioning equipment and thus have normal rear window layout.
(Nigel Eadon-Clarke)

BELOW: The almost complete form of a Dennis (G) awaiting the final attentions of the assembly staff at Tuen Mun in April 1994. *(Nigel Eadon-Clarke)*

Dennis Dragon (F) - without air-conditioning but ducted
S3N271-320

Having commenced a plan to purchase large numbers of air-conditioned buses, KMB was asked by Government to slow-down the process but, as it would remain the longer-term intention to increase the proportion of air-conditioned buses while, in the meantime requiring additional new vehicles, KMB specified a batch of Duple-Metsec bodied 11-metre Dragons that were fully-ducted for air-conditioning but were delivered without the actual plant installed.

The resultant buses have an unusual appearance from the rear in having no lower-deck rear window but are in cream livery. Neither do they have any signs of the louvres for air-intake in the back panels to indicate that these will, one day, be retro-fitted.

In most other respects the body shell is similar in appearance to the air-conditioned (and the last non-air-conditioned Dragons). Opening windows are fitted which are mounted in rubber gaskets. Opening segments in the windscreens are also a feature but in these buses, unlike the earlier-style S3BL-class vehicles, the vents and windscreen divider are black in colour, in place of bright metal. Due to the provision of high-capacity ducts, there are no external louvers in the roof, front dome or on the body-sides.

When delivered, these buses were included in the S3N-class but, presumably, will later be added to the AD-class *(qv)*, after receiving their air-conditioning equipment.

Gardner LG1200 engines and Voith D863 fully-aotomatic gearboxes are fitted.

RIGHT: An offside front view of S3N294, one of the type without air-conditioning but with the ducting fitted for later installation of the a/c plant. The body-style is similar to that of the air-conditioned Dragons of AD-class rather than earlier S3BL vehicles *(Nigel Eadon-Clarke*

BELOW: The rear of S3N294 displays clearly the appearance of the Dragon (F)s without air-conditioning which have been provided with all the necessary ducting, etc. for retro-fitting in due course. *(Nigel Eadon-Clarke*

Part Five - Section D.
Air-conditioned double-deck buses.

All-three-axle:	Class:	Page:
Leyland Olympian 11-m	AL-class	252
Dennis Dragon 11m	AD-class	254
Dennis Dragon 10m	ADS-class	256
Scania 11.5m	AS-class	257
Volvo Olympian 11m	AV-class	258
Volvo Olympian 12m	3AV-class	259

Faced with criticism that their services failed to match the quality offered on commuter services from residential estates by its competitors, KMB sought, in 1989, to provide air-conditioning on certain routes where a premimum fare would be acceptable.

Air-conditioning ('air-con' or 'air-cond' in the Hong Kong vernacular) was not new to KMB who had a fleet of Dennis Falcon single-deck airport coaches and had operated two Albion/Duple coaches and a Bedford YRQ (*qv*) so fitted. Double-deckers were another matter, however, and historically had been a stumbling block for a number of reasons, no matter where tried. The New York Atlanteans had failed to meet somewhat demanding criteria and KMB's own tentative attempts, first in 1980/81 with Alexander bodied double-deck coaches, one each on Leyland Victory and Dennis Jubilant chassis and, in 1987, in conjunction with MCW, had been less than successful. With the exception of the MCW example, early double-deck air-conditioned buses had auxiliary engines and were, thus, uneconomic.

Under pressure from its customers, KMB, in conjunction with Leyland and Alexander, introduced an 11-metre Leyland Olympian in May 1989, fitted with an Alexander 'RH- type' body, a combination also chosen by competitors Citybus and KCRC. After some fine tuning, this bus, DX2437, was found to be acceptable and KMB sought to introduce the type on the premimum fare services mentioned above. This was not so simple as KMB fell foul of the financial arrangements under which it holds its franchise, it only being permitted to return a certain sum per bus operated and premium fares would have exceeded this so-called 'permitted return'. A solution was found, but not before KMB, under mounting pressure from its customers, was forced to hire air-conditioned single-deckers under an agreement with non-franchised operator Argos Bus Services. It also took into stock large numbers of 'off-the-peg' Toyota 'Coaster' large-mini-buses and Mitsubishi midi-buses, together with a pair of MCW Metroriders and, later Dennis Darts.

When eventually approved, KMB's initial order was for 20 Leyland Olympian/Alexander, 11-metre, 3-axle buses to the same general specification as DX2437, an order which was placed in November 1988.

Dennis, in conjunction with bodybuilder Duple-Metsec, developed its own airconditioned double-deck bus, hoping to retain its share of KMB orders. KMB inspected this alternative and placed an order for a prototype in January 1990. This Dennis, registered EL5113, later fleet-numbered AD 1, entered service in April 1990. The new design had no opening windows, such was the faith in the air-conditioning equipment!

KMB confirmed a major purchase in March 1990, when a further 20 Olympian/Alexander and 20 Dennis Dragon/Duple Metsec air-conditioned buses were ordered. Of these the first 16 Leylands entered service in November 1990, just as orders were placed for yet another 20 Leyland and 20 Dennis for 1991 delivery. Numbers of air-conditioned buses rose quickly and soon reached 150 Olympians and 130 Dragons, at which point additional buses were introduced without air-conditioning but KMB publicly re-affirmed that it believed that the entire fleet should consist of air-conditioned buses as soon as possible.

Special livery.
With the introduction of air-conditioned double-deck buses, KMB adopted the white livery with grey relief hitherto reserved for its air-conditioned single-deck fleet of Dennis Falcon coaches and various mini and midi-buses. Initially no fleet numbers were carried but in 1991 this policy, reportedly made on aesthetic grounds, was reversed and fleet numbers were applied using adhesive vinyl letters and numerals in blue.

NOTE: Air-cond., air-con. and a/c are widely used in Hong Kong as abbreviations for air-conditioned/ing.

BELOW: Dennis, in conjunction with Duple (Metsec), supplied a completely-built-up air-conditioned double-decker in early 1990, with the body to a new design and having no opening front or side windows. The prototype is seen here posed for the official camera in England prior to its being shipped to Hong Kong. *(Duple-Metsec*

Leyland Olympian - 11m air-conditioned. AL-class

DX2437. Later numbered AL 1.

This precursor of what was to become the the AL-class of air-conditioned double-deckers, and later its premier member, was developed as a joint venture by Leyland Bus and Walter Alexander, the body manufacturer. Chassis was the tried and tested 11-metre, 3-axle, Leyland Olympian powered by a Cummins L10 engine rated at 220bhp to give a chassis model code ONCL10/5RZ, indicating the provision of a ZF fully automatic gearbox, in this case the 4-speed model. The chassis number, 10743, was out of sequence and in advance of others registered at that time in the series 10675-83 and to similar specification but without air-conditioning.

The Alexander RH-type body was adapted for air-conditioning from the standard design and is unique amongst the class in having opening windows - essential in the prevailing climatic conditions should the air-con. fail. Seating was H67/40+31, as opposed to H69/40+34 for a standard S3BL-class bus registered a few days previously, reflecting the extra weight and intrusion into the rear of the upper deck by the a/c equipment.

DX2437 was first registered on 12th May 1988 and was used for extensive trials of the air-conditioning equipment, entering service on residential/commuter Route 238M, between Riviera Gardens and Tsuen Wan MTR Station where it was joined in May 1990 by the prototype air-cond. Dennis, EL5113. KMB's air-cond. buses were all allocated fleet-numbers by the end of April 1991, although the two prototypes did not actually have them applied until mid-year. Livery was mainly white with grey relief and blue lining with red lettering.

VEHICLE SPECIFICATION: Leyland Olympian Air-conditioned.

Fleet numbers:	AL 1-150,
Chassis make:	Leyland Olympian ONCL10/5RZ (AL 1)
	Leyland Olympian ON3R49C18Z4 (AL2-150)
Engine:	Cummins LT10, 180kw (245bhp)
Gearbox:	ZF4HP500 4-speed fully-automatic.
Body Make:	W. Alexander, ckd.
Body layout:	H63/39+30FE,FdXS,A/C.
Date introduced:	1990-93
Length:	11100mm
Width:	2500mm
Height:	4400mm
Wheelbase:	4900mm+1600mm
Unladen weight:	13,110kg
Gross vehicle weight:	21,450kg

A/C system.

Air-conditioning equipment is by Sutrak and three small compressors are shaft-driven from the gearbox - the previous MCW air-cond. bus having a single large compressor driven from the engine which was vibration prone and unreliable.

AL2-21; 22-62 et seq.

The first 'production' AL-class (Air-conditioned Leyland) buses entered service with KMB towards the end of 1990 with the premier member, AL 2 (ER4295), being registered in October with the remaining 19 following in November to complete the first batch of 20.

The chassis and body specification followed that of the experimental DX2437 in all significant ways. All 20 entered service without fleet-numbers, as was normal until early April 1991, when these buses received the numbers AL2-21.

AL22-41 and AL42-62 were further orders for similar buses, the latter being extended to 41. AL22 entered service in February 1991 and AL42 on 19th September. After this, further orders were placed for 23, 30 and 15 AL-chassis, bringing the total of AL-class buses to 150.

AL134 was converted to a 'Studio Bus' with the classification 'Oly11M(air con)(S)' by the removal of all the upper-deck seats to make provision for the broadcasting equipment. The exercise was to promote the use of the premimum-fare air-conditioned bus services whilst providing a minor public service in the form of traffic reports. Layout for this unusual role was H00/39+30FE,FdXS.

BELOW: The prototype Leyland Olympian air-conditioned double-decker, DX2437, commenced its trial in 1990 on service 238M, a 2.8km route between Rivera Gardens and Tsuen Wan MTR station. It remained on the 238M and was seen here on the roof of Lai Chi Kok Depot in 1994, still allocated to the route. *(Clement Lau*

ABOVE: AL45 leaving Jordan Road Ferry bus terminus en-route for Tuen Mun on express Route 60x. The deeper destination screen and the black surrounds to the non-opening bonded windows make for a very smart appearance, enhanced by bright metal wheel-trims of traditional Leyland pattern. *(Mike Davis*

BELOW: AL110, followed by tram 47, in Des Voeux Road Central in September 1994. The forward staircase can be located from this aspect. *(Mike Davis*

253

Dennis Dragon 11m - air-conditioned.   AD-class.

With dual-sourcing a matter of Company policy, KMB ordered a single trial Air-conditioned Dennis Dragon for evaluation alongside the trial Olympian, DX2437, a vehicle that was registered EL5113 and entered service in April 1990, sporting a new-style of Duple (Metsec) body, featuring fixed windows. Presumably if the air-conditioning was to fail it would be taken out of service immediately. In the winter months, forced-air ventilation at ambient tempreture replaces the chilled-air from the air-con. unit. The bodywork is to the same general outline as that supplied to CMB for its air-conditioned Dennis Condor vehicles, DA1-36.

The Sutrak air-conditioning is driven by a shaft from the engine, and, like the Olympian, three small compressors are provided rather than a single large capacity unit.

Almost exactly a year after its entry into service, EL5113 received the fleet number AD 1.

VEHICLE SPECIFICATION: Airconditioned Dennis Dragon,

Fleet numbers:	AD1-245 (by 1994)
Chassis:	Dennis Dominator 6x2/Dragon
Engine:	Cummins LTA10 or Gardner LG1200
Transmission:	
Body make:	Duple Metsec
Body layout AD1:	H66/41+34FE,FdXS,A/C.
AD2-130:	H62/40+33FE,FdXS,A/C.
Date introduced: AD1:	1990
AD 2-130:	1991-93
Total:	130
Length:	11m
Width:	2.5m
Height:	4.4m
Wheelbase:	
Unladen weight:	
Gross vehicle weight:	

AD 2-130.

In March 1990, three-months after placing the order for what was to become AD1, KMB ordered another twenty air-cond. Dragons and equivalent number of Duple-(Metsec) body kits. From then on, orders were placed in tandem with orders for AL-class Leyland Olympians:

January 1990	1	PLUS:	
March 1990	20	1994	60
January 1991	41	and	55
April 1991	23		
October 1991	30		
November 1991	15		
Total	130		

Equivalent Duple (Metsec) orders tally.

All the AD-class Dragons are similar mechanically but the specification is the subject of continuous up-rating, AD1 being DDA1806, AD2-85 DDA1811, except AD53/6 which are DDA1812 with Gardner LG1200 engine. AD86 on are DDA1813.

The Duple (Metsec) bodies are similar in appearance to AD1 and are known by KMB as Dragon(A)11M(air-con) with a layout of H62/40+33FE,FdXS,A/C.

LEFT: The prototype air-conditioned Dennis Dragon, EL5113, stands in Tsuen Wan bus station in October 1990. No destination blinds had been provided for the 238M and this brand-new bus lost some of its appeal by carrying loose slip-boards. *(Nigel Eadon-Clarke)*

ABOVE: AD37, of the second 'production' batch, on lay-over between turns on the 81c in April 1994. *(Nigel Eadon-Clarke)*

RIGHT: AD108, an example of an air-conditioned bus in advertising livery. There appears to be a concious effort to keep advertising on air-cond. buses in pale colours, often based on white. *(Mike Davis)*

LOWER: AD46 of the third batch of 'air-cond' Dragons shows-off the deeper bumper bar fitted from AD42. Seen on the then newly introduced Route 64M in June 1992. *(Nigel Eadon-Clarke)*

Dennis Dragon 10m - 3-axle-air-conditioned. ADS1-30.

During 1993, KMB placed in service thirty short wheelbase Dennis Dragons of 9.983m overall length. The unusually short length on three-axles was required in order to keep the gross axle weight within limits whilst carrying the weight of the air-conditioning plant above the rear-engine.

Bodywork is the usual Duple (Metsec) for mounting on Dennis chassis and is similar to the AD-class Bonded windows are fitted to side windows and those at front and rear are rubber gasket mounted.

VEHICLE SPECIFICATION:	Air-conditioned Short Dennis Dragon.	
Fleet numbers:	ADS 1-30	
Chassis:	Dennis Dragon 10-metre	
Engine:	Cummins LT10 - 252bhp	
Transmission:	Voith D863	
Body make:	Duple (Metsec)	
Body layout:	H55/34+28FE,FdXS,DOO,A/C	
Date introduced:	1993.	Height: 4369mm
Total:	30	Wheelbase: 4117mm+1600mm
Length:	9983mm	Unladen weight: 13,310kg
Width:	2500mm	Gross vehicle weight: 21,800kg

RIGHT: The 9.98 metre version of the Dragon is actually the same length as a 2-axle Dominator, from which the Dragon was evolved. ADS27 was photographed on the roof of Lai Chi kok Depot, after licensing but prior to entering service. *(Clement Lau*

BELOW: In order to provide an air-conditioned service on routes where the geometry of the roads precludes the use of 11 or 12 metre buses, KMB purchased thirty Dennis Dragons of only 9983mm in overall length. ADS 5 was at work on Route 2 when photographed outside the Penninsula Hotel in September 1994. *(Mike Davis*

Scania N113 - Air-conditioned, 3-axle, 11.5m. AS 1 & 2.

During 1992, it was revealed that KMB had ordered two 11.5-metre, 3-axle Scania double-deckers for evaluation purposes and early in 1993 photographs revealed the prototypes at the Alexander body plant at Fiakirk ready for shipping. These two buses represent a new move by Scania into the field of 3-axle double-deck buses with air-conditioning.

The Scania designation for these two chassis is N113 DRB.

The Alexander bodywork shows a departure from the previous orders in having a one-piece curved windscreen and forward staircase

VEHICLE SPECIFICATION: Scania 3-axle - air-conditioned.	
Registration numbers:	FU 482 & FU2948.
Fleet numbers:	AS1 & 2.
Chassis:	Scania N113 DRB
Engine:	
Transmission:	
Body make:	W. Alexander (nos RH110/2092/1 & 2092/2)
Body layout:	H63/39+34FE,CXS,DOO (provisional)
Date introduced:	1993
Total:	2
Length:	11.5m
Width:	2.5m
Height:	?
Wheelbase:	?
Unladen weight:	?
Gross vehicle weight:	?

ABOVE: The first Scania, AS1, on MTR relief service 300 in Central, about to return to Prince Edward MTR Station in Kowloon shortly after entering service. The curved windscreen was a new feature on KMB double-deck buses. *(Nigel Eadon-Clarke*

RIGHT: The offside of AS2 arriving in Des Voeux Road on 18th April 1994, followed closely by Dennis Lance (FW3600) on the Airport Service A2. According to another photograph by Chan Ka Yu, AS1 also bore this livery for exhibition to the public. *(Nigel Eadon-Clarke)*

Volvo Olympian - 11m - 3-axle, Air-conditioned. AV 1-50.

This vehicle was the first KMB Olympian to be supplied by Volvo rather than Leyland and was bodied by Alexanders at the same time as the two Scanias to which it shows a family likeness, having the one-piece curved windscreen of the home market model. It was, the first one of this 11.3 metre length.

The chassis designation 53.80 is a Volvo development code and has not been used for the subsequent 49 vehicles..

Because bodywork conceals the mechanical units of a modern bus, only the front wheels give away the fact that these are Volvo-built at Irvine, Scotland, rather than Leyland-built at Workington. Inside, the Volvo steering-wheel and dash-panel give the game away.

The foot of the staircase is opposite the exit doorway, both are forward rather than centre located.

VEHICLE SPECIFICATION: Volvo Olympian 53.80

Fleet numbers :	AV1-
Chassis: AV1:	Volvo 53.80 Olympian
AV2 -11:	Volvo Olympian
Engine:	
Transmission:	
Body make:	Alexander (AV1: cbu; AV2 et seq: ckd)
Body layout:	H63/39+29FE,FdSX,DOO
Date introduced:	1994
Total:	10 (first order).
Length:	11.3m
Width:	2500mm
Height:	
Wheelbase:	
Unladen weight:	
Gross vehicle weight:	

LEFT: Nearside view of the Volvo 53.80 for KMB standing in the forecourt of Alexanders at Falkirk. This shows well the forward located exit doors. Although painted in base white, the finishing details, fleet-name, legal lettering, stripes and grey skirt were applied in Kowloon. *(David Kat*

BELOW: The first Tuen Mun assembled 11.3m Volvo Olympian, AV 2, leaves the auto-toll booth at the Kowloon end of the Cross Harbour Tunnel in September 1994. The high visibility handrails can be clearly seen through the front glazing

Volvo Olympian - 12m, 3-axle, air-conditioned. 3AV-class.

With air-conditioned buses interspersed between non-air-con. 12-metre buses, fifty 12-metre, air-conditioned, Volvo Olympians were ordered for delivery in mid-1994 - when the early 12m buses became twelve years old.

The bodywork contract was, as usual, for Olympian chassis with KMB, awarded to Walter Alexander who used the improved design introduced with the 11.3M AV-class, having the curved windscreen of the 'Royale' body sold on the home market..

For the type-code, KMB reverted to the use of the number '3' as a prefix, rather than enter the realms of long letter groups, to produce the 3AV-class.

VEHICLE SPECIFICATION:	Air-conditioned Volvo Olympian
Fleet numbers :	3AV 1-50
Chassis:	Volvo Olympian SLVYNC - 12m
Engine:	Cummins LT10 - 252bhp
Transmission:	ZF 4HP 500
Body make:	Walter Alexander
Body layout:	H68/43+26FE,FdXS,DOO,A/C
Date introduced:	1994
Total:	50
Length:	12,000mm
Width:	2500mm
Height:	4382mm
Wheelbase:	5600mm+1600mm
Unladen weight:	14,640kg
Gross vehicle weight:	23,141kg

RIGHT: 3AV2 in Queensway, Hong Kong Island. Ex-Servicemen who served in Hong Kong prior to 1980 will find it hard to recognise the site of the former Sergeants Mess on what was Queen's Road, by the sharp double-bend. 3AV2 had its back to Central and was heading towards Hennesy Road and the Cross-Harbour Tunnel when photographed here in 1994. *(Mike Davis*

BELOW: 3AV42 passes the Hong Kong Polytechnic as it approaches the toll plaza of the Cross-Harbour Tunnel in September 1994, on its way to North Point. *(Mike Davis*

Liveries (See page 270).

ABOVE: The original post-war KMB fleet livery remained unaltered until the early 1970's when the advent of One-Conductor-Operation (OCO) and Driver-Only-Operation brought variations and, eventually alterations. Here Daimler(c) AD4546 displays the original style of mainly red and cream with lining in dark green as it approaches the "Star" Ferry terminus in 1971. *(Lyndon Rees*

BELOW: The yellow stripes indicating that the bus is operated by one conductor (OCO) show well in this view and indicate to intending passengers that they should enter by the front doorwar with fares ready and pay at the cash desk. Daimler (a) No 4966, built in 1949, seen here in 1974, turning out of Jordan Road whilst on its way to Shek Kip Mei Resettlement Estate, on Route 2D. *(Mike Davis*

ABOVE: The 'half-and-half' cream-over-red livery was introduced firstly on the 'tunnel' routes which worked alongside CMB buses as Driver Only (then OMO). At first the green lining was retained but was later dropped. Daimler(e), D468, turns out of Jackson Road, Central, on Hong Kong Island in 1974 when working on the 105 from West Point to Lai Chi Kok (Amusemant Park). *(Mike Davis*

BELOW: As driver-only-operation spread route by route through the KMB system, older buses were painted into the 'half-and-half' livery, complete with 'coin-in-the-slot' signs either side of the destination screen. Daimler (a) D40, which entered service in February 1950 with a traditional exposed radiator as seen on 4966, opposite, photograph. By the time it was photographed here in January 1978, it had exchanged its original front for a 'Manchester' type as fitted to the later Daimler(b))s. *(Derek Lucas*

ABOVE: BD5118, previously Rhondda and Western Welsh No 463, was typical of the AEC Regent MkV's purchased second-hand and displays to advantage the livery chosen for the 1973 second-hand fleet. Photographed here in Jordan Road in 1974, BD5118, become 2A11 later in the year, was approaching its terminus having travelled in from the New Territories on trunk Route70. *(Mike Davis*

BELOW: Here Daimler(cb) The older types were also painted into the 1982 livery, which bore a striking resemblance to that of the second-hand fleet of a decade earlier but was probably influenced by the new Chief Engineer who had recently arrived from South Yorkshire PTE whose livery was similar. *(Clement Lau*

ABOVE: Seddon Mk 17, HK4415 stands in an unidentified KMB depot yard in the 1960's *(Julian Bowden collection*

CENTRE LEFT: AEC Regent MkV, AD7178 stands at the Kowloon "Star" Ferry terminus, sporting a colourful livery advertising 'Rodania' Swiss watches. *(Mike Davis*

BELOW LEFT: Daimler(e), D548 (AD7454) was advertising 'Titoni' watches in a contemporary 'flower power' style in 1973 when it was only about six months old. It is seen here at Causeway Bay Magistracy terminus, behind lowbridge CMB Atlantean. *(Mike Davis*

ABOVE: The use of fareboxes with a flat-fare tarif quickly became the norm, the use of 'coin-in-the-slot' signs was dropped. AD7324, a Daimler(e) was on Kowloon Urban Route 14 and would have differed from similar buses on 'tunnel' routes in having hard, moulded plastic (fibre-glass) seats rather than cushion seats. *(Mike Davis)*

CENTRE LEFT: Daimler(e), AD7450, in 1974 on a 'Tunnel' Route, showing the Government slogan 'Fight Crime'. (Later fleet numbered D544.) *(Mike Davis)*

BELOW: 1984 photograph of yet to be numbered Leyland Olympian, DB5551, 3BL27, in the 1982 livery with 1984 logo. *(Mike Davis)*

Coach Liveries.

LEFT: For its "Hong Kong in Britain Exhibition" in London's Battersea Park during September 1980, the Hong Kong Government sponsors took advantage of there being a KMB air-conditioned coach in the country at the time, awaiting shipping. Then attractively painted in yellow and off-white by Alexanders, the experimental Victory Mk2 was on show at the event. Here it was posed for the publicity camers on Westminster Bridge late one evening, together with a traditional Chinese lion dancer. *(Hong Kong Government*

LEFT: The first livery for KMB coaches was an attractive light golden-brown and cream. Coaches on Airport Coach Service 200 & 201 had an additional red panel along both sides, on which the service was promoted. *(Mike Davis*

LOWER LEFT: After the Albion/Duple coaches had been superceded, firstly by Dennis Falcon coaches nad latterly by Dennis Lances, the airport routes were revamped and renumbered in a new Airbus series commencing with the A1. Here Dennis Dart AA13 (FP1778) awaits its next trip from the Airport to Tsim Sha Tsui in May 1994. *(Nigel Eadon-Clarke*

265

Training Bus Liveries.

ABOVE: CN672, formerly 2D22 (DMS735) on the training circuit on the roof of Lai Chi Kok Depot in April 1990 in the 'old' livery. *(Nigel Eadon-Clarke)*

ABOVE RIGHT: For training duties, KMB only amended the coach livery by adding yellow arrows to the sides of Albion coaches, as illustrated in this view of BH7644 in 1984. *(John Shearman)*

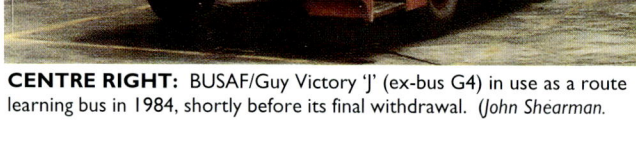

CENTRE LEFT: The ex-Singapore demonstrator, DAF engined Dennis Dominator SBS7003Y, after conversion to advanced trainer DL9735, stands alongside Metsec bodies 33ft long Fleetline BM4486 (ex-D894). *(C. Lau)*

CENTRE RIGHT: BUSAF/Guy Victory 'J' (ex-bus G4) in use as a route learning bus in 1984, shortly before its final withdrawal. *(John Shearman.)*

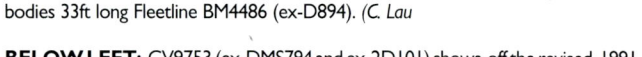

BELOW LEFT: CV9753 (ex-DMS794 and ex-2D101) shows-off the revised, 1991, Training Unit livery, similar to that of the air-conditioned buses. *(Nigel Eadon Clarke)*

BELOW RIGHT: Two Leyland 'Victory Mk2's', of which G263 (CK4013) in cream livery, was on short term loan to the Training Unit in April 1992. *(Nigel Eadon-Clarke)*

Company Insignia.

ABOVE: The KMB Company crest that was introduced circa 1950 and which continued in general use until about 1984. *(Kowloon Motor Bus*

The KMB initials as introduced on the Albion coaches in the form of an arrow; quietly dropped

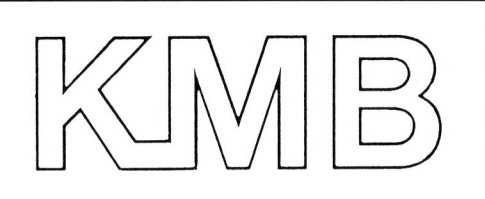

The form of the KMB initials as generally introduced in the mid 1970's.

BELOW: The current KMB insignia, introduced after the Jubilee in 1983. *(Kowloon Motor Bus*

Tickets (see page 273).

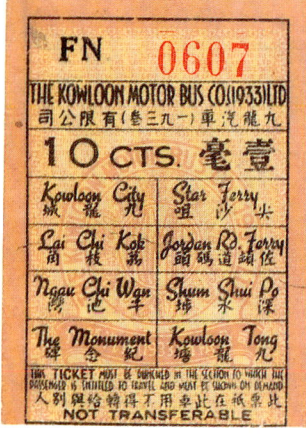
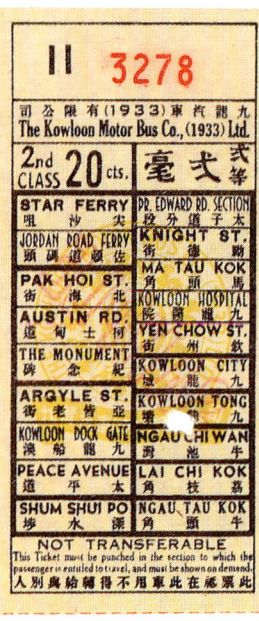

TOP ROW - 1950's/60's style 'flimsy' tickets printed on glazed, semi-waterproof, transluscent paper; **Left to Right: II 3278:** 20 cents second class of the 1950's showing Monument as a destination (It has been difficult to find an informant who can recall the location of this!). **FN 0607:** 10 cents also from the 1960's. **DPF 2747:** 30 cents dating from the 1971 fare increases, issued for a bus from L.C.K. (Lai Chi Kok); **BTW 0579:** 40 cents from the same period; **RIGHT: BQG9330:** 50 cents post-1971 ticket issued on a New Territories route with fare stages.

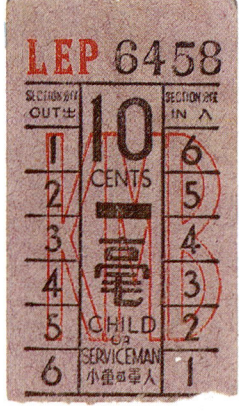
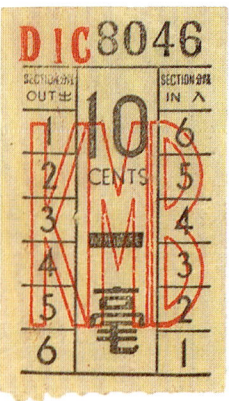
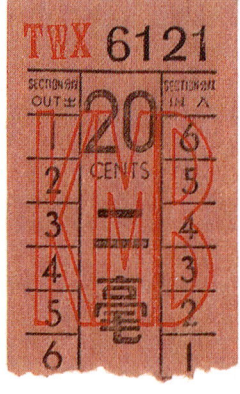
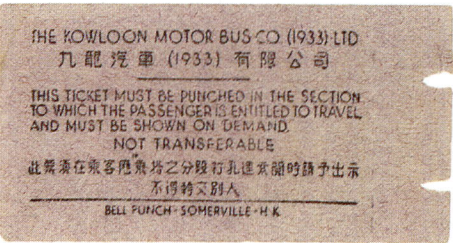

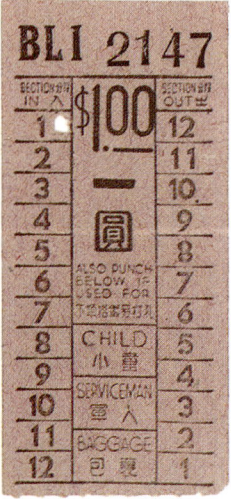

LEFT: 1960's 'Bell Punch' tickets: **LEP 6458:** 10 cents 'Child or Serviceman'; **DIC8046:** 10 cents Adult (short ride from start of route or last mile); **TWX6121:** 20 cents Adult (full distance on Kowloon routes - up to seven miles.); **BLI2147:** New Territories routes $1.00 Adult, Child, Serviceman or Baggage; **DIW0022:** Shorter distance NT routes 50 cents Adult, Child Serviceman or Baggage.

ABOVE: Reverse of 'Bell Punch' ticket.

BELOW; Autofare machine issued ticket 1980.

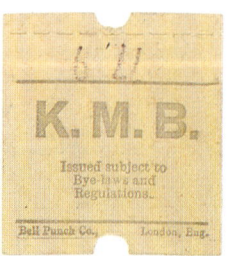

Part Six:
Other Aspects of KMB.

Franchise.

As has been briefly described in the introduction, during September 1932, the Government of Hong Kong called for tenders to operate bus services in the Colony under a new franchise system to commence on 11th June 1933. These franchises would ensure the exclusivity of operation within the region successfully bid for. These areas were to be either;
1) Hong Kong Island, or
2) Kowloon and the New Territories, or
3) The entire Colony.

According to a report in a KMB publication, five men met in November 1932 to compile a tender to operate bus services in Kowloon and the New Territories. These men were Tang Shui Kin, William Lui Shui Tak, Tam Woong Tong and Lam Ming Fan and, in January 1933, it was announced in the Government 'Gazette' that their tender had been accepted.

As a result of the tendering, separate franchises were let, one for each side of the harbour, that for Kowloon and the New Territories being won by Tang Shui Kin and his colleagues who, on 13th April 1933, formed the Kowloon Motor Bus Company (1933) Limited, an amalgamation of the original Kowloon Motor Bus Co. with the Kai Tack Motor Bus Co. (1926) Ltd. and a few small operators in the New Territories, with Tang Shui Kin (later to be Knighted as Sir Tang Shui Kin) as its chairman, a post he retained until his recent death. Not included in the amalgamation was the China Motor Bus Company which, although a Kowloon operator, had won the franchise to operate bus services on Hong Kong Island.

While CMB chose to transfer its smaller buses, those more suited to the Island's prevailing road conditions, to its new operating area, its larger buses, including the Leyland PLSC1 'Lions' and Thornycroft CD4LW 'Cygnets', were sold to the new KMB

As far as terms and conditions were concerned, KMB's franchise was similar in most respects to that of CMB's and was awarded for a period of fifteen years, extendable for periods of five years dependent on good performance by the operator. This has remained the case, despite the intervening war years and the enactment of various new items of legislation and, even today, even today the Company is required to conform to much the same terms and conditions as were introduced in 1933. The first franchise was renewed in 1948, for a period of ten years, expiring in 1958. Lengthy negotiations resulted in a new fifteen year franchise being awarded but not until 1960. A temporary extension was granted to cover the period of the negotiations from 1958-60. When the resulting fifteen year franchise expired in 1975, another period of protracted re-negotiation took place as a result of which, the franchise has since been renewed on a route by route, rather than an area, basis, initially for a period of ten years but rolled-over every two years for a further two years dependent on satisfactory performance as detirmined by the Governor-in-Council, acting on davice from Transport Department. This means that, should Government consider that the Company is under-performing on a particular route, or routes, only the particular route or routes would be forfeited, rather than the somewhat impractical, simultaneous termination of over three-hundred routes. This option has never been exercised against KMB but it was the instrument by which Government, in 1992, could offer twenty-six routes for tender on Hong Kong Island.

The terms of the franchise were amended to permit KMB and CMB buses to work on routes penetrating each other's exclusive territory following the opening of the Cross-Harbour Tunnel in 1972.

One turbulent matter affecting the exclusivity of KMB's operating area occurred in 1988 when, in readiness for the opening of the extensive light rail transit (LRT) in the large New Town of Tuen Mun in the north-west New Territories, Government sought to protect the embryo tramway from competition by creating an exclusive 'North-west Transit Service Area' (TSA). The LRT operator, the KCRC (Kowloon Canton Railway Corporation), would absorb all existing KMB bus routes which lay wholly within the TSA. Bus routes within the TSA were operated by the LRT, either as feeder-bus services to the LRT or as pre-LRT routes, subsequently to be replaced by tram services. KMB bus services with one or both terminus/i external to the TSA were restricted to setting-down on inward journeys and picking-up on outward journeys, thus preventing local journeys which would offer competition to the LRT. Lastly, there were restrictions to minibus (Public Light Bus) operations within the TSA so as to minimise their local role.

KMB was, not surprisingly, most unhappy about having to transfer so much potentially lucrative business which had been, slowly developed, probably at a loss in the early days of the population the build-up during the development of Tuen Mun and prior to the opening of the LRT with its TSA. It was not without a struggle that the Government's will was to prevail but, however, the LRT became the victim of its own success, hard-pressed to cope with the traffic it generated and so, in order to relieve the situation, KMB was released from many of the restrictions and, today is free to operate routes in the area, although, at the time of writing, has declined to do so.

RIGHT: A cutting from the leading Hong Kong English language daily newspaper, the 'South China Morning Post', for 16th January 1933.

OMNIBUS SERVICES.

Confirmation of Local Monopolies.

FOR FIFTEEN YEARS.

The reports in circulation at the week-end that the China Motor Bus Company had been awarded the franchise for the operation of motor buses on Hongkong Island, and the Kowloon Motor Bus Co. for Kowloon and the New Territories, are officially confirmed, the *Gazette* intimating the acceptance of their tenders.

Tenderers were given the option of applying, for the rights either in respect of the whole Colony, or separately in respect of Hongkong and Kowloon. The franchise extends for a period of 15 years as from June 11th next.

It is laid down that British vehicles must be used and that the majority of the partners or directors of the successful tendering firms must be British subjects.

Tenderers had to offer on the basis of a royalty by way of percentage on gross earnings. Specimen lists of fares considered reasonable for the various sections were included in the tender forms, and all tenderers had to submit itemised maximum fares. These fares are to prevail for three years after which, and from time to time subsequently, the Governor-in-Council may prescribe such fares as are deemed to be reasonable having regard both to the licensees and the general public.

Double-Deckers.

Double-deck vehicles are not allowed on the Hongkong side, while the use of such vehicles at Kowloon is subject to the advice of the Director of Public Works as to the suitability of the roads therefor.

A special condition laid down was that if any company operating a bus service so desired, the successful tenderers might be required by the Governor in Council to purchase at such time as the Governor in Council may direct, under some method of valuation to be determined by the Government, all or any suitable and effective vehicles, repair plant, and machinery, lands buildings and materials used by such company for the purpose of its undertaking prior to the 10th June, 1933.

269

KMB Fleet Livery.

During the thirty years following the end of the Pacific War, the predominant colours used to paint the buses of Hong Kong, Kowloon and Lantau (and nearby Macau) were red and cream (yellow on Lantau), red being a very fortuitous colour in the culture of China. The differences between the companies lay in the method of dividing the two main colours and in the lining-out. The style adopted by Kowloon Motor Bus was as follows:

a. Single-deck buses.

The post-war style of single-deck bus livery remained in use on the majority of such buses until the mid-1980's and on the older types until their eventual demise. Exceptions were the Albion coach fleet and some late Albion buses. Later, air-conditioned coaches and small buses were to introduce another special livery. The main livery was as follows:
Cream roof and window surrounds;
Green beading band, approx. 2.5cm (1in) wide, later omitted;
Cream moulding band, approx. 15cm (6in) wide, later omitted;
Green beading as above, also later omitted;
Red side, rear and front panels;
Black mudguards.

The Company crest was displayed on each side, between the wheel-arches, on the red side panels. Below the crest there was, at one time, a repeat of the registration number in cream; this latter served as a fleet number until 1973/4, since when fleet numbers have been applied on the cream moulding in red stencilled characters or in cream, similarly applied on red panels. In 1974, the green beading was omitted for a time but was reinstated at the request of the Chairman of the Company, only to disappear again by the early 1980's, when the double-deck livery was revised.

A minor departure from the standard practice was the painting of the front dash panels of the 1970/71 delivery of Seddon Pennine 4's and Albion Vikings (L220-269) which, for an unascertained reason, were painted cream.

A totally revised style appeared in 1976, with the introduction of the last series, L279-312, Albion Viking buses. These were basically cream with three broad bands of red around the side panels, the uppermost being below the side windows, the lowest around the skirt while the third lay centrally between the upper and lower bands. No green beading was applied. These buses were later repainted in the standard livery.

b. Double-deck buses.

From 1949-1972, all new deliveries of double-deck buses were painted in a similar style to the single-deckers but extended to these vehicles as follows:
Cream roof and upper-deck window surrounds;
Green beading;
Cream moulding beneath upper-deck windows;
Green beading;
Red upper-deck side panels and lower-deck window surrounds;
Green beading;
Cream moulding;
Green beading;
Red lower-deck side panels.

Crests, registration numbers and fleet numbers were carried as for the single-deckers, *qv*.

From 1972, with the introduction of seated conductors, three yellow bands were applied to the red upper-deck front panels to warn intending passengers of the need to have the correct fare ready and that they should board by the entrance indicated on the side panel marked by the English word "ENTRANCE" and the equivalent Chinese character.

With the subsequent introduction of Driver Only Operation (then called OMO), the livery was revised, on those buses so operating, to a 'half-and-half' livery and all upper-deck red disappeared. Green beading was retained for a time but was dropped, reinstated and finally dropped in 1981. The cream moulding was retained below the lower deck windows on older buses, but the Fleetlines, introduced in the 1970's, were in a very basic 'half-and-half' livery with red to the top of the lower-deck windows and all cream above with no lining. At this time the Company's initials first appeared on the offside on the staircase panels and, on Fleetlines, on the rear engine lid.

All-over advertisements were introduced to the KMB fleet in 1972 and was extended to buses of most types, except the oldest Daimlers of the (a) & (b) series and single-deckers.

c. Second-hand buses (1973).

From 1973 (one bus Dec '72) many double-deckers were introduced second-hand from UK operators and these were given a new, mainly cream, livery relieved only by two broad bands of red, one at skirt level and one at upper-deck floor level. In the latter years of these buses the standard 'half-and-half' livery was adopted on surviving examples, mainly AECs and Daimlers, although some rebodied Atlanteans were also so treated.

d. '1982' livery.

A new livery was adopted in 1982 where only the roof and skirt panels were red, the remainder of the bus being in a rich cream colour with KMB in large block letters between decks. For the Company's Jubilee in 1983, appropriate slogans were carried between decks together with a new logo in the form of a stylised 'KMB' (see 'Insignia'). At this stage the traditional KMB 'nine dragons' crest was dropped, together with the registration number painted on each side. The new logo was based on the Chinese character for nine, (*kau* or *gau*) - nine as in Kowloon, or, more correctly, '*Kau Leung*' - Nine Dragons.

e. Coaches - 1970's

The single-deck Bedford and Albion coaches were painted in a completely different livery devised especially for them, where the main body colour of golden-brown at the front, was changed to a sandy cream towards the rear, the change being effected by means of intermingled lines or bands. After repaints, the change from dark to light paint was achieved by a simple diagonal break. The reader is directed to photographs to help visualise both this and the original arrangement. The lower front and sides were the lighter shade right around the vehicle, while the roof, including the front upper part of the 'Bristol' dome, was white.

The two experimental double-deck coaches, one Leyland and one Dennis, introduced during 1981 and 1982, were painted in an adaptation of the single-deck coach livery.

f. Air-conditioned vehicles.

With the entry into service in 1985 of the Dennis Falcon airport coaches, in replacement of the Albions, a new all-white livery was introduced and later extended to other air-conditioned types, including midi-coaches and three-axle double-deckers.

Although the description of 'all-white' is used, relief is applied, the complete scheme being, from roof downwards:
White roof;
Black window surrounds;
white between-decks panel;
red stripe at upper-deck floor level, kicking-up towards the rear and crossing the back of the bus below the upper-deck windows;
Black lower-deck window surrounds;
White front side and rear lower panels;
12-inch grey lower skirt panels;
Black bumpers front and rear.

The words 'AIR CONDITIONED' are applied in blue lettering, in English and Chinese, above the front destination screen, towards the rear between decks, below the up-raised red band and below the nearside of the windscreen on the dash panel. The KMB logo appears on the side panels, above the exit doors and in the equivalent offside position

Insignia.

Close examination of pre-1941 photographs reveals that the Company crest was different from that in use post-war, from approximately 1950 to 1984. It has not, however, been possible to find an illustration of the earlier design.

Both pre-1941 and early post-1945, the Company name was sign-written along the sides of the buses, on the cream moulding band, with the English language version in block capitals above the Chinese equivalent. The exact date of the change to the Company crest, as illustrated, on this page, remains unclear but the first double-deckers, introduced in early 1949 did not feature it, whereas photographs taken in 1952 show only the crest.

The name Kowloon means Nine Dragons in Cantonese and these nine dragons, each represent one of the *eight* hills that surround the region. Legend has it that, many years ago, the Emperor of China, whose territory the region still was, asked of a wise man why there were only eight hills when the place was called Gau Leung (Nine Dragons). The wise man answered the Emperor, saying that the ninth dragon was the biggest and worthiest of all; the Emperor himself.

During the early nineteen-seventies, KMB sought to make its buses more distinctive beside those of China Motor Bus, prior to the latter's change from red to blue livery. KMB chose to apply its Company's initials on the sides and rear of its buses in an un-serifed block capital style in which the lower leg of the K was joined to the first leg of the M. This was, at first, applied in cream on the red paintwork, usually on the offside panels of the stair-well and the lower rear panels but was usually omitted on the nearside where there was no suitable location. Exceptions were the British assembled buses shipped complete, including the first Victory Mk2, G5, and the ECW bodied Olympians. Later, the three-axle prototype buses had the initials in red on cream panels.

With the introduction of the Albion/Duple coaches in 1974, a variation of the 'KMB' initials was tried on the first few where the 'K' was styled so as to form an arrow-head and the 'B' formed the tail. This was held by some to represent 'arrows of ill fortune' and was quietly dropped.

The two Alexander bodied double-deck coaches had 'KMB' applied in brown above the door, as well as on the offside and the rear in cream.

With the introduction of the '1982' livery, the 'KMB' initials appeared in red on both sides on the between-decks panels, in large letters which sometimes showed below advertisements, when these were superimposed.

The following year, 1983, KMB celebrated its Jubilee and added to the 'KMB' initials the legend '50th Anniversary' in Chinese on the nearside and English on the offside. The initials themselves were in a new, stylised form and the first use was then made of the new logo, based on the Chinese character representing the number nine, *'gau'* (九). While the shape of the character was clear, included in the design of the stylised version was a flyover, a road junction and three main roads.

After the jubilee year was passed, the new 'Gau' logo continued in use together with the stylised 'KMB' to become the Company's standard insignia. Later the slogan, in Chinese characters, 'KMB - getting better and better' was added below the main components of the logo. This style is now used on all KMB vehicles, buildings and paperwork.

RIGHT: Until about 1983, KMB repeated the registration number on each side, beneath the Company's crest although, in isolated instances, the fleet number appeared after 1974 but this was probably done in error. *(Mike Davis)*

KMB Registration Numbers

Prior to the Pacific War (1941-1945) registration numbers were allocated on the basis of the next available number, using a system with a maximum of three numerals, ie 1-999. No prefix letters were used and, it is assumed, the authorities reissued surrendered numbers. All plates were white with black numerals, regardless of vehicle class. Photographs show that franchised buses also carried the words, in small letters, PUBLIC VEHICLE, on the plate, above the number.

Following Liberation in August 1945, surviving buses retained their original pre-1941 number until mid-1946 when a new system was commenced and all vehicles were re-registered. Vehicles which had no tracable number or which were newly introduced were allocated a serial number by the Military Administration and it is believed that these were issued on a local, regional, basis, eg HK Island, Kai Tak, Kowloon, etc.

From mid-1946, existing and additional buses received new four-figure numbers which were issued in blocks ahead of requirement and allocated by the Company themselves, the number becoming 'issued' only when a vehicle was first registered, beginning with un-prefixed 4xxx, followed by HK4xxx and then AD4xxx, AD7xxx and AR7xxx, irrespective of the prefix, if any. (The reason for this is outlined in the annex describing the system of vehicle registration more fully.) These were, in general order of issue, 4201-50, 4591-4690, 4951-4999, HK4251-4590, AD4701-4900, AD7000-7499 and AR7500-7750.

In the early post-war days, up until the late-1950's, many of the un-prefixed numbers were reissued to new buses, the number plates having been physically removed from scrapped buses and, in many cases, stored.

Until 1962, KMB buses, like those of CMB, were fitted with number plates of a noticeably smaller size to those in use by other vehicles but in that year a larger size was introduced, being fractionally larger than those then normal in the United Kingdom.

From 1945 to 1982 all Public Omnibuses were required to carry their registration numbers on plates with a red background and with the letters and figures in white. In the case of KMB these were usually painted, having the appearance of having been stove enamelled.

As is the case with bus companies the world over, KMB buses, which have become too old for passenger service, are frequently converted to other uses, including driver training and service tenders of many types. Until at least the mid-1970's, buses converted for duties other than carrying paying passengers were required to be re-registered in a new category, ie Private Omnibus for driver training or Goods Vehicle for service cars. Not only were the registration number plates changed to have a white background with white characters, the vehicle lost its original Public Omnibus number and received in its place a new number that was the next in the sequence then being issued. This applied particularly to the Albion Victor VT17L's converted to Training buses in 1970-72 and which reverted to passenger use in 1974-75. By this time the rules had been changed, or forgotten, and the reinstated buses retained the 'new' numbers they received as trainers, the plates merely being replaced with red ones carrying the same mark in white.

In 1973-4 the Government ceased to allow the use of pre-booked numbers and the bus companies were allocated the next number in the public sequence, thus ceasing to enable KMB to use registration numbers as fleet numbers.

Fleet Numbering *(a complete list of fleet-number prefixes appears on page 80).*

Commencing in June/July 1973, the then existing fleet was allocated fleet numbers with very broadly based groups having, originally, a single letter to denote the chassis manufacturer; thus, the 34ft long AEC double-deckers became A1-210, Daimlers, regardless of chassis type, D1-665, single-deck Albions took the Leyland name to avoid confusion with the AEC's and became L1-270 while the Seddons were naturally S1-63. The second-hand fleet was separately identified by the addition of the figure '2' - for 2nd hand - so that the 30ft long AEC's, formerly with British operators, became 2A1-28, the Leylands 2L1-79 and the Daimlers 2D1-7.

When studying photographs with a view to dating them, I should be remembered that, although the numbers were allocated in mid-1973, many buses remained un-numbered well into 1974.

Additional vehicles were fleet-numbered on entry into service, Fleetlines, regardless of who made them, Daimler or Leyland, received the numbers D666-1115, but to avoid the re-use of 'D,' when the first Dennis arrived, the initial letter of the second syllable in the name 'Dennis', thus the 'Jubliants' became N1-4, with subsequent additions.

New complications arrived with the advent of the 12-metre, 3-axle buses, the first being from Metro-Cammell Weymann. This long 'Metrobus' was initially prefixed 'M' but was very soon amended to have a second prefix '3' - for 3-axle - becoming 3M1, the similar Dennis was 3N1 but the prototype example of a British Leyland 'Olympian' on three-axles broke new ground in having <u>two</u> prefix letters, 'BL', themselves prefixed by '3' to take the style '3BL1'. The prototype two-axle 'Olympians' were BL1-3.

With the introduction of 11-metre, 3-axle double-deckers, the prefix took on another letter to indicate a shorter version of the main type. To take the case of the 'Olympians' again, the new code 'S3BL' was adopted. Other types followed suite.

At one time the coach fleet, together with the air-conditioned double-deckers and mini-buses, were allocated 'paper' fleet-numbers that were never applied to the vehicles, reportedly for aesthetic reasons, and so the two Bedford coaches and 100 Albion coaches were nominally CB and CA (Coach Bedford and Coach Albion) classes respectively, in a single series from 1-102; the Bedfords being CB67 & 78. The Dennis Falcon airport coaches were, again, nominally FD1-20 but, in 1991, it was agreed with the HK Government Transport Department to allocate fleet-numbers to the coach and air-conditioned bus fleets and the Dennises received the totally different prefix letters 'AF'. Just to confuse the student of the subject, there were only 19 remaining as 'FD6' had been destroyed by fire, making every AF above 6, one lower than the previous, notional, FD number.

The fleet of air-conditioned double-deck buses and minibuses received fleet numbers at the same time as the 'Falcon' coaches. These classes were, predictably, prefixed by the letter 'A' - for Air-conditioned - followed by the initial letter of the manufacturer's name, thus Dennis Dragon became 'AD-class', the already mentioned Falcons, 'AF', while the short, three-axle, Olympians, thankfully, started a new Olympian series (not taking the cumbersome style 'AS3BL'!!), to be simply AL. The only type to gain an additional letter was 'AMR' for the two Metroriders, the Mitsubishi MP117 single-deck buses having taken the prefix 'AM'.

Upon their introduction in 1993, the short (10m) 3-axle Dragons were given ADS fleet numbers while the 11-metre Volvo Olympians became AV and 12-metre Volvos reverted to the older system to become 3AV.

I should be pointed-out that all letter series commenced at 1 but for fuller details of fleet numbers, and their allocation to types, reference should be made to the text or fleet list.

Fleet numbers have been applied by means of stencilled letters and figures from the start - red on cream surfaces and cream on red surfaces. The numbering of the Air-conditioned types has been undertaken by application of neat, self-adhesive characters, smaller than the stencils but not as small as some very recent applications which are little more than 25mm high and almost impossible to pick-out from any but the sharpest photographs.

KMB Depots.

In order to operate its very large fleet of buses, KMB utilizes large amounts of one of Hong Kong's most scarce assets; namely land. Because of its density of population and small useable land mass, Hong Kong people have been forced to accept that buildings must spread upwards; housing, factories, even churches are accommodated in high rise buildings, and, of course, bus garages are no exception, most being like oversized multi-storey car parks.

The headquarters of the Company are at 1, Po Lun Street, Lai Chi Kok, New Kowloon and is an integral part of the large, three-storey, operating depot and garage and, in the 1970's there were two adjacent open-air parking grounds.

Pre-1941 photographs show a depot at the north end of Nathan Road while other older premises at To Kwa Wan, Camp Street and Shing Lok also probably date from pre-war or early post-war. Good use was made of Shing Lok Body Works in 1972 74 when the large fleet of second-hand buses was purchased. Here they were re-built to varying degrees and painted.

To Kwa Wan was a workshop in the '70's, probably an ex-depot, and undertook much of the mechanical work required by the second-hand buses.

In the New Territories, there were small largely open air depots at Yuen Long and Shung Shui which were out stations to Lai Chi Kok and Kwun Tong until rebuilt on new sites in the 1980's.

Two of the earlier, and more spectacular, multi-storey depots were at Kwun Tong, where tthey stood on either side of a public road and, until one was recently demolished, connected one to the other at second and third-floor levels by bridges spanning the road. It was a sight, almost unremarkable in Hong Kong's cramped environment, to see a 12-metre double-deck bus pass high above one's head, but to the newcomer, or interested visitor, the experience was well worth a visit. There is also a nearby open area for overnight bus parking

Since its renewal in the 1980's the third main operating base, at Yuen Long in the less crowded New Territories, has developed into a major facility. There is also the almost wholly open air, but very modern, satellite parking and fuelling facility at Shung Shui.

A new Headquarters and engineering department was established at Kwai Chung in the 1970's and this included an apprentice training school. The Headquarters returned to Lai Chi Kok in the 1990's, occupying a new building on the original site.

By 1980, the Company had established a driver training yard at Shatin, but this was soon developed into another large, multi-storey depot in order to house the ever-expanding fleet of larger buses.

The Company also conducts various activities from numerous other premises, including the engineering and bus assembly works at Tuen Mun, but it is beyond the scope of this book to expand on this theme..

BELOW: An interesting juxtaposition: KMB owned a filling-station on the corner of its Camp Street Depot site in Sham Shui Po and here the Company is apparently fuelling the minibuses of the opposition! *(John Shearman collection*

KMB Tickets and Fares.

Tickets. *(For illustrations see page 268.)*

Until the 1960's, KMB used tickets printed on flimsy paper with a hard glazed upper surface which went some way to combat the dampening effects of the humid climate and preventing the ticket from becoming soggy when clutched in the, somewhat sweaty, fist of the hard pressed conductor, working in crowded conditions in a relative humidity which remained above 85% for months on end and with daily shade temperatures of 30°C. These tickets were stapled through their lower margin and perforated just above the staple. The normal method was for the conductor to place a block of adult fare tickets back to back with a block of child tickets and tear-off each ticket with a single deft upward movement of his thumb. Examples of these tickets remained in stock as a reserve for some years and when the restored 1949 Daimler was exhibited to the public in connection with the Company's Jubilee celebrations, visitors boarding this static vehicle on its display stand were issued with a 'period' ticket of the type described.

By the time that the writer arrived in Hong Kong, in mid 1964, KMB had introduced a different type of ticket, printed on similar quality paper to that used for the 'Bell Punch' tickets used on London buses until the introduction of the 'Gibson' automatic ticket issuing machine. The Kowloon version differed from the London version in that the tickets were stapled through the lower margin in a manner similar to that described above for the 'flimsy' type that preceded them. The paper used for these tickets was heavily inked and this helped to combat the effects of dampness on the otherwise fairly absorbent paper.

Tickets were provided for 'adults' and for 'child or serviceman'. Those were the days prior to the so-called 'cultural revolution' of 1967, when British servicemen were permitted to travel on public transport in uniform and it was a condition of the KMB franchise - and those of China Motor Bus and The Hongkong Tramways - that servicemen in uniform and the police were to be carried at the appropriate child's fare, usually as near half the adult fare as possible.

On longer routes to the New Territories, stages applied and tickets of suitable denominations were printed for these routes. Where the adult fare was reduced, normally to half, for passengers boarding at, or after the mid point of the route, a special colour ticket was issued form that of the normal 'half' child ticket of the same denomination.

With the introduction of one conductor operation (OCO) from 1972, tickets continued to be issued but the subsequent introduction of driver only operation from 1973 - from 1972 on Cross-Harbour Tunnel Routes - saw the the end for the traditional ticket. Some longer routes continued with section fares, still using buses equipped with Bell Punch "Farebox" hoppers, as on flat fare routes where no ticket was issued, but these were additionally connected to one of two types of ticket issuing equipment. These ticket issuing machines were known by KMB as 'Simple' *(sic)* or 'Complex', the difference being in the number of types of classes and values of fare that could be obtained.

The 'Farebox' works as follows: the passenger deposits the exact fare in coins into a hopper on top of the machine. These coins are then displayed behind a glass screen where the driver can visually check that the correct fare has been paid. Having satisfied himself that all is well, he presses a release handle and the money drops into the vault at the bottom of the 'Farebox'.

BELOW: The farebox intowhich passengers drop fare; no change is given. Coins and notes can be seen by the driver through a glass screen to check that the correct fare has been paid. He then releases the coins into the vault which is secures by a security lock. *(Clement Lau)*

RIGHT: An unused strip of Autofare tickets for use on routes with fare stages.

Fares and their collection.

Background.

Fares in Hong Kong are controlled by the statutory authorities who, from 1945 to 1970 had kept fares unchanged. This was partly as a result of the delicate political position in which Hong Kong found itself, lying, as it does, on the coast of what was then the second largest and powerful Communist state whose system was irritated by the relative affluence of the capitalist Colony. Instability within China had spilled-over into Hong Kong in 1966 and, to a greater extent, in 1967. The disturbances in April 1966 were allegedly brought to flash point by the impending 1st May first-class fare increase on the "Star" Ferry from 20 to 25 cents; there was no increase to the 10 cents second-class fare.

As a result of this reaction to such a minimal fare increase (with none on the deck then traditionally used by the less well-off) and the severity with which the effects of the so-called cultural revolution, across the border in China, affected the Colony, Government held-back from permitting any general fare increase on buses and the fare structure was to remain unaltered until September 1971.

The 1945 bus fares that persisted for so long were low even when introduced. At that time, there was a system of class segregation on all Hong Kong transport and the buses were no exception, most having glazed partitions dividing single-deckers into first and second-class saloons. The only form of fare increase to occur was when the two-class system was removed and the former first-class rate became the flat, classless, fare sometime in the early 1950's.

Within the urban area, two fares were charged. The lower fare was 10 cents and the length of that stage was roughly one mile. Travel exceeding this distance cost 20 cents and could be up to seven miles. In the New Territories, fares were charged on a stage basis and ranged up to one dollar (the exchange rate with the £ Sterling in 1966 being 1 shilling and 3 (old) pence sterling, or 6p, to 1 HK$).

There were half-fares for children (a minimum of 10 cents) and for policemen, postmen and servicemen in uniform.

OCO - One Conductor Operation.

Up until this point, all fares had been collected by roving conductors issuing tickets and cancelling them with a pair of clippers which doubled as a signalling device - both to the driver to start (a sharp banging on a convenient metal panel) and to chivvey along any slow passengers who were taking their time in finding coins from deep in pockets or purses - a sharp 'rap-rap' on the saloon ceiling, just behind the offender, had the necessary quickening effect - and also kept the painters in work. It should not go without mention that double-deckers had a minimum of two conductors and often three - two on the lower deck, one each to control the front and rear gates and the third upstairs.

Following the "Star" Ferry fares problem, it was generally recognised that fares would have eventually to rise, and a milestone in the Colony's process of growing into a modern economy had been passed.

With effect from 15th September 1971, a flat fare of 30 cents was introduced on 47 urban bus routes, 40 cents on two longer urban routes and a minimum 20 cents sectional fare on New Territories bus routes. On nine urban routes (three miles or less) the fare remained a flat 20 cents.

Shortly after this, KMB introduced the concept of one conductor operation (OCO) on big, crowded buses where up to three had been previously used. This was made possible by replacing with more a pliant workforce, the mass of highly unionised staff dismissed as a result of the 1967 strike, connected with the politico-cconomic disturbances which were, in their turn, the result of the cultural revolution.

The OCO system was quite simple. As KMB buses had secondary doorways at the front and (on double-deckers) staircases at the rear, the conductor was seated at a desk about one third of the way back from the front doorway. All but a few seats ahead of the conductor were removed to create a standing area where passengers could wait after boarding, prior to paying. The bus could then proceed while passengers filed forward past the cash desk and either took any available seat at the rear of the lower saloon or made their way to the upperdeck. The front half of the system worked very well but the conflict of movement on the stairs can easily be imagined. Tickets continued to be issued with the OCO system.

OMO or Driver Only Operation.

Even with these economies, fares were to rise steadily but slowly by European standards, as costs rose, even before the 1970' energy crisis and the world-wide inflation that followed. KMB looked towards driver-only-operation - then more widely referred to as One Man Operation (OMO) - as the next step in cost reduction. KMB was able to take advantage of the introduction of an entirely new network of routes (three at first) on which to try the concept. These were the Cross-Harbour Tunnel services which commenced in 1972 with the opening of the tunnel.

Fares on the tunnel routes were paid by dropping the exact $1 fare in coins (no change) into the top of a Bell Punch 'Farebox'. The no-change aspect seems harsh at face value but passengers quickly learned to 'make change' between themselves at bus stops whilst awaiting the arrival of their bus. Additionally, newspaper sellers set-up pitches near bus stops to provide change - provided that a newspaper was purchased!

Tunnel bus fares included an element to cover the toll charge and were thus higher than those on parallel routes. The result was that, after leaving the tunnel, few further passengers were willing to pay an excess fare and buses completed their journeys with fewer and fewer passengers whilst passing queues along the way. After the tunnel routes had been running for a year or so, the problem was resolved by introducing a section fare where, after passing through the tunnel, the fare reduced to only 50 cents - ie half the full journey fare but still slightly higher than fares on local services. This differential has been retained and, generally, tunnel-bus fares today are reduced to half after leaving the tunnel.

The 'Farebox' was located (on a half-cab double-decker) against the front bulkhead and the window over the bonnet was angled forward so that the driver could supervise fare payment over his left shoulder.

The success of one manning was immediate and steps were taken to extend the concept throughout the network. The tunnel buses had been built with front entrances and staircases which made them more suitable for farebox operation - passengers entering via the front door and paying almost in one motion. The older buses, Daimler (a) (b) (c) & (d) and the AEC Regents were less suitable but the economies were so great that even they were fitted with fareboxes over a protracted period of years. Many vehicles were rebuilt - some rebodied - to improve the passenger-flow. All 210 AEC's, the 20 Daimler(d)'s, most of the Daimler(c)'s and some Daimler(a)'s & (b)'s were rebuilt with front entrances and centre staircase with centre exit - the evolved optimum in ease of passenger flow. Exceptions were those Daimler(a)'s & (b)'s which were not rebodied but were, nevertheless, still fitted with fareboxes inside the front doorway while retaining the rear platform and staircase; fare-dodging was minimal.

In the 1990's all buses are DOO but, with modern vehicles, this has become more convenient for the driver who has the farebox in front of him on the platform.

On urban routes, fares are usually charged on a flat 'any distance' basis from any point to any other point in the line of route regardless of distance.

On some longer New Territories routes, section fares were catered for by attaching 'Autofare' ticket-issuing machines to the fareboxes. On the trunk Routes 50 and 70, tapering fares were introduced circa 1978 whereby a passenger boarding part-way along the route paid a reduced flat fare regardless of length of his journey. On the return, assuming the same passenger boards at the start of the route, he pays a full-journey fare. In this way a part fare is paid outward and a full fare is paid for the return. The average fare paid represents the true cost of the journeys and is thus fair - provided a two-way trip is made.

Profit Control.

Fares today reflect true costs and provide for a reasonable profit, the latter being controlled by Government in view of the virtual monopoly enjoyed by the major transport carriers in their own geographical areas. A 'Profit Control Scheme' is imposed on the bus companies, where a notionial maximum percentage profit figure is fixed. This is known as the 'permitted return' which, in the case of KMB, is 16% *(Hibbs J. pub. 1985)*. Profit in excess of this 'permitted return' must be paid into a reserve fund either to be used for development or to cover any future year's revenue shortfall.

The permitted return has limited the extent to which premimum fares can be charged for special services such as express coaches, air-conditioned bus services, etc., as the premium would otherwise cause the return to exceed the maximum laid down by Government.

Public Relations Department Vehicles

KMB's Public Relations Department operates five open-top buses, converted from Daimler(e)'s which are sponsored by various media organisations, together with a selection of miniature 'Fun Buses', together with their tender, another former Daimler(a), AD7336, which appear on page 328.

TOP: Of the five open-top Daimler(e)'s, AD7315 received the most drastic treatment and was given a pre-1930's style open staircase. It is currently painted in a livery featuring a well known cat. *(Clement Lau)*

LEFT: A challenger for the most drastic rebuild was AD7442 which was given a sharply raked roof over the driver's cab. *(Clement Lau)*

BOTTOM LEFT: As with AD7442, the other open-toppers retained their original staircase and doorway arangements as seen here on AD7452. *(Clement Lau)*

BELOW: AD7456 was yet another variation; It kept its front roof dome and side windows and was in a basically yellow livery. *(Clement Lau)*

KMB Route Development Since 1945.

It has been briefly described how KMB resurrected itself following the deverstation it suffered at the hands of the occupying Japanese from December 1941 to August 1945. Slowly, as vehicles, often no more than unconverted Naval or Military lorries, became available, routes were reintroduced following the pre-war pattern where termini had largely been the major cross-harbour ferry piers. To describe the introduction of every new route individually would require a volume of its own whereas the main purpose here is to describe the vehicles operated. It is, however, of interest to note how the network of routes in the 1990's has evolved.

As early as 1933, it had been foreseen that double-deck buses might one day be required as the population grew and there had been a move towards their introduction when the war in Europe intervened and the idea was deferred.

Soon after the Liberation of 1945, it became obvious that the immediate pre-1941 population figure would be exceeded with the rapid return of the Chinese element from their refuge in the less affected parts of their homeland. The small Bedfords and the Tilling-Stevens single-deckers ordered soon after the cessation of hostilities would not alone be sufficient to cope with post-war traffic. Double-deckers were to be the new order and, although they have never been 100% of the fleet, the proportion of single-deckers steadily dwindled from April 1949 when the Colony's first four double-deck buses entered service.

The routes upon which the 'deckers were introduced were those radiating from, in the local vernacular, the 'Kowloon-side" "Star" Ferry Pier, particularly Routes 1, 2, 3, 6 & 7; routes which remain largely unaltered today. Prior to the 1972 opening of the first Cross-Harbour road tunnel, the "Star" Ferry was the main transport focus as it was both the shortest crossing and the one that ran directly into the main business district on Hong Kong Island, known universally as 'Central'. It was also the service most favoured by the expatriate, and better off Chinese, population.

Gradual change affected the bus route network, which grew steadily around its core of 1945 services, both in the urban Kowloon area and in the slowly emerging and more remote districts of the, then, largely rural New Territories, as new roads paved the way for housing developments around old villages that had remained largely unchanged for centuries. All these rural routes were serviced by single-deck buses of various sizes and types. The beginnings of the industrialisation of New Territories towns such as Kwun Tong and Tsuen Wan brought new demands for bus services but the Government saw the still relatively low key bus service being offered by KMB as a deterrent to the population mobility required to allow workers from remote villages the opportunity of jobs in the new urban areas and planned New Towns.

Tunnel

A major influence on KMB's route structure took place in 1972, when the first road tunnel was opened linking Hong Kong Island with the southern tip of the Kowloon peninsula. The original plan by Government was to run a fleet of coaches as a shuttle-service between points near the two tunnel portals but this was subsequently changed to an initial network of three routes between major termini on either side of the harbour, running parallel to existing routes. These new Cross-Harbour Tunnel (XHT) routes were to be operated jointly with the China Motor Bus Company and route numbers commenced a new series at 101, with white lettering on a red background to help identify quickly an XHT bus service. As the 1970's progressed, cross-harbour traffic grew enormously and additional routes were added to improve the service and to try to reduce the degree of overcrowding that very quickly became a feature of these services. Despite most obvious pressure on the XHT buses, the two bus companies were constantly refused additional routes and/or short workings in order to provide a better spread of routes into new and expanding housing developments, although the later 1980's did see some relaxation on the part of Government.

It is of note that, while most XHT routes introduced since the opening of the tunnel have been operated jointly with CMB, since September the 170 has been transferred from CMB to Citybus Ltd. and from 6th June 1994, an additional XHT Route, the 171, has been added, operated jointly with Citybus from the outset.

New Territories.

During 1973, in response to a Government request, KMB embarked on a massive upgrading and renumbering programme to improve the bus services in the New Territories by the improvement of two new trunk routes between Kowloon (Jordan Road Ferry) and the New Territories, one of which was Route 16, which traversed the west coast, travelling along the Castle Peak Road to Yuen Long, while the second, Route 19, took the more central road almost due north to Tai Po and Sheung Shui via the Tai Po Road. Not only were these revised trunk routes of high frequency, they also saw the general[1] introduction of double-deck buses outside the Kowloon urban area for the first time. Under the renumbering the 16 and 19 became the 50 and 70 respectively. This introduction of a large fleet of additional vehicles not only increased capacity on the two trunk corridors quite dramatically, it also allowed the single-deck fleet to be dispersed to other rural services to provide a parallel improvement on routes serving as yet less developed areas. In turn, these latter routes have gradually been double-decked in response to the growing traffic demands of an expanding and relocating population.

Until the 1973 re-structuring of NT services, route numbers had been allocated on a 'next available number' basis, regardless of the location served. Along with the double-decking of Routes 50 and 70, KMB undertook a comprehensive revision of route numbering in the New Territories and, since then, route numbers have reflected the area which the routes serve. Under the system, new route numbers are allocated within groups according to district, as follows:
 30-49: Tsuen Wan;
 50-69: Yuen Long and Tuen Mun;
 70-89: Sheung Shui, Tai Po and Sha Tin;
 90-99: Sai Keung.

Coach Services.

A further 1970's innovation was the introduction of urban express coach services, again parallel to existing routes, not unlike the London 'Green Line' services. These ran at frequent intervals and, with no standing permitted, offered a guaranteed, comfortable, coach-style, high-backed seat, but at a premium fare. Again a dedicated route number series was introduced, commencing at 200. Of these, Nos. 200 and 201 were Airport Coach Service routes; the 200 from Central District on Hong Kong Island which ran via the XHT, and the 201 from Kowloon, Tsim Sha Tsui. These services were replaced in 1985 by a new series marketed as 'Airbus' and with the route numbers A1 and A2, later to be joined by a third, A3, again an XHT service serving hotels in the Causeway Bay area, and a fourth, A4, to connect the Airport with the China Ferry Terminal in Canton Road.

Further developments have been more in the nature of refinements than radical new services.

MTR Connecting Routes.

The opening of the Mass Transit Railway (MTR) in 1979, KMB lost a considerable amount of traffic, particularly from the Urban Coach services along the route of the new underground railway. The overall drop in passenger journeys was not as large as had been predicted, perhaps reflecting the number of potential bus passengers who had previously resigned themselves to walking to school or work, who could find a place on a bus. Bus fare levels remained generally lower than those of the MTR and this helped prevent complete defection, particularly amongst the lower paid. KMB, with a little encouragement from Government, introduced special feeder services to many MTR stations, often as a major deviation from an existing route, the new feeder gaining a suffix 'M' such as 7M, 13M, etc. An MTR logo appeared on the destination blind as an additional means of identifying an MTR feeder route.

KCR Connecting Routes.

Having introduced its MTR feeder services, KMB turned to requests that it should do likewise for the Kowloon Canton Railway, the Colony's recently electrified and modernised main line railway on which are operated an intense inner and outer suburban service and international trains to and from Guangzhou in China. A similar system was introduced so that services to a KCR station carry a suffix 'K', but no KCR logo appears.

Other categories.

Other categories include the 'R'-suffix Recreational routes which run only at weekends and on public holidays, and 'X'-suffix express bus routes, which are self-explanatory.

600 series routes.

600 series routes were introduced to indicate services via the 'Eastern Harbour Crossing', a tunnel carrying both road and rail (MTR) traffic between Lei Yue Mun, to the east of Kai Tak Airport, and Quarry Bay, on Hong Kong Island.

NOTE: [1] Prior to 1973, there had been Sunday operation of double-deckers to Sha Tin, via the Lion Rock Tunnel and also a daily double-deck route to Tsuen Wan

300 series routes.

Routes from 300 to 373 (Dec. '94) were introduced in 1991 to help relieve the MTR on the Nathan Road Corridor where demand had produced capacity saturation conditions. These services run both via the Cross-harbour Tunnel and via the Eastern Harbour Crossing and, in Dec '94, all but the 300 were peak hour only. Although originally 300 and 348 were KMB-only routes, these are now operated jointly with CMB

800 series routes.

800 series routes are special service serving the horse racing track at Sha Tin, in the New Territories, running only to serve race meetings.

Air-conditioned Services.

The introduction of air-conditioned bus services did not in itself bring about any new route number series but where a service is wholly operated by air-conditioned vehicles, the 200-series has been utilised.

Summary of Post 1973 Block Number Allocation for KMB Bus Routes.

Kowloon Routes	1 to 29
Urban New Territories Routes	30 to 49
Rural New Territories Routes	50 to 99
Cross Harbour Tunnel Routes	101 +
Coach Routes	200 +
Peak Hour Cross-Harbour Routes	300 +
Eastern Harbour Crossing Routes	600 +
Race Course Special Services	800 +

Since the introduction of the block categories, expansion of the urban area in the NT has resulted in the NT urban routes being extended into the NT Rural series, leaving only nineteen Rural Routes. With the withdrawal of many Coach Routes following the opening of the MTR, the largely unused series has been adopted for new air-conditioned bus services.

Comparison of the two route lists included here will identify many of these changes for the reader.

Route 300 which was originally introduced as a Peak Hours Only service by KMB only has evolved into an all-day service, joint with CMB.

Routes renumbered.

The following list of route number changes between 1968 and 1992 was very kindly passed to the author by **Mr. Kelvin Lai** who researched and compiled it and whom I thank. This list takes no account of routes which has termini/destinations changed as well as receiving a new route number. eg. Route 26 from Jordan Road Ferry to Yuen Long via Twisk was changed in 1973 to 51, starting from Tai Kok Tsui Ferry.

Date:	Changed from:	Operating between:
1968	15A to 19A	Jordan Road Ferry and Sheung Shui via Tai Po Road
1968	31B to 16B	Jordan Road Ferry and Kwai Chung (North)
1968	33A to 31B	Sham Shui Po Ferry and Kwai Chung (North)
1973	16 to 50	Jordan Road Ferry and Tsuen Wan Ferry
1973	16A to 30	Jordan Road Ferry and Tsuen Wan Ferry
1973	17 to 76	Sheung Shui and Yuen Long (West) via Lok Ma Chau
1973	18 to 77	Sheung Shui and Yuen Long (West) vai Kam Tin
1973	18A to 54	Yuen Long and Sheung Tsuen.
1973	19 to 70	Jordan Road Ferry and Shung Shui, via Lion Rock Tunnel
1973	20 to 78	Sheung Shui and Sha Tau Kok
1973	21 to 91	Choi Hung and Tai O Mun.
1973	22 to 92	Choi Hung and Sai Keung.
1973	23 to 74	Tai Po Market and Yuen Long (West)
1973	24 to 55	Yuen Long (East) and Lau Fau Shan.
1973	25 to 75	Tai Po Market and Tei Mei Tuk.
1973	28 to 93	Sai Keung and Tai Mong Tsai
1973	29 to ,79	Sheung Shui and Ta Kwu Ling
1973	30 to 90	Choi Hung and Rennie's Mill.
1977	82A to 73A	Sheung Shui and Man Kam To.
1978	60 to 60R	Lei Muk Shue and Ting Kau Beach. (Summer holidays only).
1978	62 to 62R	Kwai Shing (East) and Sham Tseng (Summer holidays only)
1978	70A to 70R	Jordan Road Ferry and Lek Yuen. (Winter holidays only)
1978	73A to 73R	Sheung Shui and Man Kam To (Ching Ming & Cubg Yeung Festivals only)
1978	74 to 64	Tai Po Market and Yuen Long (West).
1978	75A to 75R	Tai Po Market and Bride's Pool. (Holidays only).
1978	84 to 74	Tai Po Market and Sam Mun Tsai
1978	85 to 74R	Tai Po Railway Station and Luk Keng. (Winter holidays only)
1980	45A to 46A	Lai Yiu and Tsuen Wan Ferry.
1983	5B to 5K	Kowloon KCR Station and Kwun Tong Ferry.
1983	7A to 7K	Kowloon KCR Station and Lok Fu.
1983	12A to 12K	Kowloon KCR Station and Sham Shui Po Ferry.
1983	64 to 64K	Tai Poi KCR Station and Yuen Long (West)
1983	65 to 65K	Tai Po KCR Station and Sheung Tsuen.
1983	73A to 73K	Sheung Shui and Man Kam To
1983	74 to 74K	Tai Po KCR Station and Sam Mun Tsai.
1983	75 to 75K	Tai Po KCR Station and Tai Mei Tuk.
1983	76 to 76K	Sheung Shui and Yuen Long (West) via Mai Po.
1983	77 to 77K	Sheung Shui and Yuen Long (West) via Kam Tin.
1983	78 to 78K	Sheung Shui and Sha Tau Kok.
1983	78A to 69K	Fanling KCR Station and Luk Keng.
1983	79 to 79K	Sheung Shui and Ta Kwu Ling.
1986	200 to A2	Central (Macau Ferry) and Airport.
1986	201 to A1	Airport to Tsim Sha Tsui (Circular Route).
1987	5E to 11K	Kowloon KCR Station and Chuk Yuen Estate.
1987	44A to 43C	Tai Kok Tsui Ferry and Cheung Hong.
1987	44B to 43D	Sham Shui Po Ferry and Cheung Ching.
1987	69 to 73A	Yuen Chau Kok and Choi Yuen Estate.
1987	70M to 80M	Kowloon Tong MTR Station and Sui Wo Court.
1987	71 to 81	Jordan Road Ferry and Wo Che.
1988	111A to 111S	Causeway Bay (Victoria Park Flower Fair) and Choi Hung.
1988	112A to 112S	Causeway Bay (Victoria Park Flower Fair) and Mong Kok.
1989	38M to 238M	Tsuen Wan MTR Station and RivieraGarden.
1989 (Circular).	95A to 95M	Kwun Tong (Elegance Road) and Tsui Lam Estate
1991 Station (Circular).	42B to 42M	Ching Yan Temporary Housing Area and Tsuen Wan MTR
84X to 284X		Ravana Garden and Daimond Hill MTR Station (Circular).

KMB Routes as at 31st December 1977

A complete list of KMB's routes was most kindly presented to the author during a seven working day visit to the Company's Kwai Cheung Headquarters in January 1978 and extracts from KMB's route lists are shown on page 280/81 to illustrate the style of the time, with the route details printed in small boxes, one per route, on semi-waterproof, hard-glazed paper, essential to prevent damage to the paper by the high relative humidity, a feature of Hong Kong's climate for ten months of the year. Bus tickets were also printed on similar paper at certain times for certain routes. The following 1978 list of routes was taken from these sheets.

NOTE: For some unknown reason, for many years KMB chose to use the spelling 'Un Long' in place of the more usual 'Yuen Long'. This wass reflected on destination blinds, in time tables and on bus shelters but by the 1990' has been dropped in favour of 'Yuen Long'.

Kowloon Urban Routes.

Route	From:
1	"Star" Ferry to Wang Tau Hom
1A	"Star" Ferry to Sau Mau Ping (Central)
2	"Star" Ferry to So Uk
2A	So Uk to Ngau Tau Kok
2B	Jordan Road Ferry to Kowloon City Ferry Cancelled by Dec 77
2C	"Star" Ferry to Tai Hang Tung
2D	Pak Tin to Chuk Yuen
2E	Kowloon City Ferry to Pak Tin
2F	Cheung Sha Wan to Tsz Wan Shan (South)
3	Jordan Road Ferry to Chuk Yuen
3A	Chuk Yuen to Tsz Wan Shan
3B	Hong Hom Ferry to Tsz Wan Shan (South)
3C	Jordan Road Ferry to Tsz Wan Shan (North)
3D	Tsz Wan Shan (N) to Kwun Tong (Yue Man Sq)
4	Jordan Road Ferry to Cheung Sha Wan
4A	Jordan Road Ferry to Tai Hang Tung
5	"Star" Ferry to Choi Hung
5A	"Star" Ferry to Kowloon City Ferry (Shing Tak St)
5B	Hung Hom Railway Station to Kwun Tong Ferry
5C	"Star" Ferry to Tsz Wan Shan (South)
6	"Star" Ferry to Lai Chi Kok (Bridge)
6A	"Star" Ferry to Lai Chi Kok (Amusement Park)
6B	Lai Chi Kok (Bridge) to Chuk Yuen
6C	Lai Chi Kok (Bridge) to Kowloon City Ferry
6D	Cheung Sha Wan to Ngau Tau Kok
7	"Star" Ferry to Kowloon Tong
7A	Hong Hom Railway Station to Lok Fu
7B	Hong Hom Ferry to Wang Tau Hom
8	"Star" Ferry to Oi Man
9	"Star" Ferry to Choi Hung
10	Tai Kok Tsui Ferry to Ping Shek
11	Jordan Road Ferry to Chuk Yuen
11A	Kwun Tong Ferry to San Po Kong
11B	Kowloon City Ferry to Kwun Tong (Tsui Ping Road)
11C	Wang Tau Hom to Kwun Tong (Tsui Ping Road)
11D	Lok Fu to Kwun Tong Ferry
12	Jordan Road Ferry to Lai Chi Kok Amusement Park)
12A	Hung Hom Railway Station to Sham Shui Po Ferry
12B	Wang Tau Hom to Lai Chi Kok (Bridge)
13	Jordan Road Ferry to Choi Hung
13A	Kowloon City Ferry to Sau Mau Ping(Upper)
13B	Kwun Tong Ferry to Sau Mau Ping (Central)
13C	Kwun Tong Ferry to Sau Mau Ping (Upper)
13D	Tai Kok Tsui Ferry to Sau Mau Ping (Central)
14	Jordan Road Ferry to Yau Tong (Ko Chiu Road)
14A	Kwun Tong (Yue Man Square) to Yau Tong (Ko Chiu Road)
14B	Ngau Tau Kok to Yau Tong
14C	Kwun Tong (Yue Man Sq) to Lei Yue Mun (Sam Ka Tsuen Ferry)
15	Hung Hom Ferry to Lam Tin (North)
15A	Tsz Wan Shan (North) to Lam Tin (North)
15B	Kwun Tong Ferry to Lam Tin (North)
15C	Lam Tin (South) to Kwun Tong (Yue Man Square)
16	Tai Kok Tsui Ferry to Lam Tin (North)
17	Oi Man to Kwun Tong (Yue Man Sq)
18	Tai Kok Tsui Ferry to Oi Man/Ho Man Tin(Circular)
19	Ngok Yue Shan to Kwun Tong Ferry
19A	Kwun Tong (Yue Man Square) to Kung Lok Road (Circular)
20	Jordan Road Ferry to Oi Man
25	"Star" Ferry to Airport (Cargo Terminal)

New Territories Urban Routes

Route	From:
30	Jordan Road Ferry to Tsuen Wan Ferry
31	Shek Lei to Tsuen Wan Ferry
31A	Shek Lei to Tsuen Wan West
31D	Sham Shui Po Ferry to Shek Lei
32	Tsuen Wan Ferry to Shing Mun Reservoir
32A	Tsuen Wan Ferry to Lo Wai
33	Sham Shui Po Ferry to Tsuen Wan Ferry
33A	Sham Shui Po Ferry to Kwai Chung Central (Tai Wo Hau)
33B	Kwai Hing to Sham Shui Po Ferry
33C	Kwai Fong to Sham Shui Po Ferry
34	Kwai Shing (Central) to Tsuen Wan West
35	Tsuen Wan Ferry to Shek Yam
36	Tsuen Wan Ferry to Lei Muk Shu
36A	Sham Shui Po Ferry to Lei Muk Shu
36B	Jordan Road Ferry to Lei Muk Shu
37	Kwai Shing (Central) to Tai Kok Tsui Ferry
37A	Kwai Shing (East) to Sham Shui Po Ferry
37B	Kwai Shing (East) to Lai Chi Kok (Bridge)
38	Kwai Shing (East) to Kwun Tong (Yue Man Square)
40	Kwun Tong Ferry to Tsuen Wan Ferry
43	Cheung Ching to Tsuen Wan Ferry
44	Lai Chi Kok (Bridge) to Tsing Yi
44A	Tai Kok Tsui Ferry to Cheung Ching
45	Lai King (North) to Kowloon City Ferry
45A	Tsuen Wan Ferry to Lai King (Circular)
45H	Lai Chi Kok (Bridge) to Princess Margaret Hospital
46	Lai Yiu to Sham Shui Po Ferry

New Territories Rural Routes.

Route	From:
50	Jordan Road Ferry to Un Long (East)
51	Tai Kok Tsui Ferry to Un Long (West)
52	Tsuen Wan Ferry to Tuen Mun
52A	Tsuen Wan Ferry to Tai Hing
53	Yuen Long (East) to Castle Peak Bay
54	Un Long (West) to Shung Tsuen (Sek Kong)
55	Un Long to Lau Fau Shan
56	Un Long (West) to Tai Tong (Wong Nei Tun)
57	Un Long (East) to Pak Nei
58	Un Long (East) to Sha Kiu (Tsim Bei Tsui)
59	Tai Hing to Pak Kok
60	Lek Muk Shu to Ting Kau Beach (weather permitting) SuPH
61	Yuen Long (East) to Tai Hing
62	Kawi Shing (East) to Sham Tseng (weather permitting) SuPH
70	Jordan Road Ferry to Sheung Shui
70A	Jordan Road Ferry to Lek Yuen San Tsuen
70B	Jordan Road Ferry to Tai Po Market
71	Jordan Road Ferry to Sha Tin Market
72	Tai Kok Tsui Ferry to Tai Po Market
73	Tai Po Market to Man Kam To (San Uk Ling)
73A	Sheung Shui to Man Kam To (San Uk Ling)
74	Tai Po Market to Yuen Long (West)
75	Tai Po Market to Tai Mei Tuk
75A	Tai Po (Railway Station) to Bride's Pool (Su PH only)
76	Shung Shui to Un Long (West) - direct
77	Sheung Shui to Un Long (West) via Kam Tin Ro
78	Sheung Shui to Sha Tau Kok
78A	Luen Wo Hui to Sha Tau Kok (secondary school)
79	Sheung Shui to Ta Kwu Ling
79A	Luen Wo Hui to Ping Yuen
81	Luen Wo Hui to Luk Keng
84	Tai Po Market to Sam Mun Tsai
85	Tai Po (Railway Station) to Luk Keng
87	Sha Tin (Lek Yuen San Tsuen) to Tai Kok Tsui Ferry
88	Sha Tin Market to Siu Lek Yuen
88A	Sha Tin Market to Hin Tin
89	Lek Yuen San Tsuen to Kwun Tong (Yue Man Square)
90	Choi Hung to Rennie's Mill
91	Choi Hung to Tai O Mun
91A	Chuk Yuen to Hang Hau
92	Choi Hung to Sai Kung
93	Sai Keung to Tai Mong Tsai

Cross-Harbour Tunnel Bus Routes - Joint with CMB.

Route	From:
101	Kwun Tong (Yue Man Sq.) to Kennedy Town
102	Lai Chi Kok (Bridge) to Shau Kei Wan
103	Lok Fu to Pokfield Road
104	Pak Tin to West Point
105	Lai Chi Kok (Amusement Park) to West Point
106	Chuk Yuen to Chai Wan (East)
111	Ping Shek to Cleverly Street
111A	Causeway Bay (Magistracy) to Choi Hung (Chinese New Year only)
112	So Uk to North Point
112A	Causeway Bay (Magistracy) to Mong Kok (Chinese New Year only)
113	Choi Hung to West Point
121	Choi Hung to Cleverly Street (Night Route only)
122	So Uk to North Point (Night Route only)
170	Sha Tin Market to Aberdeen

Coach Services.

Route	From:
200	Kai Tak Airport to Central (Ice House Street)
201	"Star" Ferry to Kai Tak Airport
202	"Star" Ferry to Yau Yat Chuen
203	"Star" Ferry to Tsz Wan Shan (North)
204	Lai Chi Kok (Bridge) to Kwun Tong (Yue Man Sq)
206	"Star" Ferry to Lai Chi Kok (Bridge)
207	"Star" Ferry to Beacon Hill Road
208	"Star" Ferry to Broadcast Drive
209	"Star" Ferry to Ngau Tau Kok
210	"Star" Ferry to Caldecott Road
211	"Star" Ferry to Kwun Tong (Yue Man Sq)
216	Mei Foo Sun Chuen to Hung Hom (Railway Station)
246	"Star" Ferry to Lai Yiu
250	"Star" Ferry to Un Long (East)
291	Kowloon City Ferry to Tai O Mun (Su PH weather permitting)
293	Kowloon City Ferry to Tai Mong Tsai (Su PH weather permitting)

NOTE: Su PH = Sundays and Public Holidays Only.

1991 - MTR Relief Services.

These Mass Transit Railway Relief Services were introduced in 1991 in order to attract potential MTR users onto special bus services so as to keep directional peak hour loadings on the railway below 75,000 and, thereby, prevent the MTR from further raising its peak congestion fare, believed to be politically unacceptable. The cause of the congestion on the MTR is the trunk line between Prince Edward, in the north, and Central in the south, the 'Nathan Road Corridor', so named because of the main road beneath which the MTR runs. Almost the entire length of line north of Tsim Sha Tsui attracts residents who work in the main business district, Central, on Hong Kong Island and, quite simply, there is over-demand for the available capacity. When the MTR was first built, there was concern that it should be protected from competition by the bus companies; 12-years on and the buses are required to provide relief from the railway's success!

The routes themselves comprise a series of (mainly) morning peak, unidirectional, services from Kowloon and the New Territories to Central, mostly running along the MTR catchment corridors. All the services use air-conditioned double-deckers and are numbered in the 300-series.

Route 300 itself is unique, as it operates in both directions all day; it is worked by KMB alone. Most other routes are jointly worked with CMB.

The routes are:-

300	Prince Edward MTR to Central - all-day; two-way;
301	Hung Hom (Toll Plaza) to Central
302	Tse Wan Shan to Central
303	Yui On (Ma On Shan) to Central
305	Mei Lam to Central
307	Tai Po to Central
348	Tsing Yi (Cheung On Estate) to Central (and vice versa) **KMB only.**

BELOW: Route 300 was introduced to relieve the Nathan Road to Central section of the Mass Transit Railway. Here KMB's AD56 approaches the first bus stop in Queensway on Hong Kong Island in September 1994. Sheung Wan, shown on the blind, is a few hundred metres past the business heart of Central. *(Mike Davis)*

The Kowloon Motor Bus Co., (1933) Ltd.
九龍汽車（1933）有限公司

ROUTE NO. 3 號路線

JORDAN ROAD FERRY to CHUK YUEN: via Jordan Road, Nathan Road, Public Square Street, Shanghai Street, Waterloo Rd., Prince Edward Rd., La Salle Rd., Boundary St., Grampian Rd., Dumbarton Rd., Junction Road and Tung Tau Tsuen Road.

佐敦道碼頭至竹園　經：佐敦道，彌敦道，公衆四方街，上海街，窩打老道，太子道，喇沙利道，界限街，甘霖遁道，東實庭道，聯合道及東頭村道。

CHUK YUEN to JORDAN ROAD FERRY: via Tung Tau Tsuen Rd., Junction Rd., Prince Edward Rd., Waterloo Rd., Nathan Rd., Public Square St., Shanghai Street and Jordan Road.

竹園至佐敦道碼頭　經：東頭村道，聯合道，太子道，窩打老道，彌敦道，公衆四方街，上海街及佐敦道

TIME TABLE 時間表

Mondays to Saturdays 星期一至星期六

From Jordan Rd. Ferry 佐頓碼頭開	From Chuk Yuen 竹園開	Frequency 每次
6.10a.m. 上午 to 6.30a.m. 上午	5.40a.m. 上午 to 6.00a.m. 上午	10 minutes 分
6.30 a.m. 上午 to 10.42a.m. 上午	6.00a.m. 上午 to 10.18a.m. 上午	6 ,,
10.42 a.m. 上午 to 12.26p.m. 下午	10.18a.m. 上午 to 11.54a.m. 上午	8 ,,
12.26 p.m. 下午 to 7.56p.m. 下午	11.54a.m. 上午 to 7.18p.m. 下午	6/7 ,,
7.56p.m. 下午 to 9.16p.m. 下午	7.18p.m. 下午 to 8.46p.m. 下午	8 ,,
9.16p.m. 下午 to 11.54p.m. 下午	8.46p.m. 下午 to 11.24p.m. 下午	6/7 ,,
11.54 p.m. 下午 to 12.30 a.m. 上午	11.24p.m. 下午 to 12.00mid. 子夜	12 ,,

Sundays and Public Holidays 星期日及公衆假期

6.10a.m. 上午 to 7.10a.m. 上午	5.42a.m. 上午 to 6.42a.m. 上午	12 minutes 分
7.10a.m. 上午 to 10.57a.m. 上午	6.42a.m. 上午 to 10.28a.m. 上午	6/7 ,,
10.57 a.m. 上午 to 12.09p.m. 下午	10.28a.m. 上午 to 11.40a.m. 上午	8 ,,
12.09 p.m. 下午 to 7.51p.m. 下午	11.40a.m. 上午 to 7.22p.m. 下午	6 ,,
7.51p.m. 下午 to 9.07p.m. 下午	7.22p.m. 下午 to 8.34p.m. 下午	8 ,,
9.07p.m. 下午 to 11.28p.m. 下午	8.34p.m. 下午 to 10.58p.m. 下午	6/7 ,,
11.28 p.m. 下午 to 12.28 a.m. 上午	10.58p.m. 下午 to 11.58p.m. 下午	12 ,,

Faretable 30 cents per Single Journey 票價：單程三毫

ROUTE NO. 3A 號路線

CHUK YUEN to TSZ WAN SHAN: via Shatin Pass Rd., Fung Tak Rd., Sheung Fung St., Wan Wah St. and Tsz Wan Shan Rd.

竹園至慈雲山　經：沙田坳道，鳳德道，雙鳳街，雲華街及慈雲山道。

TSZ WAN SHAN to CHUK YUEN: via Tsz Wan Shan Rd., Wai Wah St., Wan Wah St., Shung Wah St., Sheung Fung St., Fung Tak Rd. and Shatin Pass Rd.

慈雲山至竹園　經：慈雲山道，惠華街，雲華街，崇華街，雙鳳街，鳳德道及沙田坳道。

TIME TABLE 時間表

From Chuk Yuen 竹園開	From Tsz Wan Shan 慈雲山開	Frequency 每次
5.36 a.m. 上午 to 6.12 a.m. 上午	5.44 a.m. 上午 to 6.20 a.m. 上午	6 minutes 分
6.12 a.m. 上午 to 1.50 p.m. 下午	6.20 a.m. 上午 to 1.58 p.m. 下午	3 ,,
1.50 p.m. 下午 to 3.58 p.m. 下午	1.58 p.m. 下午 to 4.06 p.m. 下午	4 ,,
3.58 p.m. 下午 to 10.10 p.m. 下午	4.06 p.m. 下午 to 10.18 p.m. 下午	3 ,,
10.10 p.m. 下午 to 10.38 p.m. 下午	10.18 p.m. 下午 to 10.46 p.m. 下午	4 ,,
10.38 p.m. 下午 to 11.38 p.m. 下午	10.46 p.m. 下午 to 11.46 p.m. 下午	3 ,,
11.38 p.m. 下午 to 12.14 a.m. 上午	11.46 p.m. 下午 to 12.22 a.m. 上午	6 ,,

Faretable 20 cents per Single Journey 票價：單程二毫

ROUTE NO. 3B 號路線

HUNG HOM FERRY to TSZ WAN SHAN (NORTH): via Gillies Avenue, Wuhu St., Ma Tau Wei Rd., To Kwa Wan Rd., Ma Tau Kok Rd., Ma Tau Chung Rd., Prince Edward Rd., Choi Hung Rd., Tai Shing St., Tung Tau Tsuen Rd., Shatin Pass Rd., Fung Tak Rd., Sheung Fung St., Wan Wah St. and Tsz Wan Shan Rd.

紅磡碼頭至慈雲山（北）　經：機利士路，蕪湖街，馬頭圍道，土瓜灣道，馬頭角道，馬頭涌道，太子道，彩虹道，大成街，東頭村道，沙田坳道，鳳德道，雙鳳街，雲華街及慈雲山道。

TSZ WAN SHAN (NORTH) to HUNG HOM FERRY: via Tsz Wan Shan Rd., Wan Wah St., Shung Wah St., Sheung Fung St., Fung Tak Rd., Shatin Pass Rd., Choi Hung Rd., Flyover, Prince Edward Rd., Ma Tau Chung Rd., Mok Chong St., To Kwa Wan Rd., Ma Tau Wei Road, Wuhu Street and Gillies Avenue.

慈雲山（北）至紅磡碼頭　經：慈雲山道，雲華街，崇華街，雙鳳街，鳳德道，沙田坳道，彩虹道，天橋，太子道，馬頭涌道，木廠街，土瓜灣道，馬頭圍道，蕪湖街及機利士路。

TIME TABLE 時間表

Mondays to Saturdays 星期一至星期六

From Hung Hom Ferry 紅磡碼頭開	From Tsz Wan Shan 慈雲山開	Frequency 每次
6.05 a.m. 上午 to 6.25 a.m. 上午	5.40 a.m. 上午 to 6.00 a.m. 上午	10 minutes 分
6.25 a.m. 上午 to 9.45 a.m. 上午	6.00 a.m. 上午 to 9.15 a.m. 上午	5 ,,
9.45 a.m. 上午 to 1.00 p.m. 下午	9.15 a.m. 上午 to 12.15 p.m. 下午	7/8 ,,
1.00 p.m. 下午 to 7.30 p.m. 下午	12.15 p.m. 下午 to 7.00 p.m. 下午	5 ,,
7.30 p.m. 下午 to 12.00 mid. 子夜	7.00 p.m. 下午 to 11.30 p.m. 下午	7/8 ,,
12.00 mid. 子夜 to 12.30 a.m. 上午	11.30 p.m. 下午 to 12.00 mid. 子夜	10 ,,

Sundays and Public Holidays 星期日及公衆假期

6.05 a.m. 上午 to 6.55 a.m. 上午	5.40 a.m. 上午 to 6.30 a.m. 上午	10 ,,
6.55 a.m. 上午 to 9.55 a.m. 上午	6.30 a.m. 上午 to 9.30 a.m. 上午	6 ,,
9.55 a.m. 上午 to 12.55 p.m. 下午	9.30 a.m. 上午 to 12.30 p.m. 下午	7/8 ,,
12.55 p.m. 下午 to 7.55 p.m. 下午	12.30 p.m. 下午 to 7.30 p.m. 下午	6 ,,
7.55 p.m. 下午 to 11.40 p.m. 下午	7.30 p.m. 下午 to 11.15 p.m. 下午	7/8 ,,
11.40 p.m. 下午 to 12.25 a.m. 上午	11.15 p.m. 下午 to 12.00 mid. 子夜	15 ,,

Faretable 30 cents per Single Journey 票價：單程三毫

ROUTE NO. 3C 號路線

JORDAN ROAD FERRY to TSZ WAN SHAN ... Square St., Shanghai Street, Mongkok Road, ... Road, Fu Mei St., Fung Mo Street, Lung C... Kwong Village Road and Tsz Wan Shan Roa...

佐敦道碼頭至慈雲山（北）　經：佐敦道，彌敦道，... 打老道，聯合道，... 道及慈雲山道。

TSZ WAN SHAN (NORTH) to JORDAN ROA... Village Road, Lung Cheung Road, Fung Mo ... Prince Edward Road, Nathan Rd., Public Squa...

慈雲山（北）至佐敦道碼頭　經：慈雲山道，蒲崗村... 子道，彌敦道，公...

TIME T...

From Jordan Road Ferry 佐頓道碼頭開	FromTsz...
6.15 a.m. 上午 to 6.37 a.m. 上午	5.36 a...
6.37 a.m. 上午 to 10.54 a.m. 上午	6.06 a...
10.54 a.m. 上午 to 12.00 noon 正午	10.30 a...
12.00 noon 正午 to 12.00 mid. 子夜	12.24 p...
12.00 mid. 子夜 to 12.42 a.m. 上午	11.20 p.m...

Faretable 30 cents ...

ROUTE NO. 3D 號路線

TSZ WAN SHAN (NORTH) to KWUN TONG (Yue ... Cheung Rd., Kwun Tong Rd., Hong Ning Road, Yu...

慈雲山（北）至觀塘（裕民坊）　經：慈雲山道，斧山道，...

KWUN TONG (Yue Man Sq.) to TSZ WAN SHAN (NO... Road, Kwun Tong Rd., Lung Cheung Rd., Ham... Road and Tsz Wan Shan Road.

觀塘（裕民坊）至慈雲山（北）　經：同仁街，裕民坊，... 及慈雲山道。

TIME TAB...

Mondays to Sat...

From Tsz Wan Shan (NORTH) 慈雲山（北）開	From K...
5.55 a.m. 上午 to 6.40 a.m. 上午	5.30
6.40 a.m. 上午 to 9.40 a.m. 上午	6 30
9.40 a.m. 上午 to 12.40 p.m. 下午	9.30
12.40 p.m. 下午 to 7.40 p.m. 下午	12.30
7.40 p.m. 下午 to 8.40 p.m. 下午	8 00
8.40 p.m. 下午 to 11.25 p.m. 下午	9.10
11.25 p.m. 下午 to 12.25 a.m. 上午	10.55

Sundays and Public H...

6.00 a.m. 上午 to 11.30 a.m. 上午	5.30
11.30 a.m. 上午 to 12.30 p.m. 下午	11.00
12.30 p.m. 下午 to 8.30 p.m. 下午	12.00
8.30 p.m. 下午 to 9.30 p.m. 下午	8.00
9.30 p.m. 下午 to 12.00 mid. 子夜	9.00
12.00 mid. 子夜 to 12.30 a.m. 上午	11.30

Faretable 30 cents pe...

ROUTE NO. 4 號路線

JORDAN ROAD FERRY to CHEUNG SHA W... Shanghai Street, Lai Chi Kok Rd., Prince Edw... St., Cheung Sha Wan Rd., Yen Chow St., Un Ch...

佐敦道碼頭至長沙灣　經：佐敦道，廣東道，... 界限街，長沙灣道，欽...

CHEUNG SHA WAN to JORDAN ROAD FER... Square Street, Shanghai Street and Jordan Ro...

長沙灣至佐敦道碼頭　經：長沙灣道，彌敦道，公...

TIME...

Mondays to...

From Jordan Road Ferry 佐敦道碼頭開	
6.10 a.m. 上午 to 6.30 a.m. 上午	
6.30 a.m. 上午 to 9.30 a.m. 上午	
9.30 a.m. 上午 to 1.00 p.m. 下午	
1.00 p.m. 下午 to 8.00 p.m. 下午	
8.00 p.m. 下午 to 12.30 a.m. 上午	

Sundays and Pu...

6.10 a.m. 上午 to 9.40 a.m. 上午	
9.40 a.m. 上午 to 10.48 a.m. 上午	
10.48 a.m. 上午 to 12.18 p.m. 下午	
12.18 p.m. 下午 to 7.40 p.m. 下午	
7.40 p.m. 下午 to 12.30 a.m. 上午	

Faretable 30 cents per S...

February, 1977.

Subject to Alterat...
「行車時間如有更...

SCHEDULE OF BUS SERVICES IN KOWLOON URBAN AREA
九龍市區巴士路線　價目及行車時間表

ROUTE 4A 號路線

JORDAN ROAD FERRY to TAI HANG TUNG: via Jordan Road, Canton Rd., Public Square St., Shanghai Street, Lai Chi Kok Rd., Prince Edward Road, Portland Street, Boundary St., Tai Hang Tund Road, Tong Yam Street and Tai Hang Tung Road.

佐敦道碼頭至大坑東　經：佐敦道，廣東道，公眾四方街，上海街，荔枝角道，太子道，砵蘭街，界限街大坑東道，棠蔭街及大坑東道。

TAI HANG TUNG to JORDAN ROAD FERRY: via Tai Hang Tung Rd., Boundary St., Knight St., Prince Edward Rd., Nathan Rd., Public Square Street, Shanghai Street and Jordan Road.

大坑東至佐敦道碼頭　經：大坑東道，界限街，勵德街，太子道，彌敦道，公眾四方街，上海街及佐敦道

TIME TABLE 時間表
Mondays to Saturdays 星期一至星期六

From Jordan Road Ferry 佐敦道碼頭開	From Tai Hang Tung 大坑東開	Frequency 每次
6.03a.m.上午 to 6.53a.m.上午	5.40a.m.上午 to 6.30a.m.上午	10 minutes 分
6.53a.m.上午 to 11.33a.m.上午	6.30a.m.上午 to 11.10a.m.上午	5 ,,
11.33a.m.上午 to 12.48p.m.下午	11.10a.m.上午 to 12.10p.m.下午	7 8 ,,
12.48p.m.下午 to 4.06p.m.下午	12.10p.m.下午 to 4.28p.m.下午	6 ,,
4.06p.m.下午 to 7.41p.m.下午	4.28p.m.下午 to 7.18p.m.下午	5 ,,
7.41p.m.下午 to 8.56p.m.下午	7.18p.m.下午 to 8.18p.m.下午	7 8 ,,
8.56p.m.下午 to 10.56p.m.下午	8.18p.m.下午 to 10.30p.m.下午	6 ,,
10.56p.m.下午 to 12.26a.m.上午	10.30p.m.下午 to 12.00mid.子夜	10 ,,

Sundays and Public Holidays 星期日及公眾假期

6.05a.m.上午 to 7.25a.m.上午	5.40a.m.上午 to 7.00a.m.上午	10 minutes 分
7.25a.m.上午 to 11.37a.m.上午	7.00a.m.上午 to 11.12a.m.上午	6 ,,
11.37a.m.上午 to 1.05p.m.下午	11.12a.m.上午 to 12.40p.m.下午	8 ,,
1.05p.m.下午 to 7.53p.m.下午	12.40p.m.下午 to 7.28p.m.下午	6 ,,
7.53p.m.下午 to 9.21p.m.下午	7.28p.m.下午 to 8.56p.m.下午	8 ,,
9.21p.m.下午 to 10.51p.m.下午	8.56p.m.下午 to 10.26p.m.下午	6 ,,
10.51p.m.下午 to 12.21a.m.上午	10.26p.m.下午 to 11.56p.m.下午	10 ,,

Faretable　30 cents per Single Journey 票價：單程三毫

ROUTE 5 號路線

STAR FERRY to CHOI HUNG: via Salisbury Rd., Chatham Road, Cheong Wan Rd., Hong Chong Road, Flyover, Chatham Road, Ma Tau Wei Rd., Ma Tau Chung Rd., Prince Edward Road and Choi Hung Road.

尖沙咀碼頭至彩虹　經：疏利士巴利道，漆咸道，暢運道，康莊道，天橋，漆咸道，馬頭圍道，馬頭涌道太子道及彩虹道。

CHOI HUNG to STAR FERRY: via Choi Hung Road, Flyover, Prince Edward Rd., Ma Tau Chung Rd., Ma Tau Wei Road, Chatham Road and Salisbury Road.

彩虹至尖沙咀碼頭　經：彩虹道，天橋，太子道，馬頭涌道，馬頭圍道，漆咸道及疏利士巴利道。

TIME TABLE 時間表
Mondays to Saturdays 星期一至星期六

From Star Ferry 尖沙咀碼頭開	From Choi Hung 彩虹開	Frequency 每次
6.03a.m.上午 to 6.33a.m.上午	5.30a.m.上午 to 6.00a.m.上午	6 minutes 分
6.33a.m.上午 to 10.33a.m.上午	6.00a.m.上午 to 10.00a.m.上午	4 ,,
10.33a.m.上午 to 11.53a.m.上午	10.00a.m.上午 to 11.20a.m.上午	5 ,,
11.53a.m.上午 to 8.05p.m.下午	11.20a.m.上午 to 7.28p.m.下午	4 ,,
8.05p.m.下午 to 9.05p.m.下午	7.28p.m.下午 to 8.28p.m.下午	6 ,,
9.05p.m.下午 to 11.57p.m.下午	8.28p.m.下午 to 11.24p.m.下午	4 ,,
11.57p.m.下午 to 12.45a.m.上午	11.24p.m.下午 to 12.12a.m.上午	6 ,,

Sundays and Public Holidays 星期日及公眾假期

6.03a.m.上午 to 7.33a.m.上午	5.30a.m.上午 to 7.00a.m.上午	6 minutes 分
7.33a.m.上午 to 10.33a.m.上午	7.00a.m.上午 to 10.00a.m.上午	4 ,,
10.33a.m.上午 to 11.53a.m.上午	10.00a.m.上午 to 11.20a.m.上午	5 ,,
11.53a.m.上午 to 8.05p.m.下午	11.20a.m.上午 to 7.28p.m.下午	4 ,,
8.05p.m.下午 to 9.05p.m.下午	7.28p.m.下午 to 8.28p.m.下午	6 ,,
9.05p.m.下午 to 11.33p.m.下午	8.28p.m.下午 to 11.00p.m.下午	4 ,,
11.33p.m.下午 to 12.45a.m.上午	11.00p.m.下午 to 12.12a.m.上午	6 ,,

Faretable　30 cents per Single Journey 票價：單程三毫

ROUTE 5A 號路線

STAR FERRY to KOWLOON CITY (SHING TAK ST.): via Salisbury Rd., Chatham Rd., Cheong Wan Road, Hong Chong Road, Flyover, Chatham Road, Ma Tau Wei Road, Ma Tau Chung Rd., Fu Ning Street and Shing Tak Street.

尖沙咀碼頭至九龍城（盛德街）經：疏利士巴利道，漆咸道，暢運道，康莊道，天橋，漆咸道，馬頭圍道，馬頭涌道，富寧街及盛德街。

KOWLOON CITY (SHING TAK ST.) to STAR FERRY: via Shing Tak Street, Ma Tau Kok Rd., Ma Tau Chung Rd., Ma Tau Wei Rd., Chatham Road and Salisbury Road.

九龍城（盛德街）至尖沙咀碼頭　經：盛德街，馬頭角道，馬頭涌道，馬頭圍道，漆咸道及疏利士巴利道

TIME TABLE 時間表
Mondays to Saturdays only 星期一至星期六
(No service on Public Holidays 星期日及公眾假期停止服務)

From Star Ferry 尖沙咀碼頭開	From Kowloon City 九龍城開	Frequency 每次
7.20a.m.上午 to 9.50a.m.上午	7.00a.m.上午 to 10.10a.m.上午	10 minutes 分
4.00p.m.下午 to 7.10p.m.下午	3.40p.m.下午 to 7.30p.m.下午	10 ,,

Faretable　30 cents per Single Journey 票價：單程三毫

Without Notice. ▶
不再另行通告」

KMB Routes and fares as at 10th August 1994

▼ Service in Morning Peak Hour Only
◊ - Service in Evening Peak Hour Only
◑ - Service in Pear Hours Only
N - Night Service Only
❋ - Sundays and Public Holidays Only
✣ - No service on Sundays and Public Holidays
▲ - Section fare

Kowloon Urban Routes.

Route No.	Route:	Non A/C Fare	A/C Fare
1	"Star" Ferry to Chuk Yuen Estate	$3.00	$4.00
	▲ Mongkok Road to Chuk Yuen Estate	$3.00	$3.50
	▲ Lok Fu (Fung Mo St.) to Chuk Yuen Estate	$2.50	$3.50
	▲ Nathan Road to Star Ferry	$3.00	$3.50
1A	"Star" Ferry to Sau Mau Ping (Central)	$3.50	$5.00
	▲ Kwun Tong (Yue Man Sq) to Star Ferry	$3.50	$4.70
	▲ Prince Edward Road East (San Po Kong) to Star Ferry	$3.50	$4.00
	▲ Nathan Road to Star Ferry	$3.00	$3.50
	▲ Mongkok Road to Sau Mau Ping	$3.50	$4.00
	▲ Kwun Tong Road (L. Ngau Tau Kok Est.) to Sau Mau Ping	$3.50	$4.00
	▲ Kwun Tong Road (Yue Man Square) to Sau Mau Ping	$2.50	$4.00
2	"Star" Ferry to So Uk	$2.50	$4.00
	▲ Nathan Road/Bute St. Junction to Star Ferry		
2A	So Uk to Lok Wah	$3.00	/
	▲ Ngau Tau Kok Road to Lok Wah	$2.50	/
2B	Jordan Road Ferry to Cheung Sha Wan	$2.50	/
2C	Yau Yat Tsuen to "Star" Ferry	$2.50	/
2D	Tung Tau Estate to Chak On Estate	$2.50	/
2E	Kowloon City Ferry to Pak Tin	$2.50	/
2F	Cheung Sha Wan to Tsz Wan Shan (South)	$3.50	/
	▲ Wong Tai Sin MTR Station to Tse Wan Shan	$2.50	/
3	Jordan Road Ferry to Diamond Hill MTR Station	$2.50	/
3A	Tsz Wan Shan (North) to Chuk Yuen Estate (Circular)	$1.90	/
3B	Hong Hom Ferry to Tsz Wan Shan (North)	$2.70	/
	▲ Ming On Street to Hung Hom Ferry		
3C	China Ferry Terminal to Tsz Wan Shan (North)	$3.00	/
	▲ Wong Tai Sin MTR Station to Tsz Wan Shan	$2.50	/
3D	Kwun Tong (Yuet Wah Street) to Tsz Wan Shan (N)	$2.20	/
3M	Tsz Wan Shan (North) to Choi Wan	$2.20	/
	▲ Clear Water Bay Road to Chio Wan	$1.90	/
▼✣ 3P	Fung Tak Road to Kwun Tong (Yuet Wah Street)	$2.50	/
4A	Jordan Roud Ferry to Tai Hang Tung	$1.80	/
5	"Star" Ferry to Choi Hung	$2.50	$3.80
◑ 5A	"Star" Ferry to Kowloon City (Shing Tak Street)	$2.20	/
5C	"Star" Ferry to Tsz Wan Shan (South)	$3.00	/
	▲ Choi hung·Road to Tsz Wan Shan	$2.50	/
5D	Hung Hom Ferry to Telford Gardens	$3.00	/
	▲ Ming On Street to Hung Hom Ferry	$2.50	/
5K	Kowloon Station to Kwun Tong Ferry	$3.30	/
6	"Star" Ferry to Mei Foo	$2.50	$4.00
	▲ Nathan Rd (Nr Prince Edward MTR) to Star Ferry	$2.50	$3.50
	▲ Nathan Rd (Nr Prince Edward MTR) to Mei Foo	$2.50	$3.50
6A	"Star" Ferry to Lai Chi Kok (Amusement Park)	$2.50	/
6B	Wong Tai Sin to Mei Foo	$3.50	/
6C	Kowloon City Ferry to Mei Foo	$3.00	$4.00
	▲ Nathan Rd (Nr Prince Edward MTR) to Mei Foo	$3.00	$3.50
	▲ Wuhu Street to Kowloon City Ferry	$3.00	$3.50
6D	Ngau Tau Kok to Cheung Sha Wan	$3.00	/
6F	Kowloon City Ferry to Lai Kok	$2.50	/
7	Lok Fu to "Star" Ferry	$2.50	/
7B	Hong Hom Ferry to Lok Fu	$3.00	/
	▲ Ming On Street to Hung Hom Ferry	$2.50	/
7M	Lok Fu to Chuk Yuen Estate (Circular)	$1.90	/
8	Jordan Road Ferry to "Star" Ferry	$2.70	/
8A	Jordan Road (Canton Road) to Whampoa Garden	$2.50	/
9	"Star" Ferry to Ping Shek	$3.00	/
10	Tai Kok Tsui Ferry to Choi Wan	$2.50	/
11	Jordan Road Ferry to Wong Tai Sin	$2.50	$3.80
11B	Kowloon City Ferry to Kwun Tong (Tsui Ping Road)	$2.50	/
11C	Chuk Yuen Estate to Sau Mau Ping (Central)	$3.00	/
	▲ Kwun Tong Rd (Nr Yue Man Sq) to Sau Mau Ping	$2.50	/
	▲ Lok Fu Bus Terminus to Chuk Yuen Estate	$2.50	/
11D	Kwun Tong Ferry to Lok Fu	$2.50	/
11K	Kowloon Station to Chuk Yuen Estate	$2.50	/
12	China Ferry Terminal to Lai Chi Kok (Amusement Park)	$2.50	/
12A	Sham Shui Po Ferry to Whampoa Garden	$2.50	/
13	Jordan Road Ferry to Choi Hung	$2.50	/
13A	Hung Hom Ferry to Sau Mau Ping(Upper)	$3.50	/
	▲ Mut Wah Street to Sau Mau Ping	$2.50	/
	▲ Ming On Street to Hung Hom Ferry	$3.50	/
13D	Tai Kok Tsui Ferry to Sau Mau Ping (Central)	$3.50	/
	▲ Mut Wah Street to Sau Mau ping	$2.50	/
13E	Kai Yip Street to Sau Mau Ping (Upper)	$2.50	/
	▲ Kwun Tong Ferry to Sau Mau Ping	$2.20	/
13M	Kwun Tong MTR Station to Sau Mau Ping(Circular)	$2.20	/
◊❋ 14X	Sau Mau Ping (Central) to "Star" Ferry	$4.00	/
	▲ Ngau Tau Kok Road to Sau Mau Ping	$2.50	/
14	China Ferry Terminal to Yau Tong (Ko Chiu Road)	$3.50	$4.70
	▲ Kwun Tong Road (L. Ngau Tau Kok Estate) to Yau Tong	$3.50	$4.00
	▲ Kwun Tong (near Yue Man Sq.) to Yau Tong	$2.50	$4.00
	▲ Prince Edward Rd. E. (San Po Kong) to China Ferry Terminal	$$3.50	$4.00
14B	Ngau Tau Kok to Lam Tin (Kwong Tin Estate)	$2.50	/
	▲ Kwun Tong Ferry to Lam Tin	$2.20	/
14C	Kwun Tong (Yue Man Sq) to Lei Yue Mun (Sam Ka Tsuen)	$2.20	/
◊❋ 14X	Jordan Road Ferry to Yau Tong (Ko Chiu Road)	$4.00	/
	▲ Kwun Tong Road (nr. Yui Man Sq.)	$2.50	/
15	Hung Hom Ferry to Lam Tin (North)	$3.50	/
	▲ Kwun Tong Road (nr. Yui Man Sq.) to Lam Tin	$2.50	/
	▲ Ming On Street to Hung Hom Ferry	$2.50	/
15A	Tsz Wan Shan (North) to Lam Tin (North)	$3.00	/
	▲ Choi Hung (Lung Chung Road) to Rsz Wan Shan	$2.50	/
	▲ Kwun Tong Ferry to Lam Tin	$2.20	/
◊❋ 15X	Tsim Sha Tsui (Hankow Rd.) to Lam Tin (Kwong Tin Est.)	$4.00	/
	▲ Kwun Tong Road (nr. Yui Man Sq.) to Lam Tin	$2.50	/
16	Tai Kok Tsui Ferry to Lam Tin (North)	$3.50	$4.70
	▲ Kwun Tong Road (Ngau Tau Kok Estate) to Lam Tin	$3.50	$4.00
	▲ Kwun Tong (nr Yue Man Sq) to Lam Tin	$2.50	$4.00
	▲ Prince Edward Road.E.(San Po Kong) to Tai Kok Tsui	$3.50	$4.00
16M	Kwun Tong MTR to Lam Tin (Hong Wah Court)(Circular)	$1.90	/
17	Oi Man to Kwun Tong (Yuet Wah Street)	$3.00	/
18	Sham Shui Po to Oi Man (Circular)	$2.50	/
21	Choi Wan to Kowloon Station	$2.50	/
23	Kwun Tong to Shun Lee (Circular)	$2.20	/
23M	Lok Wah to Shun lee (Circular)	$2.50	/
24K	Kai Yip to Mong Kok KCR Station (Circular)	$2.50	/
◑❋ 24X	Kai Yip to Tsim Sha Tsui (East) (Circular)	$4.00	/
26	Shun Tin to Kowloon Station	$3.50	$4.70
	▲ Chatham Road North to Kowloon Station	$3.50	$4.00
	▲ Prince Edward Road East to Shun Tin	$3.50	$4.00
	▲ Ping Shek (Home for the Aged) to Shun Tin	$2.50	$4.00
26M	Kwun Tong (Yuet Wah St.) to Choi Hung MTR (Circular)	$2.50	$3.80
27K	Shun Tin to Mong Kok KCR Station (Circular)	$2.50	$3.80
28	Lok Wah - Tsim Sha Tsui (Hankow Road)	$2.50	/
28A	Lok Wah to Kwun Tong Ferry	$1.90	/
29M	Shun Lee to San Po Kong (Circular)	$2.20	/

New Territories Urban Routes.

Route	From:		
30	Allway Gardens to Cheung Sha Wan	$3.30	/
	▲ Sha Tsui Road to Cheung Sha Wan	$2.50	/
	▲ Lai King to Allway Gardens	$2.50	/
30X	Allway Gardens to Whampoa Gardens	$4.50	$7.00
	▲ Mei Foo to Whampoa Gardens	$3.00	$4.50
	▲ Nathan Rd. (Nr Prince Edward MTR) to Whampoa Gdn.	$3.00	$3.50
	▲ Mei Foo to Allway Gardens	$3.50	
31	Tsuen Wan Ferry to Shek Lei (Circular)	$2.20	/
31B	Shek Lei to Sham Shui Po Ferry	$2.50	/
31M	Shek Lei (Lei Pui Street) to Kwai Fong MTR Station	$2.20	/
32	Shek Wai Kok to Tai Kok Tsui Ferry	$3.50	/
	▲ Mei Foo to Shek Wai Kok		
32B	Tsuen Wan Ferry to Cheung Wan Ferry (Circular)	$2.20	/
32M	Cheung Shan to Kwai Fong MTR Station	$2.50	/

Route	Description	Fare 1	Fare 2
33A	Tsuen Wan Ferry to Tai Kok Tsui Ferry	$3.50	/
	▲ Yen Chow Street to Tsuen Wan Ferry	$3.00	/
34	Kwai Shing (East) to Bayview Garden (Circular)	$2.60	/
34B	Sham Sheng to Tsuen Wan Ferry	$2.50	/
34M	Bayview Garden to Tsuen Wan MTR Stn (Circular)	$1.90	/
35A	On Yam to Sham Shui Po Ferry	$3.00	/
36	Lei Muk Shu to Tsuen Wan Ferry	$2.20	/
36A	Lei Muk Shu to Sham Shui Po Ferry	$3.00	/
36B	Lei Muk Shu to Jordan Road Ferry	$3.50	/
	▲ Mei Foo to Lei Muk Shue	$3.00	/
36M	Lei Muk Shue to Kwai Fong MTR Station	$2.20	/
37	Kwai Shing (Central) to Tai Kok Tsui Ferry	$3.00	/
37M	Kwai Hing to Kwai Shing (Central)	$1.90	/
38	Kwai Shing (East) to Lam Tin MTR Station	$4.50	/
	▲ Lung Cheung Road (Beacon Heights) to Lam Tin MTR	$3.50	/
	▲ Butterfly Valley Interchange to Kwai Shing (East)	$3.00	/
38A	Rivera Garden to Mei Foo	$2.50	/
	▲ Kwan Mun Hau Street to Rivera Garden	$2.20	/
▼‡ 38P	Rivera Garden to Mong Kok (Sai Yuen Choi Street)	$3.50	/
39A	Tsuen Wan Ferry to Allway Gardens (Circular)	$2.20	/
39M	Allway Gardens to Tsuen Wan MTR Station	$1.90	/
40	Tsuen Wan Ferry to Kwun Tong Ferry	$4.50	/
	▲ Mei Foo to Kwun Tong Ferry	$3.50	/
	▲ Mei Foo to Tsuen Wan Ferry	$3.00	/
40X	Lee On to Kwai Hing MTR Station	$5.00	/
	▲ Boardings at Shing Mun Tunnels Bus Interchange to Lee On	50¢	/
	▲ After Shing Mun Tunnels to Lee On	$3.50	/
	▲ After Shing Mun Tunnels to Kwai Hing MTR Stn.	$3.00	/
	▲ No charge for boardings at Shing Mun Tunnels Bus Interchange to Kwai Hing MTR Stn.		
41	Cheung Ching to Kowloon City Ferry	$4.50	/
	▲ Mei Foo to Kowloon City Ferry	$3.50	/
	▲ Mei Foo to Cheung Ching	$3.00	/
41A	Cheung On to Sham Shui Po Ferry	$3.00	/
41M	Tsing Yi Estate to Tsuen Wan MTR Station	$2.50	$3.50
●‡ 41P	Tsing Yi Ferry to Tsuen Wan MTR Station	/	$3.50
42	Cheung Ching to Shun Lee	$4.10	/
	▲ Mei Foo to Shun Lee	$3.50	/
	▲ Ping Shek (Home for the Aged) to Shun Lee	$2.50	/
	▲ Mei Foo to Chung Ching	$3.00	/
42A	Chung Hing to Jordan Road Ferry	$3.50	/
	▲ Mei Foo to Cheung Hang	$3.00	/
42C	Cheung Hang to Lam Tin MTR Station	$5.00	/
	▲ Tsuen Wan (Tsuen Fu Street) to Lam Tin MTR	$4.50	/
	▲ Lung Cheung Road (Beacon Heights) to Lam Tin MTR.	$3.50	/
	▲ Butterfly Valley Interchange to Cheung Hang	$3.50	/
42M	Tsing Yan to Tsuen Wan MTR Station (Circular)	$2.50	$3.50
43	Cheung Hong to Tsuen Wan Ferry	$2.50	/
43A	Tsing Yan to Shek Lei (Circular)	$2.50	/
43B	Cheung Ching to Tsuen Wan Ferry	$2.50	/
43C	Cheung Hong to Tai Kok Tsui Ferry	£3.50	/
	▲ Mei Foo to Cheung Hong	$3.00	/
43M	Kwai Fong MTR Stn. to Chung Ching (Circular)	$2.20	$3.00
43X	Yiu On to Tsuen Wan Ferry	$5.00	/
	▲ Boardings at Shing Mun Tunnels Bus Interchange to Yiu On	50¢	/
	▲ After Shing Mun Tunnels to Yiu On	$3.50	/
	▲ Chevalier Garden Bus Terminus to Yiu On	$2.50	/
	▲ After Shing Mun Tunnels to Tsuen Wan Ferry	$3.00	/
	▲ No charge for boardings at Shing Mun Tunnels Bus Interchange to Tsuen Wan Ferry.		
44	Tsing Yi Estate to Mong Kok KCR Station	$3.50	/
	▲ Mei Foo to Tsing Yi Estate	$3.00	/
44M	Cheung On Estate to Kwai Hing MTR Station	$2.50	/
✧‡ 44P	Mong Kok KCR Station to Tsing Yi Ferry	$3.50	/
	▲ Mei Foo to Tsing Yi Ferry	$3.00	/
45	Lai Yiu to Kowloon City Ferry	$3.60	/
	▲ Mei Foo to Lai Yiu	$3.00	/
46	Lai Yiu to Jordan Road Ferry	$3.00	/
●‡ 46P	Lai King (South) to Hin Keng	$4.50	/
	▲ After Shing Mun Tunnels to Hin Keng	$3.00	/
	▲ After Shing Mun Tunnels to Kwai Fong	$3.00	/
	▲ No charge for boardings at Shing Mun Tunnels Bus Interchange to Kwai Fong & Hin Keng.		
46X	Hin Keng to Mei Foo	$4.50	/
	▲ After Shing Mun Tunnels to Hin Keng	$3.00	/
	▲ After Shing Mun Tunnels to Mei Foo	$3.00	/
	▲ No charge for boardings at Shing Mun Tunnels Bus Interchange to Mei Foo & Hin Keng.		
47X	Chun Shek to Kwai Shing (East)	$4.50	/
	▲ After Shing Mun Tunnels to Chun Shek	$3.00	/
	▲ After Shing Mun Tunnels to Kwai Shing (East)	$3.00	/
	▲ No charge for boardings at Shing Mun Tunnels Bus Interchange to Mei Foo & Hin Keng.		
48X	Wo Che to Bayview Garden	$4.50	/
	▲ After Shing Mun Tunnels to Wo Che	$3.00	/
	▲ After Shing Mun Tunnels to Bayview Garden	$3.00	/
	▲ No charge for boardings at Shing Mun Tunnels Bus Interchange to Bayview Garden & Wo Che		
49X	Kwun Yuen to Tsing Yi Ferry	$5.00	/
	▲ Yuen Uk Road to Kwong Yuen	$.50	/
	▲ After Shing Mun Tunnels to Kwong Yuen	$3.00	/
	▲ After Shing Mun Tunnels to Tsing Yi Ferry	$3.00	/
	▲ No charge for boardings at Shing Mun Tunnels Bus Interchange to Kwun Yuen to Tsing Yi Ferry		
51	Kam Tin to Tsuen Wan Ferry	$4.50	/
	▲ Chuen Lung to Tsuen Wan Ferry	$2.50	/
	▲ Shek Kong to Kam Tin	$2.50	/
▼‡ 51P	Kam Tin (Opposite Shek Kong Swimming Pool) to Tsuen Wan Ferry	$4.50	/
	▲ Chuen Lung to Tsuen Wan Ferry	$2.50	/
52M	Prime View Garden to Kwai Fong MTR Station	$5.00	/
	▲ Tsuen Wan (Tsuen Tak Garden) to Kwai Fong MTR	$3.00	/
	▲ Tsuen Wan (Chung On St) to Prime View Garden	$4.50	/
	▲ Sam Shing to Prime View Garden	$2.50	/
52X	Tuen Mun Town Centre to Sham Shui Po Ferry	$6.20	/
	▲ Tsing Lung Tau to Sham Shui Po Ferry	$4.50	/
	▲ Mei Foo to Sham Shui Po Ferry	$2.60	/
	▲ Sham Tseng to Tuen Mun Town Centre	$4.00	/
	▲ Tsing Lung Tau to Tuen Mun Town Centre	$3.50	/
	▲ Tai Lam to Tuen Mun Town Centre	$3.00	/
	▲ Yuen Long (East) to Tsuen Wan Ferry	$5.60	/
53	Yuen Long (East) to TsuenWan Ferry	$5.60	/
	▲ Yuen Long (East) to Tsing Lung Tau	$4.50	/
	▲ Fu Tei to TsuenWan Ferry	$4.50	/
	▲ Fu Tei to Tsing Lung Tau	$3.50	/
	▲ Sam Shing to Tsing Lung Tau	$3.50	/
	▲ Tsing Lung Tau to TsuenWan Ferry	$3.50	/
	▲ Sam Shing to Yuen Long (East)	$3.50	/
	▲ Fu Tei to Yuen Long (East)	$2.70	/
57M	Shan king Estate to Lai King (South)	$5.00	/
	▲ TsuenWan (Tsuen Tak Garden) to Lai king (South)	$3.00	/
	▲ TsuenWan (Chung On Street) to Shan King Estate	$4.50	$7.00
	▲ Tuen Mun Technical Institute to Shan King Estate	$2.50	$7.00
58M	Leung King Estate to Kwai Fong MTR Station	$5.00	$7.00
	▲ TsuenWan (Tsuen Tak Garden) to Kwai Fong MTR	$3.00	$3.50
	▲ TsuenWan (Chung On Street) to Leung King Estate	$4.50	$7.00
	▲ Tuen Mun Town Centre to Leung King Estate	$2.50	$3.00
✧‡ 58P	Kwai Fong MTR Stn to Leung King Estate	$5.00	$7.00
	▲ TsuenWan (Chung On Street) to Leung King Estate	$4.50	$7.00
	▲ Tuen Mun Town Centre to Leung King Estate	$2.50	$3.00
58X	Leung King Estate to Mong Kok KCR Station	$7.00	$10.50
	▲ Mei Foo to Mong Kok KCR Station	$3.50	$4.50
	▲ Tuen Mun Town Centre to Leung King Estate	$2.50	$6.20
59A	Tuen Mun Pier Head to Sham Shui Po Ferry	$6.20	/
	▲ Goodview Garden to Sham Shui Po Ferry	$5.60	/
	▲ TsuenWan (Tsuen Tak Garden) to Sham Shui Po Ferry	$3.00	/
	▲ Cheung Sha Wan (Tai Nam West St) to Tuen Mun Pier Head	$5.60	/
	▲ TsuenWan (Chung On Street) to Tuen Mun Pier Head	$4.50	/
	▲ San Shek Wan/On Ting Estate to Tuen Mun Pier Head	$2.50	/
59M	Tuen Mun Pier Head to TsuenWan MTR Station	$4.50	$6.50
	▲ TsuenWan (Tsuen Tak Garden) toTsuenWanMTR	$2.50	$3.00
	▲ San Shek Wan to Tuen Mun Pier Head	$2.50	$3.00
59X	Tuen Mun Pier Head to Mong Kok KCR Station	$4.50	$6.50
	▲ Mei Foo to Mong Kok KCR Station	$3.00	$4.50
	▲ San Shek Wan to Tuen Mun Pier Head	$2.50	$3.00
●‡ 60	Yau Oi (South) to Cheung Sha Wan	$5.60	/
	▲ TsuenWan (Tsuen Tak Garden) to Cheung Sha Wan	$3.00	/
	▲ Tsuen Wan (Chung On Street) to Yau Oi (South)	$4.50	/
	▲ On Ting Estate to Yau Oi (South)	$2.50	/
60M	Tuen Mun Town Centre to Tsuen Wan MTR Stn.	$4.50	$6.50
	▲ Tsuen Wan (Tsuen Tak Garden) to Tsuen Wan MTR	$2.50	$3.00
	▲ Goodview Garden to Tuen Mun Town Centre	$2.50	$3.00
60M	Tuen Mun Town Center to Jordan Road Ferry	$7.00	$10.50
	▲ Mei Foo to Jordan Road Ferry	$3.00	$4.50
	▲ Nathan Road (near Prince Edward MTR to Jordan Road Ferry	$$3.00	$3.50
	▲ Goodview Garden to Tuen Mun Town Centre	$2.50	$3.00
61X	Tuen Mun Town Centre to kowloon City Ferry	$7.70	/
	▲ Beacon heights to Kowloon City Ferry	$3.50	/
	▲ Goodview Garden to Tuen mun Town Centre	$2.50	/
● 62X	Tuen Mun Town Centre to Yau Tong (Ko Chiu Road)	$8.50	$13.00
	▲ Lung Chung Road (Beacon Heights) to Yau Tong	$3.50	$4.50
	▲ Kwun Tong Road (nr. Yue Man Square) to Yau Tong	$2.50	$4.00
	▲ Goodview Garden to Tuen Mun Town Centre	$2.50	$3.00

Route	Description	Fare 1	Fare 2
◆ 64M	Tin Yau Estate to Tsuen Wan Ferry	$5.60	$7.50
	▲ Fu Tei to Tsuen Wan ferry	$4.50	$6.50
	▲ Tsuen Wan (Tsuen Tak Garden) to Tsuen Wan Ferry	$2.50	$3.00
	▲ Tuen Mun Town Centre) to Tin Yiu Estate	$3.60	$4.30
	▲ Fu Tei to Tin Yiu Estate	$2.70	$3.50
66	Tai Hing to Tai kok Tsui Ferry	$6.20	/
	▲ Tsuen Wan (Tsuen Tak Garden) to Tai Kok Tsui Ferry	$3.50	/
	▲ Cheung Sha Wan (Tai Nam West Street) to Tai Hing	$5.60	/
	▲ Tsuen Wan (Chung On Street) to Tai hing	$4.50	/
	▲ Tuen Mun Town Park to Tai Hing	$2.50	/
66M	Tai Hing to Tsuen Wan MTR Station	$4.50	$6.50
	▲ Tsuen Wan (Tsuen Tak Garden) to Tsuen Wan MTR	$2.50	$3.00
	▲ San Fat Estate to Tai Hing	$2.50	$3.00
▼✻ 66P	Tai Hing to Tsuen wan MTR Station	$4.50	$3.00
	▲ Tsuen Wan (Tsuen Tak Garden) to Tsuen Wan MTR	$2.50	$3.00
	▲ Tuen Mun Technical Institute to Tai Hing	$2.50	$3.00
66X	Tai Hing to Tai Kok Tsui Ferry	$7.00	/
	▲ Mei Foo to Tai Kok Tsui Ferry	$3.50	/
	▲ Tuen Mun Technical Institute to Tai Hing	$2.50	/
67M	Siu Hong Court to Kwai Fong MTR Station	$5.00	/
	▲ Tsuen Wan (Tsuen Tak Garden) to Kwai Fong MTR	$3.00	/
	▲ Tsuen Wan (Chung On Street) to Siu Hong Court	$4.50	/
	▲ Tuen Hing Road to Siu Hong Court	$2.50	/
67X	Siu Hong Court to Mong Kok KCR Station	$7.00	/
	▲ Mei Foo to Mong kok KCR Station	$3.50	/
	▲ Tuen Hing Road to Siu u Hong Court	$2.50	/
68	Yuen Long (East) to Jordan Road (Canton Road)	$6.20	/
	▲ FuTei to Jordan Road	$6.20	/
	▲ Tsuen Wan (Tsuen Tak Garden) to Yuen Long (East)	$5.50	/
	▲ Tsuen Wan (Chung On Street) to Yuen Long (East)	$5.60	/
	▲ Tuen Mun Town Centre to Yuen Long (East)	$3.60	/
	▲ Fu Tei to Yuen Long (East)	$2.70	/
68A	Long Ping Estate to Sham Shui Po Ferry	$7.00	/
	▲ Fu Teu to Sham Shui Po Ferry	$6.20	/
	▲ Tsuen Wan (Tsuen Tak Garden) to Sham Shuipo Ferry	$3.00	/
	▲ Tsuen Wan (Chung On Street) to Long Ping Estate	$5.60	/
	▲ Tuen Mun Town Centre to Long Ping Estate	$3.60	/
	▲ Fu Tei to long Ping Estate	£2.70	/
68M	Yuen Long (West) to Tsuen Wan MTR Station	$5.60	$7.50
	▲ Fu Tei to Tsuen Wan MTR Station	$4.50	$6.50
	▲ Tsuen Wan (Tsuen Tak Garden) to Tsuen Wan MTR	$2.50	$3.00
	▲ Tuen Mun Town Centre to Yuen Long (West)	$3.60	$4.30
	▲ Fu Tei to Yuen Long (West)	$2.70	$3.50
68X	Yuen Long (East) to JordanRoad Ferry	$8.50	$13.00
	▲ Fu Tei to Jordan Road Ferry	$7.00	$10.50
	▲ MeiFoo to Jordan Road Ferry	$3.00	$4.50
	▲ Nathan Rd.(nr. Prince Edward MTR Stn.) to Jordan Road Ferry	$3.00	$3.50
	▲ Tuen Mun Town Centre to Yuen Long(East)	$3.60	$4.30
	▲ Fu Tei to Yuen Long (East)	$2.70	$3.50
69M	Tin Shui to Lai King (South)	$6.20	$9.00
	▲ Fu Tei to Lai King (South)	$5.00	$7.00
	▲ Tsuen Wan (Tsuen Tak Garden) to Lai King (South)	$3.00	$3.50
	▲ Tsuen Wan (Chung On Street) to Tin Shui Estate	$5.60	$7.50
	▲ Tuen Mun Town Centre to Tin Shui Estate	$3.60	$4.30
	▲ Fu Tei Ti to Tin Shui Estate	$2.70	$3.50
69X	Tin Shui Estate to Kowloon Station	$8.80	$13.50
	▲ Fu Tei to Kowloon Station	$7.50	$11.50
	▲ Mei Foo to Kowloon Station	$3.50	$4.50
	▲ Nathan Road (nrPrince Edward MTR) to Kowloon Stn.	$3.00	$3.50
	▲ Tuen Hing Road to Tin Shui Estate	$3.60	$4.30
	▲ Fu Tei to Tin Shui Estate	$2.70	$3.50
70	Sheung Shui to Jordan Road Ferry	$7.00	/
	▲ Tai Po Road/Lam Kam Road Junct. to Jordan Road Ferry	$5.60	/
	▲ Cheung Shue Tan to Jordan Road Ferry	$5.00	/
	▲ AfterLionRock Tunnel to Jordan Road Ferry	$3.50	/
	▲ Cheung Shue Tan to Sheung Shui	$3.20	/
▼✻ 70P	Shung Shui to Kwun Tong (Tsui Ping Road)	$8.50	/
	▲ After Lion Rock Tunnel to KwunTong	$3.50	/
70X	Sheung Shui to Kwun Tong (Tsui Ping Road)	$8.50	/
	▲ After Lion Rock Tunnel to KwunTong	$3.50	/
	▲ Wo Hing Tsuen to Sheung Shui	$2.60	/
72	Tai Wo to TaiKok Tsui Ferry	$5.00	/
	▲ Tai Po Road (North Kln. Magistracies) to Tai Kok Tsui Ferry	$3.50	/
	▲ Cheung Shue Tan to Tai Wo	$3.00	/
72A	Tai Po Industrial Estate to Tai Wai	$4.50	/
	▲ Tai Po Industrial Estate to Cheung Shue Tan	$3.00	/
	▲ TaiWai to Cheung Shue Tan	$3.00	/
	▲ Kwong Fuk to Tai Po Industrial Estate	$2.60	/
72X	Tai Po Central to Tai Kok Tsui Ferry	$5.60	/
	▲ After Lion Rock Tunnel to Tai Kok Tsui	$3.50	/
	▲ Kwomg Fuk to Tai Po Central	$2.60	/
73A	Choi Yuen to Yuen Chau Kok	$5.60	/
	▲ Tai Po Road/Lam Kam Junct. to Yuen Chau Kok	$4.00	/
	▲ Cheung Shue Tan to Choi Yuen	$3.20	/
	▲ Wo Hing Tsuen to Choi Yuen	$2.60	/
73X	Fu Shin to Tsuen Wan Ferry	$5.60	/
	▲ After Shing MunTunnels to Tsuen Wan Ferry	$3.00	/
	▲ Boarding at Shing Mun Tunnels Bus Interchange to Fu Shin	$1.10	
	▲ Kwong Fuk to Fu Shin	$2.60	/
	▲ No charge for boarding at Shing Mun Tunnel Bus Interchange to Tsuen Wan	/	/
74X	Tai Wo to Kwun Tong Ferry	$5.60	$8.00
	▲ Belair Gardens to Kwun Tong Ferry	$3.50	$4.50
	▲ Cheung Shue Tan to Tai Wo	$3.00	$3.50
75X	Fu Shin to Kowloon City Ferry	$5.60	$8.00
	▲ After Tate's Cairn Tunnel to Kowloon City Ferry	$3.50	$4.50
	▲ Kwong Fuk to Fu Shin	$2.60	$3.50
80	Mei Lam - Kwun Tong Ferry	$3.80	/
	▲ After Lion Rock Tunnel to Mei Lam	$3.00	/
	▲ After Lion Rock Tunnel to Kwun Tong Ferry	$3.50	/
80K	Sun Chui - Yuen Chau Kok	$2.50	/
80M	Sui Wo Court - Kowloon Tong MTR Station	$3.00	$4.20
▼✻80P	Hin Keng - Kwun Tong Ferry	$3.80	/
	▲ After Lion Rock Tunnel to Hin Keng	$3.00	/
	▲ After Lion Rock Tunnel to Kwun Tong Ferry	$3.50	/
80X	Chun Shek - Kwun Tong Ferry	$4.00	/
	▲ After Tate's Caim Tunnel to Kwun Tong Ferry	$3.50	/
	▲ After Tate's Caim Tunnel to Chun Shek	$3.00	/
81	Wo Che - Jordan Road Ferry	$3.50	$4.70
	▲ Tai Po Road (near Keng Hau Road) to Wo Che	$3.00	$4.00
81C	Yiu On - Kowloon Station	$4.50	$6.50
	▲ Tai Chung Kiu Rd./Ma On Shan Rd.Junction to Kowloon Stn.	$4.00	$5.50
	▲ After Lion Rock Tunnel to Kowloon Station	$3.50	$4.50
	▲ After Lion Rock Tunnel to Yiu On	$3.50	$4.50
	▲ Chevalier Garden Bus Terminus to Yiu On	$2.50	$3.00
81K	Sun Tin Wai - Sui Wo Court	$2.50	/
81M	Sun Tin Wai - Kowloon Tong MTR Station	$2.50	/
▼✻81P	Sha Tin Wai to Tsim Sha Tsui (Chatham Rd South)	$4.00	$5.50
	▲ After Lion Rock Tunnel to Tsim Sha Tsui (Chatham Road S.)	$3.50	$4.50
82K	Mei Lam - Fo Tan KCR Station	$3.10	/
82M	Kwong Yuen - Kowloon Tong MTR Station	$3.50	$4.70
	▲ Yuen Chau Kok to Kowloon Tong MTR Station	$3.00	$4.00
	▲ After Lion Rock Tunnel to Kwong Yuen	$3.00	$4.00
82X	Ravana Garden to Wong Tai Sin [Circular]	$3.20	$4.50
83K	Wong Nai Tau to Sha Tin Central [Circular]	$2.50	/
83P	Kwong Yuen to Kowloon City Ferry	$4.00	/
	▲ Yuen Chau Kok to Kowloon City Ferry	$3.00	/
83X	Kwong Yuen - Kwun Tong Ferry	$4.00	/
	▲ After Tate's Caim Tunnel to Kwun Tong Ferry	$3.50	/
84M	Chevalier Garden - Lok Fu	$3.60	/
	Tai Chung Kiu Road/Ma On Shan Road Junction to Lok Fu	$3.00	/
85	Fo Tan (Shan Mei Street) to Kowloon City Ferry	$3.50	/
	▲ After Lion Rock Tunnel to Fo Tan	$3.00	/
85A	Yuen Chau Kok to Kowloon City Ferry	$3.50	/
	▲ After Lion Rock Tunnel to Yuen Chau Kok	$3.00	/
85B	Chun Shek - Kowloon City Ferry	$3.50	/
	▲ After Lion Rock Tunnel to Chun Shek	$3.00	/
85C	Yiu On to Hung Hom Ferry	$4.80	/
	▲ Tai Chung Kiu Rd/Ma on Shan Rd Junction to Hung Hom Ferry	$4.00	/
	▲ After Tate's Caim Tunnel to Hung Hom Ferry	$3.50	/
	▲ Ming On Street to Hung Hom Ferry	$2.50	/
	▲ After Tate's Caim Tunnel to Yiu On	$3.50	/
85K	Heng On to Sha Tin KCR Station	$2.50	/
85M	Kam Ying Court to Wong Tai Sin [Circular]	$4.50	/
	▲ After Tate's Caim Tunnel to Kam Ying Court	$3.50	/
◆✻ 85P	Sha Tin KCR Station - Chevalier Garden	$2.50	/
86	Wong Nai Tau - Mei Foo	$4.50	/
	▲ Siu Lek Yuen to Mei Foo	$4.00	/
	▲ Wo Che to Mei Foo	$3.50	/
	▲ After Lion Rock Tunnel to Wong Nai Tau	$3.00	/
86A	Sha Tin Wai - Cheung Sha Wan	$3.50	/
	▲ After Lion Rock Tunnel to Sha Tin Wai	$3.00	/
86B	Hin Keng - Mei Foo	$3.50	/
	▲ Tai Po Road (near Keng Hau Road) to Hin Keng	$3.00	/

Route		Non A/C Fare	A/C Fare
86C	Heng On - Cheung Sha Wan	$4.50	/
	▲Tai Chung Kiu Road/Ma On Shan Road Junction to Cheung Sha Wan	$4.00	/
	▲After Lion Rock Tunnel to Cheung Sha Wan	$3.50	/
	▲After Lion Rock Tunnel to Heng On	$3.50	/
86K	Kam Ying Court - Sha Tin KCR Station	$3.00	/
87	Lek Yuen - Tai Kok Tsui Ferry	$3.50	/
	▲After Lion Rock Tunnel to Lek Yuen	$3.00	/
87A	Pok Hong - Tai Kok Tsui Ferry	$3.50	$4.70
	▲After Lion Rock Tunnel to Pok Hong	$3.00	$4.00
87B	Sun Tin Wai - Tai Kok Tsui Ferry	$3.50	/
	▲After Lion Rock Tunnel to Sun Tin Wai	$3.00	/
87D	Kam Ying Court - Kowloon Station	$5.00	$7.20
	▲Ma on Shan Rd./Tai Chung Kiu Rd. Junction to Kowloon Stn	$4.00	$5.50
	▲After Lion Rock Tunnel to Kowloon Station	$3.50	$4.50
	▲After Lion Rock Tunnel to Kam Ying Count	$3.50	$4.50
87K	University KCR Station to Kam Ying Court [Circular]	$2.50	/
▼❋ 87P	Hin Keng - Tai Kok Tsui Ferry	$3.50	/
	▲After Lion Rock Tunnel to Hin Keng	$3.00	/
▼❋ 87S	Kam Ying Court to University KCR Station	$2.50	/
88K	Hin Keng - Ho Tung Lau	$2.50	/
88P	Hin Keng - Shatin Central	$2.50	/
88M	Hin Keng - Kowloon Tong MTR Station	$2.50	/
89	Lek Yuen - Kwun Tong(Yuet Wah Street)	$3.50	$4.70
	▲After Lion Rock Tunnel to Lek Yuen	$3.00	$4.00
89B	Sha Tin Wai - Kwun Tong (Elegance Road)	$3.50	/
	▲After Lion Rock Tunnel to Sha Tin Wai	$3.00	/
89C	Heng On - Kwun Tong (Tsui Ping Road)	$4.80	/
	▲Tai Chung Kiu Rd./Ma on Shan Rd. Junction to Kwun Tong	$4.00	/
	▲After Tate's Cairn Tunnel to Kwun Tong (Tsui Ping Road)	$3.50	/
	▲After Tate's Cairn Tunnel to Heng On	$3.50	/
	▲Chevalier Garden Bus Terminus to Heng on	$2.50	/
●❋ 89D	Lee On - Lam Tin MTR Station	$5.00	$7.20
	▲After Tate's Cairn Tunnel to Lam Tin MTR Station	$3.50	$4.50
	▲After Tate's Cairn Tunnel to Lee On	$3.50	$4.50
89X	Shatin KCR Station - Kwun Tong (Yuet Wah St)	$4.30	/
	▲After Tate's Cairn Tunnel to Kwun Tong (Yuet Wah Street)	$3.50	/
	▲After Tate's Cairn Tunnel to Shatin KCR Station	$3.00	/
91	Choi Hung - Clear Water Bay	$4.00	/
	▲Tseng Lan Shue to Clear Water Bay	$3.50	/
	▲Tseng Lan Shue to Choi Hung	$3.50	/
91M	Hang Hau (On Ning Gdn) - Choi Hung MTR	$3.00	/
92	Choi Hung - Sai Kung	$3.50	/
93A	Po Lam - Kwun Tong Ferry	$2.50	/
93K	Po Lam - Mong Kok KCR Station	$4.50	$6.50
	▲Sau Mau Ping (Hiu Kwong Street) to Mong Kok KCR Stn.	$3.50	$5.00
	▲Po Lam Road to Po Lam	$2.50	$3.50
	▲Kwun Tong Road (Yue Man Square) to Po Lam	$3.00	$4.50
93M	Po Lam to Lam Tin MTR Station [Circular]	$2.50	/
95	Tsui Lam - Jordan Road Ferry	$4.00	/
	▲Sau Mau Ping Road to Jordan Road Ferry	$3.50	/
	▲Po Lam Road to Tsui Lam	$2.50	/
	▲Choi Hung (Lung Cheung Road) to Tsui Lam	$3.00	/
95M	Tsui Lam - Kwun Tong (Elegance Road)	$2.50	/
98A	Hang Hau (North) - Kwun Tong (Yue Man Sq)	$3.00	$4.20
●❋ 98C	Hang Hau (North) - Mei Foo	$5.60	/
	▲Po Lam Road to Hang Hau (North)	$3.20	/
98D	Hang Hau (North) - Jordan Rd Ferry	$5.00	$7.00
	▲After Airport Tunnel to Jordan Road Ferry	$3.20	$4.50
	▲After Tseung Kwan O Tunnel to Hang Hau (Nth)	$3.20	$4.50

New Territories Rural Routes

Route No.	Route	Non A/C Fare	A/C Fare
54	Yuen Long (West) - Sheung Tsuen (Shek Kong)	$3.10	/
64K	Yuen Long (West) - Tai Po KCR Station	$4.50	$6.50
	▲Yuen Long (West) - Sheung Tsun Po	$3.10	$4.00
	▲Sheung Tsun Po - Tai Po KCR Station	$3.10	$4.00
▼❋ 64P	Yuen Long (West) - Tai Po KCR Station	$4.50	$6.50
	▲Yuen Long (West) - Sheung Tsun Po	$3.10	$4.00
	▲Sheung Tsun Po - Tai Po KCR Station	$3.10	$4.00
▼❋ 65K	Sheung Tsuen - Tai Po KCR Station	$3.10	/
	▲Tai Po Rd./Lam Kam Rd. Junction to Tai Po KCR Station	$2.60	/
70K	Wah Ming to Sheung Shui [Circular]	$2.60	/
71A	Fu Heng - Tai Po KCR Station	$2.00	/
71B	Tai Po Central to Fu Heng [Circular]	$1.00	/
71K	Tai Wo to Tai Po Market [Circular]	$2.50	/
❋ 71S	Fu Shin - Kwong Fuk	$2.20	/
73	Choi Yuen - Tai Po Industrial Estate	$4.30	/
	▲Luen Wo Hui to Tai Po Industrial Estate	$3.50	/
	▲Tai Po Rd./Lam Kam Rd. Junction to Tai Po Industrial Est.	$3.20	/
	▲Tai Po Market (Po Heung Street) to Tai Po Industrial Est.	$2.60	/
	▲Tai Po Market (Po Heung Street) to Choi Yuen	$3.20	/
73K	Sheung Shui - Man Kam to (San UK Ling)	$2.20	/
● 74K	Sam Mun Tsai - Tai Po KCR Station	$2.60	/
75K	Tai Mei Tuk - Tai Po KCR Station	$3.10	/
	▲Tai Po Industrial Estate to Tai Po KCR Sttn	$2.60	/
76K	Sheung Shui - Yuen Long (West)	$4.00	/
	▲Sheung Shui - Mai Po	$3.10	/
	▲Yuen Long(West) - Mai Po	$2.80	/
77K	Cheung Wah - Yuen Long (West)	$4.50	/
	▲Sheung Shui - Yuen Long (West)	$4.00	/
	▲Wang Toi Shan - Yuen Long (West)	$3.10	/
	▲Wang Toi Shan - Sheung Shui	$3.10	/
	▲Wang Toi Shan - Cheung Wah	$3.50	/
	▲Sheung Shui - Cheung Wah	$3.10	/
78K	Sheung Shui - Sha Tau Kok	$4.00	/
	▲Luen Wo Hui to Sha Tau Kok	$3.10	/
	▲Hung Leng to Sha Tau Kok	$2.20	/
	▲Hung Leng to Sheung Shui	$3.10	/
	▲Luen Wo Hui to Sheung Shui	$2.20	/
79K	Sheung Shui - Ta Kwu Ling	$3.10	/
94	Sai Kung - Wong Shek Pier	$3.50	$5.00
	▲Tai Mong Tsai to Wong Shek Pier	$3.10	$4.50
	▲Tai Mong Tsai to Sai Kung	$2.20	$3.00
99	Sai Kung - Nai Chung	$2.60	/
	▲Tai Wan to Nai Chung	$2.00	/
	▲Tai Wan to Sai Kung	$2.00	/

Cross Harbour Tunnel Route

Route No.	Route	Non A/C Fare	A/C Fare
101	Kwun Tong (Yue Man Square) - Kennedy Town	$6.20	$8.00
	▲Ma Tau Chung Rd. (near Mok Cheong St.) to Kennedy Town	$6.20	$7.50
	▲After Cross Harbour Tunnel to Kennedy Town/Kwun Tong	$3.10	$4.50
	▲Sheung Wan (Des Voeux Road Central) to Kwun Tong	$6.20	$7.50
102	Mei Foo - Shau Kei Wan	$6.20	$8.00
	▲Nathan Rd. (near Prince Edward MTR Stn.) to Shau Kei Wan	$6.20	$7.50
	▲After Cross Harbour Tunnel to Shau Kei Wan/Mei Foo	$3.10	$4.50
	▲King's Road (Health Garden) to Mei Foo	$6.20	$7.50
103	Chuk Yuen Estate - Pokfield Road	$6.20	$8.00
	▲Waterloo Rd./Prince Edward Rd. West Jnct to Pokfield Rd.	$6.20	$7.50
	▲After Cross Harbour Tunnel to Pokfield Rd./Chuk Yuen Est.	$3.10	$4.50
	▲Hennessy Road (near Arsenal Street) to Chuk Yuen Estate	$6.20	$7.50
104	Pak Tin - Kennedy Town	$6.20	$8.00
	▲Nathan Road to Kennedy Town	$6.20	$7.50
	▲After Cross Harbour Tunnel to Kennedy Town/Pik Tin	$3.10	$4.50
	▲Sheung Wan (Des Voeux Road Central) to Pik Tin	$6.20	$7.50
105	Lai Chi Kok (Amusement Park) - West Point	$6.20	$8.00
	▲Nathan Road (near Prince Edward MTR Stn.) to West Point	$6.20	$7.50
	▲After Cross Harbour Tunnel to West Point/Lai Chi Kok	$3.10	$4.50
	▲Sheung Wan (Des Voeux Road Central) to Lai Chi Kok	$6.20	$7.50

106	Wong Tai Sin - Chai Wan (East)	$6.20	$8.00
	▲ Mok Cheong St./Ma Tau Chung Rd. Junction to Chai Wan (E.)	$6.20	$7.50
	▲ After Cross Harbour Tunnel to Chai Wan (E.)/Wong Tai Sin	$3.10	$4.50
	▲ King's Road (Health Garden) to Wong Tai Sin	$6.20	$7.50
107	Kowloon Bay - Aberdeen	/	$9.00
	▲ After Cross Harbour Tunnel to Aberdeen	/	$5.00
	▲ After Aberdeen Tunnel to Aberdeen	/	$4.00
	▲ After Aberdeen Tunnel to Kowloon Bay	/	$7.50
	▲ After Cross Harbour Tunnel to Kowloon Bay	/	$4.50
108	Kowloon City (Shing Tak St) - Braemar Hill	$6.20	$7.50
	▲ After Cross Harbour Tunnel to Braemar Hill/Kowloon City	$3.10	$4.50
♯ 109	Ho Man Tin - Central (Macau Ferry)	$6.20	/
	▲ After Cross Harbour Tunnel to Central/Ho Man Tin	$3.10	/
110	Tsim Sha Tsui East - Sai Wan Ho Ferry	$6.20	$7.50
	▲ After Cross Harbour Tunnel to Sai Wan Ho/Tsim Sha Tsui E.	$3.10	$4.50
111	Ping Shek - Central (Macau Ferry)	$6.20	$7.50
	▲ After Cross Harbour Tunnel to Central/Ping Shek	$3.10	$4.50
112	So Uk - North Point	$6.20	$7.50
	▲ After Cross Harbour Tunnel to North Point/So Uk	$3.10	$4.50
113	Choi Hung - West Point	$6.20	/
	▲ After Cross Harbour Tunnel to West Point/Choi Hung	$3.10	/
114	Sham Shui Po Ferry - Central (Macau Ferry)	$6.20	/
	▲ After Cross Harbour Tunnel to Central/Sham Shui Po Ferry	$3.10	/
116	Tsz Wan Shan (North) - Quarry Bay	$6.20	$8.00
	▲ Ma Tau Chung Rd. (near Mok Cheong St.) to Quarry Bay	$6.20	$7.50
	▲ After Cross Harbour Tunnel to Quarry Bay/Tsz Wan Shan (N.)	$3.10	$4.50
	▲ King's Road (Health Garden) to Tsz Wan Shan (North)	$6.20	$7.50
♯ 119	Shun Lee - Central (Macau Ferry)	$6.20	$8.00
	▲ Mok Cheong Street/Ma Tau Chung Road Junction to Central	$6.20	$7.50
	▲ After Cross Harbour Tunnel to Central/Shun Lee	$3.10	$4.50
	▲ Hennessy Road (near Arsenal St) to Shun Lee	£6.20	$7.50
N 121	Ngau Tau Kok - Central (Macau Ferry)	$10.50	/
	▲ After Cross Harbour Tunnel to Central/Ngau Tau Kok	$5.30	/
	▲ No charge for boarding Route No.122 at Cross Harbour Tunnel Toll Plaza to North Point Ferry/Mei Foo		
N 122	Mei Foo - North Point Ferry	$10.50	/
	▲ After Cross Harbour Tunnel to North Point Ferry/Mei Foo	$5.30	/
	▲ No charge for boarding Route No.121 at Cross Harbour Tunnel Toll Plaza to Central/Ngau Tau Kok		
170	Sha Tin KCR Station - Wah Fu (Central)	/	$13.00
	▲ After Lion Rock Tunnel to Wah Fu (Central)	/	$12.00
	After Cross Harbour Tunnel to Wah Fu (Central)	/	$7.00
	▲ After Aberdeen Tunnel to Wah Fu (Central)	/	$4.50
	▲ After Aberdeen Tunnel to Sha Tin KCR Station	/	$12.00
	▲ After Cross Harbour Tunnel to Sha Tin KCR Station	/	$8.00
	▲ Kowloon Tong (near Suffolk Road) to Sha Tin KCR Station	/	$5.00
♯ 171	Cheung Sha Wan - South Horizon	/	$9.00
	▲ After Cross Harbour Tunnel to South Horizon	/	$5.00
	▲ After Aberdeen Tunnel to South Horizon	/	$4.00
	▲ After Aberdeen Tunnel to Cheung Sha Wan	/	$7.50
	▲ After Cross Harbour Tunnel to Cheung Sha Wan	/	$4.50
♯ 182	City One Shatin - Central (Macau Ferry)	$10.00	$13.00
	▲ After Lion Rock Tunnel to Central (Macau Ferry)	$6.20	$8.00
	▲ Cross Harbour Tunnel Toll Plaza to Central (Macau Ferry)	$6.20	$7.50
	▲ After Cross Harbour Tunnel to Central (Macau F.)	$3.10	$4.50
	▲ After Cross Harbour Tunnel to City One Shatin	$6.20	$8.00
	▲ Kowloon Tong (near Suffolk Road) to City One Shatin	$3.70	$5.00
300	Prince Edward MTR Station - Sheung Wan	/	$7.50
	▲ After Cross Harbour Tunnel to Sheung Wan/ ▲ Prince Edward	/	$4.50
▼♯ 301	Hung Hom (Cross Harbour Tunnel Toll Plaza) to Sheung Wan	/	$8.00
	▲ After Cross Harbour Tunnel to Sheung Wan	/	$4.50
▼♯ 302	Tsz Wan Shan (North) to Sheung Wan	/	$8.50
	▲ After Eastern Harbour Crossing to Sheung Wan	/	$4.50
▼♯ 303	Yiu On to Sheung Wan	/	$15.50
	▲ City One Shatin to Sheung Wan/Eastern Harbour		
	▲ Crossing Toll Plaza to Sheung Wan	/	$8.00
	▲ After Eastern Harbour Crossing to Sheung Wan	/	$4.50
▼♯ 305	Mei Lam to Sheung Wan	/	$13.00
	▲ After Lion Rock Tunnel to Sheung Wan	/	$8.00
	▲ After Cross Harbour Tunnel to Sheung Wan	/	$4.50
▼♯ 307	Tai Po Central to Sheung Wan	/	$17.50
	▲ Tate's Caim Tunnel Toll Plaza to Sheung Wan	/	$13.00
	▲ Eastern Harbour Crossing Toll Plaza to Sheung Wan	/	$8.00
	▲ After Eastern Harbour Crossing to Sheung Wan	/	$4.50
▼♯ 334	Bayview Garden to Sheung Wan	/	$13.00
	▲ After Cross Harbour Tunnel to Sheung Wan	/	$4.50
▼♯ 336	Lei Muk Shue to Sheung Wan	/	$11.00
	▲ After Cross Harbour Tunnel to Sheung Wan	/	$4.50
▼♯	Admiralty East to Lei Muk Shue	/	$10.50
	▲ After Cross Harbour Tunnel to Lei Muk Shue	/	$5.50
▼♯ 337	Kwai Shing (East) to Sheung Wan	/	$13.00
	▲ After Cross Harbour Tunnel to Sheung Wan	/	$4.50
▼♯	Central (Macau Ferry) to Kwai Shing (East)	/	$11.00
	▲ After Cross Harbour Tunnel to Kwai Shing East	/	$5.50
▼♯ 348	Cheung On to Sheung Wan	/	$13.00
	▲ After Cross Harbour Tunnel Cheung On	/	$8.50
	▲ Central (Macau Ferry) to Cheung On	/	$11.00
	▲ After Cross Harbour Tunnel to Sheung Wan	/	$4.50
▼♯ 368	Yuen Long (West) to Sheung Wan	/	$20.00
	▲ Tate's Caim Tunnel Toll Plaza to Sheung Wan	/	$13.00
	▲ Eastern Harbour Crossing Toll Plaza to Sheung Wan	/	$8.00
	▲ After Eastern Harbour Crossing to Sheung Wan	/	$4.50
▼♯ 369	Tin Shui to Sheung Wan	/	$20.00
	▲ Tate's Caim Tunnel Toll Plaza to Sheung Wan	/	$13.00
	▲ Eastern Harbour Crossing Toll Plaza to Sheung Wan	/	$8.00
	▲ After Eastern Harbour Crossing to Sheung Wan	/	$4.50
▼♯ 373	Sheung Shu to Sheung Wan	/	$19.00
	▲ Tate's Caim Tunnel Toll Plaza to Sheung Wan	/	$13.00
	▲ Eastern Harbour Crossing Toll Plaza to Sheung Wan	/	$8.00
	▲ After Eastern Harbour Crossing to Sheung Wan	/	$4.50
601	Sau Mau Ping (Central) - Admiralty (East)	$6.20	$8.00
	▲ After Eastern Harbour Crossing to Admiralty/Sau May Ping	$3.10	$4.50
	▲ Hennessy Road (near Fenwick Street) to Sau Mau Ping (Central)	$6.20	$7.50
▼♯ 603	Lam Tin (North) to Central	/	$8.50
	▲ After Eastern Harbour Crossing to Central	/	$4.50
606	Choi Wan - Siu Sai Wan Estate	$6.20	/
	▲ After Eastern Harbour Crossing to Siu Sai Wan Estate/Choi Wan	$3.10	/
	▲ After East Harbour Crossing to Central/Kai Yip	$3.10	$4.50
◐♯ 641	Kai Yip to Central	$7.00	$8.50
	▲ After Eastern Harbour Crossing to Kai Yip	$3.10	$4.50
680	Lee On - Central (Macau Ferry)	$11.00	$15.50
	▲ City One Shatin to Central (Macau Ferry)	$6.20	$8.00
	▲ After Eastern Harbour Crossing to Central (Macau Ferry)	$3.10	$4.50
	▲ After Eastern Harbour Crossing to Lee On	$6.20	$8.00
	▲ After Tate's Caim Tunnel to Lee On	$3.50	$4.50
690	Hong Sing Garden - Central (Macau Ferry)	$9.00	$13.00
	▲ After Tseung Kwan O Tunnel to Central (Macau Ferry)	$6.20	$9.50
	▲ Eastern Harbour Crossing Toll Plaza to Central (Macau Ferry)	$6.20	$8.00
	▲ After Eastern Harbour Crossing to Central/ Hong Sing Garden	$3.10	$4.50
▼♯690P	Hang Hau (North) to Central (Macau Ferry)	/	$13.00
	▲ After Tseung Kwan O Tunnel to Central (Macau Ferry)	/	$9.50
	▲ Eastern Harbour Crossing Toll Plaza to Central (Macau Ferry)	/	$8.00
	▲ After Eastern Harbour Crossing to Central (Macau Ferry)	/	$4.50

Air-Conditioned Routes:

Route No.	Route	A/C Fare
A1	Airport to Tsim Sha Tsui [Circular]	$11.00
A2	Airport - Central (Macau Ferry)	$16.00
	▲ On leaving Airport to Central	$12.00
	▲ After Cross Harbour Tunnel to Central (Macau Ferry)	$6.00
	▲ After Cross Harbour Tunnel to Airport	$11.00

A3	Airport to Causeway Bay [Circular]	$16.00
	▲ On leaving Airport to Causeway Bay	$12.00
	▲ Causeway Bay to Airport	$16.00
	▲ After Cross Harbour Tunnel to Airport	$11.00
A5	Airport to Tai Koo Shing [Circular]	$16.00
	▲ On leaving Airport to Tai Koo Shing	$12.00
	▲ After Cross Harbour Tunnel to Airport	$16.00
	After Eastern Harbour Tunnel to Airport	$11.00
A7	Airport to Kowloon Tong MTR Station [Circular]	$6.00
	▲ Fu Mei Street to Airport	$4.50
▼※235X	On Yam to Star Ferry	$7.00
	▲ Butterfly Valley Road to Star Ferry	$4.50
	▲ Lai Chi Kok Road to Star Ferry	$3.50
238M	Riviera Garden - Tsuen Wan MTR Station	$2.50
238X	Riviera Garden - China Ferry Terminal	$7.00
	▲ Mei Foo to China Ferry Terminal	$4.50
	▲ Nathan Rd.(nr Prince Edward MTR) to China Ferry Terminal	$3.50
	▲ Mei Foo to Riviera Garden	$4.50

BELOW: Although the Airbus services are normally intended for single-deck operation using dedicated single-deck Dennis Dart and Lance coaches in Airbus livery, some journeys have been undertaken using air-conditioned double-deckers, as seen here on Friday 9th September 1994 as AD86 made its way along Queensway on the A2, past Pacific Plaza (formerly Victoria and Murray Barracks) to Kai Tak International Airport, across the harbour. (Mike Davis

90	Tiu Keng Leng - Choi Hung	$4.50
●※	Tiu Keng Leng to Hang Hau [Circular]	$3.50
203	Yau Yat Chuen to Tsim Sha Tsui East [Circular]	$6.50
	▲ Nathan Road/Soy Street to Tsim Sha Tsui East	$3.50
	▲ Salisbury Road (outside Shangrila Hotel) to Yau Yat Chuen	$6.50
	▲ Nathan Road/Gascoigne Road Junction to Yau Yat Chuen	$3.50
203E	Jordan Road Ferry - Fu Shan (King Tung Street)	$4.00
	▲ Choi Hung Road to Jordan Road Ferry	$3.50
208	Broadcast Drive to Tsim Sha Tsui East [Circular]	$6.50
	▲ Waterloo Road/Argyle Street Jnct to Tsim Sha Tsui East	$3.50
	▲ Salisbury Road (Shangrila Hotel) to Broadcast Drive	$6.50
	▲ Waterloo Road/Soares Avenue Jnct to Broadcast Drive	$3.50
211	Tsui Chuk Garden to Wong Tai Sin MTR Stn [Circular]	$2.60
	▲ Chuk Yuen Road to Tsui Chuk Garden	$1.60
212	Whampoa Gardens - Sham Shui Po Ferry	$4.00
	▲ Nathan Road (near Fife Street) to Whampoa Gardens	$3.50
	▲ Lai Chi Kok Rd.(near Canton Road) to Sham Shui Po Ferry	$3.50
216M	Lam Tin (Kwong Tin Est.) to Lam Tin MTR Stn. [Circular]	$2.30
N 216S	Lam Tin (Kwong Tin Estate) - Kowloon Station	$10.00
	▲ Argyle Street to Lam Tin	$8.50
	▲ Pink Shek (Home for the Aged) to Lam Tin	$4.50
▼※219M	Laguna City to Kwun Tong (Yue Man Square) [Circular]	$2.50
219X	Laguna City to Tsui Sha Tsui [Circular]	$5.30
224M	Telford Gardens to Kowloon Bay [Circular]	$2.20
▼※230X	Allway Gardens to Whampoa Gardens	$7.50
	▲ Mei Foo to Whampoa Gardens	$4.50
	▲ Lai Chi Kok Road to Whampoa Gardens	$3.50
234A	Sea Crest Villa - Tsuen Wan Ferry	$4.20
	▲ Tsuen Tak Garden to Tsuen Wan Ferry	$3.00
	▲ Sham Tseng to Sea Crest Villa	$3.00
234X	Bayview Garden - Tsim Sha Tsui (Hankow Road)	$7.50
	▲ Mei Foo to Tsim Sha Tsui (Hankow Road)	$4.50
	▲ Nathan Rd.(nr Prince Edward MTR) to Tsim Sha Tsui	$3.50
	▲ Mei Foo to Bayview Garden	$4.50
235	On Yam to Tsuen Wan [Circular]	$3.00
235M	On Yam to Kwai Fong MTR Station [Circular]	$2.70
N 241S	Cheung Hang - Kowloon Station	$13.00
	▲ Mei Foo - Kowloon Station	$9.00
	▲ Mei Foo to Cheung Hang	$6.00
▼※242X	Cheung Hang to Star Ferry	7.00
	▲ Mei Foo to Star Ferry	$4.50
	▲ Lai Chi Kok Road to Star Ferry	$3.50
243M	Mayfair Garden - Tsuen Wan MTR Station	$3.20
▼※243P	Mayfair Garden to Tsuen Wan MTR Station	$3.20
▼※244X	Tsing Yi Ferry to Star Ferry	$7.00
	▲ Mei Foo to Star Ferry	$4.50
	▲ Lai Chi Kok Road to Star Ferry	$3.50
▼※252B	Handsome Court to Star Ferry	$11.00
	▲ Mei Foo to Star Ferry	$4.50
	▲ Nathan Road. (near Prince Edward MTR) to Star Ferry	$3.50
▼※258B	Leung King Estate to Star Ferry	$11.50
	▲ Mei Foo to Star Ferry	$4.50
	▲ Lai Chi Kok Road to Star Ferry	$3.50
▼※258C	Kin Sang Estate to Star Ferry	$11.50
	▲ Mei Foo to Star Ferry	$4.50
	▲ Lai Chi Kok Road to Star Ferry	$3.50
●※258D	Leung King - Lam Tin MTR Station	$13.00
	▲ Lung Cheung Road (Beacon Heights) to Lam Tin MTR Stn.	$4.50
	▲ Kwun Tong Road (Yue Man Square) to Lam Tin MTR Stn.	$4.00
	▲ Tuen Mun Technical Institute to Leung King Estate	$3.00
▼※259B	Tuen Mun Pier Head to Star Ferry	$11.50
	▲ Mei Foo to Star Ferry	$4.50
	▲ Lai Chi Kok Road to Star Ferry	$3.50
▼※259C	Sun Tuen Mun Centre to Star Ferry	$11.50
	▲ Mei Foo to Star Ferry	$4.50
	▲ Lai Chi Kok Road to Star Ferry	$3.50
▼※259D	Tuen Mun Pier Head to Lam Tin MTR Station	$13.00
	▲ Lung Cheung Road (Beacon Heights) to Lam Tin MTR Stn.	$4.50
	▲ Kwun Tong Road (Yue Man Square) to Lam Tin MTR Stn.	$4.00
▼※260B	Tuen Mun Town Centre to Star Ferry	$11.50
	▲ Mei Foo to Star Ferry	$4.50
	▲ Lai Chi Kok Road to Star Ferry	$3.50

	Route		Fare
▼✳ 260C	Yau Oi (South) to Kwai Fong MTR Station		$7.00
	▲ Tsuen Tak Garden to Kwai Fong MTR Station		$3.50
◆✳261M	Sam Shing - Tsuen Wan MTR Station		$6.50
	▲ Tsuen Tak Garden to Tsuen Wan MTR Station		$3.00
	▲ Goodview Garden to Sam Shing		$3.00
▼✳ 261P	Sam Shing to Tsuen Wan MTR Station		$6.50
	▲ Tsuen Tak Garden to Tsuen Wan MTR Station		$3.00
▼✳ 267S	Siu Hong Court to Star Ferry		$11.50
	▲ Mei Foo to Star Ferry		$4.50
	▲ Lai Chi Kok Road to Star Ferry		$3.50
▼✳ 268B	Yuen Long (East) to Star Ferry		$13.50
	▲ Fu Tei to Star Ferry		$11.50
	▲ Mei Foo to Star Ferry		$4.50
	▲ Lai Chi Kok Road to Star Ferry		$3.50
▼✳ 269C	Tin Shui to Lam Tin MTR Station		$15.00
	▲ After Tate's Cairn Tunnel to Lam Tin MTR Station		$4.50
270	Tin Ping - Tsui Lai Garden		$2.50
271	Fu Heng - Tsim Sha Tsui (Canton Road)		$8.00
	▲ After Lion Rock Tunnel to Tsim Sha Tsui (Canton Rd)		$4.50
	▲ Kwong Fuk to Fu Heng		$3.50
N 271S	Fu Heng - Kowloon Station		$13.50
	▲ Yuen Wo Road to Kowloon Station		$8.00
	▲ After Lion Rock Tunnel to Kam Ying Court		$8.00
273	Wah Ming to Fanling KCR Station [Circular]		$2.50
276	Tin Shui - Sheung Shui		$6.50
	▲ Castle Peak Road/Fung Cheung Road Junction to Tin Shui		$3.00
276P	Tin Shui - Sheung Shui		$6.50
	▲ Castle Peak Road/Fung Cheung Road Junction to Tin Shui		$3.00
▼✳ 280P	Siu Wo Court to Star Ferry		$6.50
	▲ After Lion Rock Tunnel to Star Ferry		$4.50
▼ 281P	Kwong Yuen to Kowloon Station		$6.50
	City One Shatin to Kowloon Station		$5.50
	▲ After Lion Rock Tunnel to Kowloon Station		$4.50
N 281S	Kam Ying Court - Kowloon Station		$12.00
	▲ Ma On Shan Rd/Tai Chung Kiu Rd. Jnct to Kowloon Sttn		$8.00
	▲ After Lion Rock Tunnel to Kam Ying Court		$5.80
282	Shatin Central to Sun Tin Wai [Circular]		$2.60
284	Ravana Garden - Sha Tin Central		$2.60
290	Choi Hung -Tiu Keng Leng		$4.50
▼✳293P	Po Lam to Kowloon City (Prince Edward Road East)		$5.50
N 293S	Hang Hau (North) - Mongkok		$12.00
	▲ Argyle Street - Hip Wo Street		$6.50
	▲ Mut Wah Street to Hang Hau (North)		$6.50
	▲ Hang Hau (North) to Kwun Tong (Yue Man Square)		$6.50
298	Lam Tin MTR Station to HK University of Science & Technology		$4.8
	▲ Po Fung Road to HK University of Science & Technology		$4.00
	▲ Po Ning Road to Lam Tin MTR Station		$4.00
299	Sha Tin Central - Sai Kung		$7.00
	▲ Wu Kai Sha (Youth Camp) to Sai Kung		$6.00
	▲ Nai Chung Bus Terminus to Sai Kung		$7.00
	▲ Shap Sz Heung Bus Terminus to Sha Tin Central		$7.00
	▲ Wu Kai Sha (Youth Camp) to Shatin Central		$6.00
	▲ Ma On Shan Rd/Tai Chung Kui Rd. Junct to Shatin Central		$2.60

Special Service Routes:
Lunar New Year/Mid-Autumn Festival/Christmas Only

Route No.	Route	Non A/C Fare	A/C Fare
A8	Airport to Prince Edward MTR Station [Circular]	/	$6.00
N 6S	Star Ferry to Mei Foo	$4.00	/
N 26S	Shun Tin to Kowloon Station	$6.50	/
N 43S	Tsuen Wan MTR Stn to Mayfair Gdn [Circular]	$5.30	$6.50
N 60S	Tsuen Wan MTR Stn to Tuen Mun Town Centre	$8.50	$12.00
	▲ Goodview Garden to Tuen Mun Town Centre	$4.00	$6.00
N 68S	Jordan Road Ferry to Yuen Long (East)	$11.00	$17.00
	▲ Goodview Garden to Yuen Long (East)	$5.30	$8.00
	▲ Fu Tei to Yuen Long (East)	$4.20	$6.50
N 80S	Sui Wo Court to Kowloon Tong MTR Station	$6.50	/
N 82S	Kwong Yuen to Kowloon Tong MTR Station	$6.50	/
N 85S	Kam Ying Court to Shatin KCR Station	$5.30	/
N 111S	Choi Hung to Causeway Bay (Victoria Park Flower Fair)	$10.50	/
	▲ After Cross Harbour Tunnel to Choi Hung	$5.30	/
N 112S	Prince Edward MTR Station - Causeway Bay (Victoria Park Flower Fair)	$10.50	/
	▲ After Cross Harbour Tunnel to Prince Edward MTR Stn	$5.30	/

Special Services:

Route No.	Route	Non A/C Fare	A/C Fare
39R	Tsuen Wan MTR Station to Tsuen Wan Plaza [Circular]	$2.00	/
44S	Cheung Ching to Mongkok KCR Station	$6.50	/
59S	Tuen Mun Pier Head to Mongkok (Sai Yeung Choi Street)	£10.00	/

Sundays & Public Holidays Only:

Route No.	Route	Non A/C Fare	A/C Fare
91R	Clear Water Bay to Wong Tai Sin (June to Sept Only)	$8.00	/
96R	Choi Hung to Wong Shek Pier	$10.00	/
	▲ Sai Kung to Wong Shek Pier	$7.00	/
	▲ Tai Mong Tsai to Wong Shek Pier	$5.00	/
	▲ Pak Tam Chung to Wong Shek Pier	$3.50	/
	▲ Pak Tam Chung to Choi Hung	$9.00	/
	▲ Tai Mong Tsai to Choi Hung	$8.00	/
	▲ Sai Kung to Choi Hung	$6.00	/
263R	▲ Tuen Mun Town Centre to Shatin Central	/	$15.00
	▲ Tsuen Tak Garden to Shatin Central	/	$10.00
	▲ Shing Mun Tunnels Interchange to Shatin Central	/	$2.50
	▲ Shing Mun Tunnels Interchange to Tuen Mun Town Centre	/	$12.00
	▲ After Shing Mun Tunnels to Tuen Mun Town Centre	/	$10.00
291R	Tai Au Mun to Tai Hang Tun (Nov. to May only)	/	$2.80

Occasion of Ching Ming Festival & Chung Yeung Festivals Only:

Route No.	Route	Non A/C Fare	A/C Fare
3S	Diamond Hill MTR Stn. to Diamond Hill Cemetery [Circular]	/	$3.40
14R	Yau Tong (Ko Chiu Rd.) to Tseung Kwan O Chinese Cemetery Circular]	/	$3.50
38S	Kwaii Fong MTR Stn. to Tsuen Wan Chinese Cemetery [Circular]	$3.30	/
70R	Wo Hop Shek - Fanling KCR Station	$3.80	/
70S	Wo Hop Shek - Jordan Road Ferry	$9.50	/
74R	Wo Hop Shek - Lam Tin MTR Station	$10.00	/

Racecourse Route:

Route No.	Route	Non A/C Fare	A/C Fare
101R	Happy Valley Race Course to Kwun Tong (Yue Man Square)	$10.50	/
	▲ After Cross Harbour Tunnel to Kwun Tong (Yue Man Square)	$5.30	/
102R	Happy Valley Race Course to Mei Foo	$10.50	/
	▲ After Cross Harbour Tunnel to Mei Foo	$5.30	/
802	Shatin Race Course to Shau Kei Wan	/	$30.00
	▲ After Eastern Harbour Crossing to Shau Kei Wan	/	$5.00
811	Shatin Race Course to Central (Macau Ferry)	/	$30.00
	After Eastern Harbour Crossing to central	/	$5.00
848	Shatin Race Course - Kwai Fong MTR Stn	$21.00	/
	▲ Shing Mun Tunnels Bus Interchange to Shatin Race Course	$16.00	/
868	Shatin Race Course - Tuen Mun Town Centre	$28.00	/
	▲ After Tuen Mun Road to Shatin Race Course	$21.00	/
	▲ Shing Mun Tunnels Bus Interchange to Shatin Race Course	$16.00	/
872	Shatin Race Course to Tai Po Central	$13.00	/
885	Shatin Race Course - Heng On	$13.00	/
886	Shatin Race Course to Mei Foo	$21.00	/
887	Shatin Race Course - Tai Kok Tsui Ferry	$13.00	/
888	Shatin Race Course - Sshatin KCR Station	$8.00	/
889	Shatin Race Course - Kwun Tong Ferry	$21.00	/
891	Shatin Race Course - Kowloon City Ferry	$13.00	/

ABOVE: KMB's operating area varies from the dense urban high-rise residential housing blocks of Kwun Tong to the windswept hills behind. Both these exyremes are experienced by passengers on Route 90 - more recently the 290, since it has been allocated air-conditioned Toyota Coaster midibuses. Because of both the light traffic and the narrow winding roads encountered towards the extreme end of this route to Rennie's Mill, smaller than average buses have always been allocated. Reference to the main text will reveal that small Ford Thames Traders and Albion Chieftains once plyed this road. *(G. Law)*

BELOW: A far cry from either the high density housing or the open hillsides found on the 290 is Kowloon's famed 'Golden Mile' - the tourist shopping district better known as Nathan Road. A three-axle MCW Metrobus, S3M16, leads a pair of Dennis Dragons as they head for the "Star" Ferry. *(Ron Phillips collection)*

Part Seven:
Fleet List - Kowloon Motor Bus Co. (1933) Ltd.
Known buses from 1945.

The following Fleet List records in tabular form the fleet, registration and chassis numbers of all known post-1945 buses owned and/or operated by the *KOWLOON MOTOR BUS COMPANY (1933) LIMITED*.

The most obvious of the known omissions are those of the immediate post-war period when numerous unidentified vehicles, mostly ex-military lorries and restored pre-1941 buses recovered from the defeated Japanese, were operated.

Readers are directed to the section describing *KMB FLEET NUMBERING* for further background information and to the main text.

The fleet list is arranged in order of first introduction of the premier vehicle of each type, or the lowest number allocated at the time of fleet-numbering in 1974. For example D666 was introduced many years after A1, but D1, introduced in the 1950's, commenced the D series; thus D-class precedes A-class.

Fleet list - Contents:

1. Single-deck buses and coaches from 1945   Page 290
2. Double-deck buses to 1980   Page 298
3. Double-deck buses purchased second-hand   Page 310
4. Double-deck buses from 1980.   Page 313
5. Air-conditioned double-deck buses   Page 322
6. Training buses and service vehicles   Page 325

NOTE: Throughout these lists the following abbreviations are used:
Sc. - scrapped
Trg - Training bus

UNPREFIXED REGISTRATION NUMBERS ALLOCATED 1946-8 by KMB to a variety of mostly untraceable vehicles.

NOTE: Many numbers were later reallocated to new vehicles from about 1948/49 - these are shown in the right-hand column.

Regn. number	Approximate year introduced	Vehicle type type if known	Regn. no subsequently to	Regn. number	Approximate year introduced	Vehicle type type if known	Regn. no subsequently to
4201	1945	?	Daimler(a) B'ham	4241	1946	?	Dr(a) B'ham
4202	1945	?	Dr(a) B'ham	4600	1947	?	Dr(b)
4203	1945	?	Dr(a) B'ham	4601	?	?	Dr(b)
4204	1945	?	Dr(a) Trad	4602	?	?	Dr(b)
4205	1945	?	Dr(a) B'ham	4603	?	?	Dr(b)
4206	1945	?	Dr(a) B'ham	4604	?	?	Dr(b)
4207	1945	?	Dr(a) B'ham	4605	?	?	Dr(b)
4208	1945	Commer	Commer(a)	4606	?	?	Dr(b)
4209	1945	?	Dr(a) B'ham	4607	?	?	Dr(b)
4210	1945	?	Dr(a) B'ham	4608	?	?	Dr(b)
4211	1945	?	Dr(a) B'ham	4609	?	?	Dr(b)
4212	1945	?	Dr(a) B'ham	4612	?	Commer 2-door	Dr(b)
4213	1945	?	Dr(a) B'ham	4613	?	?	Dr(b)
4214	1945	?	Dr(b)	4614	?	?	Dr(b)
4215	1945	?	Dr(b)	4615	?	?	Dr(b)
4216	1945	Dodge military chassis	Dr(b)	4624	?	?	Dr(b)
4217	1945/6	?	Dr(a) B'ham	4625	?	?	Dr(b)
4218	1945/6	?	Dr(a) B'ham	4626	1947	?	Dr(b)
4219	1945/6	?	Dr(a) B'ham	4634	1947	?	Dr(a) B'ham
4220	1945/6	?	Dr(a) B'ham	4635	1947	?	Dr(a) B'ham
4221	1945/6	?	Dr(a) B'ham	4663	1947	?	Dr(b)
4222	1945/6	?	Dr(a) B'ham	4664	?	?	Dr(b)
4223	1945/6	?	Dr(a) B'ham	4666	?	?	Dr(b)
4224	1945/6	?	Dr(a) B'ham	4669	?	?	Dr(b)
4225	1945/6	?	Dr(a) B'ham	4670	?	?	Dr(b)
4228	1945/6	?	Dr(a) B'ham	4672	?	?	Dr(b)
4229	1945/6	?	Dr(a) B'ham	4952	?	?	Dr(b)
4240	1946	?	Dr(a) B'ham				

Single-deck Buses from 1945

BEDFORD OB
Total 30; no fleet numbers.

Regn. number	Chassis number	Date first registered	Date withdrawn
4226		1946	1959
4227		1946	1959
4230		1946	1959
4231		1946	1959
4232		1946	1959
4233		1946	1959
4234		1946	1959
4235		1946	1959
4236		1946	1959
4237		1946	1959
4238		1946	1959
4239		1946	1959
4242		1946	1959
4243		1946	1959
4244		1946	1959
4245		1946	1959
4246		1946	1959
4247		1946	1959
4248		1946	1959
4249		1946	1959
4250		1946	1959
4591		1946	1959
4592		1946	1959
4593		1946	1959
4594		1946	1959
4595		1946	1959
4596		1946	1959
4597		1946	1959
4598		1946	1959
4599		1946	1959

All registration numbers were transferred to Daimler(b) double-deckers in 1959/60.

NOTE: All were introduced in 1946. 18 withdrawn in April 1959 12 withdrawn in October 1959 order unknown.

CHASSIS NUMBERS: The following chassis were exported by Bedford to Hong Kong in 1946. These and others are **PROBABLY** related to the vehicles listed here but there is no record as to which chassis number received which registration number:-

26062	32306
26158	34864
27086	36949
27157	37227
28217	57770
28242	57851
28928	59407
32073	59475

TILLING-STEVENS K5LA7
Total 50; no fleet numbers. KMB TIMBER FRAMED BODIES

Regn. number	Chassis number	Date first registered	Date withdrawn
4610	?	1947	?
4616	?	1947	?
4617	?	1947	?
4618	?	1947	?
4619	?	1947	?
4620	?	1947	?
4621	?	1947	?
4622	?	1947	?
4623	?	1947	?
4627	?	1947	?
4628	?	1947	?
4629	?	1947	?
4636	?	04.48	?
4637	?	04.48	?
4638	?	04.48	?
4639	?	05.48	?
4640	?	05.48	?
4641	?	05.48	?
4642	?	05.48	?
4643	?	05.48	?
4644	?	05.48	?
4645	?	07.48	?
4646	?	07.48	?
4647	?	07.48	?
4648	?	07.48	?
4649	?	07.48	?
4650	?	07.48	?
4651	?	07.48	?
4653	?	08.48	?
4654	?	08.48	?
4655	?	08.48	?
4656	?	08.48	?
4657	?	08.48	?
4673	?	10.48	?
4674	?	10.48	?
4675	?	10.48	?
4677	?	10.48	?
4678	?	10.48	?
4679	?	10.48	?
4680	?	10.48	?
4681	?	11.48	?
4682	?	11.48	?
4683	?	11.48	?
4684	?	11.48	?
4685	?	12.48	?
4686	?	12.48	?
4687	?	12.48	?
4688	?	12.48	?

> **BUSES CONVERTED TO DRIVER TRAINING** - operated until about 1972. These buses were re-registered as 'Private Omnibuses' but it is not known which original number relates to which later number.
> **Private Omnibuses:**
> AJ7144 AP7884
> AJ7145 AP7885
> AJ7146 AP7886
> AP7879 AP7887
> AP7880 AP7888
> AP7881 AP7889
> AP7882 AP7890
> AP7883 AP7892
> AP7893

> **BUS USED BY H.K. GOVERNMENT CIVIL AID SERVICES**
> for training volunteers as bus drivers in case of emergency situations arising in the Colony. Of these, one Tilling-Stevens at least was re-registered in the Government's own reserved AM-series to become AM5018

> **WATER TANKER**
> AA6147 delicensed 17.06.74
> Subsequently used within Lai Chi Kok depot until the mid 1980's. Chassis number unknown..

The actual dates of withdrawal are not known, but c1967-70 seems most likely.

COMMER - model unknown;
Origianl total: Uncertain. - BODIES OF UNKNOWN ORIGIN

Regn. number	Chassis number	Date first registered	Withdrawn	
4208	?	12.48	?	Number **probably** reallocated from earlier vehicle
4612	?	?	?	Photo shows 4612 to be 2-door
4613	?	?	?	Commer Commando chassis nos. 17A0703-6, built in
4614	?	?	?	1947, went to Hong Kong - the only ones to do so.
4615	?	?	?	These **MAY** have been 4612-4615
4630	?	?.48	01.65	
4631	?	?.48	11.65	
4632	?	?.48	11.65	
4633	?	?.48	11.65	
4652	?	07.48	?	
4658	?	09.48	?	
4659	?	09.48	?	
4660	?	09.48	?	
4661	?	09.48	?	
4662	?	09.48	?	
4665	?	09.48	?	
4667	?	09.48	?	
4668	?	09.48	?	
4671	?	09.48	?	
4689	?	12.48	?	
4690	?	12.48	?	
4951	?	01.49	?	
4953	?	01.49	?	
4954	?	01.49	?	
4955	?	01.49	?	
4956	?	01.49	?	
4957	?	01.49	?	
4984	?	08.49	?	Became service truck
4985	?	08.49	?	Became service truck
4986	?	08.49	?	Became service truck
4987	?	08.49	?	Became service truck

It is believed that all these early Commers were withdrawn in 1964 & 65.

COMMER SUPERPOISE
Total 30; no fleet numbers. SPARSHATTS BODIES

Regn. number	Chassis number	Date first registered	Date withdrawn	Disposal
HK4016	?	?.50	?	Converted to service truck
HK4017	?	?.50	?	Converted to service truck
HK4018	?	?.50	?	Converted to service truck
HK4019	?	?.50	?	Converted to service truck
HK4020	?	?.50	?	
HK4021	?	?.50	?	
HK4022	?	?.50	?	
HK4023	?	?.50	?	

COMMER SUPERPOISE (contd.)
Total 30; no fleet numbers. SPARSHATTS BODIES

Regn. number	Chassis number	Date first registered	Date withdrawn	Disposal
HK4024	?	?.50	?	
HK4025	?	?.50	?	
HK4082	?	06.52	?	
HK4083	?	06.52	?	
HK4084	?	06.52	?	
HK4085	?	06.52	?	
HK4086	?	06.52	?	
HK4087	?	06.52		
HK4088	?	06.52		
HK4089	?	06.52		
HK4120	?	10.52		
HK4121	?	10.52		
HK4122	?	10.52		
HK4123	?	10.52		
HK4124	?	10.52		
HK4125	?	10.52		
HK4126	?	10.52		
HK4127	?			
HK4128	?			
HK4129	?			
HK4130	?			
HK4131	?			

> **NOTE:** The only information regarding withdrawal dates that has become available concerns all the Commers generally, including the earlier, types as follows:-
> 1964 - 3 buses
> 1965 - 36 buses
> 1966 - 18 buses

> **A number of Commers** were converted to service trucks, ie. licensed as 'Goods Vehicles'. There is no known information to show which bus became which Goods Vehicle. These became:-
>
> AA5406 AF4064 (4004??)
> AA5710 AF4963 AF5190
> AA5747 AF7493 AF8023
> AA5758 AG8807
> AA5759 AG8896 AG8968
> AA5760 AG9323 AH4410
> AA5759 AH4410
> AA6147 AH4411 AS ?
>
> **NOTE:#** AA5760, AF7493 & AG8968 were rebuilt with angular 'home-made' bonnets, thereby rendering them unrecognisable as Commers but they remained normal-control.

DENNIS 'PAX' - SPARSHATTS BODIES.
Total 30; no fleet numbers.

Regn. number	Chassis number	Date first regisered	Date withdrawn	Disposal
HK4090	?	07.52	?.65	Sold to Lantao Motor Bus Co.
HK4091	?	07.52		
HK4092	?	07.52		
HK4093	?	07.52		
HK4094	?	07.52		
HK4095	?	07.52		
HK4096	?	07.52		
HK4097	?	07.52		
HK4098	?	07.52		
HK4099	?	07.52		
HK4100	?	07.52		
HK4101	?	07.52		
HK4102	?	07.52		
HK4103	?	07.52		
HK4104	?	07.52	?.65	Sold to Lantao Motor Bus Co.
HK4105	?	07.52		
HK4106	?	07.52		
HK4107	?	07.52		
HK4108	?	07.52		
HK4109	?	07.52		
HK4110	? 08.52		?.65	Sold to Lantao Motor Bus Co.
HK4111	?	08.52		
HK4112	?	08.52	?.65	Sold to Lantao Motor Bus Co.
HK4113	?	08.52		
HK4114	? 08.52		?.65	Sold to Lantao Motor Bus Co.
HK4115	?	08.52		
HK4116	?	08.52		
HK4117	?	08.52		
HK4118	?	08.52		
HK4119	?	08.52		Service truck AP7892.

> **Actual dates of withdrawal** not known except that annual totals withdrawn were: :
> 1965 5
> 1968 1
> 1969 24

> An **unidentified Pax** became AP7892 in the KMB service fleet and then with the HK A.A. for immobile use as an office under the Marsh Road flyover in Wanchai.

> **Buses sold** to Lantao Motor Bus Co.
> AG7942
> AG7943
> AG7944
> AG7945
> AG7946
> **NOT** necessarily in order.

> **Chassis numbers** were as listed below and were re-registered: but it is not known which chassis had what number.
> 1218D2, 1221D2, 1222D2, 1223D2, 1224D2,
> 1225D2, 1225D2, 1226D2, 1227D2, 1228D2,
> 1229D2, 1230D2, 1231D2, 1232D2, 1233D2,
> 1234D2, 1235D2, 1236D2, 1237D2, 1238D2,
> 1239D2, 1240D2, 1241D2, 1244D2, 1245D2,
> 1246D2, 1247D2, 1248D2, 1249D2, 1250D2, 1251D2.

BEDFORD SBO

Total 30; No fleet numbers. BODIES OF UNCERTAIN ORIGIN

Regn. number	Chassis number	Date first registered	Date withdrawn
HK4272		1956	1970
HK4283	See	1956	1970
HK4284	note	1956	1970
HK4285	on	1956	1970
HK4286	right	1956	1970
HK4287		1956	1970
HK4288		1956	1970
HK4289		1956	1970
HK4290		1956	1970
HK4291		1956	1970
HK4292		1956	1970
HK4293		1956	1970
HK4294		1956	1970
HK4295		1956	1970
HK4296		1956	1970
HK4297		1956	1970
HK4298		1956	1970
HK4299		1956	1970
HK4300		1956	1970
HK4301		1956	1970
HK4302		1956	1970
HK4303		1956	1970
HK4304		1956	1970
HK4305		1956	1970
HK4306		1956	1970
HK4307		1956	1970
HK4308		1956	1970
HK4309		1956	1970
HK4310		1956	1970
HK4311		1956	1970

These vehicles carried chassis numbers as shown below but which bus carried which chassis number remains unknown. (11/90)

41484, 41552, 41582, 41655, 41798, 41838, 41856, 41868, 41952, 41982, 42180, 42310, 42340, 42370, 42769, 42856, 43126, 43154, 43181, 43252, 43277, 43289, 43315, 43341, 43427, 43453, 43479, 43569, 43621, 43668.

NOTE: All believed *delivered* (ie not date into service) 5/56 to 7/56

SEDDON MK 17/M/3

Total 100; No fleet numbers. METAL SECTIONS BODIES

Regn. number	*Probable Chassis number	Date first registered	Date withdrawn
HK4352	32298	1957	?
HK4353	32317	1957	?
HK4354	32318	1957	?
HK4355	32319	1957	?
HK4356	32320	1957	?
HK4357	32327	1957	?
HK4358	32328	1957	?
HK4359	32329	1957	?
HK4360	32330	1957	?
HK4361	32331	1957	?
HK4362	32346	1957	?
HK4363	32347	1957	?
HK4364	32350	1957	?
HK4365	32351	1957	?
HK4366	32352	1957	?
HK4367	32353	1957	?
HK4368	32354	1957	?
HK4369	32355	1957	?
HK4370	32356	1957	?
HK4371	32360	1957	?
HK4372	32385	1957	?
HK4373	32386	1957	?
HK4374	32387	1957	?
HK4375	32388	1957	?
HK4376	32389	1957	?
HK4377	32390	1957	?
HK4378	32391	1957	?
HK4379	32392	1957	?
HK4380	32393	1957	?
HK4381	32394	1957	?
HK4382	32402	1957	?
HK4383	32403	1957	?
HK4384	32404	1957	?
HK4385	32405	1957	?
HK4386	32406	1957	?
HK4387	32418	1957	?
HK4388	32419	1957	?
HK4389	32420	1957	?
HK4390	32421	1957	?
HK4391	32422	1957	?
HK4392	32425	1957	?
HK4393	32426	1957	?
HK4394	32427	1957	?
HK4395	32428	1957	?
HK4396	32429	1957	?
HK4397	32430	1957	?
HK4398	32431	1957	?
HK4399	32432	1957	?
HK4400	32433	1957	?
HK4401	32434	1957	?
HK4402	32456	1958	?
HK4403	32457	1958	?
HK4404	32459	1958	?
HK4405	32458	1958	?
HK4406	32460	1958	?
HK4407	32461	1958	?
HK4408	32462	1958	?
HK4409	32463	1958	?
HK4410	32464	1958	?
HK4411	32465	1958	?
HK4412	34272	1958	?
HK4413	32473	1958	?
HK4414	34274	1958	?
HK4415	32475	1958	?
HK4416	32476	1958	?
HK4417	32477	1958	?
HK4418	32478	1958	?
HK4419	32479	1958	?
HK4420	32480	1958	?

***NOTE:** Chassis numbers are correct as a group but **MAY** be in a different order.

SEDDON MK 17/M/3 (contd.).

No fleet numbers. METAL SECTIONS BODIES

Regn. number	*Probable Chassis number	Date first registered	Date withdrawn
HK4421	32481	1958	?
HK4422	32496	1958	?
HK4423	32497	1958	?
HK4424	32498	1958	?
HK4425	32499	1958	?
HK4426	32509	1958	?
HK4427	32510	1958	?
HK4428	32511	1958	?
HK4429	32512	1958	?
HK4430	32513	1958	?
HK4431	32514	1958	?
HK4432	32525	1958	?
HK4433	32526	1958	?
HK4434	32527	1958	?
HK4435	32528	1958	?
HK4436	32529	1958	?
HK4437	32530	1958	?
HK4438	32542	1958	?
HK4439	32543	1958	?
HK4440	32548	1958	?
HK4441	32549	1958	?
HK4442	32591	1958	?
HK4443	32592	1958	?
HK4444	32593	1958	?
HK4445	32594	1958	?
HK4446	32595	1958	?
HK4447	32600	1958	?
HK4448	32601	1958	?
HK4449	32602	1958	?
HK4450	32603	1958	?
HK4451	32604	1958	?

NOTE: Dates withdrawn by years:-
Premature	1963	1
April	1969	30
October	1969	1
April	1970	12
July	1970	15
November	1970	17
March	1971	16

TROJAN - 14-seat private-hire coaches; Total 8;

No fleet numbers. BODIES OF UNCERTAIN MAKE - possibly Trojan.

Regn. number	Chassis number	Date first registered	Date withdrawn
HK4452		26.02.60	31.01.66
HK4453		26.02.60	31.01.66
HK4454		26.02.60	31.01.66
HK4455		26.02.60	31.01.66
HK4456		26.02.60	31.01.66
HK4457		26.02.60	31.01.66
HK4458		26.02.60	31.01.66
HK4459		26.02.60	31.01.66

Engine number when new: 1109739

FORD THAMES TRADER Small single-deck buses

Total 10; No fleet numbers. BODIES OF UNCERTAIN MAKE possibly BACO.

Regn. number	Chassis number	Date first registered	Date withdrawn
AD4713		1961	28.02.69
AD4714		1961	28.02.69
AD4715		1961	28.02.69
AD4716		1961	28.02.69
AD4787		1962	28.02.69
AD4788		1962	28.02.69
AD4819		1963	28.02.69
AD4820		1963	28.02.69
AD4821		1963	28.02.69
AD4822		1963	28.02.69

ALBION VICTOR VT17AL-350 - BRITISH ALUMINIUM CO BODIES

Fleet numbered in 1973/4:

L1-85 - plus L270-6, & 278 (L277 omitted) Total 101

Fleet number	Regn. number	Chassis number	Date first registered	Rebody date	Withdrawn	Disposal
L1	HK4502	79608L	05.01.61		10.01.80	Demonstrator - arrived 03/60
L2	HK4503	79621A	05.01.61	11.10.76	26.10.83	
L3	HK4504	79621B	05.01.61		25.04.85	
L4	HK4505	79621C	05.01.61		25.04.85	
L5	HK4506	79621D	05.01.61		25.04.85	
L6	HK4507	79621E	05.01.61	11.10.76	25.10.83	
L7	HK4508	79621F	05.01.61		23.11.82	
L8	HK4509	79621J	05.01.61		23.11.82	
L9	HK4510	79621H	05.01.61		15.07.82	
L10	HK4511	79622D	30.01.61		26.04.84	
L11	HK4512	79622A	30.01.61		11.11.80	Trg fleet
L12	HK4513	79622F	30.01.61		15.07.82	
L13	HK4514	79622E	30.01.61		25.04.85	
L14	HK4515	79622H	30.01.61		25.04.85	
L15	HK4516	79621L	30.01.61	29.11.76	25.10.83	
L16	HK4517	79621K	30.01.61		25.04.85	
L17	HK4518	79622B	30.01.61		25.04.85	
L18	HK4519	79623E	07.03.61		25.04.85	
L19	HK4520	79623D	07.03.61		30.01.81	Trg fleet
L20	HK4521	79623B	07.03.61		25.04.85	
L21	HK4522	79623J	07.03.61		25.04.85	
L22	HK4523	79623K	13.03.61		09.01.80	Trg fleet
L23	HK4524	79623H	13.03.61		25.08.83	
L24	HK4525	79623C	13.03.61		22.12.80	
L25	HK4526	79623F	13.03.61		25.04.85	
L26	HK4527	79624K	24.04.61		20.02.81	
L27	HK4528	79624J	24.04.61		08.10.86	
L28	HK4529	79625A	24.04.61		02.01.81	Trg fleet
L29	HK4530	79624L	24.04.61		03.03.81	
L30	HK4531	79625E	27.04.61		22.12.80	
L31	HK4532	79625B	27.04.61		02.01.81	Trg fleet
L32	HK4533	79625C	27.04.61		09.01.80	Trg fleet
L33	HK4534	79625D	27.04.61		17.01.80	Trg fleet
L34	HK4535	79625J	12.05.61		13.02.86	

ALBION VICTOR VT17AL(contd.). *BRITISH ALUMINIUM BODIES*

Fleet number	Regn. number	Chassis number	Date first registered	Withdrawn	Disposal	Trg. bus regn. no.	Returned to PSV	Withdrawn
L35	HK4536	79626C	12.05.61	21.02.80	Trg fleet			
L36	HK4537	79625K	12.05.61	03.03.80	Trg fleet			
L37	HK4538	79625H	12.05.61	26.11.87				
L38	HK4539	79626B	12.05.61	21.02.80	Trg fleet			
L39	HK4540	79625F	12.05.61	23.03.81	Trg fleet			
L40	HK4541	79626A	12.05.61	29.05 84				
L41	HK4542	79626D	12.05.61	29.05.84				
L42	HK4543	79626H	30.05.61	26.11.87				
L43	HK4544	79626J	30.05.61	13.04.86				
L44	HK4545	79626E	30.05.61	07.10.87				
L45	HK4546	79626F	30.05.61	25.11.00	Trg fleet			
L46	HK4547	79627A	30.05.61	29.05.84				
L47	HK4548	79627B	30.05.61	13.04.84				
L48	HK4549	79627C	30.05.61	26.11.87				
L49	HK4550	79626L	30.05.61	01.04.81				
L50	HK4551	79627J	19.06.61	14.05.81				
L51	HK4552	79627L	19.06.61	16.07.87				
L52	HK4553	79627K	19.06.61	14.05.81				
L53	HK4554	79628A	19.06.61	24.03.81				
L54	HK4555	79628B	19.06.61	29.05.84				
L55	HK4556	79629C	19.06.61	29.05.84				
L56	HK4557	79628D	19.06.61	12.11.79	Trg fleet			
L57	HK4558	79628E	19.06.61	29.05.84				
L58	HK4559	79628H	12.07.61	25.10.83				
L59	HK4560	79629H	12.07.61	25.10.83				
L60	HK4561	79629L	12.07.61	29.10.86				
L61	HK4562	79628L	12.07.61	12.11.79	Trg fleet			
L62	HK4563	79629K	12.07.61	15.01.80	Trg fleet			
L63	HK4564	79629J	12.07.61	22.12.80				
L64	HK4565	79628K	12.07.61	16.12.81				
L65	HK4566	79628J	12.07.61	17.11.81				
L66	HK4567	79630C	27.07.61	19.07.79				
L67	HK4568	79630B	27.07.61	20.10.86				
L68	HK4569	79630D	27.07.61	13.02.86				
L69	HK4570	79630A	27.07.61	31.05.80				
L70	HK4571	79630H	27.07.61	16.07.87				
L71	HK4572	79630F	27.07.61	20.03.87				
L72	HK4573	79630J	27.07.61	28.06.84				
L73	HK4574	79630E	27.07.61	19.03.87				
L74	HK4575	79631J	18.08.61	05.12.80	Trg fleet			
L75	HK4576	79631C	18.08.61	16.07.87				
L76	HK4577	79631A	18.08.61	12.11.79	Trg fleet			
L77	HK4578	79631E	18.08.61	20.02.79	first to be scrapped			
L78	HK4579	79631D	18.08.61	31.05.80				
L79	HK4580	79631H	18,08.61	16.07.87				
L80	HK4581	79631F	18.08.61	31.05.80				
L81	HK4582	79631B	18.08.61	29.05.84				
L82	HK4583	79632C	19.09.61	31.05.80				
L83	HK4584	79632D	19.09.61	25.04.85				
L84	HK4585	79633B	19.09.61	20.04.82				
L85	HK4586	79632E	19.09.61	25.04.85				
L86	HK4587	79632F	19.09.61	25.04.85				
L274	HK4588	79632J	09 .61	17.12.71	Trg fleet	AX 1209	06.08.74	16.12.81
P	HK4589	79632L	10 .61	12.01.72	Trg fleet	AX 3283	-	?
L273	HK4590	79633D	10 .61	17.12.71	Trg fleet	AX 1210	15.07.71	14.03.80**T**
L270	AD4701	79632H	10 .61	17.12.71	Trg fleet	AX 1211	19.05.75	14.05.71
L271	AD4702	79632K	10 .61	17.12.71	Trg fleet	AX 1213	29.11.73	31.07.84
P	AD4703	79634A	10 .61	17.11.71	Trg fleet	AW8728	-	?
L272	AD4704	79634B	10 .61	17.12.71	Trg fleet	AX 1214	09.11.73	15.07.82
P	AD4705	79634C	10 .61	05.11.71	Trg fleet	AW7672	-	?
L275	AD4706	79633E	10 61	17.11.71	Trg fleet	AW8727	12.08.74	09.03.81**T**
P	AD4707	79633C	11 .61	05.11.71	Trg fleet	AW7671	-	?
P	AD4708	79634D	11 .61	21.10.71	Trg fleet	AW6340	-	?
L276	AD4709	79634F	11 .61	08.09.74	Trg fleet	AW5196	25.04.75	
P	AD4710	79634K	11 .61	10.06.70	Trg fleet	AE 5272	-	?
L270	AD4711	79634H	11 .61	06.07.70	Trg fleet	AS 6125	26.11.73	17.11.79
P	AD4712	79634J	11 .61	06.07.70	Trg fleet	AS 6126	-	?

NOTE 1: HK4588-90 and AD4701-12 were withdrawn from PSV duties and re- reistered as Private Omnibuses (at that time, white plates with black letters and numerals) on the dates shown for duties as training buses. Of these, 8 were subsequently returned to PSV duties but retained their new registration numbers carried as trainers. The plates were repainted red and white, as required for 'Public Omnibuses'.

NOTE 2: **'P'** in the fleet number column indicates a bus retained as a '**P**rivate Omnibus' after 1984.

NOTE 3: **T** L273 & 275 returned once again to the training fleet in 1980/1, before final withdrawal.

ALBION VICTOR VT23L-370 - *METAL SECTIONS BODIES*

Fleet-numbered from 1973/4: L87-185 & 277.

Fleet number	Regn. number	Chassis number	Date first registered	Rebody date	Withdrawn or scrapped	Disposal
L87	AD7000	78400H	06.08.63	? 78	22.03.82	
L88	AD7001	78400K	13.08.63	-	30.01.82	
L89	AD7002	78401B	13.08.63		30.04.82	
L90	AD7003	78401A	13.08.63	? 78	22.03.82	
L91	AD7004	78401H	13.08.63		30.01.82	
L92	AD7005	78400L	13.08.63		30.01.82	
L93	AD7006	78400J	13.08.63		16.12.81	
L94	AD7007	78402D	13.08.63		24.03.81	
L 95	AD7008	78400F	13.08.63		30.01.82	
L 96	AD7009	78400E	13.08.63		30.04.82	
L 97	AD7010	78401K	13.08.63		23.06.80	
L 98	AD7011	78402H	13.08.63		30.04.82	
L 99	AD7012	78401J	13.08.63		20.02.79	
L100	AD7013	78402B	13.08.63		22.03.82	
L101	AD7014	78401L	13.08.63		23.06.80	
L102	AD7015	78402J	03.09.63		30.04.82	
L103	AD7016	78404D	03.09.63	? 78	22.03.82	
L104	AD7017	78402H	03.09.63		23.08.82	

ALBION VICTOR VT23L(contd.). *METAL SECTIONS BODIES*

Fleet number	Regn. number	Chassis number	Date first registered	Rebody date	Withdrawn or scrapped	Disposal
L105	AD7018	78404C	03.09.63		16.12.81	
L106	AD7019	78400D	03.09.63		25.08.83	
L107	AD7020	78404F	03.09.63		30.01.82	
L108	AD7021	78400B	03.09.63		25.08.83	
L109	AD7022	78400C	03.09.63	09.11.77	28.03.84	
L110	AD7023	78402C	03.09.63		20.02.79	
L111	AD7024	78404B	03.09.63		30.01.82	
L112	AD7025	78401D	28.09.63		23.08.82	
L113	AD7026	78404J	28.09.63		23.08.82	
L114	AD7027	78401F	28.09.63		30.01.82	
L115	AD7028	78404H	28.09.63	? .78	29.10.83	
L116	AD7029	78404F	28.09.63		29.10.83	
L117	AD7030	78401E	28.09.63		20.02.79	
L118	AD7031	78402E	28.09.63		25.08.83	
L119	AD7032	78401C	28.09.63		10.08.83	
L120	AD7033	78402F	28.09.63		12.01.84	
L121	AD7034	78404A	28.09.63		25.08.83	
L122	AD7035	78402L	08.11.63		10.08.83	
L123	AD7036	78403E	08.11.63		21.02.84	
L124	AD7037	78403K	08.11.63		16.10.82	
L125	AD7038	78403J	08.11.63		10.08.83	
L126	AD7039	78404L	08.11.63		10.08.83	
L127	AD7040	78403L	08.11.63		10.08.83	
L108	AD7041	78403H	08.11.63		25.08.83	
L129	AD7042	78403C	08.11.63		16.10.82	
L130	AD7043	78403F	08.11.63		16.10.82	
L131	AD7044	78403D	08.11.63		30.07.79	
L132	AD7045	78403B	08.11.63		16.10.82	
L133	AD7046	78403A	28.11.63	04.04.78	16.10.82	
L134	AD7047	78405B	28.11.63		16.10.82	
L135	AD7048	78405D	28.11.63		23.11.82	
L136	AD7049	78405A	28.11.63		25.08.83	
L137	AD7050	78405C	28.11.63		23.11.82	
L138	AD7051	78404K	28.11.63		29.10.83	
L139	AD7052	78405J	28.11.63		23.11.82	
L140	AD7053	78405K	28.11.63		25.08.83	
L141	AD7054	78402K	28.11.63		25.08.83	
L142	AD7055	78405L	28.11.63		23.11.83	
L143	AD7056	78400A	18.03.64		21.02.84	
L144	AD7057	78405E	18.03.64		16.12.81	
L145	AD7058	78405F	18.03.64		16.12.81	
L146	AD7059	78405H	18.03.64		21.02.84	
L147	AD7060	78406E	18.03.64		21.02.84	
L148	AD7061	78406F	18.03.64	19.12.77	02.03.84	
L149	AD7062	78406H	18.03.64		16.12.81	
L150	AD7063	78406J	18.03.64		19.07.79	
L151	AD7064	78407A	18.03.64		21.02.84	
L152	AD7065	78407B	18.03.64		16.12.81	
L153	AD7066	78406L	05.05.64	03.03.78	21.02.84	
L154	AD7067	78407C	05.05.64		23.03.83	
L155	AD7068	78406C	05.05.64		30.01.82	
L156	AD7069	78406K	05.05.64		30.01.82	
L157	AD7070	78406D	05.05.64	06.04.78	16.12.81	
L158	AD7071	784064	05.05.64	? 78	22.03.82	
L159	AD7072	78406B	05.05.64	07.04.78	16.12.81	
L160	AD7073	78407D	05.05.64		23.03.83	
-	AD7074	78407F	05.05.64		pre 1973	see L277 below and box
L161	AD7075	78407E	05.05.64		30.01.82	
L162	AD7076	78407J	19.06.64	? 78	30.04.82	
L163	AD7077	78409E	19.06.64		23.03.83	
L164	AD7078	78408E	19.06.64		23.03.83	
L165	AD7079	78408D	19.06.64		23.03.83	
L166	AD7080	78408H	19.06.64	01.02.78	12.01.84	
L167	AD7081	78408J	19.06.64		29.10.83	
L168	AD7082	78408F	19.06.64		23.03.83	
L169	AD7083	78408C	30.06.64		22.03.82	
L170	AD7084	78408B	30.06.64		23.03.83	
L171	AD7085	78408A	30.06.64		22.03.83	
L172	AD7086	78409A	30.06.64		22.03.83	
L173	AD7087	78409B	30.06.64		23.03.83	
L174	AD7088	78409K	30.06.64		22.03.82	
L175	AD7089	78408L	30.06.64	11.01.78	25.08.83	
L176	AD7090	78409H	22.07.64		30.04.82	
L177	AD7091	78409K	22.07.64		23.08.82	
L178	AD7092	78407L	22.07.64	07.03.77	21.02.84	
L179	AD7093	78409L	22.07.64		25.08.83	
L180	AD7094	78409J	22.07.64		30.04.82	
L181	AD7095	78409D	03.09.64		16.12.81	
L182	AD7095	78409F	03.09.64		? 78	22.03.82
L183	AD7097	78407H	03.09.64		23.08.82	
L184	AD7098	78407K	03.09.64		23.08.82	
L185	AD7099	78409C	03.09.64		25.08.83	
L277	BH3001	78407F	05.05.64	17.01.75	30.01.82	to Goods Vehicle.

NOTE: L277 - BH3001 was originally registered AD7074, but following a serious accident, was withdrawn for write-off but was subsequently rebuilt and reinstated as a bus. As the time it had been de-registered spanned the period of fleet numbering this bus was not allowed for in the VT23L series and was allocated the next available fleet number when it was returned to traffic in 1975.

ALBION CHIEFTAIN CH13AXL

Fleet numbered from 1973/4: Nos: L186-219 - *METSEC BODIES*

Fleet number	Regn. number	Chassis number	Date first registered	Rebody date	Withdrawn	Disposal
-	AD7202	77023F	27.10.65			to truck AV5269
L186	AD7203	77023H	27.10.65	01.02.78	25.04.85	
L187	AD7204	77023J	27.10.65	?.78	08.82	Goods veh. 8.82
L188	AD7205	77023E	27.10.65		21.09.81	Goods veh. 21.9.81
L189	AD7206	77027L	27.10.65		15.07.82	
L190	AD7207	77028A	27.10.65		15.07.82	
L191	AD7208	77027J	27.10.65		15.07.82	
L192	AD7209	77027K	27.10.65	10.03.78	29.10.83	

ALBION CHIEFTAIN CH13AXL (contd.). METSEC BODIES

Fleet number	Regn. number	Chassis number	Date first registered	Rebody date	Withdrawn	Disposal
L193	AD7210	77029D	27.10.65	04.07.77	26.11.87	
L194	AD7211	77033H	27.10.65		15.07.82	
L195	AD7212	77033L	27.10.65		30.07.82	Goods veh. 30.7.82
L196	AD7213	77029A	27.10.65	? 78	21.06.82	Goods veh. 21.6.82
L197	AD7214	77033J	27.10.65		22.09.87	
L198	AD7215	77033K	27.10.65		21.09.81	Goods veh. 21.9.81
L199	AD7216	77029B	27.10.65		24.06.82	Goods veh. 24.6.82
L200	AD7217	77029C	27.10.65		16.09.81	Goods veh. 16.9.81
L201	AD7218	77035E	27.10.65	? .78	27.01.87	
L202	AD7219	77035D	27.10.65		10.08.83	
L203	AD7220	77035F	27.10.65		28.09.82	Goods veh. 28.09.82
L204	AD7221	77035H	27.10.65	07.04.78	27.01.87	
L205	AD7222	77040A	22.11.65		27.10.81	Goods veh 27.10.81
L206	AD7223	77039L	22.11.65	07.04:78	20.10.83	
L207	AD7224	77039K	22.11.65		16.10.87	
L208	AD7225	77039C	22.11.65		16.10.87	
L209	AD7226	77039D	22.11.65		26.11.87	
L210	AD7227	77037D	22.11.65	? .78	27.01.87	
L211	AD7228	77037F	22.11.65		26.10.87	
L212	AD7229	77037E	22.11.65		16.10.87	
L213	AD7230	77046J	22.11.65		20.10.87	
L214	AD7231	77046L	22.11.65		25.08.83	
L215	AD7232	77047A	22.11.65		25.08.83	
L216	AD7233	77046K	22.11.65		16.09.81	Goods veh 16.9.81
L217	AD7234	77046L	22.11.65	? .78	15.07.82	
L218	AD7235	77048H	22.11.65		29.10.83	
L219	AD7236	77048E	22.11.65	09.11.77	29.05.81	

> **NOTE:** AD7202 was withdrawn prior to fleet numbering and at a time when it remained the practice to re-register former buses when they were reclassified as Goods Vehicles. The later conversions retained their orignal numbers.

ALBION VIKING EVK41XL. - Leyland nameplate.
Fleet numbered from 1973/4: L220-269.   METSEC BODIES

Fleet number	Regn. number	Chassis number	Date first registered	Date withdrawn
L220	AR7701	53115D	23.10.70	17.07.87
L221	AR7702	53116B	23.10.70	05.09.87
L222	AR7703	53115C	23.10.70	05.09.87
L223	AR7704	53115B	23.10.70	05.09.87
L224	AR7705	53115E	23.10.70	05.09.87
L225	AR7706	53116D	06.11.70	05.09.87
L226	AR7707	53117B	06.11.70	05.09.87
L227	AR7708	53117A	06.11.70	05.09.87
L228	AR7709	53116E	06.11.70	05.09.87
L229	AR7710	53117D	06.11.70	05.09.87
L230	AR7711	53116H	06.11.70	05.09.87
L231	AR7712	53116C	06.11.70	05.09.87
L232	AR7713	53115F	06.11.70	05.09.87
L233	AR7714	53115H	06.11.70	05.09.87
L234	AR7715	53116K	06.11.70	05.09.87
L235	AR7716	53120J	04.11.70	05.09.87
L236	AR7717	53119F	04.12.70	05.09.87
L237	AR7719	53119L	04.12.70	05.09.87
L238	AR7719	53119K	04.12.70	26.11.87
L239	AR7720	53120C	04.12.70	05.09.87
L240	AR7721	53118E	18.12.70	05.09.87
L241	AR7722	53118F	18.12.70	26.11.87
L242	AR7723	53119B	18.12.70	05.09.87
L243	AR7724	53119D	18.12.70	26.11.87
L244	AR7725	53118D	16.04.71	21.01.87
L245	AR7726	53116J	16.04.71	08.03.85
L246	AR7727	53117E	16.04.71	21.01.88
L247	AR7728	53116F	16.04.71	08.03.85
L248	AR7729	53117H	16.04.71	25.04.85
L249	AR7730	53118B	27.04.71	29.04.87
L250	AR7731	53119C	27.04.71	23.04.87
L251	AR7732	53120B	27.04.71	05.09.87
L252	AR7733	53120K	27.04.71	29.04.87
L253	AR7734	53121A	27.04.71	19.01.88
L254	AR7735	53119E	27.04.71	05.09.87
L255	AR7736	53115A	27.04.71	24.04.87
L256	AR7737	53120A	27.04.71	29.04.87
L257	AR7738	53117J	27.04.71	22.04.88
L258	AR7739	53119H	27.04.71	22.04.88
L259	AR7740	53118A	06.08.71	17.07.87
L260	AR7741	53120L	06.08.71	17.07.87
L261	AR7742	53117K	06.08.71	16.06.88
L262	AR7743	53116L	06.08.71	
L263	AR7744	53117F	06.08.71	17.07.87
L264	AR7745	53119J	06.08.71	12.11.84
L265	AR7746	53119A	06.08.71	
L266	AR7747	53118C	06.08.71	12.11.84
L267	AR7748	53117L	06.08.71	09.11.90
L268	AR7749	53117C	06.08.71	26.04.90
L269	AR7750	53121B	06.08.71	09.11.71

ALBION VIKING EVK55CL
Fleet numbered from 1973/4: L279-308   - UNION AUTO BODIES (HK)

Fleet number	Regn. number	Chassis number	Date first registered	Date withdrawn
L279	BK4969	51750B	05.05.76	
L280	BK4970	51750A	05.05.76	20.12.90
L281	BK4971	51749B	05.05.76	30.03.90
L282	BK4972	51749E	05.05.76	11.08.86
L283	BK4973	51749E	05.05.76	30.03.90
L284	BK4974	51749C	05.05.76	20.12.89

ALBION VIKING EVK55CL (contd.) - UNION AUTO BODIES (HK)

Fleet number	Regn. number	Chassis number	Date first registered	Date withdrawn	
L285	BK4975	51727L	05.05.76	20.12.89	
L286	BK8210	51749K	18.06.76	25.02.77	Sek Kong accident victim
L287	BK8211	51750C	18.06.76	03.01.92	
L288	BK8212	51750D	18.06.76	30.03.90	
L289	BK8213	51750H	18.06.76		
L290	BK8214	51750J	18.06.76	30.03.90	
L291	BK8215	51751B	18.06.76	20.12.89	
L292	BK8216	51751C	18.06.76	20.11.86	
L293	BL4318	51749J	07.09.76	30.03.90	
L294	BL4319	51749L	07.09.76		
L295	BL4320	51750F	07.09.76	01.05.90	
L296	BL5038	51751A	16.09.76		
L297	BL5039	51750L	16.09.76		
L298	BL5040	51750K	16.09.76		
L299	BL5041	51750E	16.09.76		
L300	BL5042	51749H	16.09.76		
L301	BL5043	51749D	16.09.76		
L302	BL5044	51727F	16.09.76	01.05.90	
L303	BM 243	51727J	15.11.76		
L304	BM 244	51727K	15.11.76		
L305	BM 245	51728C	15.11.76	03.01.92	
L306	BM 246	51728B	15.11.76		
L307	BM 247	51727H	15.11.76		
L308	BM 248	51728A	15.11.76	03.01.92	

ALBION VIKING EVK41L
Fleet numbered from new: L309-312   - UNION AUTO BODIES (HK)

Fleet number	Regn. nun6er	Chassis number	Date first registered	Date withdrawn
L309	BM3028	53191L	13.12.76	
L310	BM3029	53191K	13.12.76	
L311	BM3031	53193F	13.12.76	
L312	BM3032	53193E	13.12.76	18.11.88

SEDDON 510-V8 PENNINE 4   PENNINE BODIES
Fleet numbered as S-class 1973/4.   Total built: 100 (AR7601-7700)

Regn. number	Fleet number	Chassis number	Date first registered	Date withdrawn	
AR7601	S69	47138	22.05.70	03.09.80	
AR7602	S1	47140	22.05.70	14.11.77	
AR7603	S2	47141	22.05.70	19.07.79	
AR7604	S3	47142	22.05.70	14.11.77	
AR7605	S75	47136	22.05.70	19.07.79	
AR7606	S67	47139	22.05.70	02.06.80	
AR7607	S91	47137	22.05.70	01.05.80	
AR7608	S4	47143	05.06.70	08.12.29	
AR7609	S5	47144	05.06.70	19.07.79	
AR7610	-	47145	05.06.70	13.02.73	
AR7611	S82	47146	05.06.70	19.07.79	
ARF612	S6	47147	12.06.70	22.09.79	
AR.7613	S7	47148	12.06.70	14.11.77	
AR7614	S82	47149	12.06.70	06.08.80	
AR7615	S8	47150	24.06.70	03.09.80	(One time coach; then KMB class Pennine(B)
AR7616	S9	47151	24.06.70	22.09.79	
AR7617	S10	47152	24.06.70	08.12.79	
AR7618	S11	47153	24.06.70	02.06.80	
AR2619	-	47154	24.06.70	03.03.72	
AR7620	S24	47155	07.07.70	19.07.79	
AR7621	S12	47156	07.07.70	04.07.80	
AR7622	S13	47159	09.07.70	18.05.77	
AR7623	S14	47158	09.07.70	30.01.80	
AR7624	S15	47163	09.07.70	04.07.80	
AR7625	S16	47160	09.07.70	19.07.79	
AR7626	S17	471?	20.07.70	08.12.79	
AR7627	S80	47157	20.07.70	01.05.80	
AR7628	S18	47162	20.07.70	20.02.79	
AR7629	S19	47161	20.07.70	10.10.77	
AR7630	S20	47136	20.07.70	08.12.79	
AR7631	-	46561	20.07.70	23.12.72	
AR7632	S78	46579	20.07.70	08.12.79	
AR7633	S21	47128	20.07.70	19.07.79	
AR2634	S68	47131	20.07.70	19.07.79	
AR7635	S22	47132	20.07.70	01.05.80	
AR7636	S23	47135	20.07.70	08.12.79	
AR7637	S24	46581	24.07.70	08.08.77	
AR7638	S25	46594	24.07.70	22.09.79	
AR7639	-	46597	24.07.70	30.11.71	Became water tank BE7470 19.12.73
AR7640	S26	46599	24.07.70	08.12.79	
AR7641	S27	46134	24.07.70	08.12.79	
AR7642	S28	47133	24.07.70	24.04.79	
AR7643	S29	47120	24.07.70	13.09.79	
AR7644	S30	46578	31.07.70	13.09.79	
AR7645	S31	46596	31.07.70	08.12.79	
AR7646	S32	47129	07.07.70	08.12.79	
AR7647	-	46583	07.08.70	13.02.73	
AR7648	S33	46598	07.08.70	04.12.73	
AR7649	S34	46600	07.08.70	08.12.79	
AR7650	S84	47123	07.08.70	08.12.79	
AR7651	S35	46592	18.08.70	19.07.79	
AR7652	S36	47126	18.08.70	08.12.79	
AR7653	-	47121	18.08.70	28.02.73	
AR7654	S37	47124	18.08.70	14.11.77	
AR7655	S66	47125	18.08.70	19.07.79	
AR7656	S72	47127	18.08.70	14.11.77	
AR7657	S38	47563	21.08.70	22.09.79	
AR7658	S88	47573	21.08.70	22.09.79	
AR7659	S39	47574	21.08.70	06.08.80	

SEDDON 510-V8 PENNINE 4 (contd.) PENNINE BODIES
S-class. (AR7601-7700)

Regn. number	Fleet number	Chassis number	Date first registered	Date withdrawn
AR7660	S40	46584	21.08.70	08.12.79
AR7661	S86	46589	21.08.70	08.12.79
AR7662	S41	46587	? ? ?	09.11.73
AR7663	S42	46568	01.09.70	15.12.76
AR7664	S43	46556	01.09.70	23.06.80
AR7665	S44	46569	01.09.70	19.07.79
AR7666	S74	46575	01.09.70	08.03.80
AR7667	S45	46595	01.09.70	22.09.79
AR7668	S76	46585	01.09.70	18.01.00
AR7669	S65	46586	18.09.70	22.09.79
AR7670	S64	46590	18.09.70	15.08.79
AR7671	S46	46576	18.09.70	14.09.77
AR7672	S77	47126	18.09.70	19.07.79
AR7673	S47	47122	18.09.70	13.11.73
AR7674	S48	46591	18.09.70	22.09.79
AR7675	S89	46557	18.09.70	19.07.79
AR7676	S49	46558	18.09.70	14.11.77
AR7677	S87	46560	02.12.70	19.07.79
AR7678	S79	46570	02.12.70	19.07.79
AR7679	S50	46572	02.12.70	22.09.79
AR7680	S51	46559	02.12.70	19.07.79
AR7681	S52	46533	02.12.70	03.08.73
AR7682	S53	46571	02.12.70	21.03.77
AR7683	S92	46562	24.12.70	08.03.70
AR7684	S90	46558	24.12.70	03.09.80
AR7685	S71	46547	24.12.70	19.07.79
AR7686	S54	46564	24.12.70	14.09.77
AR7687	-	46566	24.12.70	14.04.73
AR7688	S55	46554	24.12.70	08.12.79
AR7689	S56	46552	19.01.71	17.11.79
AR7690	S57	46567	19.01.71	14.11.77
AR7691	S58	46550	19.01.71	08.03.80
AR7692	S59	46584	19.01.71	24.04.79
AR7693	S81	47130	14.10.71	19.07.79
AR7694	S60	46555	14.10.71	08.12.79
AR7695	S85	46551	14.10.71	20.02.79
AR7696	S61	46549	12.11.71	03.09.80
AR7697	S62	46563	12.11.71	22.09.79
AR7698	S63	46577	14.12.71	23.06.80
AR7699	S93	46582	17.05.72	23.06.80
AR7700	S73	46580	17.05.72	08.03.80

NOTE: Many of these buses were out of service at the time of fleet-numbering, awaiting premature scrapping due to a multitude of fires and accidents. The remaining 63 operational buses were given the numbers S1-63 in registration number order. Subsequently a further thirty were restored and returned to service, when they received the next available 'S' number, regardless of registration numbers. Six were completely written-off and one, AR7639, became a water-tank car, was re-registered as a goods vehicle and allocated the new number BE7470.

ALBION VIKING EVK55CL - DUPLE BODIES
Single-deck coaches. Fleet numbers not carried

Regn. number	Chassis number	Date first registered	Date of withdrawal	Disposal
BH3002	51698H	17.01.75	13.10.83	Trg.bus
BH3004	51698J	17.01.75	13.10.83	Trg.bus
BH3005	51700B	17.01.75	13.10.83	Trg.bus
BH3006	51700D	17.01.75	13.03.84	Scrapped
BH3007	51700E	17.01.75	13.10.83	Trg.bus
BH3008	51700F	17.01.75	13.10.83	Trg.bus
BH3009	51700J	17.01.75	13.11.83	Trg.bus
BH3010	51700K	17.01.75	28.11.83	Trg.bus
BH3011	51700H	17.01.75	28.11.83	Trg.bus
BH3012	51702D	17.01.75	29.09.85	Scrapped
BH3013	51702C	17.01.75	28.11.83	Trg.bus
BH3014	51702J	17.01.75	09.09.86	Scrapped
BH3015	51702K	17.01.75	13.03.84	Scrapped
BH3016	51703A	17.01.75	30.11.84	To China
BH3706	51707L	06.02.75	05.06.84	Scrapped
BH3707	51708B	06.02.75	12.09.83	To Goods Veh
BH3708	51708C	06.02.75	22.05.85	Scrapped
BH3709	51708F	06.02.75	05.06.84	Scrapped
BH3710	51708H	06.02.75	12.07.84	To Goods Vehl
BH3711	51708J	06.02.76	18.03.87	Trg. bus
BH3712	51700C	06.02.75	30.11.84	To China
BH3713	51708K	06.02.75	30.11.84	To China
BH3714	51708G	06.02.75	09.09.86	Scrapped
BH3715	51709B	06.02.75	30.11.84	To China
BH3716	51710C	06.02.75	30.11.84	To China
BH3717	51708D	06.02.75	12.07.84	To Goods Veh
BH3718	51698F	06.02.75	07.06.84	To China
BH3719	51700A	06.02.75	07.06.84	To China
BH3720	51702L	06.02.75	20.06.85	To Goods Veh.
BH3721	51703L	06.02.75	13.03.84	Scrapped
BH3722	51704A	06.02.75	30.11.84	To China
BH3723	51704B	06.02.75	18.03.87	To Trg.bus
BH3724	51704D	06.02.75	30.11.84	To China
BH3725	51705B	06.02.75	04.10.84	To Goods Veh.
BH3726	51707C	06.02.75	27.12.84	To China
BH3727	51705A	06.02.75	27.12.84	To China
BH3728	51707D	06.02.75	27.12.84	To China
BH3729	51707E	06.02.75	09.10.84	To Goods Veh.
BH3730	51707F	06.02.75	30.11.84	To China
BH3731	51707H	06.02.75	27.12.84	To China
BH3732	51707J	06.02.75	27.12.84	To China
BH3733	51707K	06.02.75	30.11.84	To China
BH3734	51709J	06.02.75	04.02.85	To China
BH5088	51708D	26.03.75	12.04.85	To China
BH5089	51708E	26.03.75	18.04.88	To Goods Veh.
BH5090	51708A	26.03.75	12.04.85	To China
BH5091	51708L	26.03.75	23.02.88	To Goods Veh.
BH7629	51709E	02.06.75	27.12.84	To China

ALBION VIKING EVK55CL - DUPLE BODIES
Single-deck coaches.

Regn. number	Chassis number	Date first registered	Date of withdrawal	Disposal
BH7630	51709F	02.06.75	27.03.85	To China
BH7631	51709H	02.06.75	27.03.85	To China
BH7632	51709K	02.06.75	30.11.84	To China
BH7633	51709L	02.06.75	02.03.84	To Goods Veh
BH7634	51710A	02.06.75	25.10.84	Scrapped
BH7635	51710B	02.06.75	16.10.85	Scrapped
BH7636	51710C	02.06.75	12.03.88	To Goods Veh.
BH7637	51710D	02.06.75	30.04.90	Sold to Road Safety Assn.
BH7638	51710E	02 06 75	04 02 85	Scrapped
BH7639	51710F	02 06 75	13 11 85	Scrapped
BE7640	51710J	02 06 75	11 04 88	To Goods Veh
BH7641	51710K	02 06 75	12 03 88	To Goods Veh
BH7642	51710L	02 06 75	27 04 84	Scrapped
BH7643	51711C	02 06 75		
BH7644	51711D	02 06 75	24 05 83	Trg bus
BH7645	51711F	02 06,75	16 10 85	Scrapped
BH7646	51719B	02 06 75	16 10 85	Scrapped
BH8196	51710H	12 06 75	23 05 86	To Goods Veh
BJ1862	51711H	01 09 75	25 11 83	To Goods Veh
BJ1863	51711J	01 09 75	09 09 86	Scrapped
BJ1864	51711E	01 09 75	11 02 84	To Goods Veh
BJ1973	51719C	01 09 75		
BJ2628	51723D	15 09 75	16 10 86	Scrapped
BJ2629	51733F	15 09 75	19 09 88	Goods Vehicle
BI2630	51733H	15 09 75	09 09 86	Scrapped
BJ2631	51733J	15 09 75	09 09 86	Scrapped
BJ2632	51733L	15 09 75	22 05 85	Scrapped
BJ2633	51734H	15 09 75	16 10 85	Scrapped
BJ4810	51723C	15 09 75	16 10 86	Scrapped
BJ4811	51733K	15 09 75	28 02 88	To Goods Veh
BJ4812	51734B	15 09 75	09 09 86	Scrapped
BJ4813	51734C	15 09 75	20 06 85	Scrapped
BJ4814	51734E	15 09 75	02 02 88	Scrapped
BJ4815	51734F	15 09 75	02 02 88	To Goods Veh
BJ4816	51734J	15 09 75	09 09 86	Scrapped
BJ4817	51734K	15 09 75	20 06 85	Scrapped
BJ4818	51735A	15 09 75	16 10 85	Scrapped
BJ8953	51722E	23.01.76	09.09.86	Scrapped
BJ8954	51722J	23.01.76	16.10.86	Scrapped
BJ8955	51722L	23.01.76	09.09.86	Scrapped
BJ8956	51723B	23.01.76	28.12.89	Sold to Malawi (UTM)
BJ8957	51734A	23.01.76	16.10.86	Scrapped
BL1822	51719A	04.08.76	01.11.86	To Goods Veh.
BL1823	51723A	04.08.76	28.12.89	Sold Malawi (UTM)
BL1824	51722D	04.08.76	28.12.89	Sold Malawi (UTM)
BL1825	51734D	04,08.76	19.06.90	Sold to ??
BL1826	51723E	04.08.76	16.10.85	Scrapped
BL1827	51722F	04.08.76	22.05.85	Scrapped
BL1828	51722H	04.08.76	22.05.85	Scrapped
BL1829	51722K	04.08.76	28.12.89	Sold Malawi (UTM)
BL1830	51718L	04.08.76	28.12.89	Scrapped
BL1831	51734L	04.08 76	09.09.86	Scrapped

NOTES: Although the Albion coaches never carried fleet-numbers on the actual vehicles, they were in fact allocated internally within the Company. These were, CA1-66, 68-77 & 79-102, the two gaps being taken up by the two Bedford coaches which were also allocated the same 'CA' class-letters, despite their presumably indicating 'Coach Albion'.

BEDFORD YRQ - WILLOWBROOK AND DUPLE BODIES
Single-deck coaches. Fleet numbers not carried.

Regn. number	Chassis number	Date first registered	Date of withdrawal	Disposal
BH8315	YRQ2T475484	16.06.75	14.02.80	Believed to have
BJ2726	YRQ202CW453734	16.09.79	14.02.80	been sold to China

NOTES: 1. These coaches were allocated the fleet numbers CB67 & 78 but, like the Albions, these were never carried on the Vehicles.
2. Both these Bedford coaches were manufacturer's demonstrators on loan to KMB from 1974 but were not used until the following year, when they were purchased.

DENNIS FALCON HC Air-conditioned single-deck coaches.
KMB class: All to Falcon(A) 1986 DUPLE(METSEC) BODIES
Fleet numbered 1991.

Fleet number	Regn. number	Chassis number	Date first registered	Date of withdrawal	
AF 1	DH 743	SDA418/165	14.11.85		
AF 2	DH1523	SDA418/161	14.11.85		
AF 3	DH 972	SDA418/164	02.12.85		
AF 4	DH1577	SDA418/163	02.12.85		
AF 5	DH1634	SDA418/166	02.12.85		
AF 6	DH1644	SDA418/167	02.12.85		
-	DH1700	SDA418/160	02.12.85	09.01.89	scrapped (fire).
AF 7	DH2134	SDA418/174	02.12.85		
AF 8	DH2192	SDA418/169	02.12.85		
AF 9	DH3323	SDA418/168	10.12.85		
AF10	DH3087	SDA418/170	16.12.85		
AF11	DH4881	SDA418/172	30.12.85		
AF12	DH5034	SDA418/171	30.12.85		
AF13	DH5921	SDA418/173	03.01.86		
AF14	DH8180	SDA418/175	20.01.86		Falcon(A) as new
AF15	DH9046	SDA418/179	07.02.86		Falcon(A) as new
AF16	DJ 107	SDA418/176	07.02.86		Falcon(A) as new
AF17	DJ 112	SDA418/162	07.02.86		Falcon(A) as new
AF18	DJ 167	SDA418/177	07.02.86		Falcon(A) as new
AF19	DJ 258	SDA418/178	07.02.86		Falcon(A) as new

NOTES: 1: All 15 Falcon converted to Falcon(A) during 1986.
2: Falcons allocated fleet numbers FD1-20 but not applied to vehicles.
3: Fleet numbers AF1-19 allocated and applied to remaining 19 Falcons 1991.
4: Converted to buses 1994.

TOYOTA COASTER - ARAKAWA AUTO 24 SEAT MIDIBUS BODIES.
Fleet numberer from 1991: AT1-91.

Fleet number	Regn. number	Chassis number HB30-000	Date first registered	Withdrawn	Disposal
AT 1	DS8752	5556	09.09.87	09.92	?
AT 2	DS9892	5561	09.09.87	09.92	Speedybus - Wuzhou
AT 3	DS9894	5566	10.09.87	09.92	?
AT 4	DT 139	5571	10.09.87	09.92	?
AT 5	DS8549	5569	14.09.87	09.92	?
AT 6	DS9790	5576	14.09.87	09.92	?
AT 7	DT 956	5559	14.09.87	09.92	Speedybus - Wuzhou
AT 8	DT2373	5573	14.09.87	09.92	?
AT 9	DT1445	5570	16.09.87	09.92	?
AT10	DV 644	5983	.01.88	01.93	?
AT11	DV 647	5976	.01.88	13.11.91	Trg fleet
AT12	DV 699	5982	.01.88	13.11.91	Trg fleet.
AT13	DV1312	5985	.01.88	01.93	?
AT14	DV1309	5981	.01.88	01.93	?
AT15	DV3822	5987	.01.88	01.93	?
AT16	DV2802	5988	.01.88	02.93	?
AT17	DV2996	5986	.01.88	01.93	?
AT18	DX1424	6414	.05.88		
AT19	DX4533	6296	.06.88		
AT20	DX5231	6297	.06.88		
AT21	DX5406	6308	.06.88		
AT22	DX5813	6300	.06.88		Converted to (A) 5/91
AT23	DX6289	6303	.06.88		
AT24	DX8893	6318	06.07.88	By 6/94	Speedybus - Wuzhou
AT25	DX8907	6294	06.07.88		
AT26	DX9280	6310	06.07.88		
AT27	DX9537	6401	06.07.88	By 6/94	Speedybus - Wuzhou
AT28	DX9699	6322	06.07.88	By 6/94	Speedybus - Wuzhou
AT29	DY 337	6311	06.07.88		
AT30	DZ4947	6867	30.09.88		
AT31	DZ5379	6862	30.09.88		
AT32	DZ5559	6871	30.09.88		
AT33	DZ7231	6864	05.10.88	By 6/94	Speedybus - Wuzhou
AT34	EA5269	6870	17.11.88		Toyota(A)
AT35	EA6127	6868	17.11.88		Toyota(A)
AT36	EA6976	7026	01.12.88		
AT37	EA8057	7033	01.12.88		
AT38	EB5189	7151	12.01.89		
AT39	EB5694	7137	12.01.89	By 6/94	Speedybus - Wuzhou
AT40	EB5697	7183	12.01.89		
AT41	EB7408	7190	19.01.89		
AT42	EB6537	7198	20.01.89		
AT43	EC 514	7145	01.02.89		
AT44	EC 790	7196	01.02.89		
AT45	EC1575	7195	01.02.89		
AT46	EG4635	7902	02.10.89		
AT47	EG4742	7913	02.10.89		
AT48	EG8431	7984	06.10.89		
AT49	EH 293	7988	16.10.89		
AT50	EG6752	8134	10.10.89		
AT51	EH 149	8131	16.10.89		
AT52	EH 766	8132	23.10.89		
AT53	EH2443	8140	23.10.89		
AT54	EH2620	8136	01.11.89		
AT55	EH2959	8143	01.11.89		
AT56	EH3534	8147	01.11.89		
AT57	EH3730	8139	01.11.89		
AT58	EH5495	8149	14.11.89		
AT59	EH6302	8145	14.11.89		
AT60	EH5970	8133	16.11.89		
AT61	EH5296	8150	01.11.89		
AT62	EJ4491	7993	04.01.90		
AT63	EJ4740	7995	04.01.90		Toyota(A)
AT64	EJ5883	7998	04.01.90		Toyota(A)
AT65	EJ6276	8035	04.01.90		Toyota(A)
AT66	EJ7909	8461	24.01.90		
AT67	EK 502	8507	15.02.90		
AT68	EK 539	8502	15.02.90		
AT69	EK3738	8497	22.02.90		
AT70	EK4546	8524	28.02.90		
AT71	EK5062	8479	28.02.90		
AT72	EK5068	8496	28.02.90		
AT73	EK6195	8491	28.02.90		
AT74	EK9655	8513	12.03.90		
AT75	EL 452	8510	12.03.90		
AT76	EK9579	8514	16.03.90		
AT77	EL3913	8546	03.04.90		
AT78	EL2636	8547	09.04.90		
AT79	EL5138	8556	18.04.90		
AT80	EL6500	8553	18.04.90	11.92	?
AT81	EL4705	8532	26.04.90		
AT82	EL7795	8559	07.05.90	By 6/94	Speedybus - Wuzhou
AT83	EL8144	8560	07.05.90		
AT84	EM5808	8572	20.06.90		
AT85	EM6057	8565	20.06.90		
AT86	EM8997	8577	10.07.90		
AT86	EM9421	8578	10.07.90		
AT88	EN 387	8573	10.07.90		
AT89	EN 401	8571	10.07.90		
AT90	EN4271	8586	23.07.90		
AT91	EN6160	8585	.08.90		

MCW METRORIDER - MF154/7 or MF154/018 (AMR2)
Fleet numbered from 1991. - MCW MIDIBUS BODIES

Fleet number	Regn. number	Chassis number	Date first registered	Disposal
AMR 1	DY6050	MB9786	03.08.88	Ranger (UK)
AMR 2	EA4591	MB10041	17.11.88	

NOTE: AMR2 used as demonstrator prior to purchase by KMB in Nov. 1988.

HINO RK176K - POC (TAIWAN) BODIES.
Coaches & Buses hired from Argos Bus Services Co.

KMB fleet number	Argos fleet number	Regn. number	Chassis number	Date first registered	Date returned to Argos.
Low-deck-model.					
AH 1	103	EA9323	40319	09.12.88	01.93
AH 2	104	EA9383	40327	09.12.88	01.93
AH 8	107	EB2749	40346	29.12.88	01.93
AH10	109	EB3009	40344	29.12.88	01.93
AH14	117	EB4195	40369	29.12.88	01.93
AH15	119	EB4409	40367	29.12.88	01.93
High-deck model.					
AH 3	208	EB 155	40324	09.12.88	01.93
AH 4	209	EB 523	40352	19.12.88	01.93
AH 5	215	EB1603	40334	19.12.88	01.93
AH 6	220	EB2438	40343	19.12.88	01.93
AH 7	223	EB2567	40345	29.12.88	01.93
AH 9	225	EB2886	40366	23.12.88	01.93
AH11	226	EB3247	40355	29.12.88	
AH12	231	EB3570	40335	23.12.88	01.93
AH13	234	EB3937	40347	29.12.88	01.93
AH16	201	EA7222	40317	01.12.88	
AH17	202	EA9367	40326	09.12.88	
AH18	206	EA9838	40325	09.12.88	
AH19	207	EA9903	40323	09.12.88	
AH20	212	EB1077	40354	19.12.88	
AH21	218	EB1801	40353	19.12.88	
AH22	222	EB2473	40339	19.12.88	01.93
AH23	230	EB3552	40358	29.12.88	
AH24	235	EB4035	40376	29.12.88	
AH25	239	EB4780	40364	09.01.89	

FUSO-MITSUBISHI MK117J - Midibuses.
Fleet numbered from 1991: AM 1-143 - MITSUBISHI BODIES

Fleet number	Regn. number	Chassis number MK117J-	Date first registered	Withdrawn	Disposal
AM 1	EL8611	92788	11.05.90		
AM 2	EL9048	92783	11.05.90		
AM 3	EL9171	92786	11.05.90		
AM 4	EL9337	92751	11.05.90		
AM 5	EL9970	92749	11.05.90		
AM 6	EM 312	92785	11.05.90		
AM 7	EL8933	92787	14.05.90		
AM 8	EM 263	92750	14.05.90		
AM 9	EL8779	92812	17.05.90		
AM10	EL8873	92822	17.05.90		
AM11	EL9025	92791	17.05.90		
AM12	EL9254	92815	17.05.90		
AM13	EL9310	92755	17.05.90		
AM14	EL9314	92757	17.05.90		
AM15	EL9549	92790	17.05.90		
AM16	EL9639	92821	17.05.90		
AM17	EL9700	92834	17.05.90		
AM18	EL9701	92810	17.05.90		
AM19	EL9780	92811	17.05.90		
AM20	EL9807	92824	17.05.90		
AM21	EL9808	92817	17.05.90		
AM22	EM 125	92820	17.05.90		
AM23	EM4983	92758	20.06.90		
AM24	EM5061	92818	20.06.90		
AM25	EM5288	92823	20.05.90		
AM26	EM5396	92784	20.06.90		
AM27	EM5451	92756	20.06.90		
AM28	EM5783	92813	20.06.90		
AM29	EM5788	92836	20.06.90		
AM30	EM5804	92816	20.06.90		
AM31	EM6023	92754	20.06.90		
AM32	EM6130	92789	20.06.90		
AM33	EM6236	92792	20.06.90		
AM34	EM6244	92753	20.06.90		
AM35	EM8699	92752	10.07.90		
AM36	EM8782	92819	10.07.90		
AM37	EM9814	92833	10.07.90		
AM38	EN 182	92814	10.07.90		
AM39	EN 342	92832	10.07.90		
AM40	EN 419	92835	10.07.90		
AM41	EU6518	20256	10.05.91		
AM42	EU7075	20249	10.05.91		
AM43	EU7311	20253	10.05.91		
AM44	EU7330	20252	10.05.91		
AM45	EU7966	20251	10.05.91		
AM46	EU7970	20255	10.05.91		
AM47	EU8152	20270	10.05.91		
AM48	EU8184	20248	10.05.91		
AM49	EU6804	20254	13.05.91		
AM50	EU7309	20250	13.05.91		
AM51	EU8298	20247	13.05.91		
AM52	EU9302	20274	24.05.91		
AM53	EU9493	20276	24.05.91		
AM54	EU9869	20271	24.05.91		
AM55	EV 349	20275	24.05.91		
AM56	EU8798	20278	29.05.91		
AM57	EU8970	20277	29.05.91		
AM58	EU9116	20273	29.05.91		
AM59	EU9325	20294	29.05.91		
AM60	EU9596	20279	29.05.91		
AM61	EV 341	20293	29.05.91		
AM62	EV 487	20272	29.05.91		
AM63	EV2921	20295	12.06.91		
AM64	EV3589	20298	12.06.91		
AM65	EV4210	20296	12.06.91		
AM66	EV4454	20329	12.06.91		
AM67	EV2661	20320	13.06.91		
AM68	EV3417	20301	13.06.91		
AM69	EV3791	20319	13.06.91		
AM70	EV3872	20317	13.06.91		

FUSO-MITSUBISHI MKII7J (contd.). — MITSUBISHI BODIES

Fleet number	Regn. number	Chassis number MKII7J-	Date first registered	Withdrawn	Disposal
AM71	EV2603	20322	20.06.91		
AM72	EV2681	20318	20.06.91		
AM73	EV3179	20300	20.06.91		
AM74	EV3221	20330	20.06.91		
AM75	EV3449	20299	20.06.91		
AM76	EV4070	20331	20.06.91		
AM77	EV4148	20315	20.06.91		
AM78	EV4472	20321	20.06.91		
AM79	EV4573	20314	25.06.91		
AM80	EV4952	20316	25.06.91		
AM81	EV5027	20302	25.06.91		
AM82	EV5427	20313	25.06.91		
AM83	EV6483	20297	25.06.91		
AM84	EV4921	20333	27.06.91		
AM85	EV5285	20332	27.06.91		
AM86	EV5667	20335	27.06.91		
AM87	EV6500	20334	27.06.91		
AM88	EV4897	20336	01.07.91		
AM89	EV5322	20337	01.07.91		
AM90	EV7175	20338	02.07.91		
AM91	EV7544	20343	02.07.91		
AM92	EV9305	20349	16.07.91		
AM93	EV9573	20365	16.07.91		
AM94	EV8783	20367	17.07.91		
AM95	EV8949	20369	17.07.91		
AM96	EV9913	20363	17.07.91		
AM97	EV9984	20346	17.07.91		
AM98	EV8606	20366	17.07.91		
AM99	EV9686	20371	18.07.91		
AM100	EW 225	20362	18.07.91		
AM101	EW 468	20347	18.07.91		
AM102	EV8912	20361	19.07.91		
AM103	EV9853	20345	19.07.91		
AM104	EV9905	20368	19.07.91		
AM105	EW 347	20370	19.07.91		
AM106	EW 733	20350	22.07.91		
AM107	EW 849	20344	22.07.91		
AM108	EW 918	20364	22.07.91		
AM109	EW1131	20351	22.07.91		
AM110	EW1568	20352	22.07.91		
AM111	EW4094	20348	06.08.91		
AM112	EW4617	20372	15.08.91		
AM113	EW4947	20376	15.08.91		
AM114	EW5660	20373	15.08.91		
AM115	EW6173	20378	15.08.91		
AM116	EW4517	20379	16.08.91		
AM117	EW4906	20375	16.08.91		
AM118	EW4934	20381	16.08.91		
AM119	EW5191	20380	16.08.91		
AM120	EW5360	20374	16.08.91		
AM121	EW5510	20377	16.08.91		
AM122	FF6647	20866	20.08.92		
AM123	FF7510	20874	20.08.92		
AM124	FF8496	20871	20.08.92		
AM125	FG 665	20867	01.09.92		
AM126	FG1156	20868	01.09.92		
AM127	FG1189	20870	01.09.92		
AM128	FG1356	20872	01.09.92		
AM129	FG2055	20869	01.09.92		
AM130	FG1954	20884	02.09.92		
AM131	FG2479	20877	02.09.92		
AM133	FG4597	20879	14.09.92		
AM133	FG4804	20878	14.09.92		
AM134	FG5583	20882	17.09.92		
AM135	FG4552	20873	17.09.92		
AM136	FG5027	20876	17.09.92		
AM137	FG5402	20883	17.09.92		
AM138	FG5767	20865	17.09.92		
AM139	FG6084	20880	17.09.92		
AM140	FJ2624	20885	11.11.92		
AM141	FJ3786	20886	11.11.92		
AM142	FJ4139	20881	11.11.92		
AM143	FJ4266	20875	11.11.92		
AM144	FN6619	21117	03.05.93		
AM145	FN6924	21064	03.05.93		
AM146	FN7236	21115	03.05.93		
AM147	FN7313	21063	03.05.93		
AM148	FN7362	21090	03.05.93		
AM149	FN7382	21066	03.05.93		
AM150	FN7475	21119	03.05.93		
AM151	FN7657	21085	03.05.93		
AM152	FN7719	21091	03.05.93		
AM153	FN7890	21118	03.05.93		
AM154	FN8250	21065	03.05.93		
AM155	FN8469	21092	03.05.93		
AM156	FP4511	21094	01.06.93		
AM157	FP4543	21093	01.06.93		
AM158	FP4790	21061	01.06.93		
AM159	FP4975	21087	01.06.93		
AM160	FP5039	21088	01.06.93		
AM161	FP5118	21089	01.06.93		
AM162	FP5465	21062	01.06.93		
AM163	FP5496	21060	01.06.93		
AM164	FP5595	21089	01.06.93		
AM165	FP6073	21086	01.06.93		
AM166	FP6263	21057	01.06.93		
AM167	FS8774	21058	01.09.93		
AM168	FS9389	21116	01.09.93		

MITSUBISHI MP618N Demonstrator bus.
Temporary fleet number: AP I. - MITSUBISHI BODY

Fleet number	Regn. number	Chassis number	Date first registered	Withdrawn	Disposal
AP 1	FL5622	10010	15.02.93	.02.94	Passed on to Citybus, 1205, and then KCRC, 318, on demonstration.

DENNIS 'DART' 9SDL - 9m   Total 2
Fleet nos: AA 1 & 2 - CARLYLE BODIES

Fleet number	Regn. number	Chassis number	Date first registered	Withdrawa	Disposal
	EP5213	9SDL-132	24.09.90	? .11.90	Purchased outright (see below)

NOTE: EP5213 remained in the ownership of Dennis Specialist Vehicles Ltd. and was returned to them following the trial period. It was then with KCRC for a trial period but was later purchased by KMB together with the second demonstrator, EP1863, that had been with KCRC since 08.90. Both vehicles were added to the KMB fleet to become AA1 & 2.

Fleet number	Regn. number	Chassis number	Date to KMB 9SDL-3002	Withdrawa	Disposal
AA1	EP5213	132	10.92		
AA2	EP1863	115	10.92		

DENNIS 'DART' 98SDL3014 - 9.8m - Total 20
Fleet nos: AA 3-22.   ALEXANDER BODIES

Fleet number	Regn. number	Chassis number 98SDL3014-	Date first registered	Withdrawn	Disposal
AA 3	FN1332	1059	02.04.93		
AA 4	FN1854	1050	02.04.93		
AA 5	FN1967	1019	02.04.93		
AA 6	FN6824	1031	03.05.93		
AA 7	FN6879	1020	03.05.93		
AA 8	FN7282	1051	03.05.93		
AA 9	FN7995	1052	03.05.93		
AA10	FN8039	1017	03.05.93		
AA11	FN8391	1030	03.05.93		
AA12	FP 915	1012	17.05.93		

DART (A) Airbus   ALEXANDER BODIES
AA13	FP 957	1061	17.05.93		
AA14	FP 997	1010	17.05.93		
AA15	FP1202	1028	17.05.93		
AA16	FP1330	1049	17.05.93		
AA17	FP1678	1029	17.05.93		
AA18	FP1695	1060	17.05.93		
AA19	FP1778	1062	17.05.93		
AA20	FP1891	1013	17.05.93		
AA21	FP2107	1018	17.05.93		
AA22	FP6003	1011	01.06.93		

METSEC BODIES ASSEMBLED BY WADHAM STRINGER
		98SDL3039-			
AA23	GD3852	1962	09.09.94		
AA24	GD3952	2010	09.09.94		
AA25	GD5114	1999	13.09.94		
AA26	GD5192	1961	13.09.94		
AA27	GD4555	2009	16.09.94		
AA28	GD5813	1963	16.09.94		
AA29					
AA30					
AA31					
AA32					
AA33					
AA34					
AA35					
AA36					
AA37					
AA38					
AA39					
AA40					
AA41					
AA42					

DENNIS 'LANCE' 11SDA3104 - 1.7m
Fleet nos: AN 1-24.   Dual-door bus - ALEXANDER BODIES

Fleet number	Regn. number	Chassis number 11SDA3104-	Date first registered	Withdrawn	Disposal
AN 1	FP2584	151	25.05.93		
AN 2	FP3934	153	25.05.93		
AN 3	FP2756	145	27.05.93		
AN 4	FP2943	155	27.05.93		
AN 5	FP4320	135	27.05.93		
AN 6	FP2738	139	28.05.93		
AN 7	FP3732	156	28.05.93		
AN 8	FP4583	144	01.06.93		
AN 9	FP5902	143	03.06.93		
AN10	FP5327	149	04.09.93		

LANCE 11.7(A) single-door Airbus - ALEXANDER BODIES.
AN11	FP6762	129	11.06.93		
AN12	FR 667	134	01.07.93		
AN13	FR 755	154	01.97.93		
AN14	FR9173	142	26.07.93		
AN15	FS6584	152	23.08.03		
AN16	FS6896	157	23.08.93		
AN17	FS6904	148	23.08.93		
AN18	FS7610	150	23.08.93		
AN19	FS7776	135	23.08.93		
AN20	FS7880	140	23.08.93		
AN21	FS7603	141	24.08.93		
AN22	FU5663	146	28.10.93		
AN23	FU5072	147	28.10.93		
AN24	FW3600	158	29.12.93		

Double-deck Buses Purchased New -1949-1980

DAIMLER CVG5 - 26ft long new with *'Traditional'* radiator.

KMB class: Daimler(a). METAL SECTIONS BODIES

Fleet number	Regn. number	Chassis number	Date first registered	Rebody date	Withdrawn scrapped	Other Disposal	Final type
-	4958	?	13.04.49	-	12.70		a
-	4959	?	13.04.49	-	22.02.73		a
-	4960	?	13.04.49	-	12.70		a
D26	4961	16309	13.04.49	-	30.01.80	Trg bus	a
-	4962	16299	01.05.49	-	02.05.80	Trg bus	a
D28	4963	16310	01.05.49	-	01.03.80	Trg bus	a
D29	4964	16312	01.05.49	-	03.03.80	Trg bus	a
D30	4965	16313	01.05.49	-	23.04.80	Trg bus	a
D31	4966	16300	09.05.49	-	30.01.80	Trg bus	a
-	4967	?	09.05.49	-	22.03.73		a
-	4968	?	?		.12.70		a
D32	4969	16302	01.06.49	-	11.02.81		a
-	4970	?	01.06.49	-	22.02.73		a
-	4971	?	01.06.49	-	04.04.73		a
-	4972	?	01.06.49	-	22.02.73		a
D33	4973	16314	01.06.49	-	15.02.82		a
-	4974	?	?		12.70		a
-	4975	?	?		12.70		a
D34	4976	16306	18.06.49	-	15.01.82		a
D35	4977	16311	08.07.49	-	08.10.79		a
D36	4992	14712	.02.50	-	23.01.78		a
-	4993	?	07.02.50	-	30.04.73		a
-	4994	?	?		12.70		a
D37	4995	14724	07.02.50	-	10.03.80		a
-	4996	?	?		12.70		a
-	4997	?	07.02.50	-	30.04.73		a
D38	4998	14713	07.02.50	-	14.02.80		a
D39	4999	14711	07.02.50	-	17.02.80		a
D40	HK4001	14722	07.02.50	-	19.03.80	Trg bus	a
D41	HK4002	14720	07.02.50	-	17.02.80		a
D 4	**4204**	14725	28.03.50	-	17.07.82		a
D42	HK4003	?	28.03.50	-	24.05.74		a
D43	HK4004	14733	28.03.50	-	17.02.82		a
-	HK4005	?	28.03.50	-	12.70		a
D44	HK4006	14717	28.03.50	-	21.02.80	Trg bus	a
-	HK4007	?	28.03.50	-	30.04.73		a
-	HK4008	?	28.03.50	-	12.70		a
D45	HK4009	14715	28.03.50	-	16.09.80	Trg bus	a
D46	HK4010	14729	28.03.50	-	23.04.80	Trg bus	a
-	HK4011	?	28.03.50	-	22.02.73		a
-	HK4012	?	08.05.50	-	22.02.73		a
D47	HK4013	14731	08.05.50	-	12.08.82		a
-	HK4014	?	08.05.50	-	12.70		a
-	HK4015	?	08.05.50	-	12.70		a
-	HK4026	?	26.06.50	-	04.04.73		a
-	HK4027	?	26.06.50	-	04.04.73		a
-	HK4028	?	26.06.50	-	27.02.73		a
D48	HK4029	17355	26.06.50	-	02.05.80		a
-	HK4030	?	26.06.50	-	04.04.73		a
D49	HK4031	17349	26.06.50	-	18.08.82	To PDS.	a
D50	HK4032	?	26.06.50	-	28.01.76		a
D51	HK4033	?	26.06.50	-	12.08.82		a
-	HK4034	?	26.06.50	-	12.70		a
-	HK4035	?	26.05.50	-	04.04.73		a
D52	HK4036	17360	26.06.50	-	20.03.80		a
D53	HK4037	17359	22.08.50	-	24.02.82		a
-	HK4038	?	22.08.50	-	12.70		a
D54	HK4939	17365	22.08.50	-	26.06.81		a
-	HK4040	?	22.08.50	-	12.70		a
-	HK4041	17351	22.08.50	-	15.09.81		a
D56	HK4042	17352	22.08.50	-	22.01.81		a
-	HK4043	?	22.08.50	-	04.04.73		a
D57	HK4044	17361	22.08.50	-	12.08.82		a
D58	HK4045	17362	22.08.50	-	01.04.82		a
-	HK4046	?	22.08.50	-	30.04.73		a
-	HK4047	?	22.08.50	-	12.70		a
	HK4048	?	29.09.50	-	22.02.73		a
D59	HK4049	17358	29.09.50	-	17.02.82		a
-	HK4050	?	29.09.50	-	04.04.73		a
D60	HK4051	14714	13.10.51	-	08.05.80		a
-	HK4052	?	08.05.51	-	12.70		a
-	HK4053	?	08.05.51	-	12.70		a
D61	HK4054	17696	08.05.51	-	22.09.80	Trg bus	a
D62	HK4055	17693	08.05.51	-	04.05.81		a
-	HK4056	?	08.05.51	-	12.70		a
-	HK4057	?	08.05.51	-	12.70		a
-	HK4058	?	08.05.51	-	12.70		a
D63	HK4059	17358	08.05.51	-	04.05.77	R.S.Assn.	a
D64	HK4060	17695	08.05.51	-	20.04.82		a
D65	HK4061	?	08.05.51	-	20.09.77		a
D66	HK4062	17703	27.06.51	-	08.06.81		a
D67	HK4063	17702	27.06.51	-	01.04.82		a
-	HK4064	?	27.06.51	-	04.04.73		a
-	HK4065	?	27.06.51	-	27.02.73		a
-	HK4066	?	27.06.51	-	12.70		a
D68	HK4067	?	27.06.51	-	30.08.77		a
-	HK4068	?	27.06.51	-	22.02.73		a
D69	HK4069	17710	27.06.51	-	25.04.80		a
-	HK4070	?	27.06.51	-	22.02.73		a
D7O	HK4071	17706	27.06.51	-	04.05.81		a
-	HK4072	?	29.08.51	-	12.70		a
-	HK4073	?	29.08.51	-	12 70		a
-	HK4074	?	29.08.51	-	03.01.73		a
-	HK4075	?	29.08.51	-	12.70		a
D71	HK4076	17719	29.08.51	-	12.08.82		a
-	HK4077	?	29.08.51	-	12.70		a
D72	HK4078	17721	29.08.51	-	25.02.82		a
D73	HK4079	17716	29.08.51	-	22.12.80		a
-	HK4080	?	29.08.51	-	12.70		a
-	HK4081	?	10.10.51	-	12.70		a
D74	HK4132	18156	17.11.52	-	07.08.81		a
D75	HK4133	18140	17.11.52	-	24.02.82		a
D76	HK4134	18146	17.11.52	-	02.05.80		a
D77	HK4135	18153	17.11.52	-	07.08.81		a
D78	HK4136	18152	17.11.52	-	25.02.82		a

Many of the first 125 chassis (with 'Traditional radiators') cannot be directly matched to individual registration numbers but the whole batch was as follows:
14711-735 (25) - 1950*
16295-314 (20) - 1949*
17347-371 (25) - 1950
17687-696 (10) - 1951
17702-721 (20) - 1951
18136-160 (25) - 1952/3
NB: * Higher numbers built before lower numbers.

DAIMLER CVG5 with *'Traditional'* radiator (contd.).

KMB class: Daimler(a). METAL SECTIONS BODIES

Fleet number	Regn. number	Chassis number	Date first registered	Rebody date	Withdrawn scrapped	Disposal	Final type
D79	HK4137	18145	17.11.52	-	25.02.82		a
D80	HK4138	18150	17.11.52	-	20.04.82		a
D81	HK4139	18148	01.12.52	-	24.02.82		a
-	HK4140	?	01.12.52	-	12.70		a
-	HK4141	?	01.12.52	-	12.70		a
-	HK4142	?	24.12.52	-	21.05.73		a
D82	HK4143	18155	24.12.52	-	27.05.81		a
D83	HK4144	18139	24.12.52	-	19.06.81		a
-	HK4145	?	24.12.52	-	30.04.73		a
-	HK4146	?	24.12.52	-	04.04.73		a
-	HK4147	?	24.01.53	-	04.04.73		a
D84	HK4148	?	24.01.53	-	22.09.77		a
-	HK4149	?	24.01.53	-	12.69	a - First to be scrapped	
-	HK4150	?	24.01.53	-	12.70		a
D85	HK4251	18160	24.01.53	-	16.12.81		a
-	HK4252	?	24.01.53	-	12.70		a
-	HK4253	?	24.01.53	-	12.70		a
D86	HK4254	18158	24.01.53	-	01.04.82		a
D87	HK4255	18138	27.01.53	-	29.07.82		a
D88	HK4256	?	24.01.53	-	24.05.74		a

NOTES:
1) These buses were allocated fleet numbers as shown in 1974.
2) Buses scrapped prior to fleet numbering are denoted by a dash
3) Chassis numbers were not required by the licencing authority until the mid-1970's and it is thus that those buses withdrawn prior to that time remain anonymous as the Company had no need to retain this information.

DAIMLER CVG5 26ft; new with *'Birmingham'* grilles.

KMB class: Daimler(a). METAL SECTIONS BODIES

Fleet number	Regn. number	Chassis number	Date first registered	Rebody date	Withdrawn scrapped	Other disposal	Final type
D 1	4201	18586	24.04.54	-	15.02.82	TSW(HK)Ltd	a
D 2	4202	?	24.04.54	-	13.11.73		a
D 3	4203	?	24.04.54	-	30.05.74		a
D 5	4205	?	24.04.54	-	24.05.74		a
D 6	4206	18585	24.04.54	-	24.02.82		a
D 7	4207	18594	15.05.54	-	17.07.82		a
D 8	4209	18591	15.05.54	-	08.10.79		a
D 9	4210	18593	15.05.54	-	02.01.81	Trg bus	a
D10	4211	18592	15.05.54	-	10.07.80		a
D11	4212	18590	15.05.54	-	16.10.82	Preservation- Speedybus Ltd.	a
D12	4213	18598	08.06.54	-	10.07.80		a
D13	4217	?	08.06.54	-	24.04.74		a
D14	4218	?	08.06.54	-	25.10.73		a
D15	4220	18595	08.06.54	-	25.08.83		a
D16	4221	18596	08.06.54	-	03.04.78		a
D17	4222	18600	05.07.54	-	23.03.83		a
D18	4223	18601	05.07.54	-	15.10.79		a
D19	4224	18603	05.07.54	-	15.02.81		a
D20	4225	18604	05.07.54	-	30.12.82		a
D21	4228	18602	05.07.54	-	15.02.81		a
D22	4229	18672	01.11.54	-	10.07.80		a
D23	4240	18668	01.11.54	-	15.10.80		a
D24	4241	18669	01.11.54	-	30.07.80	Speedybus	a - 06.12.88
-	4634	18670	01.11.54	-	22.02.73		a
D25	4635	18671	01.11.54	-	15.02.82		a
D89	HK4257	18673	27.11.54	-	24.02.82		a
D90	HK4258	18677	27.11.54	05.10.77	25.08.83		ab
D91	HK4259	18676	27.11.54	05.10.77	20.06.83		ab
D92	HK4260	18674	27.11.54	04.10.76	07.08.80		aa
D93	HK4261	18675	27.11.54	-	24.01.80		a
-	HK4262	18679	01.01.55	-	15.09.72		a
D94	HK4263	18678	01.01.55	-	25.02.82		a
D95	HK4264	18681	01.01.55	-	23.06.80		a
D96	HK4265	18682	01.01.55	-	24.01.80		a
D97	HK4266	18680	01.01.55	-	05.12.80	Trg bus	a
D98	HK4267	18685	13.01.55	-	22.12.80		a
D99	HK4268	18686	13.01.55	-	24.01.83		a
D100	HK4269	18684	13.01.55	-	24.05.74		a
D101	HK4270	18683	22.03.55	-	25.08.83		a
D102	HK4271	18687	22.03.55	-	07.05.82		a
-	HK4273	18957	03.04.56	-	15.09.72		a
D103	HK4274	18958	03.04.56	-	07.05.82		a
D104	HK4275	18959	03.04.56	-	25.08.83		a
D105	HK4276	18956	03.04.56	-	17.02.82		a
D106	HK4277	18955	03.04.56	-	24.01.80		a
D107	HK4278	18964	23.04.56	-	23.03.83		a
D108	HK4279	18960	23.04.56	-	24.01.83		a
D109	HK4280	18961	23.04.56	-	30.07.80		a
D110	HK4281	18963	23.04.56	-	20.04.82		a
-	HK4282	18962	23.04.56	-	15.09.72		a
D111	HK4312	18956	01.08.56	-	25.02.82	TSW(HK)Ltd	a
D112	HK4313	18969	01.08.56	-	10.12.79		a
D113	HK4314	18971	01.08.56	-	24.01.83		a
D114	HK4315	18966	01.08.56	-	23.09.82	Speedybus 06.12.88	a
D115	HK4316	18967	01.08.56	-	13.11.81		a
D116	HK4317	18972	25.08.56	-	27.10.81		a
D117	HK4318	18974	25.08.56	-	13.04.81		a
D118	HK4319	18973	25.08.56	-	16.02.83		a
D119	HK4320	18970	25.08.56	-	23.11.82		a
D120	HK4321	18968	25.08.56	-	13.04.81		a
D121	HK4322	19108	18.09.56	-	17.02.82		a
D122	HK4323	19109	18.09.56	-	12.01.81		a
D123	HK4324	19107	18.09.56	-	29.07.82		a

The 90 chassis/numbers with 'Birmingham fronts'for KMB :
18585-604 (20) - 1954
18668-687 (20) - 1954/5
18955-974 (20) - 1956
19105-134 (30) - 1956/7
NB: Two similar chassis for the Manila Motor Coach Co. were: 18689/90 in 1954

PDS: Police Driving School. TSW : Transport Supplies Worldwide (HK) Ltd.
R.S.Assn: Road Safety Association. Trg bus: Driver training vehicle

DAIMLER CVG5 with *'Birmingham'* grilles (contd.).

KMB class: Daimler(a) - *METAL SECTIONS BODY*

Fleet number	Regn. number	Chassis number	Date first registered	Rebody date	Withdrawn	Disposal	Final type
D124	HK4325	19105	18.09.56	-	25.08.83		a
D125	HK4326	19106	18.09.56	-	27.11.80	Trg bus	a
D126	HK4327	19114	19.10.56	-	24.02.82		a
D127	HK4328	19110	01.10.56	-	17.11.79		a
D128	HK4329	19111	01.10.56	-	17.07.82		a
D129	HK4330	19112	19.10.56	-	29.07.82		a
D130	HK4331	19113	19.10.56	-	30.12.82		a
D131	HK4332	19116	10.11.56	-	11.06.81		a
D132	HK4333	19117	10.11.56	-	23.06.80		a
D133	HK4334	19115	10.11.56	-	30.07.79		a
D134	HK4335	19119	10.11.56	-	11.09.80		a
D135	HK4336	19118	10.11.56	-	25.02.82		a
D136	HK4337	19123	10.12.56	-	24.02.82		a
D137	HK4338	19120	10.12.56	-	23.03.83		a
D138	HK4339	19124	10.12.56	-	24.05.74		a
D139	HK4340	19121	10.12.56	-	12.05.75		a
D140	HK4341	19122	07.01.57	-	30.12.82		a
D141	HK4342	19128	07.01.57	-	07.05.82		a
D142	HK4343	19127	07.01.57	-	24.01.83		a
D143	HK4344	19125	07.01.57	-	28.06.82		a
D144	HK4345	19126	07.01.57	-	29.07.82		a
D145	HK4346	19129	07.01.57	-	23.09.81		a
D146	HK4347	19132	23.04.57	-	29.07.82		a
D147	HK4348	19130	23.04.57		21.02.81		a
D148	HK4349	19134	27.04.57	-	27.05.82		a
D149	HK4350	19133	23.04.57	07.07.76	25.08.83		aa
D150	HK4351	19131	23.04.57	-	11.02.81		a

NOTE: **1)** HK4351 (D150) Temporary Training bus 12.71 to 04.74.
2) Total Daimler(a) (both types): 215; Total at fleet numbering: 150; ie 65 scrapped.

DAIMLER CVG5DD 27ft. *'Manchester'* front as built.

KMB class: Daimler(b) - *METAL SECTIONS BODIES*

Fleet number	Regn number	Chassis number	Date first registered	Rebody date	Withdrawn	Disposal	Final type
D151	4214	19597	21.10.59	-	25.08,76		b
D152	4215	19601	31.10.59	09.11.77	16.01.84		ba
D153	4216	19596	31.10.59	-	03.05.82	Trg bus	b
D154	4219	19605	31.10.59	-	24.01.83		b
D155	4226	19599	31.10.59	-	17 11.79		b
D156	4227	19602	31.10.59	?	27.06.81		bb
D157	4230	19604	31.10.59	-	21.08.79		b
D158	4231	19603	31.10.59	-	01.04.81		b
D159	4232	19600	31.10.59	.04.78	22.03.82	Trg bus	ba
D160	4233	19598	31.10.59	?	16.01.84		ba
D161	4234	19612	25.11.59	07.78	20.04.82		bb
D162	4235	19607	25.11.59	04.78	22.03.82	Trg bus	ba
D163	4236	19611	2i.11.59	-	22.01.82	Trg bus	b
D164	4237	19610	25.11.59	-	25.08.83		b
D165	4238	19606	25.11.59	24.05.77	23.11.82		bb
D166	4239	19608	25.11.59	-	24.11.80		b
D167	4242	19613	25.11.59	?	16.01.84		ba
D168	4243	19614	25.11.59	78	10.12.79		ba
D169	4244	19609	25.11.59	-	19.04.82	Trg bus	b
D170	4245	19615	25.11.59	-	10.12.82		b
D171	4246	19627	22.12.59	?	29.07.82		ba
D172	4247	19616	22.12.59	-	22.09.79		b
D173	4248	19618	22.12.59	-	16.01.84		ba
D174	4249	19619	22.12.59	-	22.09.77		b
D175	4250	19622	22.12.59	09.03.77	16.12.81		ba
D176	4591	19621	22.12.59	27.07.77	22.12.80		ba
D177	4592	19620	22.12.59	22.04.77	28.03.84		bb
D178	4593	19617	22.12.59	09.08.77	31.07.84	***Last Daimler(b)***	bb
D179	4594	19626	22.12.59	-	03.02.81	Tree cutter	b
D180	4595	19624	22.12.59	05.78	19.04.82	Trg bus	ba
D181	4596	19623	21.01.60	-	22.09.79		b
D182	4597	19634	21.01.60	05.78	13.02.82	Trg bus	ba
D183	4598	19630	21.01.60	.78	22.12.80		ba
D184	4599	19631	21.01.60	-	28.11.81	Trg bus	b.
D185	4600	19629	21.02.60	05.78	25.03.82	Trg bus	ba
D186	4601	19633	21.01.60	-	13.04.81		b
D187	4602	19635	21.01.60	-	27.04.74	Tree cutter	b
D188	4603	19628	21.01.60	04.78	22.01.82	Trg bus	ba
D189	4604	19625	21.01.60	04.78	12.02.82	Trg bus	ba
D190	4605	19632	21.01.60	-	07.05.83		b
D191	4606	19642	11.02.60	22.04.77	28.03.84		bb
D192	4607	19639	11.02.60	-	01.04.82	Trg bus	b
D193	4608	19643	11.02.60	-	19.01.82	Trg bus	b
D194	4609	19640	11.02.60	-	22.01.82	Trg bus	b
D195	4612	19636	11.02.60	-	.01.04.81		b
D196	4613	19638	27.02.60	-	13.03.82	Trg bus	b
D197	4614	19644	27.02.60	-	12.02.82	Trg bus	b
D198	4615	19637	27.02.60	27.02.77	30.04.84		bb
D199	4624	19645	27.02.60	-	16.12.76		b
D200	4625	19641	27.02.60	16.08.77	22.03.82	Trg bus	ba
D201	4626	19693	25.07.60	-	18.01.80		b
D202	4663	19694	25.07.60	-	15.10.79		b
D203	4664	19695	25.07.60	-	28.04.81		b
D204	4666	19690	25.07.60	-	15.03.82	Trg bus	b
D205	4669	19698	25.07.60	-	18.01.80		b
D206	4670	19696	25.07.60	-	22.03.82	Trg bus	b
D207	4672	19689	25.07.60	-	23.06.80		b
D208	4952	19697	25.07.60	-	13.04.81		b
D209	4978	19692	25.07.60	-	12.08.81		b
D210	4979	19691	25.07.60	-	11.03.82	Trg bus	b
D211	4980	19708	12.08.60	-	07.04.82	Trg bus	b
D212	4981	19707	12.08.60	-	01.04.82	Trg bus	b
D213	4982	19704	12.08.60	24.12.77	08.10.82		ba
D214	4983	19705	12.08.60	24.12.77	24.09.81		ba
D215	4988	19706	12.08.60	-	07.04.82	Trg bus	b
D216	4989	19702	09.09.60	07.10.77	21.02.81		bb
D217	4990	19703	09.09.60	01.78	16.01.84		ba
D218	4991	19701	09.09.60	02.78	13.02.82	Trg bus	ba
D219	HK4460	19700	09.09.60	-	07.08.81		b
D220	HK4461	19699	09.09.60	-	10.03.82		b
D221	HK4462	19739	10.11.60	-	08.10.79		b
D222	HK4463	19740	10.11.60	05.78	20.04.82		bb

DAIMLER CVG5DD 27ft long with *'Manchester'* front.

KMB class: Daimler(b) (contd.) - *METAL SECTIONS BODY*

Fleet number	Regn number	Chassis number	Date first registered	Rebody date	Withdrawn	Disposal	Final type
D223	HK4464	19738	10.11.60	-	14.07.81		b
D224	HK4465	19737	10.11.60	09.12.77	01.08.83		ba
D225	HK4466	19741	24.11.60	-	08.10.82		b
D226	HK4467	19744	24.11.60	-	08.10.79		b
D227	HK4468	19746	24.11.60	-	13.10.77		b
D228	HK4469	19742	24.11.60	-	24.01.84		b
D229	HK4470	19743	24.11.60	?.77	08.11.80		b
D230	HK4471	19745	20.12.60	-	25.10.83		b
D231	HK4472	19740	20.12.60	-	25.02.82		b
D232	HK4473	19750	20.12.60	-	24.03.81		b
D233	HK4474	19749	20.12.60	-	30.12.81	Trg bus	b
D234	HK4475	19747	20.12.60	23.11.77	01.04.82	Trg bus	ba
D235	HK4476	19751	20.12.60	-	14 05.81		b
D236	HK4477	19764	12.01.61	-	29 11.79		b
D237	HK4478	19767	12.01.61	24.11.77	26.04.82	Trg bus	ba
D238	HK4479	19766	12.01.61	-	12.01.81		b
D239	HK4480	19765	12.01.61	-	26.09.80		b
D240	HK4481	19763	12.01.61	-	25.10.83		b
D241	HK4482	19754	08.02.61	-	25.10.73		b
D242	HK4483	19759	08.01.61	-	25.10.73		b
D243	HK4484	19760	08.02.61	-	10.11.81		b
D244	HK4485	19762	08.02.61	-	07.08.81		b
D245	HK4486	19752	08.02.61	-	16.12.81		b
D246	HK4487	19753	03.03.61	-	14.12.81	Trg bus	b
D247	HK4488	19757	03.03.61	22.04.77	28.03.84		bb
D248	HK4489	19761	03.03.61	-	18.01.80		b
D249	HK4490	19758	03.03.61	-	23.03.83		b
D250	HK4491	19755	03.03.61	-	22.03.82	Trg bus	b
D251	HK4492	19770	21.03.61	-	10.03.82		b
D252	HK4493	19776	21.03.61	-	22.03.82	Trg bus	b
D253	HK4494	19771	21.03.61	-	22.09.79		b
D254	HK4495	19775	21.03.61	-	28.04.81		b
D255	HK4496	19774	21.03.61	-	14.04.82	Trg bus	b
D256	HK4497	19756	03.05.61	-	24.01.83		b
D257	HK4498	19769	03.05.61	-	30.04.82	Trg bus	b
D258	HK4499	19773	03.05.61	-	23.01.82	Trg bus	b
D259	HK4500	19772	03.05.61	-	20.04.82		b
D260	HK4501	19768	03.05.61	-	15 03.78		b

NOTES: **1)** Tree cutting buses : 4602 & 4594)
2) Original total: 110
3) 1974 total : 110 (at time of fleet numbering).

DAIMLER CVG6-30 - 30ft long x 8ft wide D261-33.

KMB class: Daimler(c) - *METAL SECTIONS BODIES*

Fleet number	Regn. number	Chassis number	Date first registered	Converted to centre staircase	Withdrwan	Disposal	Final type
D261	AD4717	30101	25.01.62	?	31.12.84		cb
D262	AD4718	30099	25.01.62	?	31.12.84		cb
D263	AD4719	30104	25.01.62	?	31.12.84		cb
D264	AD4720	30106	25.01.62	-	31.12.84		c
D265	AD4721	30103	25.01.62	-	31.12.84		c
D266	AD4722	30107	25.01.62	?	21.01.85		cb
D267	AD4723	30100	25.01.62	?	31.12.84		cb
D268	AD4724	30105	25.01.62	?	21.01.85		cb
D269	AD4725	30102	25.01.62	?	16.11.84		cb
D270	AD4726	30108	25.01.62	?	12.02.85		cb
D271	AD4727	30110	07.03.62	-	18.03.85		c
D272	AD4728	30112	07.03.62	?	04.02.86		cb
D273	AD4729	30109	07.03.62	-	06.06.85		c
D274	AD4730	30114	07.03.62	-	11.03.86		c
D275	AD4731	30111	07.03.62	?	03.03.86		cb
D276	AD4732	30116	07.03.62	24.05.77	25.04.85		cc
D277	AD4733	30115	23.03.62	24.06.77	11.08.86		cc
D278	AD4734	30113	23.03.62	?	12.02.85		ca
D279	AD4735	30117	23.03.62	?	12.02.85		cc
D280	AD4736	30123	23.03.62	?	23.04.86		cc
D281	AD4737	30125	04.04.62	17.05.77	25.04.85		cc
D282	AD4738	30122	04.04.62	?	03.03.86		cb
D283	AD4739	30119	04.04.62	?	19.05.86		cb
D284	AD4740	30120	04.04.62	09.06.77	02.06.87		cc
D285	AD4741	30126	04.04.62	?	28.03.84		cb
D286	AD4742	30118	09.04.62	04.82	18.03.86	***(BRITISH ALUMINIUM BODY)***	ca
D287	AD4743	30121	19.04.62	?	14.03.86		cb
D288	AD4744	30124	19.04.62	23.04.77	12.11.84		cb
D289	AD4745	30098	19.04.62	26.10.77	06.10.86		cc
D290	AD4746	30127	19.04.62	?	11.03.86		cb
D291	AD4747	30154	03.07.62	-	13.05.83		c
D292	AD4748	30147	03.07.62	-	13.02.86		c
D293	AD4749	30151	03.07.62	?	05.05.82		cb
D294	AD4750	30150	03.07.62	-	28.04.86		c
D295	AD4751	30146	05.07.62	?	12.11.84		cb
D296	AD4752	30148	03.07.62	-	20.07.84	tree cutter	`c
D297	AD4753	30149	03.07.62	01.12.77	12.11.84		cc
D298	AD4754	30155	03.07.62	?	26.06.86		cb
D299	AD4755	30160	03.07.62	-	26.06.86		c
D300	AD4756	30152	03.07.62	-	20.04.86		c
D301	AD4757	30164	06.07.62	-	11.08.86		c
D302	AD4758	30156	06.07.62	01.12.77	20.11.84		cc
D303	AD4759	30162	06.07.67	-	30.01.86		c
D304	AD4760	30153	06.07.62	-	23.04.86		cb
D305	AD4761	30161	06.07.62	05.82	21.07.86		cb
D306	AD4762	30158	13.07.62	20.09.77	11.08.86		cb
D307	AD4763	30163	13.07.62	22.08.77	26.07.85		cb
D308	AD4764	30165	13.07.62	05.10.77	06.10.86		cc
D309	AD4765	30159	13.07.62	?	20.01.86		cb
D310	AD4766	30170	13.07.62	?	26.06.86		cb
D311	AD4767	30166	25.07.62	06.04.78	18.03.85		cc
D312	AD4768	30167	25.07.62	?	10.07.86		ca
D313	AD4769	30157	25.07.62	?	10.07.86		ca
D314	AD4770	30168	15.07.62	?	06.06.85		ca
D315	AD4771	30169	25.07.62	?	11.03.86		cb
D316	AD4772	30173	08.08.62	?	11.03.86		ca
D317	AD4773	30180	08.08.62	20.09.77	11.08.86		ca
D318	AD4774	30179	08.08.62	07.09.77	11.08.86		ca
D319	AD4775	30174	08.08.62	16.11.77	12.11.86		cc

NOTE: ?: Conversion dates to centre staircase: Detailed information has not become available showing exact dates of conversions but all markrd with a question mark were convertesbetween 9/74 and 3/77

DAIMLER CVG6-30 (contd.) D261-330;
KMB class: Daimler(c) - METAL SECTIONS BODIES

Fleet number	Regn. number	Chassis number	Date first registered	Converted to centre staircase	Withdrawn	Disposal	Final type
D320	AD4776	30172	08.08.62	?	22.05.85		ca
D321	AD4777	30184	21.08.62	?	27.05.84		ca
D322	AD4778	30178	21.08.62	?	11.08.86		ca
D323	AD4779	30176	21.08.62	12.09.77	19.05.86		ca
D324	AD4780	30183	21.08.62	-	25.10.84		c
D325	AD4781	30177	21.08.62	22.09.77	11.08.86		cb
D326	AD4782	30175	31.08.62	?	26.06.86		ca
D327	AD4783	30182	31.08.62	?	11.08.86		ca
D328	AD4784	30185	31.08.62	?	25.10.84		ca
D329	AD4785	30171	31.08.62	?	10.09.87		ca
D330	AD4786	30181	31.08.62	22.09.77	07.03.83		cb

> **AEC REGENT MkV:** It should be noted that, chronologically, these should appear next, the list continues with the Daimlers; the AEC's commence on page 304.

DAIMLER CVG6LX-34 - 34ft long - METAL SECTIONS BODIES
KMB class: Daimler(d) - later became Daimler(d)CS

Fleet number	Regn. number	Chassis number	Date first registered	Converted to 'CS'	Withdrawn	Disposal
D331	AD7237	30428	07.02.67	02.05.80	22.04.88	
D332	AD7238	30429	07.02.67	27.05.80	16.06.88	
D333	AD7239	30430	07.02.67	05.80*	22.04.88	
D334	AD7240	30431	30.04.67	19.05.80	22.04.88	
D335	AD7241	30432	30.04.67	19.06.80	23.06.88	
D336	AD7242	30433	30.04.67	08.08.80	23.06.88	
D337	AD7243	30434	30.04.67	18.08.80	23.06.88	
D338	AD7244	30435	31.05.67	20.08.80	16.03.88	
D339	AD7245	30446	31.05.67	23.11.79	07.10.87	
D340	AD7246	30437	31.05.67	13.06.80	29.04.88	
D341	AD7247	30438	31.05.67	26.09.80	25.10.84	
D342	AD7248	30439	31.05.67	26.09.80	21.08.87	Speedybus - GZ -1304
D343	AD7249	30440	30.06.67	11.79	16.06.88	
D344	AD7250	30441	30.06.67	12.02.80	21.01.88	
D345	AD7251	30442	31.10.67	30.10.80	26.11.87	
D346	AD7252	30443	31.10.67	12.01.80	14.12.87	
D347	AD7253	30444	31.10.67	22.04.80	23.03.88	
D348	AD7254	30427	31.12.67	25.04.80	23.03.88	
D349	AD7255	30445	31.12.67	09.80	30.11.84	
D350	AD7256	30446	31.12.67	08.04.80	23.03.88	

> **NOTES:**
> i. *AD7239 was firstly converted, in September 1979, to DOO with rear staircase retained and the forward doorway repositioned in the first nearside body bay, adjacent to the bulkhead. AD7253 was also partially convetred in this was but it seems unlikely that it was completed before the final arrangement was arrived at, ie, with centre staircase and exit.
> ii. Both were finally rebuilt to conform with the remainder of rebuilt class and all were reclassified as 'Daimler(d)CS' -where, in KMB parlance, 'CS' stands for 'Centre Staircase'.

DAIMLER CVG6LX-34 34ft long, Front Stairs. METAL SECTIONS BODIES
KMB class: Daimler(e) (D464 /90: Centre Stairs).

Fleet number	Regn. number	Chassis number	Date first registered	Withdrawn	Disposal
D351	AD7257	30450	18.10.69	15.09.87	Speedybus - Guangzhou No 1
D352	AD7258	30452	18.10.69	15.09.87	Speedybus - Chongqing 14.02.88
D353	AD7259	30456	18.10.69	15.09.87	Speedybus - Guangzhou No 1
D354	AD7260	30460	18.10.69	11.08.86	
D355	AD7261	30462	18.10.69	11.08.86	
D356	AD7262	30468	18.10.69	15.09.87	Speedybus - Guangzhou No 1
D357	AD7263	30471	18.10.69	15.09.87	Speedybus - Guangzhou No 1
D358	AD7264	30458	22.10.69	15.09.87	Speedybus - Guangzhou No 1
D359	AD7265	30451	22.10.69	15.09.87	
D360	AD7266	30453	22.10.69	15.09.87	
D361	AD7267	30459	11.11.69	16.10.87	
D362	AD7268	30479	11.11.69	16.10.87	
D363	AD7269	30470	11.11.69	16.10.87	
D364	AD7270	30473	11.11.69	29.10.86	
D365	AD7271	30467	11.11.69	29.10.86	
D366	AD7272	30477	11.11.69	29.10.86	
D367	AD7273	30457	11.11.69	29.10.86	
D368	AD7274	30469	11.11.69	29.10.86	
D369	AD7275	30464	11.11.69	29.10.86	
D370	AD7276	30472	11.11.69	29.10.86	
D371	AD7277	30466	24.11.69	29.10.86	
D372	AD7278	30465	24.11.69	29.10.86	
D373	AD7279	30463	24.11.69	29.10.86	
D374	AD7280	30476	24.11.69	29.10.86	
D375	AD7281	30455	24.11.69	29.10.86	
D376	AD7282	30474	24.11.69	29.10.86	
D377	AD7283	30461	24.11.69	29.10.86	
D378	AD7284	30475	24.11.69	29.10.86	
D379	AD7285	30478	24.11.69	29.10.86	
D380	AD7286	30454	09.12.69	26.11.86	Speedybus - Guangzhou No 1
D381	AD7287	30538	24.05.71	07.06.88	
D382	AD7288	30537	24.05.71	07.06.88	
D383	AD7289	30535	24.05.71	07.06.88	
D384	AD7290	30533	24.05.71	07.06.88	
D385	AD7291	30536	24.05.71	07.06.88	
D386	AD7292	30532	01.06.71	07.06.88	
D387	AD7293	30534	01.06.71	07.06.88	

> **Guangzhou No 1:** Guangzhou No 1 Bus Co. (via Speedybus.)

DAIMLER CVG6LX-34 - 34ft;, front stairs - D351-550.
KMB class: Daimler(e) METAL SECTIONS BODIES

Fleet number	Regn. number	Chassis number	Date first registered	Withdrawn	Disposal
D388	AD7294	30530	01.06.71	07.06.88	
D389	AD7295	30539	01.06.71	07.06.88	
D390	AD7296	30531	01.06.71	07.06.88	
D391	AD7297	30548	15.06.71	07.06.88	
D392	AD7298	30555	15.06.71	07.06.88	
D393	AD7299	30554	15.06.71	07.06.88	
D394	AD7300	30545	15.06.71	07.06.88	
D395	AD7301	30559	15.06.71	07.06.88	
D396	AD7302	30558	15.06.71	07.06.88	
D397	AD7303	30550	15.06.71	07.06.88	
D398	AD7304	30549	15.06.71	28.06.88	
D399	AD7305	30541	15.06.71	28.06.88	
D400	AD7306	30557	15.06.71	07.06.88	
D401	AD7307	30558	15.06.71	28.06.88	
D402	AD7308	30553	07.07.71	28.06.88	
D403	AD7309	30556	07.07.71	28.06.88	
D404	AD7310	30543	07.07.71	19.01.88	
D405	AD7311	30546	07.07.71	21.06.90	Stagecoach Malawi - No.?
D406	AD7312	30542	07.07.71	03.06.87	
D407	AD7313	30547	07.07.71	28.07.88	
D408	AD7314	30544	07.07.71	28.07.88	
D409	AD7315	30552	07.07.71	Temp.03.90	Reinstated -30.08.90 - see page 301
D410	AD7316	30540	07.07.71	03.08.89	Stagecoach Malawi - 1042
D411	AD7317	30617	25.11.71	16.10.87	
D412	AD7318	30624	25.11.71	26.08.87	
D413	AD7319	30623	25.11.71	16.10.87	
D414	AD7320	30615	25.11.71	19.11.88	Stagecoach Malawi -1021
D415	AD7321	30622	25.11.71	19.11.88	Speedybus - Wuzhou
D416	AD7322	30620	25.11.71	19.11.88	Stagecoach Malawi - 1031
D417	AD7323	30621	25.11.71	Temp 01.89	Reinstated - 13.07.90
D418	AD7324	30618	25.11.71	Temp 01.89	Reinstated - 03.08.90
D419	AD7325	30616	25.11.71	19.11.88	Stagecoach Malawi - 1022
D420	AD7326	30619	27.11.71	19.11.88	Speedybus - Wuzhou
D421	AD7327	30611	01.02.72	18.03.89	Stagecoach Malawi - 1045
D422	AD7328	30610	01.02.72	19.01.88	
D423	AD7329	30612	01.02.72	10.04.89	Stagecoach Malawi - 1032
D424	AD7330	30614	01.02.72	18.03.89	Stagecoach Malawi - 1043
D425	AD7331	30613	01.02.72	10.02.89	Stagecoach Malawi - No. ?
D426	AD7332	30634	01.02.72	03.03.89	Stagecoach Malawi - 1030
D427	AD7333	30633	01.02.72	10.02.89	Stagecoach Malawi - No. ?
D428	AD7334	30628	01.02.72	03.03.89	Stagecoach Malawi - 1026
D429	AD7335	30627	01.02.72	10.02.89	Stagecoach Malawi - No. ?
D430	AD7336	30626	01.02.72	Temp. 03.90	Reinstated - 09.07.90 - see page 301
D431	AD7337	30631	02.02.72	10.02.89	Stagecoach Malawi - No. ?
D432	AD7338	30629	02.02.72	31.12.90	
D433	AD7339	30646	02.02.72	10.02.89	Stagecoach Malawi - No. ?
D434	AD7340	30636	02.02.72	10.04.89	Stagecoach Malawi - 1051
D435	AD7341	30641	02.02.72	18.03.89	Stagecoach Malawi - 1040
D436	AD7342	30640	02.02.72	10.04.89	Stagecoach Malawi - 1041
D437	AD7343	30642	02.02.72	03.03.89	Stagecoach Malawi -1029
D438	AD7344	30632	02.02.72	03.03.89	Stagecoach Malawi - 1034
D439	AD7345	30625	02.02.72	03.03.89	Stagecoach Malawi - 1033
D440	AD7346	30630	02.02.72	10.04.89	Stagecoach Malawi - 1027
D441	AD7347	30651	09.02.72	31.12.90	
D442	AD7348	30645	09.02.72	18.03.89	Stagecoach Malawi - 1055
D443	AD7349	30638	09.02.72	19.01.88	
D444	AD7350	30644	09.02.72	19.01.88	
D445	AD7351	30646	09.02.72	05.11.87	
D446	AD7352	30643	09.02.72	19.01.88	
D447	AD7353	30648	09.02.72	19.01.88	
D448	AD7354	30653	09.02.72	19.01.88	
D449	AD7355	30639	09.02.72	19.01.88	
D450	AD7356	30650	09.02.72	19.01.88	
D451	AD7357	30660	17.03.72	16.02.88	
D452	AD7358	30669	17.03.72	16.02.88	
D453	AD7359	30668	17.03.72	16.02.88	
D454	AD7360	30661	17.03.72	16.02.88	
D455	AD7361	30652	17.03.72	27.01.87	
D456	AD7362	30665	17.03.72	03.03.88	
D457	AD7363	30662	17.03.72	05.11.87	
D458	AD7364	30705	17.03.72	05.11.87	
D459	AD7365	30695	17.03.72	21.01.87	
D460	AD7366	30701	17.03.72	16.02.88	
D461	AD7367	30713	04.04.72	22.07.88	Speedybus - Chendu (PCCC)*
D462	AD7368	30719	04.04.72	06.04.88	
D463	AD7369	30725	04.04.72	06.04.88	
D464 CS	AD7370	30715	04.04.72	22.07.88	Speedybus - Chendu (PCCC)*
D465	AD7371	30649	04.04.72	27.08.88	Speedybus - Guilin
D466	AD7372	30724	04.04.72	27.07.88	Speedybus - Chendu (PCCC)*
D467	AD7373	30656	04.04.72	06.04.88	
D468	AD7374	30722	04.04.72	06.04.88	
D469	AD7375	30702	04.04.72	22.07.88	Speedybus - Guilin
D470	AD7376	30731	04.04.72	22.07.88	Speedybus - Chendu (PCCC)*
D471	AD7377	30700	04.05.72	22.04.88	
D472	AD7378	30709	04.05.72	22.04.88	
D473	AD7379	30692	04.05.72	22.04.88	
D474	AD7380	30673	04.05.72	22.04.88	
D475	AD7381	30691	04.05.72	22.04.88	
D476	AD7382	30670	04.05.72	22.04.88	
D477	AD7383	30706	04.05.72	22.04.88	
D478	AD7384	30666	04.05.72	29.04.88	
D479	AD7385	30738	04.05.72	29.04.88	
D480	AD7386	30737	04.05.72	29.04.88	
D481	AD7387	30747	01.08.72	28.07.89	
D482	AD7388	30740	01.08.77	28.07.89	Stagecoach Malawi - 1049
D483	AD7389	30730	01.08.72	03.08.89	Stagecoach Malawi - 1048
D484	AD7390	30745	01.08.72	28.07.89	Stagecoach Malawi - 1025
D485	AD7391	30735	01.08.72	03.08.89	Stagecoach Malawi - 1047

> **Temp:** Indicates temporary withdrawal during overhaul, repairs, or a decision on a bus's future.

> ***PCCC:** Public Communications Company, Chendu.

DAIMLER CVG6LX-34 - 34ft (contd.)
KMB class: Daimler(e) METAL SECTIONS BODIES

Fleet number	Regn. number	Chassis number	Date first registered	Withdrawn	Disposal
D486	AD7392	30739	01.08.72	28.07.89	Stagecoach Malawi - 1039
D487	AD7393	30743	01.08.72	30.03.90	Stagecoach Malawi - 1052
D488	AD7394	30742	01.08.72	29.04.88	
D489	AD7395	30749	01.08.72	30.03.90	Stagecoach Malawi - 1053
D490	AD7396	30760	01.08.72	30.03.90	Stagecoach Malawi - 1056
D491	AD7397	30741	01.08.72	03.08.89	Stagecoach Malawi - 1044
D492	AD7398	30754	01.08.72	30.03.90	
D493	AD7399	30731	01.08.72	01.11.86	Speedybus - not used
D494 CS	AD7400	30746	01.08.72	30.03.90	Stagecoach Malawi - 1053
D495	AD7401	30755	01.08.72	03.08.89	Stagecoach Malawi - 1038
D496	AD7402	30654	01.08.72	03.08.89	Stagecoach Malawi - 1050
D497	AD7403	30748	01.08.72	01.09.89	Stagecoach Malawi - 1043
D498	AD7404	30761	01.08.72	25.07.88	
D499	AD7405	30637	01.08.72	25.07.88	
D500	AD7406	30644	01.08.72	25.07.88	Speedybus - Dalian
D501	AD7407	30659	01.08.72	25.07.88	Speedybus - Hangzhou
D502	AD7408	30756	01.08.72	25.07.88	Speedybus - Hangzhou
D503	AL7409	30755	01.08.72	25.07.88	Stagecoach Malawi - 1007
D504	AD7410	30754	01.08.72	25.07.88	Speedybus - Hangzhou
D505	AD7411	30657	01.08.72	25.07.88	Speedybus - Hangzhou
D506	AD7412	30663	01.08.72	25.07.88	Speedybus - Hangzhou
D507	AD7413	30658	01.08.72	25.07.88	Stagecoach Malawi - 1023
D508	AD7414	30751	01.08.72	25.07.88	Stagecoach Malawi - 1006
D509	AD7415	30671	01.08.72	25.07.88	Stagecoach Malawi - 1001
D510	AD7416	30635	01.08.72	25.07.88	Stagecoach Malawi - 1020
D511	AD7417	30757	09.08.72	25.07.88	Stagecoach Malawi - 1008
D512	AD7418	30752	09.08.72	25.07.88	Stagecoach Malawi - 1010
D513	AD7419	30750	09.08.72	25.07.88	Speedybus - Chungchun
D514	AD7420	30667	09.08.72	25.07.88	Speedybus - Guilin
D515	AD7421	30672	09.08.72	25.07.88	Speedybus - Hangzhou
D516	AD7422	30674	09.08.72	25.07.88	Speedybus - Hangzhou
D517	AD7423	30694	09.08.72	25.07.88	
D518	AD7424	30690	09.08.72	25.07.88	
D519	AD7425	30764	09.08.72	25.07.88	
D520	AD7426	30762	09.08.72	25.07.88	
D521	AD7427	30729	02.10.72	22.09.88	Stagecoach Malawi - 1004
D522	AD7428	30759	02.10.72	22.09.88	Stagecoach Malawi - 1009
D523	AD7429	30758	02.10.72	13.05.87	
D524	AD7430	30763	02.10.72	27.09.88	Stagecoach Malawi - 1018
D525	AD7431	30734	02.10.72	27.09.88	Stagecoach Malawi - 1005
D526	AD7432	30736	02.10.72	27.09.88	Speedybus - Wuzhou
D527	AD7433	30712	02.10.72	27.09.88	Stagecoach Malawi - 1003
D528	AD7434	30732	02.10.72	27.09.88	Speedybus - Chungchun
D529	AD7435	30701	02.10.72	27.09.88	Speedybus - Wuzhou
D530	AD7436	30714	02.10.72	27.09.88	Stagecoach Malawi - 1015
D531	AD7437	30718	02.11.72	19.11.88	Stagecoach Malawi - 1016
D532	AD7438	30720	02.11.72	19.11.88	Stagecoach Malawi - 1012
D533	AD7439	30696	02.11.72	19.11.88	Stagecoach Malawi - 10129
D534	AD7440	30695	02.11.72	19.11.88	Stagecoach Malawi - 1011
D535	AD7441	30699	02.11.72	14.12.87	
D536	AD7442	30703	02.11.72	Temp 01.89	Reinstated -13.07.90
D537	AD7443	30693	02.11.72	19.11.88	Stagecoach Malawi - 1024
D538	AD7444	30760	02.11.72	19.11.88	Stagecoach Malawi - 1002
D539	AD7445	30735	02.11.72	19.11.88	Stagecoach Malawi - 1035
D540	AD7446	30711	02.11.72	19.11.88	Stagecoach Malawi - 1014
D541	AD7447	30716	03.11.72	19.11.88	Stagecoach Malawi - 1037
D542	AD7448	30704	03.11.72	17.07.87	
D543	AD7449	30723	03.11.72	19.11.88	Stagecoach Malawi - 1046
D544	AD7450	30728	03.11.72	19.11.88	Stagecoach Malawi - 1017
D545	AD7451	30726	10.11.72	14.12.88	
D546	AD7452	30727	10.11.72	Temp	Reinstated - 09.07.90
D547	AD7453	30707	10.11.72	19.11.88	Stagecoach Malawi - 1013
D548	AD7454	30697	10.11.72	19.11.88	Stagecoach Malawi - 1036
D549	AD7455	30717	10.11.72	Temp	Reinstated - 13.07.90
D550	AD7456	30744	10.11.72	Temp	Reinstated - 09.07.90

> **NOTE:** Buses sold to Malawi were purchased by 'United Transport Malawi Ltd' - part of the 'Stagecoach' empire. Shipped to Durban, RSA, then driven overland.

KMB class: Daimler(e) (contd) - METAL SECTIONS BODIES
Reinstated Buses:

Following either 'Withdrawal' (dates as above) or 'Temporary suspension of licence' (dates NOT shown above), the following Daimler(e)'s were reinstated for service to cover vehicle shortages. Final withdrawal dates shown below.

Fleet number	Regn. number	Date reinstated	Second withdrawal	Second Disposal.
D409	AD7315	30.08.90	?	To open-top 02.04.92 - see below
D417	AD7323	23.07.90	?	?
D418	AD7324	03.08.90	?	Restored by KMB (? awaiting Scrapping 12/94 ?)
D430	AD7336	13.07.90	17.09.91	Special Purpose Vehicle (Carrier for miniature buses).
D536	AD7442	17.07.90	?	To open-top 10.07.92 - see below
D546	AD7452	13.07.90	?	To open-top ? .92 - see below
D549	AD7455	23.07.90	?	To open-top 02.02.93 - see below
D550	AD7456	09.07.90	?	To open-top 27.05.92 - see below

> **NOTES:** 1) Certain buses above were **CONVERTED TO OPEN-TOP** for tourist and promotional work.
> 2) All appeared in public on trade plates on a number of occasions prior to being re-licensed on the date shown above.
> 3) Fleet numbers not carried as open-top buses.

DAIMLER CVG6LX-30 - 31ft long, front stairs.
KMB class: Daimler(f) METAL SECTIONS BODIES

Fleet number	Regn. number	Chassis number	Date first registered	Withdrawn	Disposal
D551	AD7486	30489	19.12.69	12.11.86	
D552	AD7487	30481	19.12.69	12.11.86	
D553	AD7488	30482	19.12.69	12.11.86	
D554	AD7489	30483	19.12.69	12.11.86	
D555	AD7490	30484	19.12.69	12.11.86	
D556	AD7491	30485	19.12.69	16.10.82	
D557	AD7492	30486	19.12.69	12.11.86	
D558	AD7493	30487	19.12.69	12.11.86	
D559	AD7494	30488	19.12.69	12.11.86	
D560	AD7495	30490	19.12.69	12.11.86	
D561	AD7496	30491	29.12.69	12.11.86	
D562	AD7497	30492	29.12.69	12.11.86	
D563	AD7498	30493	29.12.69	12.11.86	
D564	AD7499	30494	29.12.69	12.11.86	
D565	AR7500	30495	29.12.69	12.11.86	
D566	AR7501	30496	29.12.69	12.11.86	
D567	AR7502	30497	29.12.69	12.11.86	
D568	AR7503	30498	29.12.69	12.11.86	
D569	AR7504	30499	29.12.69	12.11.86	
D570	AR7505	30502	29.12.69	12.11.86	
D571	AR7506	30489	12.01.70	18.12.86	
D572	AR7507	30500	12.01.70	18.12.86	
D573	AR7508	30501	12.01.70	18.12.86	
D574	AR7509	30503	12.01.70	11.03.86	
D575	AR7510	30504	12.01.70	20.10.87	
D576	AR7511	30505	12.01.70	07.10.87	
D577	AR7512	30506	12.01.70	14.12.87	
D578	AR7513	30507	12.01.70	03.11.87	
D579	AR7514	30508	12.01.70	21.01.88	
D580	AR7515	30509	12.01.70	11.03.86	
D581	AR7516	30510	27.01.70	18.12.86	
D582	AR7517	30511	27.01.70	10.02.88	
D583	AR7518	30512	27.01.70	20.10.87	
D584	AR7519	30513	27.01.70	21.09.87	
D585	AR7220	30514	27.01.70	18.12.86	
D586	AR7521	30515	27.01.70	18.07.86	
D587	AR7522	30516	27.01.70	20.10.87	
D588	AR7523	30517	27.01.70	10.02.88	
D589	AR7524	30518	27.01.70	10.02.88	
D590	AR7525	30519	27.01.70	20.10.87	
D591	AR7526	30520	03.02.70	23.12.87	
D592	AR7527	30521	03.02.70	27.01.87	
D593	AR7528	30522	03.02.70	03.11.87	
D594	AR7529	30523	03.02.70	11.03.86	
D595	AR7530	30524	03.02.70	10.02.88	
D596	AR7531	30525	03.02.70	10.02.88	
D597	AR7532	30526	03.02.70	04.02.05	
D598	AR7533	30527	03.02.70	27.01.87	
D599	AR7534	30528	03.02.70	27.01.87	
D600	AR7535	30529	03.02.70	27.01.87	
D601	AR7536	30580	28.12.70	03.11.87	
D602	AR7537	30588	28.12.70	20.10.87	
D603	AR7538	30586	28.12.70	03.11.87	
D604	AR7539	30573	28.12.70	03.11.87	
D605	AR7540	30565	28.12.70	07.10.87	
D606	AR7541	30569	18.01.71	21.09.87	
D607	AR7542	30561	18.01.71	03.11.87	
D608	AR7543	30564	18.01.71	20.10.87	
D609	AR7544	30560	18.01.71	20.10.87	
D610	AR7545	30602	22.01.71	03.11.87	
D611	AR7546	30603	22.01.71	07.10.87	
D612	AR7547	30601	22.01.71	02.09.87	
D613	AR7548	30604	22.01.71	30.01.86	
D614	AR7549	30596	22.01.71	20.10.87	
D615	AR7550	30599	22.01.71	20.10.87	
D616	AR7551	30597	22.01.71	19.03.87	
D617	AR7552	30593	25.01.71	20.10.87	
D618	AR7553	30594	25.01.71	20.10.87	
D619	AR7554	30606	25.01.71	08.10.85	
D620	AR7555	30587	25.01.71	07.10.87	
D621	AR7556	30598	25.01.71	20.10.87	
D622	AR7557	30581	03.02.71	27.01.87	
D623	AR7558	30568	03.02.71	21.08.87	
D624	AR7559	30575	03.02.71	23.05.87	
D625	AR7560	30562	07.02.71	03.11.87	
D626	AR7561	30595	15.02.71	23.12.87	
D627	AR7562	30592	15.02.71	03.11.87	
D628	AR7563	30577	02.03.71	14.12.87	
D629	AR7564	30600	02.03.71	26.11.87	
D630	AR7565	30584	02.03.71	21.08.87	Speedybus - not used
D631	AR7566	30590	02.03.71	23.03.88	
D632	AR7567	30589	02.03.71	09.01.86	
D633	AR7568	30609	02.03.71	10.02.88	
D634	AR7569	30608	02.03.71	10.02.88	
D635	AR7570	30607	02.01.71	27.08.83	
D636	AR7571	30605	02.03.71	05.06.84	
D637	AR7572	30571	02.03.71	26.11.87	
D638	AR7573	30579	24.03.71	21.08.87	
D639	AR7574	30585	24.03.71	23.03.88	
D640	AR7575	30574	24.03.71	10.02.88	
D641	AR7576	30591	24.03.71	16.07.87	
D642	AR7577	30572	24.03.71	23.03.88	
D643	AR7578	30566	24.03.71	10.07.88	
D644	AR7579	30570	24.03.71	10.02.88	
D645	AR7580	30576	24.03.71	10.02.88	
D646	AR7581	30583	24.03.71	23.03.88	
D647	AR7582	30567	02.04.71	19.03.87	
D648	AR7583	30578	02.04.71	23.03.88	

DAIMLER CVG6LX-30 - 31ft long, (contd.)

KMB sub-class: Daimler(f) METAL SECTIONS BODIES

D649	AR7584	30582	02.04.71	23.03.88	
D650	AR7585	30563	02.04.71	23.04.88	
D651	AR7586	30682	02.04.71	19.12.87	
D652	AR7587	30686	14.12.71	19.12.87	
D653	AR7588	30688	14.12.71	19.12.87	
D654	AR7589	30687	14.12.71	19.12.87	
D655	AR7590	30689	14.12.71	19.12.87	
D656	AR7591	30675	14.12.71	07.01.88	
D657	AR7592	30676	05.01.72	23.12.87	
D658	AR7593	30677	05.01.72	07.01.88	
D659	AR7594	30678	05.01.72	19.11.87	Speedybus - Chongquing 1st PTC**
D660	AR7595	30679	05.01.72	07.01.88	
D661	AR7596	30680	05.01.72	07.01.88	
D662	AR7597	30681	05.01.72	19.11.87	Speedybus - Chendu (PCCC)*
D663	AR7598	30683	05.01.72	07.01.88	
D664	AR7599	30684	05.01.72	23.12.87	
D665	AR7600	30685	05.01.72	07.01.88	

** Chongquing First Public Transport Cpn.

DAIMLER FLEETLINE CRG6 - 33ft long.

KMB sub-class: Fleetline CS. METAL SECTIONS BODIES

Fleet number	Regn. number	Chassis number	Date first registered	Withdrawn	Disposal	Final KMB sub-type
D666	BG5124	68798	26.06.74	18.11.88	Sc	CS
D667	BG5125	68803	26.06.74	07.06.89	Sc	CS
D669	BG5127	68801	26.06.74	07.06.89	Speedybus/Guangzhou No 2 Bus Co	CS
D668	BG5126	68802	26.06.74	13.02.88	Sc	CS
D669	BG5127	68801	26.06.74	07.06.89	Speedybus/Guangzhou No 2 Bus Co	CS
D670	BG5128	68800	26.06.74	28.04.88	Sc	CS
D671	BG5129	68799	26.06.77	10.07.88	Sold to China	CS
D672	BG6398	68808	25.07.74	24.06.87	Sold to China	CS
D673	BG6399	68807	25.07.74	11.08.89	Sc	CS
D674	BG6400	68804	25.07.74	12.07.89	Sc	CS
D675	BG6401	68806	25.07.74	03.03.89	Sc	CS
D676	BG6402	68809	25.07.74	15.05.89	Sc	CS
D677	BG6403	68810	25.07.74	11.08.89	Sc	CS
D678	BG7693	68821	28.08.74	10.01.90	Trg. bus.	CS
D679	BG7694	68819	28.08.74	11.08.89	Sc	CS
D680	BG7695	68818	28.08.74	13.12.89	Trg bus	CS
D681	BG7696	68817	28.08.74	11.08.89	Sc	CS
D682	BG7697	68813	28.08.74	13.12.89	Trg bus	CS
D683	BG7968	68811	03.09.74	19.10.87	Sold to China	CS
D684	BG7969	68815	03.09.74	23.06.88	Sc	CS
D685	BG7970	68816	03.09.74	19.10.87	Sold to China	CS
D686	BG7971	68820	03.09.74	01.09.89	Sc	CS
D687	BG8761	68822	20.09.74	19.09.89	Stagecoach Malawi - No. ?	CS
D688	BG8762	68814	20.09.74	19.10.87	Sold to China	CS
D689	BG8763	68805	20.09.74	19.10.87	Sold to China	CS
D690	BG8764	68812	20.09.74	01.09.89	Sc	CS
D691	BG9300	68826	04.10.74	30.10.87	Sold to China	CS
D692	BG9301	68824	04.10.74	22.10.87	Sold to China	CS
D693	BG9302	68825	01.10.74	30.10.87	Sold to China	CS
D694	BH1293	68823	21.11.74	22.10.87	Sold to China	CS
D695	BH2596	68828	06.01.75	30.07.80	Sc	CS
D696	BH2597	68827	06.01.75	19.09.89	Sc	CS
D697	BH2598	68832	06.01.75	04.12.87	Sold to China	CS
D698	BH2599	68831	06.01.75	30.12.87	Sold to China	CS
D699	BH2600	68829	06.01.75	30.12.87	Sold to China	CS
D700	BH3529	68833	03.02.75	30.12.87	Sold to China	CS
D701	BH3530	68835	03.02.75	30.12.87	Sold to China	CS
D702	BH3531	68838	03.02.75	07.05.92	Trg bus	CS
D703	BH3532	68840	03.02.75	30.12.87	Sold to China	CS
D704	BH3533	68841	03.02.75	19.08.88	Sc	CS
D705	BH3534	68843	03.02.75	22.02.90	Speedybus/Guangzhou No 2 Bus Co	CS
D706	BH3536	68844	03.02.75	23.04.90	Trg bus	CS
D707	BH3537	68845	03.02.75	23.04.90	Trg bus	CS
D708	BH3538	68847	03.02.75	10.02.89	Sc	CS
D709	BH4289	68837	03.03.75	22.02.90	Sc	CS
D710	BH4290	68836	03.03.75	30.03.90	Sc	CS
D711	BH4291	68830	03.03.75	09.01.89	Sc	CS
D712	BH4292	68834	03.03.75	28.04.88	Sc	CS
D713	BH4586	68851	11.03.75	22.02.90	Sc	CS
D714	BH4587	68842	11.03.75	30.03.90	Sc	CS
D715	BH4588	68852	11.03.75	02.02.88	Sc	CS
D716	BH4589	68839	11.03.75	14.02.92	Trg bus	CS
D717	BH4590	68848	11.03.75	28.12.90	Sc	CS
D718	BH4792	68853	18.03.75	30.03.90	Sc	CS
D719	BH4793	68856	18.03.75	09.01.89	Sc	CS
D720	BH4794	68857	18.03.75	28.12.90	Sc	CS
D721	BH5601	68865	10.04.75	02.02.88	Sc	CS
D722	BH5605	68864	10.04.75	14.02.92	Trg bus	CS
D723	BH5606	68849	10.04.75	13.02.88	Sc	CS
D724	BH5607	68858	10.04.75	22.03.89	Speedybus - Tianjin	CS
D725	BH5608	68860	10.04.75	03.03.88	Sc	CS
D726	BH5848	68862	16.04.75	07.05.92	Trg bus	CS
D727	BH5849	68846	16.04.75	30.03.90	Sc	CS
D728	BH5850	68855	16.04.75	26.11.90	Sc	CS
D729	BH5851	68861	16.04.75	26.11.90	Sc	CS
D730	BH6483	68859	01.05.75	24.04.90	Sc	CS
D731	BH6631	68850	05.05.75	24.04.90	Sc	CS
D732	BH6632	68854	05.05.75	22.04.88	Sc	CS
D733	BH6634	68863	05.05.75	30.05.92	Sc	CS
D734	BH6635	68866	05.05.75	28.04.88	Sc	CS
D735	BH6636	68867	05.05.75	29.05.90	Sc	CS
D736	BH6637	68868	05.05.75	13.05.92	Trg bus	CS
D737	BH6638	68871	05.05.75	22.04.88	Sc	CS
D738	BH6639	68873	05.05.75	15.11.85	Sc	CS
D739	BH7627	68880	02.06.75	29.10.89	Trg bus	CS
D740	BH7628	68875	02.06.75	29.05.89	Speedybus/Guangzhou No 2 Bus Co	CS

DAIMLER FLEETLINE CRG6 - 33ft long.

KMB sub-class: Fleetline CS. METAL SECTIONS BODIES

Fleet number	Regn. number	Chassis number	Date first registered	Withdrawn	Disposal	Final KMB sub-type
D741	BH7738	68869	03.06.75	29.05.89	Speedybus - Tianjin	CS
D742	BH7739	68870	03.06.75	29.05.89	Speedybus - Tianjin	CS
D743	BH7740	68877	03.06.75	30.05.92	Sc	CS
D744	BH7741	68882	03.06.75	30.05.92	Sc	CS
D745	BH7742	68884	03.06.75	05.05.88	Sc	CS
D746	BH7743	68885	03.06.75	01.06.89	Trg bus	CS
D747	BH8197	68874	12.06.75	30.05.92	Sc	CS
D748	BH8198	68879	12.06.75	12.07.89	Sc	CS
D749	BH8199	68887	12.06.75	15.05.89	Sc	CS
D750	BH8518	68872	20.06.75	22.04.88	Sc	CS
D751	BH8519	68878	20.06.75	01.06.89	Trg bus	CS
D752	BH8520	68883	20.06.75	01.06.89	Trg bus	CS
D753	BH8521	68886	20.06.75			CS
D754	BH9048	68876	02.07.75	12.07.89	Sc	CS
D755	BH9049	68881	02.07.75	10.05.88	Sc	CS
D756	BH9439	68888	10.07.75	03.03.88	Sc	CS
D757	BH9440	68891	10.04.75	06.02.88	Sc	CS
D758	BH9471	68893	10.04.75	03.03.88	Sc	CS
D759	BJ 687	68890	05.08.75	12.02.90	Trg bus	CS
D760	BJ 688	68892	05.08.75	05.07.88	Sc	CS
D761	BJ 689	68889	05.08.75	01.09.89	Sc	CS
D762	BJ 690	68894	05.08.75	08.07.89	Trg Bus > Speedybus - Dalian	CS
D763	BJ 691	68895	05.08.75	01.09.89	Sc	CS
D764	BJ 692	68896	05.08.75	01.09.89	Sc	CS
D765	BJ 693	68897	05.08.75	01.09.89	Sc	CS
D766	BJ2562	68899	16.09.75	09.92	?	CS
D767	BJ2563	68900	16.09.75	24.05.88	Sc	CS
D768	BJ2564	68901	16.09.75	09.92	?	CS
D769	BJ2565	68903	16.09.75	09.92	?	CS
D770	BJ2566	68902	16.09.75	11.06.92	Trg bus (Private Bus)	CS
D771	BJ2567	68904	16.09.75	28.04.88	Sc	CS
D772	BJ3988	68898	16.10.75	26.08.88	Sc	CS
D773	BJ3989	68906	16.10.75	17.09.91	Sc	CS
D774	BJ3990	68907	16.10.75	17.08.88	Sc	CS
D775	BJ3991	68908	16.10.75	17.08.87	Sc	CS
D776	BJ3992	68909	16.10.75	02.02.87	Sc	CS
D777	BJ3994	68910	16.10.75	18.09.86	Sc	CS
D778	BJ3995	68912	16.10.75	19.01.88	Sc	CS
D779	BJ3996	68916	16.10.75	22.04.88	Sc	CS
D780	BJ3997	68917	16.10.75	18.09.86	Sc	CS
D781	BJ4801	68905	03.11.75	22.09.87	Sc	CS
D782	BJ4802	68914	03.11.75	17.08.88	Sc	CS
D783	BJ4803	68915	03.11.75	15.11.91	Sc	CS
D784	BJ4804	68918	07.11.75		Speedybus - Dalian	CS
D785	BJ4805	68919	03.11.77	28.11.91	Sc	CS
D786	BJ4806	68920	03.11.75	28.12.90	Sc	CS
D787	BJ4807	68921	03.11.75	15.11.91	Sc	CS
D788	BJ4808	68924	03.11.75	28.11.91	Sc	CS
D789	BJ4809	68926	03.11.75	02.93	?	CS
D790	BJ5434	68911	13.11.75			CS
D791	BJ5435	68913	13.11.75			CS
D792	BJ5436	68927	13.11.75	01.93	?	CS
D793	BJ6926	68936	16.12.75			CS
D794	BJ6927	68933	16.12.75			CS
D795	BJ6928	68934	16.12.75	11.05.92	Sc	CS
D796	BJ6929	68935	16.12.75	02.93	?	CS
D797	BJ7631	68923	02.01.76	15.11.91	Sc	CS
D798	BJ7632	68925	02.01.76	12.92	?	CS
D799	BJ7633	68939	02.01.76	06.11.91	Sc	CS
D800	BJ7634	68940	02.01.76	22.09.88	Sc	CS
D801	BJ7635	68941	02.01.76	12.92	?	CS
D802	BJ8527	68928	16.01.76	12.10.88	Sc	CS
D803	BJ8528	68937	16.01.76	22.09.88	Sc	CS
D804	BJ9113	68930	27.01.76	28.12.90	Speedybus/Guangzhou No 2 Bus Co	CS
D805	BJ9114	68932	27.01.76	03.10.88	Sc	CS
D806	BJ9115	68938	27.01.76	17.10.88	Sc	CS
D807	BJ9264	68929	28.01.76	18.01.90	Speedybus 25.01.90	CS
D808	BJ9265	68931	28.01.76	12.92	Speedybus/Urban Traffic Co. Yantai.	CS
D809	BK 962	7600505	01.03.76	22.02.90	Sc	CS
D810	BK 963	7600507	01.03.76			CS
D811	BK 964	7600757	01.03.76	22.02.90	Speedybus/Guangzhou No 2 Bus Co	CS
D812	BK1163	7600504	03.03.76	22.02.90	Sc	CS
D813	BK1164	7600758	03.03.76	09.05.90	Trg bus	CS
D814	BK1165	7600506	03.03.76			CS

Tianjin

Fleetline(A) - PROTOTYPE BRITISH ALUMINIUM CO. BODY.

D815	BK3183	68922	02.04.76	21.12.90	Sc	**(A)**

FleetlineCS (contd.) - METAL SECTIONS BODIES.

D816	BK5278	7600616	10.05.76	31.02.90	Sc	CS
D817	BK5279	7600617	10.05.76			CS
D818	BK5280	7600669	10.05.76	31.05.90	Sc	CS
D819	BK5281	7600670	10.05.76			CS
D820	BK5282	7600671	10.05.76			CS
D821	BK5283	7600672	10.05.76	31.05.91	Sc	CS
D822	BK5284	7601317	10.05.76			CS
D823	BK5286	7601614	10.05.76	26.11.90	Sc	CS
D824	BK5287	7601616	10.05.76			CS
D825	BK5470	7600894	12.05.76			CS
D826	BK5471	7600895	12.05.76	09.05.91	Speedybus/Guangzhou No 2 Bus Co	CS
D827	BK5472	7601316	12.05.76			CS
D828	BK5473	7601486	12.05.76	31.05.91	Sc	CS
D829	BK5474	7601615	12.05.76			CS
D830	BK5475	7601799	12.05.76	26.11.90	Sc	CS
D831	BK7089	7601801	03.06.76	15.05.89	Speedybus/Guangzhou No 2 Bus Co	CS
D832	BK7090	7601789	03.06.76			CS
D833	BK7091	7601617	03.06.76	26.07.90	Sc	CS

LEYLAND FLEEETLINE FE33AGR.

KMB sub-classs: Fleetline CS - *METAL SECTIONS BODIES*

Fleet number	Regn. number	Chassis number	Date first registered	Withdrawn	Disposal	Final KMB sub-type
D834	BK7476	7601797	09.06.76	26.11.90	Solt (to ??)	CS
D835	BK7477	7601800	09.06.76	26.11.90	Sc	CS
D836	BK7478	7602259	09.06.76	28.12.90	Sc	CS
D837	BK7479	7602112	09.06.76			CS
D838	BK9197	7602113	02.07.76	29.05.89	Speedybus/Guangzhou No 2 Bus Co	CS
D839	BK9188	7602258	02.07.76	29.05.89	Sc	CS
D840	BK9199	7602376	02.07.76	07.06.89	Speedybus/Guangzhou No 2 Bus Co	CS
D841	BK9200	7602378	02.07.76	07.06.89	Sc	CS
D842	BK9701	7602379	02.07.76	12.07.89	Sc	CS
D843	BK9202	7602523	02.07.76	07.06.89	Speedybus/Guangzhou No 2 Bus Co	CS
D844	BK9768	7603244	09.07.76	12.07.89	Sc	CS
D845	BK9769	7603210	09.07.76	07.06.89	Sc	CS
D846	BK9770	7602924	09.07,76			CS
D847	BL1697	7603209	03.08.76	11.08.89	Sc	CS
D848	BL1698	7603234	03.08.76	21.07.89	Sc	CS
D849	BL1699	7603235	03.08.76			CS
D850	BL1700	7603489	03.08.76	29.06.89	Speedybus/Guangzhou No 2 Bus Co	CS
D851	BL1821	7603522	04.08.76			CS
D852	BL2382	7603503	11.08.76			CS
D853	BL2383	7603520	11.08.76			CS
D854	BL2384	7603521	11.08.76	03.08.89	Sc	CS
D855	BL2385	7603540	11.08.76	11.08.89	Sc	CS
D856	BL3690	7603539	01.09.76			CS
D857	BL3691	7603769	01.09.76	11.08.89	Sc	CS
D858	BL3692	7603790	01.09.76	28.09.89	Speedybus - Tianjin	CS
D859	BL3693	7604069	01.09.76	09.92	?	CS
D860	BL3694	7604084	01.09.76			CS
D861	BL5488	7602377	22.09.76	18.10.89	Sc	CS
D862	BL5489	7603502	22.09.76	18.10.89	Sc	CS
D863	BL5490	7607051	22.09.76			CS
D864	BL5491	7604141	22.09.76	10.08.88	Sc	CS
D865	BL6224	7602526	01.10.76		Speedybus/Urban Traffic Co. Yantai.	CS
D866	BL6225	7602923	01.10.76	02.93	?	CS
D867	BL6227	7607768	01.10.76			CS
D868	BL6228	7603789	01.10.76			CS
D869	BL6229	7604018	01.10.76			CS
D870	BL7446	7602524	14.10.76			CS
D871	BL7448	7603621	14.10.76			CS
D872	BL7449	7604017	14.10.76			CS
D873	BL7450	7604142	14.10.76			CS
D874	BL8955	7602525	02.11.76	07.11.89	Sc	CS
D875	BL8956	7603810	02.11.76	07.11.89	Sc	CS
D876	BL8957	7604030	02.11.76			CS
D877	BL8958	7604187	02.11.76	07.11.89	Sc	CS
D878	BL8959	7604190	02.11.76			CS
D879	BM1268	7604675	23.11.76	18.11.89	Sc	CS
D880	BM1269	7604676	23.11.76	18.11.89	Sc	CS
D881	BM1270	7604732	23.11.76	18.11.89	Sc	CS
D882	BM2458	7604188	06.12.76			CS
D883	BM2459	7604029	06.12.76			CS
D884	BM2460	7604050	06.12.76			CS
D885	BM2461	7604083	06.12.76	01.93	?	CS
D886	BM2462	7604731	06.12.76			CS
D887	BM2463	7604864	06.12.76			CS

Fleetline(B) - *KMB BODY*

D888	BM2684	7601485	08.12.76	18.01.90	**Sc**	**(B)**

FleetlineCS (contd.) - *METAL SECTIONS BODIES.*

D889	BM2907	7605292	10.12.76			CS
D890	BM2908	7605270	10.12.76	17.10.88	Sc	CS
D891	BM2909	7605370	10.12.76			CS
D892	BM4484	7603519	03.01.77			CS
D893	BM4485	7604070	03.01.77			CS
D894	BM4486	7605582	03.01.77	03.05.86	Trg bus	CS
D895	BM4487	7605591	03.01.77	04.03.91		CS
D896	BM4489	7605269	03.01.77			CS
D897	BM4490	7605581	03.01.77	28.12.90	Sc	CS
D898	BM5539	7605592	13.01.77			CS
D899	BM5540	7605593	13.01.77			CS
D900	BM5541	7606466	13.01.77	01.93	?	CS
D901	BM7053	7606279	01.02.77			CS
D902	BM7054	7606557	01.02.77	18.01.91	Sc	CS
D903	BM7055	7606298	01.02.77	21.12.90	Sc	CS
D904	BM7056	7606440	01.02.77			CS
D905	BM7057	7606595	01.02.77			CS
D906	BM7058	7606558	01.02.77	21.12.90	Sc	CS
D907	BM7691	7606367	08.02.77	28.12.90	Sc	CS
D908	BM7739	7606379	09.02.77			CS
D909	BM7740	7606956	09.02.77			CS
D910	BM8688	7606909	16.02.77			CS
D911	BM8689	7606729	16.02.77			CS
D912	BM8690	7606728	16.02.77	18.01.91	Sc	CS
D913	BM8691	7606705	16.02.77			CS
D914	BM8692	7607343	16.02.77	21.12.90	Sc	CS
D915	BM9960	7606887	10.03.77			CS
D916	BN1403	7606704	21.03.77	31.05.91	Sold (to ??)	CS
D917	BN1404	7606727	21.03.77			CS
D918	BN2419	7607153	01.04.77			CS
D919	BN2420	7606465	01.04.77	28.12.90	Speedybus/Guangzhou No 2 Bus Co	CS
D920	BN2421	7607335	01.04.77	23.01.91	Sc	CS
D921	BN3007	7606888	07.04.77			CS
D922	BN3008	7607128	07.04.77			CS
D923	BN3009	7607129	07.04.77			CS
D924	BN3010	7607154	07.04.77			CS
D925	BN3011	7607385	07.04.77	17.05.91	Sc	CS
D926	BN5071	7607334	03.05.77			CS
D927	BN5072	7607558	03.05.77	19.02.91	Sc	CS
D928	BN5073	7607559	03.05.77			CS
D929	BN5074	7607582	03.05.77	06.03.90	Sc	CS
D930	BN5075	7607583	03.05.77			CS

Fleetline(B) - *KMB BODY*

D931	BN5707	7605641	09.05.77			**(B)**

FleetlineCS (contd.) - *METAL SECTIONS BODIES.*

D932	BN5877	7608466	10.05.77	17.05.91	Speedybus/Guangzhou No 2 Bus Co	CS
D933	BN6643	7606910	18.05.77	30.03.90	Speedybus/Guangzhou No 2 Bus Co	CS
D934	BN6645	7607522	18.05.77			CS

LEYLAND FLEEETLINE FE33AGR.

KMB sub-classs: Fleetline CS *METAL SECTIONS BODIES*

Fleet number	Regn. number	Chassis number	Date first registered	Withdrawn	Disposal	Final KMB sub-type
D935	BN6646	7608444	18.05.77			CS
D936	BN6647	7608467	18.05.77	26.02.90	Sc	CS
D937	BN7692	7608489	01.06.77	04.06.91	Sc	CS
D938	BN7693	7700201	01.06.77	25.07.91	Speedybus/Guangzhou No 2 Bus Co	CS
D939	BN8321	7607344	06.06.77	02.07.91	Sc	CS
D940	BN8322	7700221	06.06.77			CS
D941	BN8323	7700558	06.06.77	15.05.89	Speedybus - Tianjin	CS
D942	BN9168	7608449	14.06.77	05.06.89	Sc	CS
D943	BN9169	7700537	14.06.77	05.06.89	Sc	CS
D944	BN9170	7700538	14.06.77	10.06.91	Sc	CS
D945	BN9171	7700557	20.06.77			CS
D946	BN9638	7700556	20.06.77	15.05.91	Speedybus/Guangzhou No 2 Bus Co	CS
D947	BP1802	7701514	07.07.77	25.07.91	Sc	CS
D948	BP1803	7700827	07.07.77	26.06.89	Speedybus - Wuzhou	CS
D949	BP1804	7700968	07.07.77	29.06.89	Speedybus - Wuzhou	CS
D950	BP1805	7707083	07.07.77	04.06.91	Sc	CS
D951	BP2384	7700826	13.07.77	06.03.90	Sc	CS
D952	BP2385	7701353	13.07.77	17.07.91	Sc	CS
D953	BP3027	7700825	19.07.77	21.07.89	Sc	CS
D954	BP3028	7701219	19.07.77	02.07.91	Sc	CS
D955	BP4633	7700823	04.08.77	09.09.91	Trg bus	CS
D956	BP4891	7700924	08.08.77	17.07.91	Sc	CS
D957	BP5148	7701220	10.08.77	17.07.91	Sc	CS
D958	BP5425	7701354	12.08.77	28.08.91	Speedybus - Dalian	CS
D959	BP5426	7701221	12.08.77	13.09.89	Speedybus - Tianjin	CS
D960	BP5872	7701484	12.08.77	21.07.89	Sc	CS
D961	BP5873	7701517	12.08.77	25.07.91	Speedybus - Dalian	CS
D962	BP6329	7701382	22.08.77	17.07.91	Sc	CS
D965	BP8772	7701550	14.09.77	09.09.91	Trg bus	CS
D966	BP8773	7701515	14.09.77	09.09.91	Trg bus	CS
D967	BP8774	7701290	17.09.77			CS

KMB sub-class: Fleetline(C) *BRITISH ALUMINIUM CO. BODIES*

D963	BP6496	7700824	23.08.77			(C)
D964	BP6903	7700969	26.08.77	17.07.91	Sc	(C)
D968	BR3691	7701886	27.10.77	28.08.91	Speedybus - Dalian	(C)
D969	BR3692	7701770	27.10.77	22.08.91	Speedybus/Guangzhou Trolleybus Co.	(C)
D970	BR4119	7701911	01.11.77	01.09.89	Sc	(C)
D971	BR4120	7701742	01.11.77	22.08.91	Sc	(C)
D972	BR4473	7701702	03.11.77	28.08.91	Speedybus/Guangzhou Trolleybus Co.	(C)
D973	BR4475	7701910	03.11.77	28.08.91	Sc	(C)
D974	BR6227	7702077	17.11.77	31.08.91	Speedybus/Urban Traffic Co. Yantai.	(C)
D975	BR7473	7702148	01.12.77			(C)
D976	BR7742	7702078	02.12.77	28.08.91	Speedybus/Guangzhou Trolleybus Co.	(C)
D977	BR8323	7702138	08.12.77	11.10.89	Sc	(C)
D978	BR8324	7702115	08.12.77	11.10.89	Sc	(C)
D979	BS 479	7702045	28.12.77	28.08.91	Sc	(C)
D980	BS1066	7702712	03.01.78	31.08.91	Speedybus/Guangzhou Trolleybus Co.	(C)
D981	BS1067	7702139	03.01.78	31.08.91	Speedybus/Guangzhou Trolleybus Co.	(C)
D982	BS2395	7702116	10.01.78	21.10.91	Sc	(C)
D983	BS2396	7702748	10.01.78			(C)
D984	BS4209	7702763	12.01.78	31.08.91	Sc	(C)
D985	BS5280	7702711	02.02.78	28.12.91	Sc	(C)
D986	BS5281	7702684	02.02.78	28.12.91	Sc	(C)
D987	BS5400	7702683	03.02.78	08.10.91	Sc	(C)
D988	BS5401	7702764	03.02.78	08.10.91	Sc	(C)
D989	BT 207	7702864	22.07.78	20.02.92	Sc	(C)
D990	BT 208	7702810	22.03.78	28.12.91	Sc	(C)
D991	BT 209	7703159	22.03.78	28.12.91	Sc	(C)
D992	BT 533	7702749	23.03.78			(C)
D993	BT 534	7703135	23.03.78	18.12.91	Sc	(C)
D994	BT3269	7702811	17.04.78			(C)
D995	BT3270	7702981	17.04.78		Speedybus - Dalian	(C)
D996	BT4341	7702904	26.04.78			(C)
D997	BT4342	7702903	26.04.78	19.12.91	Sc	(C)
D998	BT4344	7702957	26.04.78	16.04.92	Sc	(C)
D999	BT6973	7703728	18.05.78			(C)
D1000	BT6974	7703904	18.05.78	13.05.92	Sc	(C)
D1001	BT7274	7703727	22.05.78		Speedybus - 389 - ?	(C)
D1002	BT7275	7707807	22.05.78			(C)
D1003	BT7276	7704161	22.05.78	13.05.92	Speedybus/Urban Traffic Co. Yantai.	(C)
D1004	BT8336	7703969	01.06.78	29.06.92	Sc	(C)
D1005	BU 571	7703970	14.06.78	02.02.88	Sc	(C)
D1006	BU2787	7800030	03.07.78			(C)

Rebodied as Fleetline'CS' - *METAL SECTIONS BODY (23.04.82)*

D1007	BU3648	7704160	06.07.78			(C)

Fleetline(C) (contd.) - BRITISH ALUMINIUM CO. BODIES

D1008	BU3649	7704745	06.07.78	06.07.92	Speedybus/Urban Traffic Co. Yantai.	(C)
D1009	BU4894	7800423	14.07.78	06.07.92	Speedybus/Urban Traffic Co. Yantai.	(C)
D1010	BU4895	7800474	14.07.78	06.07.92	Speedybus/Urban Traffic Co. Yantai.	(C)
D1011	BU4896	7800797	17.07.78	29.06.92	Sc	(C)
D1012	BU5579	7800438	19.07.78	06.07.92	Speedybus/Urban Traffic Co. Yantai.	(C)
D1013	BU5774	7800391	20.07.78	29.06.92	Speedybus/Urban Traffic Co. Yantai.	(C)
D1014	BU6019	7704666	21.07.78	06.07.92	Sc	(C)
D1015	BU6020	7704648	21.07.78	06.07.92	Sc	(C)
D1016	BU7119	7800353	01.08.78			(C)
D1017	BU8677	7702765	11.08.78			(C)
D1018	BU8678	7705233	11.08.78	17.07.92	Sc	(C)
D1019	BU9308	7800511	15.08.78	17.07.92	Sc	(C)
D1020	BU9540	7704623	16.08.78	17.07.92	Sc	(C)
D1021	BU9541	7704696	16.08.78			(C)
D1022	BU9542	7705292	16.08.78			(C)
D1023	BV 307	7704697	21.08.78	06.07.92	Speedybus/Urban Traffic Co. Yantai.	(C)
D1024	BV2642	7703940	04.09.78	23.04.87	Sc	(C)
D1025	BV2643	7704667	04.09.78			(C)
D1026	BV2644	7704744	04.09.78			(C)
D1027	BV2645	7705033	04.09.78			(C)
D1028	BV3257	7800029	07.09.78			(C)
D1029	BV3658	7705293	11.09.78			(C)
D1030	BV3659	7705622	11.09.78			(C)
D1031	BV5003	7800707	21.09.78	25.07.88	Sc	(C)
D1032	BV5004	7800625	21.09.78	24.05.88	Sc	(C)
D1033	BV5006	7800352	21.09.78	11.92	?	(C)
D1034	BV5007	7705675	21.09.78	09.92	?	(C)
D1035	BV5729	7704783	27.09.78	09.92	?	(C)
D1036	BV5730	7800475	27.09.78	09.92	?	(C)
D1037	BV5731	7800496	27.09.78	10.92	?	(C)
D1038	BV5732	7800594	27.09.78	09.92	?	(C)

LEYLAND FLEEETLINE FE33AGR (ccntd.)

KMB sub-class: Fleetline(C) *BRITISH ALUMINIUM CO. BODIES*

Fleet number	Regn. number	Chassis number	Date first registered	Withdrawn	Disposal	Final KMB sub-type
D1039	BV6145	7800796	02.10.78	10.92	?	(C)
D1040	BV6146	7800494	02.10.78	10.92	?	(C)
D1041	BV6147	7800275	02.10.78	?	Speedybus - Dalian	(C)
D1042	BV6148	7800439	02.10.78	11.92	?	(C)
D1043	BV6692	7800668	04.10.78	11.92	?	(C)
D1044	BV6693	7800626	04.10.78	11.92	?	(C)
D1045	BV8072	7705621	13.10.78	10.92	?	(C)
D1046	BV8073	7800916	13.10.78	10.92	?	(C)
D1047	BV8255	7800667	16.10.78	11.92	?	(C)
D1048	BV8709	7800706	18.10.78	11.92	?	(C)
D1049	BV8710	7800917	18.10.78	11.92	?	(C)
D1050	BW1005	7800829	01.11.78			(C)
D1051	BW1006	7801788	01.11.78			(C)
D1052	BW1611	7800079	03.11.78			(C)
D1053	BW1612	7800208	03.11.78	12.92	?	(C)
D1054	BW1613	7805082	03.11.78			(C)
D1055	BW3024	7801387	13.11.78	26.08.88	Sc	(C)
D1056	BW3025	7801417	13.11.78	10.92	?	(C)
D1057	BW3026	7801438	13.11.78	10.92	?	(C)
D1058	BW3027	7800495	13.11.78	11.92	?	(C)
D1059	BW3028	7800828	13.11.78	10.92	?	(C)
D1060	BW3489	7801300	15.11.78	10.92	?	(C)
D1061	BW3490	7805439	15.11.78			(C)
D1062	BW3855	7800971	17.11.78	02.93	Speedybus/Urban Traffic Co. Yantai.	(C)
D1063	BW3856	7800190	17.11.78	10.92	?	(C)
D1064	BW4229	7800250	21.11.78			(C)
D1065	BW4230	7800885	21.11.78	26.06.89	Sc	(C)
D1066	BW4231	7801002	21.11.78			(C)
D1067	BW4347	7800795	22.11.78	02.93	Speedybus/Urban Traffic Co. Yantai.	(C)
D1068	BW4641	7801042	24.11.78	11.92	?	(C)
D1069	BW4642	7800249	24.11.78	12.92	?	(C)
D1070	BW4643	7800057	24.11.78	12.92	?	(C)
D1071	BW4644	7705119	24.11.78	11.92	Speedybus/Urban Traffic Co. Yantai.	(C)
D1072	BW4792	7800354	27.11.78	12.92	?	(C)
D1073	BW5697	7705168	04.12.78	10.92	?	(C)
D1074	BW5698	7800080	04.12.78	12.92	?	(C)
D1075	BW5699	7801440	04.12.78	11.92	Speedybus/Urban Traffic Co. Yantai.	(C)
D1076	BW5700	7801041	04.12.78	25.11.91	Sc	(C)
D1077	BW6029	7801759	06.12.78	11.92	Speedybus/Urban Traffic Co. Yantai.	(C)
D1078	BW6030	7801576	06.12.78	12.92	Speedybus - Dalian	(C)
D1079	BW6031	7800919	06.12.78	12.92	?	(C)
D1080	BW6032	7800056	06.12.78	11.92	Speedybus/Urban Traffic Co. Yantai.	(C)
D1081	BW6432	7801445	08.12.78	12.92	Speedybus/Urban Traffic Co. Yantai.	(C)
D1082	BW6825	7801327	12.12.78	12.92	Speedybus/Urban Traffic Co. Yantai.	(C)
D1083	BW7404	7705106	15.12.78			(C)
D1084	BW7698	7800830	19.12.78	12.92	?	(C)
D1085	BW7699	7801465	19.12.78	23.05.87	Sc	(C)
D1086	BW8346	7800031	27.12.78			(C)
D1087	BW8347	7801628	27.12.78	12.92	Speedybus - Dalian	(C)
D1088	BW8660	7705562	02.01.79	01.93	?	(C)
D1089	BW8661	7800032	02.01.79	12.92	Speedybus/Guangzhou Trolleybus Co.	(C)
D1090	BW9831	7705273	08.01.79			(C)
D1091	BW9832	7800627	08.01.79	12.92	Speedybus - Dalian	(C)
D1092	BW9833	7801389	08.01.79	?	Speedybus/Guangzhou Trolleybus Co.	(C)
D1093	BX 648	7800351	10.01.79	02.93	Speedybus - Dalian	(C)
D1094	BX 649	7801599	10.01.79		Speedybus - Dalian	(C)
D1095	BX1239	7704742	12.01.79			(C)
D1096	BX1240	7705623	12.01.79			(C)
D1097	BX1241	7801459	12.01.79			(C)
D1098	BX1713	7800276	16.01.79	02.93	Speedybus - Dalian	(C)
D1099	BX1714	7800575	16.01.79	02.93	?	(C)
D1100	BX1715	7801466	16.01.79		Speedybus - Dalian	(C)
D1101	BX1716	7801664	16.01.79	03.03.89	Speedybus 06.05.89	(C)
D1102	BX2347	7800595	19.01.79			(C)
D1103	BX2348	7801439	19.01.79	01.93	?	(C)
D1104	BX2349	7801554	19.01.79	12.92	Speedybus/Guangzhou Trolleybus Co.	(C)
D1105	BX2538	7800972	22.01.79		Speedybus - Dalian	(C)
D1106	BX2539	7801447	22.01.79	12.92	Speedybus/Guangzhou Trolleybus Co.	(C)
D1107	BX2540	7801573	22.01.79	12.92	?	(C)
D1108	BX2812	7800730	23.01.79			(C)
D1109	BX2813	7800759	23.01.79	01.93	Speedybus - Dalian	(C)
D1110	BX2814	7800760	23.01.79	01.93	?	(C)
D1111	BX2815	7800761	23.01.79			(C)
D1112	BX2816	7801629	23.01.79	01.93	Speedybus/Guangzhou Trolleybus Co.	(C)
D1113	BX3117	7800866	27.01.79			(C)
D1114	BX3115	7800886	24.01.79	01.93	?	(C)
D1115	BX3116	7801069	24.01.79	06.12.91	Sc	(C)

AEC REGENT MKV - Type 2D2RA

KMB sub-class: AEC(a)/30 *BRITISH ALUMINIUM CO. BODIES*

Fleet number	Regn. number	Chassis number	Date first registered	Body rebuilt	Withdrawn	Disposal
A 1	AD4789	-1174	21.01.63	01.82	21.09.84	
A 2	AD4790	-1166	21.01.63	05.81	25.10.84	
A 3	AD4791	-1175	21.01.63	09.81	14.11.84	
A 4	AD4792	-1161	21.01.63	09.81	27.12.84	
A 5	AD4793	-1165	21.01.63	10.81	27.12.84	
A 6	AD4794	-1170	21.01.63	03.82	18.02.85	
A 7	AD4795	-1168	21.01.63	01.82	23.01.85	
A 8	AD4796	-1171	26.02.63	01.82	23.01.85	
A 9	AD4697	-1169	26.02.63	10.81	27.02.84	
A10	AD4798	-1173	26.02.63	06.81	29.06.84	
A11	AD4799	-1172	26.02.63	01.82	11.02.85	
A12	AD4800	-1162	26.02.63	01.82	11.02.85	
A13	AD4801	-1184	28.03.63	03.82	12.02.85	
A14	AD7802	-1183	28.03.63	05.81	29.06.84	
A15	AD4803	-1182	28.03.63	03.82	18.02.85	
A16	AD4804	-1167	28.03.63	06.82	22.05.85	
A17	AD4805	-1190	28.03.63	03.82	18.02.85	
A18	AD4806	-1164	28.03.63	03.82	04.03.85	
A19	AD4807	-1188	06.05.63	04.82	25.04.85	Preserved by KMB

AEC REGENT MKV - Type 2D2RA (contd.)

KMB sub-class AEC(a)/30 *BRITISH ALUMINIUM CO. BODIES*

Fleet number	Regn. number	Chassis number	Date first registered	Body rebuilt	Withdrawn	Disposal
A20	AD4808	-1185	06.05.63	03.82	25.04.85	
A21	AD4809	-1186	06.05.63	06.81	27.06.84	
A22	AD4810	-1189	06.05.63	05.82	22.05.85	
A23	AD4811	-1177	06.05.63	04.81	17.04.84	
A24	AD4812	-1187	06.05.63	05.82	22.05.85	
A25	AD4813	-1178	12.06.63	04.81	31.07.84	
A26	AD4814	-1179	12.06.63	05.82	22.05.85	
A27	AD4815	-1180	12.06.63	05.81	31.07.84	
A28	AD4816	-1176	12.06.63	05.82	22.05 85	
A29	AD4817	-1163	12.06.63	06.81	21.09.84	
A30	AD4818	-1181	12.06.63	06.81	01.10.84	

KMB sub-class: AEC(a)/40 - A31-70

METAL SECTIONS BODIES

Fleet number	Regn. number	Chassis number	Date first registered	Body rebuilt	Withdrawn	Disposal
A31	AD4823	-1473	06.02.64	11.80	13.03.84	
A32	AD4824	-1475	06.02.64	11.80	25.01.83	
A33	AD4825	-1476	06.02.64	11.80	11.01.87	
A34	AD4826	-1479	06.02.64	11.80	17.03.86	
A35	AD4827	-1488	06.02.64	10.80	30.01.86	
A36	AD4828	-1481	26.02.64	10.80	04.02.86	
A37	AD4829	-1480	26.02.64	12.80	21.01.85	
A38	AD4830	-1471	26.02.64	12.80	21.01.85	
A39	AD4831	-1474	26.02.64	02.81	28.01.85	
A40	AD4832	-1472	26.02.64	01.81	28.01.85	
A41	AD4833	-1483	03.04.64	12.80	11.03.86	
A42	AD4834	-1485	03.04.64	01.81	17.03.86	
A43	AD4835	-1489	03.04.64	01.81	12.02.85	
A44	AD4836	-1478	03.04.64	08.78	28.02.85	
A45	AD4837	-1486	03.04.64	01.81	17.03.86	
A46	AD4838	-1510	23.04.64	01.81	17.03.86	
A47	AD4839	-1508	23.04.64	01.81	28.01.85	
A48	AD4840	-1506	23.04.64	06.75	17.03.86	
A49	AD4841	-1509	23.04.64	02.81	18.03.86	
A50	AD4842	-1484	23.04.64	11.80	13.11.84	
A51	AD4843	-1495	29.05.64	01.81	26.04.86	
A52	AD4844	-1505	29.05.64	03.81	18.03.86	
A53	AD4845	-1477	29.05.64	02.81	26.04.86	
A54	AD4846	-1493	29.05.64	01.81	25.07.83	
A55	AD4847	-1507	29.05.64	01.81	12.11.84	
A56	AD4848	-1487	29.05.64	01.81	14.03.85	
A57	AD4849	-1504	29.05.64	02.81	25.01.83	
A58	AD4850	-1482	29.05.64	02.81	01.04.86	
A59	AD4851	-1490	09.06.64	02.81	01.04.86	
A60	AD4852	-1503	09.06.64	03.81	19.08.86	
A61	AD4853	-1494	28.09.64	10.80	25.10.83	
A62	AD4854	-1502	28.09.64	11.80	25.10.84	
A63	AD4855	-1497	28.09.64	10.80	01.10.83	
A64	AD4856	-1491	28.09.64	10.80	12.11.84	
A65	AD4857	-1500	28.09.64	10.80	09.10.84	
A66	AD4858	-1498	16.10.64	10.80	01.10.84	
A67	AD4859	-1492	16.10.64	10.80	09.10.84	
A68	AD4860	-1499	16.10.64	11.80	25.10.84	to Australia
A69	AD4861	-1501	16.10.64	11.80	31.10.84	
A70	AD4862	-1496	16.10.64	11.80	25.10.84	

KMB sub-class: AEC(b)

METAL SECTIONS BODIES

Fleet number	Regn. number	Chassis number	Date first registered	Body rebuilt	Withdrawn	Disposal
A71	AD4863	-1628	06.01.65	10.81	10.11.86	
A72	AD4864	-1632	09.01.65	11.80	27.12.84	
A73	AD4865	-1633	09.01.65	10.81	10.12.86	
A74	AD4866	-1612	09.01.65	10.81	10.12.86	
A75	AD4867	-1637	09.01.65	11.81	29.12.86	
A76	AD4868	-1629	09.01.65	11.81	16.01.86	
A77	AD4869	-1608	09.01.65	11.81	29.12.86	
A78	AD4870	-1620	09.01.65	11.81	16.01.86	
A79	AD4871	-1623	09.01.65	11.81	29.12.86	
A80	AD4872	-1635	09.01.65	11.81	24.02.86	
A81	AD4873	-1621	25.01.65	11.81	27.11.84	
A82	AD4874	-1630	25.01.65	11.81	09.12.8	
A83	AD4875	-1634	25.01.65	11.81	09.12.85	
A84	AD4876	-1638	25.01.65	11.81	29.12.86	
A85	AD4877	-1639	25.01.65	11.81	30.12.86	
A86	AD4878	-1631	16.02.65	01.82	26.04.86	
A87	AD4879	-1636	16.02.65	07.80	30.04.84	
A88	AD4880	-1643	16.02.65	01.82	27.01.87	
A89	AD4881	-1644	16.02.65	01.82	01.04.86	
A90	AD4882	-1647	16.02.65	01.82	31.01.86	
A91	AD4883	-1640	16.02.65	01.82	24.02.86	
A92	AD4884	-1641	16.02.65	01.82	27.02.84	
A93	AD4885	-1642	16.03.65	01.82	27.02.84	
A94	AD4886	-1645	16.03.65	01.82	27.02.84	
A95	AD4887	-1646	16.03.65	02.82	13.03.83	
A96	AD4888	-1610	16.03.65	02.82	27.02.84	
A97	AD4889	-1614	16.03.65	01.82	27.02.84	
A98	AD4890	-1613	16.03.65	02.82	27.02.84	
A99	AD4891	-1615	16.03.65	06.82	27.06.84	
A100	AD4892	-1617	16.03.65	02.82	27.02.84	
A101	AD4893	-1627	14.04.65	06.80	27.06.84	
A102	AD4894	-1624	14.04.65	02.82	13.03.84	
A103	AD4895	-1625	14.04.65	04.82	30.04.84	
A104	AD4896	-1626	14.04.65	07.81	29.10.86	
A105	AD4897	-1622	14.04.65	02.82	29.03.84	
A106	AD4898	-1611	14.04.65	02.82	29.03.84	
A107	AD4899	-1616	14.04.65	02.82	27.02.84	
A108	AD4900	-1618	14.04.65	03.82	29.03.84	
A109	AD4100	-1619	14.04.65	03.82	29.03.84	
A110	AD7101	-1609	14.04.65	04.82	29.03.84	

A.E.C. REGENT MKV Type 2D2RA (contd.)
KMB class: AEC(c) - METAL SECTIONS BODIES

Fleet number	Regn. number	Chassis number	Date first registered	Date rebuilt	Withdrawn	Disposal
A111	AD7102	-1707	25.11.65	08.80	17.08.87	
A112	AD7103	-1705	25.11.65	10.80	16.07.87	
A113	AD7104	-1706	25.11.65	08.80	31.07.84	
A114	AD7105	-1708	25.11.65	11.80	17.08.87	
A115	AD7106	-1709	25.11.65	10.80	16.07.87	
A116	AD7107	-1710	25.11.65	11.80	17.08.87	
A117	AD7108	-1702	25.11.65	10.80	26.08.87	
A118	AD7109	-1700	25.11.65	10.80	21.08.87	
A119	AD7110	-1699	25.11.65	11.80	09.07.84	
A120	AD7111	-1703	25.11.65	11.80	26.09.86	
A121	AD7112	-1718	29.12.65	12.80	21.12.84	
A122	AD7113	-1714	29.12.65	11.80	14.11.86	
A123	AD7114	-1721	29.12.65	12.80	29.10.86	
A124	AD7115	-1701	29.12.65	12.80	31.12.84	
A125	AD7116	-1704	29.12.65	11.80	29.10.86	
A126	AD7117	-1715	29.12.65	11.80	21.09.84	
A127	AD7118	-1720	29.12.65	11.80	21.09.84	
A128	AD7119	-1717	29.12.65	11.80	21.08.87	
A129	AD7120	-1722	29.12.65	12.80	16.07.87	
A130	AD7121	-1716	29.12.65	11.80	21.08.87	
A131	AD7122	-1711	29.12.65	12.80	24.07.87	
A132	AD7123	-1712	29.12.65	11.80	25.10.84	
A133	AD7124	-1713	29.12.65	12.80	14.11.86	
A134	AD7125	-1719	29.12.65	11.80	21.08.87	
A135	AD7126	-1740	29.12.65	11.80	24.07.87	
A136	AD7127	-1739	12.01.66	10.81	30.12.86	
A137	AD7128	-1737	12.01.66	11.80	14.11.86	
A138	AD7129	-1738	12.01.66	11.80	31.12.84	
A139	AD7130	-1736	12.01.66	11.80	24.07.87	
A140	AD7131	-1735	12.01.66	12.80	21.08.86	
A141	AD7132	-1734	19.01.66	12.80	31.12.84	
A142	AD7133	-1742	19.01.66	12.80	31.12.84	
A143	AD7134	-1728	19.01.66	12.80	26.08.87	
A144	AD7135	-1726	19.01.66	10.80	31.12.84	
A145	AD7136	-1743	19.01.66	09.80	30.12.86	
A146	AD7137	-1725	29.01.66	12.80	26.08.87	
A147	AD7138	-1746	29.01.66	12.80	26.08.87	
A148	AD7139	-1724	29.01.66	01.81	17.08.87	
A149	AD7140	-1727	29.01.66	01.81	24.07.87	
A150	AD7141	-1723	29.01.66	10.81	30.12.86	
A151	AD7142	-1744	23.02.66	02.81	28.01.85	
A152	AD7143	-1747	23.02.66	12.81	27.01.87	
A153	AD7144	-1741	23.02.66	12.81	27.01.87	
A154	AD7145	-1745	23.02.66	01.82	27.01.87	
A155	AD7146	-1748	23.02.66	12.81	27.01.87	
A156	AD7147	-1753	23.02.66	02.82	14.02.87	
A157	AD7148	-1752	23.02.66	04.82	02.04.87	
A158	AD7149	-1754	23.02.66	01.82	31.01.86	
A159	AD7150	-1751	23.02.66	01.82	27.01.87	
A160	AD7151	-1759	23.02.66	03.82	24.02.87	
A161	AD7152	-1750	23.02.66	04.82	02.04.87	
A162	AD7153	-1771	18.04.66	04.02	02.04.87	
A163	AD7154	-1767	18.04.66	01.81	07.06.86	
A164	AD7155	-1772	18.04.66	03.82	19.03.87	
A165	AD7156	-1769	18.04.66	03.82	19.03.87	Exported to UK 6.87
A166	AD7157	-1768	18.04.66	08.80	16.07.87	
A167	AD7158	-1730	18.04.66	08.80	24.07.87	
A168	AD7159	-1733	18.04.66	08.81	29.10.86	
A169	AD7160	-1729	18.04.66	05.82	13.05.87	
A170	AD7161	-1732	18.04.66	02.81	24.07.87	
A171	AD7162	-1760	11.05.66	03.81	01.04.86	
A172	AD7163	-1756	11.05.66	01.81	01.04.86	
A173	AD7164	-1749	11.05.66	02.81	11.03.86	
A174	AD7165	-1773	11.05.66	05.82	23.04.87	
A175	AD7166	-1774	11.05.66	04.81	02.04.87	
A176	AD7167	-1755	11.05.66	08.81	14.11.86	
A177	AD7168	-1757	11.05.66	04.82	23.04.87	
A178	AD7169	-1758	11.05.66	04.82	23.04.87	
A179	AD7170	-1770	11.05.66	04.82	23.04.87	
A180	AD7171	-1731	11.05.66	01.81	10.04.86	
A181	AD7172	-1763	11.05.66	02.82	10.04.86	
A182	AD7173	-1766	11.05.66	08.81	14.11.86	
A183	AD7174	-1765	11.05.66	08.80	24.07.87	
A184	AD7175	-1762	11.05.66	02.81	01.04.86	
A185	AD7176	-1785	30.06.66	08.81	26.09.86	
A186	AD7177	-1782	30.06.66	10.80	21.09.84	
A187	AD7178	-1786	30.06.66	05.82	13.05.87	
A188	AD7179	-1783	30.06.66	06.82	13.05.87	
A189	AD7180	-1795	30.06.66	05.82	13.05.87	
A190	AD7181	-1794	30.06.66	03.82	24.02.87	
A191	AD7182	-1784	30.06.66	05.82	22.05.87	
A192	AD7183	-1761	18.06.66	02.81	07.06.86	
A193	AD7184	-1764	18.06.66	05.82	02.06.87	
A194	AD7185	-1788	20.09.66	08.81	29.10.86	
A195	AD7186	-1787	20.09.66	05.09.86		
A196	AD7187	-1790	20.09.66	02.82	26.04.86	
A197	AD7188	-1791	20.09.66	08.81	29.10.86	
A198	AD7189	-1789	20.09.66	08.80	07.06.86	
A199	AD7190	-1792	20.09.66	02.81	05.09.86	
A200	AD7191	-1793	20.09.66	02.82	14.02.87	
A201	AD7192	-1776	20.09.66	09.80	21.05.86	
A202	AD7193	-1796	20.09.66	02.81	05.09.86	
A203	AD7194	-1798	07.10.66	02.81	05.09.86	
A204	AD7195	-1797	07.10.66	09.80	26.08.87	
A205	AD7196	-1781	07.10.66	09.80	09.10.84	
A206	AD7197	-1780	07.10.66	02.81	26.09.86	
A207	AD7198	-1775	29.10.66	03.81	26.09.86	
A208	AD7199	-1777	29.10.66	03.81	26.09.86	
A209	AD7200	-1778	29.10.66	04.81	26.09.86	
A210	AD7201	-1779	29.10.66	07.81	26.09.86	

GUY VICTORY-J - BUSAF Prototype Buses
KMB Class: 'Victory J'   G 1-4.   BUSAF BODIES

Fleet number	Regn. number	Chassis number JVTB-	Date first registered	Withdrawn	Disposal
G 1	BJ9266	470262	28.01.76	11.03.82	To driving school
G 2	BJ9267	470263	28.01.76	23.01.82	To driving school
G 3	BJ9268	470276	28.01.76	18.02.82	To driving school
G 4	BJ9269	470277	28.01.76	19.01.82	To driving school

LEYLAND (GUY) VICTORY Mk2 Series 2
KMB Class: 'Leyland Victory Mk2'. - ALEXANDER BODIES

Fleet Note:	Regn. number	Chassis number JVTB-	Date first registered	Withdrawn	Disposal
G 5	BY8415	870199	25.05.79		
G 6	BZ9777	870332	10.08.79		
G 7	BZ9778	870390	10.08.79		
G 8	BZ9780	870287	10.08.79		
G 9	BZ9781	870284	10.08.79		
G10	CA2071	870331	21.08.78		
G11	CA2072	870389	21.08.79		
G12	CA2785	870387	24.08.79		
G13	CA2786	870290	24.08.79		
G14	CA2787	870391	24.08.79		
G15	CA2788	870269	24.08.79		
G16	CA4434	870334	05.09.79		
G17	CA4435	870396	05.09.79		
G18	CA4436	970010	05.09.79		
G19*	CA4437	870157	05.08.79		
G20	CA5648	970099	13.09.79		
G21	CA5649	970042	13.09.79		
G22	CA5650	870250	13.09.79		
G23	CA5651	870268	13.09.79		
G24	CA6070	870231	17.09.79		
G25	CA6071	970096	17.09.79		
G26	CA6072	970125	17.09.79		
G27	CA6073	970134	17.09.79		
G28	CA6074	870283	17.09.79		
G29	CA6075	970062	17.09.79		
G30	CA6076	970006	17.09.79		
G31	CA7628	870388	25.09.79		
G32	CA7629	870393	25.09.79		
G33	CA7630	970009	25.09.79		
C34	CA7631	970183	25.09.79		
G35	CA7632	970214	25.09.79		
G36	CA7949	970215	27.09.79		
G37	CA7950	970236	27.09.79		
G38	CA7951	970094	27.09.79		
G39	CA7952	970128	27.09.79		
G40	CA7953	970301	27.09.79		
G41	CA7954	970298	27.09.79		
G42	CA8145	970001	28.09.79		
G43	CA8146	870271	28.09.79		
G44	CA8147	970066	28.09.79		
G45	CA8148	970130	28.09.79		
G46	CA8149	970139	28.09.79		
G47	CA8150	970294	28.09.79		
G48	CA8151	970324	28.09.79		
G49	CA9233	970091	04.10.79		
G50	CA9234	970340	04.10.70		
G51	CA9235	970211	04.10.79		
G52	CB1650	970325	04.10.79		
G53	CB1651	970327	04.10.79		
G54	CB2605	870233	23.10.79		
G55	CB2606	870395	23.10.79		
G56	CB5066	970101	08.11.79		
G57	CB5071	970002	08.11.79		
G58	CB5135	970239	08.11.79		
G59	CB5168	970004	08.11.79		
G60	CB5183	970040	08.11.79		
G61	CB5264	970121	08.11.79		
G62	CB5377	970035	08.11.79		
G63	CB5309	870204	09.11.79		
G64	CB5352	870286	09.11.79		
G65	CB5448	970060	08.11.79	29.12.92	Scrapped
G66	CB5507	870266	08.11.79		
G67	CB5560	870392	08.11.79		
G68	CB5580	970064	08.11.79		
G69	CB5685	870427	08.11.79		
G70	CB5991	970089	12.11.79		
G71	CB6883	970038	20.11.79		
G72	CB6955	870264	20.11.79		
G73	CB8843	970242	29.11.79		
G74	CB8851	970177	29.11.79		
G75	CB8853	870213	29.11.79		
G76	CB8876	870394	29.11.79		
G77	CB7937	970209	30.11.79		
G78	CB7964	970240	30.11.79		
G79	CB9767	970284	06.12.79		
G80	CB9780	970331	06.12.79		
G81	CB9795	970103	06.12.79		
C82	CB9807	970245	06.12.79		
G83	CB9825	979180	06.12.79		
G84	CB9838	970175	06.12.79		
G85	CB9356	070328	07.12.79		
G86	CB9357	970296	07.12.79		
G87	CB9564	970184	07.12.79		
G88	CC 486	970178	07.12.79		
G89	CC1037	970138	14.12.79		
G90	CC1142	970246	14.12.79		
G91	CC1186	970326	14.12.79		

> **NOTE:** G19* Driver's cab air-conditioned retrospectively as pilot for a batch of 20 - G131-150.

LEYLAND (GUY) VICTORY Mk2 Series 2 (contd.).
KMB Class: 'Leyland Victory Mk2' (contd.) - ALEXANDER BODY

Fleet number	Regn. number	Chassis number JVTB-	Date first registered	Withdrawn	Disposal	Note:
G92	CC5179	970286	04.01.80	03.05.85		Trg bus
G93	CC5325	970107	04.01.80			
G94	CC5478	970289	04.01.89	08.06.82		
G95	CC5819	970333	04.01.80		Re-bodied & **Re-registered CY4529** - 03.83.	
G96	CC5274	970181	09.01.80			
G97	CC5426	970335	09.01.80			
G98	CC5476	970250	09.01.80			
G99	CC5621	970237	09.01.80			
G100	CC5920	970251	09.01.80			
G101	CC7441	970217	14.01.80			
G102	CC7490	970288	14.01.80			
G103	CC7533	970252	14.01.80			
G104	CC7545	970243	14.01.80			
G105	CC7584	970390	14.01.80			
G106	CC7599	970248	14.01.80			
G107	CC7763	970291	17.01.80			
G108	CC8069	970254	18.01.80			
G109	CC8215	970292	18.01.80			
G110	CC8266	970325	18.01.80			
G111	CD2761	970383	07.02.80			
G112	CD2865	970389	07.02.80			
G113	CD2912	970461	07.02.80			
G114	CD4226	970338	12.02.80			
G115	CD4227	970337	12.02.80			
G116	CD4471	970455	12.02.80			
G117	CD4472	970341	12.02.80			
G118	CD4476	970392	12.02.80			
G119	CD4601	970530	12.02.80			
G120	CD4602	970525	12.02.80			
G121	CD4665	970523	12.02.80			
G122	CD4666	970463	12.02.80			
G123	CD4315	970399	14.02.80			
G124	CD4317	970336	14.02.80			
G125	CD4318	970136	14.02.80	02.93		Trg bus
G126	CD4319	970293	14.02.80			
G127	CD4320	970387	14.02.80			
G128	CD4373	970384	14.02.80			
G129	CD4374	970386	14.02.80			
G130	CD4375	970044	14.02.80	02.93		Trg bus
G131*	CE4269	970770	15.04.80			
G132*	CE4270	970518	15.04.80			
G133*	CE4329	970453	15.04.80			
G134*	CE4375	970534	15.04.80			
G135*	CE4429	970767	15.04.80			
G136*	CE4430	970532	15.04.80			
G137*	CE3603	970398	16.04.80			
G138*	CE3629	970769	16.04.80			
G139*	CE3630	970400	16.04.80			
G140*	CE4228	970736	16.04.80			
G141*	CE4229	970520	16.04.80			
G142*	CE4277	970527	16.04.80			
G143*	CE4278	970396	16.04.80			
G144*	CE4386	970522	16.04.80			
G145*	CE4387	970531	16.04.80			
G146*	CE4560	970533	16.04.80			
G147*	CE5343	970766	18.04.80			
G148*	CE5830	970613	18.04.80			
G149*	CE5970	970465	18.04.80			
G150*	CE6080	970517	29.04.80			
G151	CG 519	970771	14.07.80			
G152	CG 546	970764	14.07.80			
G153	CG 550	970765	14.07.80			
G154	CG 226	970768	15.07.80			
G155	CG 413	970772	15.07.80			

*Driver's cab air-conditioned from new - removed 10.82.

KMB sub-class: 'Leyland Victory Mk2(A)'

Fleet number	Regn. number	Chassis number	Date first registered
G156	CG2078	970924	21.07.80
G157	CG2663	970920	21.07.80
G158	CG3023	800098	24.07.80
G159	CG3026	800091	24.07.80
G160	CG3138	970919	24.07.80
G161	CG3144	800100	24.07.80
G162	CG3147	970922	24.07.80
G163	CG3291	800099	24.07.80
G164	CG3266	800105	28.07.80
G165	CG3267	971015	28.07.80
G166	CG3719	800101	28.07.80
G167	CG3867	970921	28.07.80
G168	CG3868	970936	28.07.80
G169	CG4051	970935	30.07.80
G170	CG4552	800083	30.07.80
G171	CG4564	800081	30.07.80
G172	CG4803	970926	30.07.80
G173	CG4029	970931	01.08.80
G174	CG4341	970927	01.08.80
G175	CG4548	970938	01.08.80
G176	CG4577	800094	01.08.80
G177	CG4600	970928	01.08.80
G178	CG4868	971016	01.08.80
G179	CG4871	800082	01.08.80
G180	CG6589	800109	08.08.80
G181	CG6837	800107	08.08.80
G182	CG6838	800111	08.08.80
G183	CG6890	800112	08.08.80
G184	CG6940	970932	08.08.80
G185	CG6941	800117	08.08.80

LEYLAND (GUY) VICTORY Mk2 Series 2 (contd.).
KMB sub-class: 'Leyland Victory Mk2(A)' (contd.) - ALEXANDER BODY

Fleet number	Regn. number	Chassis number JVTB-	Date first registered	Withdrawn	Disposal	Note:
G186	CG6578	800087	13.08.80			
G187	CG6628	800095	13.08.80			
G188	CG6986	800097	13.08.80			
G189	CG7045	800084	15.08.80			
G190	CG7046	800090	15.08.80			
G191	CG7791	800102	15.08.80			
G192	CG8171	971014	20.08.80			
G193	CG8464	970937	20.08.80			
G194	CG8564	971017	20.08.80			
G195	CG8757	970933	21.08.80			
G196	CG8920	970923	21.08.80			
G197	CG8922	970934	21.08.80			
G198	CG9014	800096	26.08.80			
G199	CG9777	970929	26.08.80			
G200	CG9778	800085	26.08.80			
G201	CH 366	970925	26.08.80			
G202	CH 367	970930	26.08.80			
G203	CG9010	800093	28.08.80			
G204	CG9061	800114	28.08.80			
G205	CG9162	800106	28.08.80			
G206	CG9690	800086	28.08.80			
G207	CH 744	800089	28.08.80			
G208	CG9771	800104	29.08.80			
G209	CG9772	800092	29.08.80			
G210	CH1729	800088	03.09.80			
G211	CH1741	800287	03.09.80			
G212	CH1841	800270	03.08.80			
G213	CH1056	800118	08.09.80			
G214	CH1057	800120	08.09.80			
G215	CH1711	800113	08.09.80			
G216	CH2047	800285	09.09.80			
G217	CH2048	800273	09.09.80			
G218	CH2054	800364	09.09.80			
G219	CH2394	800365	09.09.80			
G220	CH2395	800276	09.09.80			
G221	CH2770	800272	10.09.80			
G222	CH2779	800289	10.09.80			
G223	CH2924	800363	10.09.80			
G224	CH3299	800103	17.09.80			
G225	CH3542	800363	17.09.80			
G226	CH3990	800288	17.09.80			
G227	CH5645	800274	26.09.80			
G228	CJ2088	800271	28.10.80			
G229	CJ2246	800275	28.10.80			
G230	CJ2549	800616	28.10.80			
G231	CJ2550	800279	28.10.80			
G232	CJ2201	800617	29.10.80			
G233	CJ2364	800116	29.10.80			
G234	CJ2644	800618	31.10.80			
G235	CJ2656	800280	31.10.80			
G236	CJ3947	800615	03.11.80			
G237	CJ3283	800278	05.11.80			
G238	CJ3284	800621	05.11.80	09.92		?
G239	CJ3394	800286	05.11.80			
G240	CJ3395	800282	05.11.80			
G241	CJ3645	800119	05.11.80			
G242	CJ3741	800283	05.11.80			
G243	CJ3792	800623	05.11.80			
G244	CJ3893	800108	05.11.80			
G245	CJ4515	800281	07.11.80			
G246	CJ4516	800366	07.11.80			
G247	CJ4580	800611	07.11.80	10.92		?
G248	CJ4815	800110	07.11.80			
G249	CJ4568	800610	12.11.80			
G250	CJ4579	800277	12.11.80			
G251	CJ4608	800284	12.11.80			
G252	CK2644	800639	15.12.80			
G253	CK2789	800630	15.12.80			
G254	CK2024	800643	18.12.80			
G255	CK2025	800627	18.12.80			
G256	CK3276	800629	23.12.80			
G257	CK3824	800633	23.12.80			
G258	CK3832	800612	23.12.80			
G259	CK3854	800645	23.12.80			
G260	CK3884	800614	23.12.80			
G261	CK3885	800628	23.12.80			
G262	CK3895	800376	23.12.80			
G263	CK4013	800638	05.01.81			
G264	CK4059	800625	05.01.81			

Fleet number	Sub-type	Regn. number	Chassis number	Date first registered	Withdawn	Disposal
G265	(B)	CK4220	800680	05.01.81		
G266	(A)	CK4232	800622	05.01.81		
G267	(A)	CK4558	800632	05.01.81		
G268	(A)	CK4602	800626	05.01.81	12.92	?
G269	(A)	CK4707	800624	05.01.81		
G270	(B)	CK5456	800686	07.01.81		
G272	(A)	CK5457	800676	07.01.81		
G273	(B)	CK5591	800774	07.01.81		
G274	(B)	CK5601	800683	07.01.81		
G275	(A)	CK5627	800679	07.01.81		
G276	(A)	CK6553	800380	09.01.81		
G277	(A)	CK6554	800678	09.01.81		
G278	(A)	CK8715	800677	22.01.81		
G279	(A)	CK8912	800674	22.01.81		
G280	(A)	CK9255	800671	22.01.81		

The change from Victory(A) to (B) was not neat as the last (A)'s were intermixed with the early (B)'s. To aid identification, the sub-type is added between the fleet and registration number columns for these vehicles only.

LEYLAND (GUY) VICTORY Mk2 Series 2 (Contd.).
KMB sub-class: 'Leyland Victory Mk2(B)' - ALEXANDER BODY

Fleet number	Sub-type	Regn. number JVTB-	Chassis number	Date first registered	Withdrawn	Disposal
G281	(B)	CK8821	800785	26.01.81		
G282	(B)	CK8927	800787	26.01.81		
G283	(A)	CK9046	800673	26.01.81		
G284	(B)	CK9447	800681	26.01.81		
G285	(A)	CK9586	800642	26.01.81		
G286	(B)	CL 449	800792	28.01.81		
G287	(B)	CL 514	800682	28.01.81		
G288	(B)	CL 638	800784	28.01.81		
G289	(B)	CL 761	800782	28.01.81		
G290	(B)	CL 908	800791	28.01.81		
G291	(B)	CL1026	800789	28.01.81		
G292	(B)	CL1228	800780	28.01.81		
G293	(B)	CL1625	800790	28.01.81		
G294	(A)	CL1626	800672	28.01.81		
G295	(B)	CL1650	800685	28.01.81		
G296	(B)	CL1710	800783	28.01.81		
G297	(B)	CL1719	800788	28.01.81		
G298	(A)	CL1836	800675	30.01.81	21.12.82	
G299	(B)	CL2204	800793	12.02.81		
G300	(B)	CL2253	800781	12.02.81	01.93	
G301	(B)	CL2401	800795	12.02.81	12.92	?
G302	(B)	CL2402	800797	12.02.81		
G303	(B)	CL2907	800684	12.02.81		
G304	(B)	CL2908	800779	12.02.81		
G305	(B)	CM2157	800805	01.04.81		
G306	(B)	CM2221	800889	01.04.81		
G307	(B)	CM2675	800800	01.04.81		
G308	(B)	CM3848	800801	03.04.81		
COACH		CM3879	970607	03.04.81		To bus G544 04.05.83
G309	(B)	CM3945	800803	03.04.81		
G310	(B)	CM4241	800898	10.04.81		
G311	(B)	CM4245	800990	10.04.81		
G312	(B)	CM4348	800804	10.04.81		
G313	(B)	CM4450	800893	10.04.81		
G314	(B)	CM4484	800802	10.04.81		
G315	(B)	CM4492	800983	10.04.81		
G315	(B)	CM4642	800888	10.04.81		
G316	(B)	CM4645	800886	10.04.81		
G318	(B)	CM4647	800892	10.04.81		
G319	(B)	CM4078	800890	13.04.81		
G320	(B)	CM4623	800980	13.04.81		
G321	(B)	CM5111	800985	15.04.81		
G322	(B)	CM5261	800794	15.04.81		
G323	(B)	CM5591	800798	15.04.81		
G324	(B)	CM5716	800887	15.04.81	02.93	Speedybus -minus engine/axles,etc.
G325	(B)	CM5736	800786	15.04.81		
G326	(B)	CM5746	800796	15.04.81		
G327	(B)	CM5525	800896	21.04.81		
G328	(B)	CM5606	800982	21.04.81		
G329	(B)	CM5770	800989	21.04.81		
G331	(B)	CM5867	800799	21.04.81		
G332	(B)	CM6276	800986	24.04.81		
G333	(B)	CM7425	800894	01.05.81		
G334	(B)	CM7680	800991	01.05.81		
G335	(B)	CM7723	800984	01.05.81	11.92?	To (BB) type 26.08.83
G336	(B)	CM8349	800992	05.05.81		
G337	(B)	CM8788	800979	05.05.81		
G338	(B)	CM8841	800897	05.05.81		
G339	(B)	CM8979	800981	05.05.81		
G340	(B)	CM8991	800895	05.05.81		
G341	(B)	CM8024	800998	07.05.81		
G342	(B)	CM8506	800987	07.05.81		
G343	(B)	CM8904	800988	07.05.81		
G344	(B)	CN4704	800993	01.06.81		
G345	(A)	CN3715	800115	05.06.81		

KMB sub-class: Leyland Vicictory Mk2 (C)

Fleet number	Regn. number	Chassis number	Date first registered
G346	CN3715	810298	01.10.81
G347	CR6516	810090	01.10.81
G348	CR7761	810169	12.10.81
G349	CR8267	810167	14.10.81
G350	CS1385	810094	28.10.81
G351	CS1558	810115	28.10.81
G352	CS1857	810296	28.10.81
G353	CS2392	810114	04.11.81
G354	CS2829	810166	04.11.81
G355	CS2884	810276	04.11.81
G356	CS3235	810158	06.11.81
G357	CS3236	810108	06.11.81
G358	CS3478	810091	06.11.81	17.07.91
G359	CS3550	810095	09.11.81
G360	CS3586	810109	09.11.81
G361	CS3596	810104	09.11.81
G362	CS4129	810792	12.11.81
G363	CS4328	810761	12.11.81
G364	CS5938	810796	20.11.81
G365	CS6589	810908	24.11.81
G366	CS6743	810764	24.11.81
G367	CS6896	810797	24.11.81
G368	CS6380	810284	25.11.81
G369	CS6470	810088	25.11.81
G370	CS6488	810118	25.11.81
G371	CS6489	810168	25.11.81
G372	CS6846	810079	25.11.81
G373	CS6970	810087	25.11.81
G374	CS7071	810096	01.12.81

LEYLAND (GUY) VICTORY Mk2 Series 2 (Contd.).
KMB sub-class: Leyland Victory Mk2 (C) - ALEXANDER BODY

Fleet number	Regn. number	Chassis number JVTB-	Date first registered	Withdrawn	Disposal
G375	CS7120	810097	01.12.81		
G376	CS7269	810092	01.12.81		
G377	CS7320	810089	01.12.81		
G378	CS7287	810099	03.12.81		
G379	CS7344	810287	03.12.81		
G380	CS7396	810080	03.12.81		
G381	CS7908	810101	03.12.81		
G382	CS8353	810290	07.12.81		
G383	CS8809	810106	07.12.81		
G384	CS8010	810107	09.12.81	29.10.91	sc
G385	CS8156	810116	09.12.81		
G386	CS8177	810159	09.12.81		
G387	CS8241	810083	09.12.81		
G388	CT 144	801102	10.12.81		
G389	CT 205	810102	10.12.81		
G390	CT 279	810102	10.12.81		
G391	CT 596	810117	01.12.81		
G392	CT 851	810086	16.12.81		
G393	CT 856	810288	16.12.81		
G394	CT 964	810111	16.12.81		
G395	CT 985	810109	16.12.81		
G396	CT1648	810299	18.12.81		
G397	CT1795	810081	18.12.81		
G398	CT1171	810113	22.12.81		
G399	CT1643	810085	22.12.81		
G400	CT1784	810292	22.12.81		
G401	CT2297	810264	30.12.81		
G402	CT2433	810093	30.12.81		
G403	CT2588	810155	30.12.81		
G404	CT2641	810110	30.12.81		
G405	CT3147	810289	05.01.82		
G406	CT3438	810157	05.01.82		
G407	CT3497	810082	05.01.82		
G408	CT5086	810156	15.01.82		
G409	CT5137	810105	15.01.82		
G410	CU5017	810984	17.03.82		
G411	CU5472	810954	17.03.82		
G412	CU6145	810981	17.03.82		
G413	CU7966	810956	22.03.82		
G414	CU8347	810982	22.03.82		
G415	CU8470	810965	22.03.82		
G416	CU8522	810963	29.03.82		
G417	CU8640	810799	29.03.82		
G418	CU9737	810084	29.03.81		
G419	CU9050	810297	01.04.82		
G420	CU9402	810972	01.04.82		
G421	CV 342	810979	01.04.82		
G422	CV1613	810977	14.04.02		
G423	CV1963	810771	14.04.82		
G424	CV2198	810978	14.04.82		
G425	CV 954	810980	19.04.82		
G426	CV1072	810983	19.94.82		
G427	CV1899	810825	19.04.82		
G428	CV3324	810969	20.04.82		
G429	CV3424	810964	20.04.82		
G430	CV4346	810967	20.04.82		
G431	CV2668	810293	30.04.82		
G432	CV3554	810824	30.04.82		
G433	CV3808	810760	30.04.82		
G434	CV4555	810778	03.05.82		
G435	CV5502	810294	03.05.82		
G436	CV6317	810777	03.05.82		
G437	CV5302	810285	12.05.82		
G438	CV5908	810765	12.05.82		
G439	CV5952	810779	12.05.82		
G440	CV4933	810291	17.05.82	20.06.83	
G441	CV5641	810300	17.05.82		
G442	CV5977	810755	17.05.82		
G443	CV6754	810802	24.05.82		
G444	CV6803	810768	24.05.82		
G445	CV8324	810772	24.05.82		
G446	CV7128	810766	28.05.82		
G477	CV7186	810794	28.05.82		
G448	CV7666	810770	28.05.82		
G449	CV7145	810798	03.06.82		
G450	CV7645	810819	03.06.82		
G451	CV7946	810805	03.06.82		
G452	CV8914	810795	09.06.82		
G453	CV9063	810975	09.06.82		
G454	CV9943	810767	09.06.82		
G455	CV8584	810817	17.06.82		
G456	CV9531	810910	17.06.82		
G457	CV9742	810295	17.06.82		
G458	CW 509	810769	24.06.82		
G459	CW1186	810820	24.06.82		
G460	CW1958	810971	24.06.82		
G461	CW1110	810773	02.07.82		
G462	CW1322	810803	02.07.82		
G463	CW1382	810912	02.06.82		
G464	CW 708	810800	07.07.82		
G465	CW 953	810821	07.07.82		
G466	CW1897	810286	07.07.82		
G467	CW1121	810763	12.07.82		
G468	CW2079	810762	12.07.82		
G469	CW2410	810774	12.07.82		
G470	CW4123	810940	22.07.82		
G471	CW4324	810818	22.07.82		
G472	CW4403	810911	22.07.82		

307

LEYLAND (GUY) VICTORY Mk2 Series 2 (contd.).
KMB sub-class: Leyland Victory Mk2(C) - ALEXANDER BODIES

Fleet number	Regn. number	Chassis number	Date first registered	Withdrawn	Disposal	Note:
		JVTB-				
G473	CW2668	810828	28.07.82			
G474	CW3530	810959	28.07.82			
G475	CW3525	810776	03.08.82			
G476	CW4502	810925	10.08.82			
G477	CW5446	810970	17.08.82			
G478	CW5847	810926	17.08.82			
G479	CW6187	810833	18.08.82			
G480	CW4517	810831	23.08.82			
G481	CW5659	810957	23.08.82			
G482	CW8218	810775	26.08.82			
G483	CW8210	810832	01.09.82			
G484	CW8308	810913	01.09.82			
G485	CW7021	810827	06.09.82			
G486	CW7082	810953	06.09.82			
G487	CW9935	810948	13.09.82			
G488	CX 445	810919	13.09.82			
G489	CW8528	810922	20.09.82			
G490	CW9356	810923	20.09.82			
G491	CW9545	810952	20.09.82			
G492	CW9973	810815	20.09.82			
G493	CX 264	810961	20.09.82			
G494	CX2437	810947	01.10.82			
G495	CX1476	810801	05.10.82			
G496	CX2231	810804	05.10.82			
G497	CX3230	810915	15.10.82			
G498	CX3816	810830	15.10.82	10.87	Sold to New Lantao Bus Co.	
G499	CX4015	810950	15.10.82			
G500	CX2750	810816	22.10.82			
G501	CX4008	810955	22.10.82			
G502	CX3923	810920	26.10.82			
G503	CX2978	810918	01.11.82			
G504	CX3543	810974	01.11.82			
G505	CX4011	810962	01.11.82	10.87	Sold to New Lantao Bus Co.	
G506	CX4046	810966	01.11.82			
G507	CX5136	810960	10.11.82			
G508	CX6495	810976	10.11.82			
G509	CX4673	810968	18.11.82			
G510	CX5062	810826	18.11.82			
G511	CX7112	810916	22.11.82			
G512	CX7891	810951	25.11.82			
G513	CX6673	810917	01.12.82			
G514	CX6821	810907	01.12.82			
G515	CX7427	810986	06.12.82			
G516	CX8416	810924	06.12.82			
G517	CX8977	810814	15.12.82			
G518	CX9096	810985	15.12.82			
G519	CX9923	810822	15.12.82			
G520	CY 399	810921	15.12.82			
G521	CY 401	810958	21.12.82			
G522	CX8564	811019	23.12.82			
G523	CX8911	810793	23.12.82			
G524	CX8677	811026	04.01.83			
G525	CX8956	810834	04.01.83			
G526	CX9511	811020	04.01.83	03.85	Sold to New Lantao Bus Co.	
G527	CY 057	811028	12.01.83	03.85	Sold to New Lantao Bus Co.	
G528	CY1428	811018	12.01.83			
G529	CY 899	811017	13.01.83			
G530	CY1048	810829	17.01.83			
G531	CY2203	811021	17.01.83			
G532	CY1657	811023	20.01.83			
G533	CY2557	811022	01.02.83			
G534	CY3220	811027	01.02.83			
G535	CY4246	811031	01.02.83			
G536	CY4331	810823	01.02.83			
G537	CY2851	811029	07.02.83			
G538	CY2934	810973	07.02.83			
G539	CY3558	811032	07.02.83			
G540	CY3573	811030	07.02.83			
G541	CY4308	811033	07.02.83			
G542	CY3188	811024	09.02.83	01.91	Sold to New Lantao Bus Co.	
G543	CY3089	811025	11.02.83	01.91	Sold to New Lantao Bus Co.	

KMB sub-class: Leyland Vicictory Mk2(D) - ALEXANDER BODY
Formerly COACH.

Fleet number	Regn. number	Chassis number	Date first registered			
G544	CM3897	970607	04.05.83	(date licensed as bus).		

DENNIS JUBILANT - N 1-4
KMB sub-class: Dennis FS - KMB-BUILT BODIES

Fleet number	Regn. number	Chassis number	Date first registered	Withdrawn	Disposal
N 1	BR4474	DD201-104	03.11.77		
N 2	BS6429	-101	20.02.78		
N 3	BS9154	-103	14.03.78		
N 4	BT7277	-102	22.05.78		

DENNIS JUBILANT
KMB sub-class: Dennis(A) - ALEXANDER BODIES

Fleet number	Regn. number	Chassis number	Date first registered	Withdrawn	Disposal
N 5	BY6592	DD204-109	11.05.79		
N 6	BY6593	-112	11.05.79		
N 7	BY6594	-116	11.05.79		
N 8	BY6595	-111	11.05.79		
N 9	BY8945	-107	01.06.79		
N10	BY8946	-110	01.06.79		

DENNIS JUBILANT (contd.)
KMB sub-class: Dennis(A) - ALEXANDER BODIES

Fleet number	Regn. number	Chassis number	Date first registered	Withdrawn	Disposal
N11	BY8947	-123	01.06.79		
N12	BY9932	-108	07.06.79		
N13	BY9934	-119	07.06.79		
N14	BZ 841	-114	11.06.79		
N15	BZ 842	-113	11.06.79		
N16	BZ1368	-117	13.06.79		
N17	CA3777	-155	03.09.79		
N18	CA3778	-149	03.09.79		
N19	CA3779	-150	03.09.79		
N20	CA3780	-152	03.09.79		
N21	CA3781	-153	03.09.79		
N22	CA3782	-154	03.09.79		
N23	CA7625	-120	25.09.79		
N24	CA7626	-133	25.09.79		
N25	CA7627	-147	25.09.79		
N26	CA7947	-146	27.09.79		
N27	CA7948	-148	27.09.79		
N28	CA8144	-128	28.09.79		
N29	CB5093	-127	08.11.79		
N30	CB5372	-145	08.11.79		
N31	CB5766	-151	13.11.79		
N32	CB5826	-125	13.11.79		
N33	CB5876	-143	13.11.79		
N34	CB6853	-134	20.11.79		
N35	CB7165	-121	27.11.79		
N36	CC 534	-115	12.12.79		
N37	CC 690	-136	12.12.79		
N38	CC 938	-129	12.12.79		
N39	CC2035	-130	18.12.79		
N40	CC2367	-132	18.12.79		
N41	CC7615	-118	17.01.80		
N42	CC7662	-137	17.01.80		
N43	CC7713	-131	17.01.80		
N44	CC8311	-140	18.01.80		
N45	CC8360	-126	18.01.80		
N46	CC8462	-122	18.01.80		
N47	CD4228	-138	12.02.80		
N48	CD4362	-144	14.02.80		
N49	CD4363	-135	14.02.80		
N50	CD4364	-124	14.02.80		
N51	CD4771	-139	14.02.80		
N52	CD4772	-142	14.02.80		
N53	CD4773	-141	14.02.80		
COACH	CF4180	DD204-106	11.06.80		

KMB sub-class: Dennis(B) ALEXANDER BODIES

Fleet number	Regn. number	Chassis number	Date first registered		
N54	CF6061	DD206-214	19.06.80		
N55	CF6137	-193	23.06.80		
N56	CF6167	-196	19.06.80		
N57	CF6169	-197	19.06.80		
N58	CF6167	-204	19.06.80		
N59	CF6362	-199	19.06.80		
N60	CF6151	-188	23.06.80		
N61	CF6157	-194	23.06.80		
N62	CF6159	-192	23.06.80		
N63	CF6220	-190	23.06.80		
N64	CF6807	-186	23.06.80		
N65	CF6822	-203	23.06.80		
N66	CF7627	-207	27.06.80		
N67	CF7707	-200	27.06.80		
N68	CF7719	-195	27.06.80		
N69	CF7753	-210	27.06.80		
N70	CF7754	-187	27.06.80		
N71	CF7770	-206	27.06.80		
N72	CF7790	-191	27.06.80		
N73	CF7798	-212	27.06.80		
N74	CF7876	-211	27.06.80		
N75	CF7881	-201	27.06.80		
N76	CF7882	-209	27.06.80		
N77	CF7949	-208	27.06.80		
N78	CH4759	-217	12.09.80		
N79	CH4842	-205	12.09.80		
N80	CH4868	-202	12.09.80		
N81	CH3801	-198	19.09.80		
N82	CH4145	-213	19.09.80		
N83	CH4605	-215	19.09.80		
N84	CH5143	-229	26.09.80		
N85	CH5197	-233	26.09.80		
N86	CH5291	-253	26.09.80		
N87	CH5302	-230	26.09.80		
N88	CH5519	-254	26.09.80		
N89	CH5994	-238	26.09.80		
N90	CH5162	-260	29.09.80		
N91	CH5164	-235	29.09.80		
N92	CH5213	-231	29.09.80		
N93	CH5260	-256	29.09.80		
N94	CH5468	-249	29.09.80		
N95	CH5576	-232	29.09.80		
N96	CH6011	-258	02.10.80		
N97	CH6163	-246	02.10.80		
N98	CH6175	-252	02.10.80		
N99	CH6447	-257	02.10.80		
N100	CH6470	-245	02.10.80		
N101	CH6863	-259	02.10.80		
N102	CH7035	-189	08.10.80		
N103	CH7083	-216	08.10.80		
N104	CH7412	-243	08.10.80		
N105	CH7454	-219	08.10.80		
N106	CH7515	-221	08.10.80		
N107	CH7599	-234	08.10.80		
N108	CH7623	-218	08.10.80		
N109	CH8755	-224	08.10.80		
N110	CH7763	-222	10.10.80		
N111	CH8440	-228	10.10.80		
N112	CH8423	-248	13.10.80		
N113	CH8630	-250	13.10.80		
N114	CH8843	-255	13.10.80		
N115	CJ 605	-236	23.10.80		
N116	CJ1063	-242	23.10.80		

DENNIS JUBILANT (contd.)
KMB sub-class: Dennis(B) (contd.) - ALEXANDER BODIES

Fleet number	Regn. number	Chassis number	Date first registered	Withdrawn	Disposal
N117	CJ1204	-244	23.10.80		
N118	CJ1254	-223	23.10.80		
N119	CJ1458	-239	23.10.80		
N120	CJ1538	-225	23.10.80		
N121	CJ2563	-227	29.10.80		
N122	CJ2811	-251	29.10.80		
N123	CJ2990	-237	29.10.80		
N124	CJ4816	-240	07.11.80		
N125	CJ4842	-220	07.11.80		
N126	CJ4843	-247	07.11.80		
N128	CJ6726	-226	19.11.80		
N129	CJ6728	-241	19.11.80		

KMB sub-class: Dennis(C) - ALEXANDER BODIES

Fleet number	Regn. number	Chassis number	Date first registered	Withdrawn	Disposal
N127	CJ6720	DD206-261	19.11.80		
N130	CJ6757	-274	19.11.80		
N131	CJ6833	-276	19.11.80		
N132	CJ6922	-262	19.11.80		
N133	CK1801	-275	08.12.80		
N134	CK1764	-280	08.12.80		
N135	CK 198	-269	09.12.80		
N136	CK 699	-273	09.12.80		
N137	CK1595	-267	09.12.80		
N138	CK 147	-291	11.12.80		
N139	CK 183	-283	11.12.80		
N140	CK 194	-270	11.12.80		
N141	CK 237	-278	11.12.80		
N142	CK 294	-292	11.12.80		
N143	CK 344	-272	11.12.80		
N144	CK 356	-289	11.12.80		
N145	CK 398	-266	11.12.80		
N146	CK 406	-264	11.12.80		
N147	CK 422	-285	11.12.80		
N148	CK 735	-290	11.12.80		
N149	CK 788	-279	11.12.80		
N150	CK 798	-282	11.12.80		
N151	CK1027	-265	11.12.80		
N152	CK1269	-277	11.12.80		
N153	CK1618	-284	11.12.80		
N154	CK2790	-286	15.12.80		
N155	CK3051	-271	19.12.80		
N156	CK3461	-268	23.12.80		
N157	CK3570	-288	23.12.80		
N158	CK3867	-287	23.12.80		
N159	CK3870	-281	23.12.80		
N160	CK3156	-295	30.12.80		
N161	CK3158	-263	30.12.80		
N162	CK3163	-294	30.12.80		
N163	CK3298	DD208-296	30.12.80		

KMB sub-class: Dennis(D) - ALEXANDER BODIES

Fleet number	Regn. number	Chassis number	Date first registered	Withdrawn	Disposal
N164	CL7949	DD208-304	03.03.81		
N165	CL8298	-300	06.03.81		
N166	CL8775	-298	06.03.81		
N167	CL8026	-315	09.03.81		
N168	CL8075	-299	09.03.81		
N169	CL8106	-305	09.03.81		
N170	CL8260	-302	09.93.81		
N171	CL8306	-310	09.03.81		
N172	CL8405	-303	09.03.81		
N173	CL8603	-297	09.03.81		
N174	CL8753	-313	09.03.81		
N175	CL8802	-309	09.03.81		
N176	CL8909	-306	09.03.81		
N177	CL8910	-318	09.03.81		
N178	CL8303	DD206-293	10.03.81		
N179	CL8857	DD208-326	10.03.81		
N180	CL9296	-317	13.03.81		
N181	CL9319	-307	13.03.81		
N182	CM 318	-308	13.03.81		
N183	CL9352	-316	16.03.81		
N184	CL9959	-314	16.03.81		
N185	CL9317	-327	18.03.81		
N186	CL9711	-312	18.03.81		
N187	CL9712	-334	18.03.81		
N188	CM 107	-328	18.03.81		
N189	CM 108	-319	18.03.81		
N190	CM 281	-323	18.03.81		
N191	CM 283	-332	18.03.81		
N192	CM 466	-311	81.03.81		
N193	CM 469	-329	18.03.81		
N194	CM1186	-324	19.03.81		
N195	CM1194	-331	19.03.81		
N196	CM1385	-322	19.03.81		
N197	CM1446	-325	19.03.81		
N198	CM1845	-320	19.03.81		
N199	CM1879	-330	19.03.81		
N200	CM1413	-345	24.03.81		
N201	CM1572	-301	24.03.81		
N202	CM2133	-337	26.03.81		
N203	CM2166	-339	26.03.81		
N204	CM2419	-344	26.03.81		
N205	CM2741	-338	26.03.81		
N206	CM2829	-341	26.03.81		
N207	CM2978	-321	26.03.81		
N208	CM2477	-336	27.03.81		
N209	CM2691	-342	27.03.81		
N210	CM2694	-335	27.03.81		
N211	CM2740	-343	27.03.81		
N212	CM3531	-333	03.04.81		
N213	CM3644	-340	03.04.81		

KMB sub-class: Dennis(E) - DUPLE-METSEC BODIES

Fleet number	Regn. number	Chassis number	Date first registered	Withdrawn	Disposal
N214	CM9794	DD207-346	12.05.81		
N215	CN 347	-363	22.05.81		
N216	CN 371	-358	22.05.81		
N217	CN 449	-364	22.05.81		
N218	CN 450	-354	22.05.81		

DENNIS JUBILANT (contd.)
KMB sub-class: Dennis(E) - DUPLE-METSEC BODIES

Fleet number	Regn. number	Chassis number	Date first registered	Withdrawn	Disposal
N219	CN1696	DD 207-361	22.05.81		
N220	CN1837	-353	22.05.81		
N221	CN1843	-355	22.05.81		
N222	CN1844	-348	22.05.81		
N223	CN1864	-362	22.05.81		
N224	CN1877	-360	22.05.81		
N225	CN1946	-350	22.05.81		
N226	CN1947	-351	22.05.81		
N227	CN1950	356	22.05.81		
N228	CN2145	-359	26.05.81		
N229	CN2531	-352	26.05.81		
N230	CN2827	-367	26.05.81		
N231	CN2829	-369	26.05.81		
N232	CN2901	-349	26.05.81		
N233	CN2907	-347	26.05.81		
N234	CN3297	-366	01.06.81		
N235	CN3357	-382	01.06.81		
N236	CN3373	-381	01.06.81		
N237	CN3374	-357	01.06.81		
N238	CN4414	-365	01.06.81		
N239	CN4564	-370	01.06.81		
N240	CN4978	-383	01.06.81		
N241	CN4989	-368	01.06.81		
N242	CN3568	-377	04.06.81		
N243	CN3875	-384	04.06.81		
N244	CN4369	-372	04.06.81		
N245	CN3871	-385	05.06.81		
N246	CN3979	-380	05.06.81		
N247	CN3980	-379	05.06.81		
N248	CN5479	-378	10.06.81		
N249	CN5480	-389	10.06.81		
N250	CN5076	-378	10.06.81		
N251	CN5009	-371	12.06.81		
N252	CN5011	-373	12.06.81		
N253	CN5007	-386	15.06.81		
N254	CN5037	-374	15.06.81		
N255	CN5053	-376	15.06.81		
N256	CN6213	-422	19.06.81		
N257	CN6505	-424	19.06.81		
N258	CN6533	-420	19.06.81		
N259	CN7363	-392	26.06.81		
N260	CN9137	-399	09.07.81		
N261	CN9346	-408	09.07.81		
N262	CP 640	-423	13.07.81		
N263	CP 819	-407	13.07.81		
N264	CP3612	-400	24.07.81		
N265	CP3827	-402	24.07.81		
N266	CP3161	-403	27.07.81		
N267	CP3357	-404	27.07.81		
N268	CP3074	-375	28.07.81		
N269	CP3391	-395	28.07.81		
N270	CP3392	-396	28.07.81		
N271	CP3644	-405	28.07.81		
N272	CP4037	-388	04.08.81		
N273	CP4709	-397	04.08.81		
N274	CP8108	-418	26.08.81		
N275	CP8523	-451	26.08.81		
N276	CP9040	-447	27.08.81		
N277	CP9513	-406	01.09.81		
N278	CP9531	-390	01.09.81		
N279	CP9623	-438	01.09.81		
N280	CP9645	-435	01.09.81		
N281	CP9818	-394	01.09.81		
N282	CR1251	-441	04.09.81		
N283	CR2406	-459	04.09.81		
N284	CR2505	-458	09.09.81		
N285	CR2707	-446	09.09.81		
N286	CR2358	-449	11.09.81		
N287	CR2459	-436	11.09.81		
N288	CR2478	-401	11.09.81		
N289	CR3289	-452	17.09.81		
N290	CR3290	-461	17.09.81		
N291	CR3421	-450	17.09.81		
N292	CR3481	-460	17.09.81		
N293	CR3514	-457	17.09.81		
N294	CR3615	-442	17.09.81		
N295	CR4170	-454	21.09.81		
N296	CR4313	-428	21.09.81		
N297	CR4314	-444	21.09.81		
N298	CR4022	-432	23.09.81		
N299	CR4934	-429	23.09.81		
N300	CR4513	-427	24.09.81		
N301	CR4613	-456	24.09.81		
N302	CR4907	-437	24.09.81		
N303	CR5443	-433	28.09.81		
N304	CR6515	-416	01.10.81		
N305	CR6710	-417	01.10.81		
N306	CR6819	-398	01.10.81		
N307	CR6851	-391	01.10.81		
N308	CR6873	-393	01.10.81		
N309	CR6948	-445	01.10.81		
N310	CR7291	-439	08.10.81		
N311	CR7342	-419	08.10.81		
N312	CR7504	-443	12.10.81		
N313	CR8016	-434	14.10.81		
N314	CR9170	-440	16.10.81		
N315	CR9574	-431	16.10.81		
N316	CS 711	-426	16.10.81		
N317	CS 712	-425	16.10.81		
N318	CR9010	-409	19.10.81		
N319	CR9258	-421	19.10.81		
N320	CS 420	-418	23.10.81		
N321	CS 423	-430	23.10.81		
N322	CS1705	-415	27.10.81		
N323	CS1906	-412	27.10.81		
N324	CS1352	-414	28.10.81		
N325	CS2890	-410	02.11.81		
N326	CS3209	-411	06.11.81		
N327	CS3418	-413	06.11.81		
N328	CS3580	-473	09.11.81		

DENNIS JUBILANT (contd.)
KMB sub-class: Dennis(E) DUPLE-METSEC BODIES

Fleet number	Regn. number	Chassis number	Date first registered	Withdrawn	Disposal
N329	CS3842	-489	09.11.81		
N330	CS4673	-468	12.11.81		
N331	CS4819	-453	12.11.81		
N332	CS4936	-462	12.11.81		
N333	CS6562	-471	24.11.81		
N334	CT5805	-492	14.01.82		
N335	CT5855	-493	14.01.82		
N336	CT5876	-490	14.01.82		
N337	CT5013	-465	15.01.82		
N338	CT5035	-491	15.01.82		
N339	CT5087	-487	15.01.82		
N340	CT5138	-488	15.01.82		
N341	CT5190	-484	15.01.82		
N342	CT5266	-469	15.01.82		
N343	CT5288	-494	15.01.82		
N344	CT5339	-495	15.01.82		
N345	CT5512	-464	15.01.82		
N346	CT6188	-485	20.01.82		
N347	CT6189	-455	20.01.82		
N348	CT6258	-467	20.01.82		

DENNIS JUBILANT (contd.)
KMB sub-class: Dennis(E) DUPLE-METSEC BODIES

Fleet number	Regn. number	Chassis number	Date first registered	Withdrawn	Disposal
N349	CT9348	-466	11.02.82		
N350	CT9579	-472	11.02.82		
N351	CT9682	-463	11.02.82		
N352	CU1168	-468	18.02.82		
N353	CU1171	-474	18.02.82		
N354	CU1591	-483	18.02.82		
N355	CU3861	-480	25.02.82		
N356	CU3959	-475	25.02.82		
N357	CU4010	-470	04.03.82		
N358	CU4510	-478	04.03.82		
N359	CU4787	-482	04.03.82		
N360	CU4920	-479	04.03.82		
N361	CU5672	-476	17.03.82		
N362	CU7674	-477	26.03.82		
N363	CV5040	-481	07.05.82		

KMB sub-class: Dennis(F) - FORMERLY COACH.

Fleet number	Regn. number	Chassis number	Date first registered	Withdrawn	Disposal
N364	CF4180	DD204-106	11.06.80		ALEXANDER BODY

Buses Acquired Second Hand from UK Dealers.

LEYLAND TITAN PD3/5 - NEW 1961 to Ribble Motor Services Ltd. 2L 1-8

Fleet number	HK reg number	Chassis number	UK fleet number	UK reg number	Date reg in HK	Rebody Date	Withdrawn
2L 1	AD7470	603384	1748	PCK381	03.07.73	15.02.77	20.07.81
2L 2	AD7471	603149	1716	PCK357	24.05.73	03.01.77	
2L 3	AD7472	603188	1721	PCK362	03.07.73	-	23.09.75*
2L 4	AD7473	603416	1748	PCK389	03.07.73	-	29.09.76
2L 5	AD7474	603129	1713	PCK354	11.07.73	03.01.77	20.07.81
2L 6	AD7475	603496	1716	PCK393	11.07.73	13.08.76	20.07.81
2L 7	AD7476	603591	1755	PCK396	13.09.73	12.01.77	13.04.81
2L 8	AD7477	603474	1751	PCK392	13.12.73	-	11.10.75*

NOTE: *Decision to scrap not taken until 21.02.79.

LEYLAND ATLANTEAN PDR1/1 - New to Ribble & Standerwick - 'Gay Hostess' 2L9-21 & 61

Fleet number	HK reg number	Chassis number	UK fleet number	UK reg number	Date reg in HK	Rebody date	Withdrawn	KMB sub-type
2L 9	AD7457	592630	1287 (a)	NRN613	22.12.72	-	06.01.76	G/H
2L10	AD7458	592354	22 (c)		02.02.73	-	06.01.76	G/H
2L11	AD7459	592427	1289 (a)	SFV420	25.06.73	-	24.06.75	G/H
2L12	AD7460	592402	19 (e)	SFV415	24.05.73	-	12.07.76	G/H
2L13	AD7461	610692	28 (b)	VRF370	03.07.73	-	27.06.77	G/H
2L14	AD7462	592364	21 (d)	SFV417	03.07.73	-	28.04.76	G/H
2L15	AD7463	592629	1286 (a)	NRN612	25.06.73	08.03.76	19.07.79	G/H & G/H(C)
2L16	AD7464	592412	18 (e)	SFV414	26.06.73	-	06.01.76	G/H
2L17	AD7465	592385	1258 (f)	NRN607	13.07.73	-	26.05.76	G/H(B)
2L18	AD7466	610708	30 (f)	VRF372	13.07.73	-	23.03.76	G/H(B)
2L19	AD7467	592355	1256 (g)	NRN605	17.08.73	-	31.03./4	G/H
2L20	AD7468	592628	1263 (f)	NRN611	11.09.73	-	26.05.76	G/H(B)
2L21	AD7469	592404	1288 (a)	SFV413	24.09.73	-	01.10.75	G/H
2L61	AD7485	592365	20 (h)	SFV416	28.12.73	-	18.08.75	G/H

> **NOTES:** Vehicles registered in UK with VFR & SFV registration numbers NEW to W. C. STANDERWICK.
> Vehicles registered in UK with NRN registration numbers NEW to RIBBLE MOTOR SERVICES LTD.
> Some 'Gay Hostess' coaches were transferred between Ribble and Standerwick and back again, while others were, for a while, a part of the 'Scout' company, another group member; this applies to NRN611/2/3
>
> **Last UK operators** of 'Gay Hostess' coaches were (see code beside UK fleet number):-
> (a) Ribble Motor Services.   (d) Abbey, Selby.   (g) Phipps, London WC1.
> (b) W. C. Standerwick.   (e) Continental Pioneer, Richmond.   (h) Kirby Coaches, Kirby.
> (c) Thorn, Bubwith.   (f) City Coach Lines, London NW1.

LEYLAND ATLANTEAN PDR1/1 - New to Various UK Operators. 2L22-60 & 62-79

Fleet number	HK reg number	Chassis number	UK fleet number	UK reg number	Date reg in HK	Rebody date	Withdrawn	KMB sub-type
2L22	BD3904	591074	1618 (a)	NCK359	25.06.73	-	31.03.74	H/B
2L23	BD3905	591185	D138 (b)	138BRR	26.06.73	-	12.11.74	L/B
2L24	BD3906	591499	83 (g)	KCN183	25.06.73	-	19.06.75	H/B
2L25	BD5105	590989	1608 (a)	NCK349	05.07.74	-	31.03.74	H/B
2L26	BD5106	590719	432 (d)	ORC666	05.07.73	-	18.08.75	H/B
2L27	BD5107	590818	431 (d)	ORC665	05.07.73	-	23.09.75	H/B
2L28	BD5108	591589	243 (h)	CFT643	05.07.73	-	26.07.74	H/B
2L29	BD5109	592683	1927 (f)	927GPT	05.07.73	-	18.08.75	H/B
2L30	BD5110	591096	1622 (a)	NCK363	05.07.73	-	31.03.74	H/B
2L31	BD5111	590787	430 (d)	ORC664	05.07.73	-	18.08.75	H/B
2L32	BD5112	591623	1895 (f)	895EUP	05.07.73	-	20.01.76	H/B
2L33	BD5113	591301	D137 (b)	137BRR	05.07.73	-	31.03.74	L/B
2L34	BD5114	591187	D134 (b)	134BRR	05.07.73	-	01.08.75	L/B
2L35	BD5116	591302	D139 (b)	139BRR	05.07.73	-	01.08.75	L/B
2L36	BD5117	591186	D131 (b)	131BRR	05.07.73	-	06.11.76	L/B
2L37	BD5104	590775	462 (d)	ORC660	05.07.73	-	31.03.74	H/B
2L38	BD5621	591580	305 (i)	605EUP	11.07.73	-	01.08.75	H/B(B)
2L39	BD5831	592619	1923 (f)	923GPT	13.07.73	-	06.01.76	H/B(C)
2L40	BD6203	591498	87 (g)	KCN187	20.07.73	-	18.08.75	H/B(D)
2L41	BD6487	601116	921 (c)	5921W	25.07.73 -	01.08.75		H/B
2L42	BD8041	592620	1924 (f)	924GPT	13.07.73	-	12.11.74	H/B(C)
2L43	BD8783	591547	302 (i)	602EUP	28.07.73	-	07.04.75	H/B
2L44	BD9730	591842	439 (d)	RRC 73	05.09.73	-	04.12.75	H/B(C)
2L45	BE 446	603021	1703 (a)	PCK336	11.09.73	-	01.08.75	L/B
2L46	BE1580	591207	1634 (a)	NCK364	21.09.73	-	26.05.76	H/B

LEYLAND ATLANTEAN PDR1/1 - New to Various UK Operators (contd.)   2L22-60 & 62-79

Fleet number	HK reg number	Chassis number	UK fleet number	UK reg number	Date reg in HK	Rebody date	Withdrawn	KMB sub-type
2L47	BE1722	590819	429 (d)	ORC663	24.09.73	-	28.04.76	H/B
2L48	BE2419	590997	1609 (a)	NCK350	01.10.73	-	02.03.77	H/B
2L49	BE3615	591220	1638 (a)	NCK627	15.10.73	-	01.-8.75	H/B
2L50	BE3616	590988	1607 (a)	NCK348	15.10.73	-	31.03.74	H/B
2L51	BE4165	591964	455 (d)	RRC 89	23.10.73	-	07.04.75	H/B(D)
2L52	BE4166	601119	924 (c)	5924W	23.10.73	30.12.76	20.08.79	H/B no change
2L53	BE4684	590818	438 (d)	RRC 72	01.11.73	-	01.08.75	H/B(E)
2L54	BE4685	591079	1619 (a)	NCK360	01.11.73	-	16.03.76	H/B
2L55	BE5071	591209	1636 (a)	NCK625	05.11.73	-	31.03.74	H/B
2L56	BE5347	591219	1637 (a)	NCK626	08.11.73	-	18.09.76	H/B
2L57	BE 953	591598	1893 (f)	893EUP	17.09.73	-	31.07.74	H/B
2L58	BE6219	591923	449 (d)	RRC 83	22.11.73	-	31.03.74	H/B(B)
2L59	BE6499	591805	437 (d)	RRC 71	28.11.73	-	04.03.77	H/B(D)
2L60	BE6935	603018	1704 (a)	PCK337	06.12.73	-	18.10.76	L/B
2L62	BE7945	591950	454 (d)	RRC 88	02.01.74	-	27.02.75	H/B(D)
2L63	BE7946	592007	DH535(e)	535HKJ	02.01.74	-	04.02.77	H/B
2L64	BE7974	601529	DH563(e)	563HKJ	02.01.74	-	24.11.77	H/B
2L65	BE7948	591976	DH530(e)	530HKJ	02.01.74	-	13.05.76	H/B
2L66	BE8273	592576	DH541(e)	541HKJ	08.01.74	-	25.04.77	H/B
2L67	BE8656	591742	DH527(e)	527HKJ	15.01.74	05.01.77	19.07.79	H/B & H/B(F)
2L68	BE8657	601050	917 (c)	5917W	15.01.74	-	28.06.76	H/B
2L69	BE8658	592060	DH540(e)	540HKJ	15.01.74	-	04.03.77	H/B
2L70	BE8656	592077	DH542(e)	542HKJ	15.01.74	-	18.09.76	H/B
2L71	BE8983	592308	DH549(e)	549HKJ	21.01.74	08.01.74	20.08.79	H/B & H/B(A)
2L72	BE8984	592137	DH545(e)	545HKJ	21.01.74	-	31.01.77	H/B
2L73	BE9616	591957	DH529(e)	529HKJ	12.02.74	-	15.09.76	H/B
2L74	BE9617	591231	1642 (a)	NCK631	12.02.74	-	28.04.76	H/B
2L75	BE9618	592095	DH543(e)	543HKJ	12.02.74	-	16.08.76	H/B
2L76	BE9755	601378	DH554(e)	554LKP	15.02.74	-	26.01.77	H/B
2L77	BE9756	601365	DH552(e)	552LKP	15.02.74	-	26.01.77	H/B
2L78	BG 655	601503	DH557(e)	557LKP	01.03.74	-	*31.03.74	H/B
2L79	BG1820	591600	1894 (f)	894EUP	01.04.74	-	23.07.76	H/B
-	-	591548	304 (i)	604EUP	Cannibalised	-	19.02.76	H/B

*Withdrawn after only a few days service.

THE VEHICLES LISTED ABOVE were **NEW** to: (see code letter in brackets after the UK fleet number in the table above)
- a. Ribble Motor Services.
- b. East Midlands Motor Services.
- c. Sheffield Corporation.
- d. Trent Motor Traction.
- e. Maidstone & District Motor Services.
- f. Northern General Transport.
- g. Gateshead & District
- h. Tynemouth & District
- i. Sunderland District

General Note
Many of the above Leyland 'Atlantean' buses were registered in Hong Kong using the chassis number actually carried on the chassis and this differed from that originally issued by Leyland, due to the chassis number-plates being fixed to a removable part of the rear chassis extension. They had inadvertently been swapped around during major component changes, prior to export.

AEC REGENT MkV 2D3RA or 2D2RA - New to Various UK Operators   2A1-28

Fleet number	HK reg number	Chassis number	UK fleet number	UK regd. number	Date regd. in HK	Rebody date	Withdrawn	KMB sub-type	
2A 1	BD 925	2D3RA 909	94 (a)	FDH121	25.05.73	05.78	24.03.81	(M)	
2A 2	BD3900	2D3RA 926	450 (b)	450GTX	22.06.73	05.78	10.03.81	(N)	
2A 3	BD3901	2D3RA 984	462 (b)	462GTX	22.06.73	02.78	16.10.81	(N)	
2A 4	BD3907	2D3RA 908	93 (a)	FDH120	25.06.73	02.78	11.06.81	(M)	
2A 5	BD3908	2D3RA 906	91 (a)	FDH118	25.06.73	09.76	11.06.81	(M)	
2A 6	BD3909	2D3RA 483	70 (a)	DHD181	25.06.73	06.78	28.04.81	(M)	
2A 7	BD3910	2D3RA 924	448 (b)	448GTX	25.06.73	*	05.05.82	(N)	
2A 8	BD6204	2D3RA 928	452 (b)	452GTX	20.07.73	08.77	11.08.82	(N)	
2A 9	BD6479	2D3RA 922	446 (b)	446GTX	14.07.73	09.76	22.07.81	(N)	
2A10	BD6780	2D3RA 986	464 (b)	464KTG	27.07.73	05.76	10.03.82	(N)	
2A11	BD5118	2D3RA 958	463 (b)	463KTG	05.07.73	02.78	25.09.81	(N)	
2A12	BD5119	2D3RA 981	459 (b)	459KTG	05.07.73	03.75	17.11.81	(N)	
2A13	BD5120	2D3RA 976	454 (b)	454KTG	05.07.73	03.75	24.08.81	(N)	
2A14	BD5121	2D3RA 929	453 (b)	453GTX	05.07.73	06.78	24.08.81	(N)	
2A15	BD5122	2D3RA 923	447 (b)	447GTX	05.07.73	10.76	22.07.81	(N)	
2A16	BD5123	2D3RA 749	82 (a)	EHD447	05.07.73	02.77	25.01.83	(N)	
2A17	BD5124	2D3RA 977	455 (b)	455KTG	05.07.73	02.78	16.10.81	(N)	
2A18	BD5125	2D3RA 921	445 (b)	445GTX	05.07.73	02.78	25.09.81	(N)	
2A19	BD5125	2D3RA 925	449 (b)	449GTX	05.07.73	11.76	05.03.81	(N)	
2A20	BD8782	2D3RA 920	375 (cb)	NCP475	22.08.73	11.76	11.06.81	(S)	
2A21	BE 726	2D3RA 764	11 (c)	LJX 11	13.09.73	07.78	14.05.81	(S)	
2A22	BE4682	2D3RA 980	458 (b)	458KTG	01.11.73	02.77	22.06.81	(N)	
2A23	BE4683	2D3RA 757	212 (ca)	LJX212	01.11.73	05.76	10.03.82	(S)	
2A24	BE6793	2D3RA 765	12 (c)	LJX 12	04.12.73	02.77	25.01.83	(S)	
2A25	BG 657	2D2RA1525	531 (d)	350HWE	01.03.74	08.78	07.05.82	(N)	
2A26	BG 658	2D2RA1526	532 (d)	351HWE	01.03.74	**TC**	12.05.78	(N)	
2A27	BG4139	2D2RA 950	194 (e)	SCX194	03.06.74	12.75	30.07.80	(M)	
2A28	BG8544	2D3RA 907	92 (a)	FHD119	16.09.74	-	09.03.77	(N)	Written-off 27.02.79

NOTE: Code letters after the UK fleet number indicate the original owner:
- a. Yorkshire Woollen District
- b. Rhondda via Western Welsh - same fleet numbers
- c. Halifax 'A' fleet - Halifax Corporation.
- ca. Halifax 'B' fleet - Halifax Joint Committee Calderdale Joint Omnibus Ctte.
- cb. Hebble Motor Services.
- d. Sheffield Joint Omnibus Committee. (British Railways fleet)
- e. Huddersfield Corporation.

OTHER NOTES: * 2A7 - KMB records show no period in workshops long enough to be re-bodied but the bus saw service until 1982, unusual for an original body.

TC 2A26 - converted to Tree Cutting bus 12.05.78; de-registered 23.04.80 but not broken-up until April 1981.

DAIMLER CCG6-27 - New to Chesterfield Corporation.   2D1-3

Fleet number	HK reg number	Chassis number	UK fleet number	UK reg number	Date reg in HK	Rebody date	Withdrawn
2D 1	AD7478	20017	258	3258NU	24.05.73	18.11.76	12.01.84
2D 2	AD7479	20015	256	3256NU	03.07.73	-	06.01.76
2D 3	AD7480	20012	253	3253NU	03.07.73	-	12.07.76

DAIMLER CVG6-30 - New to Bolton Corporation.   2D4-7

Fleet number	HK reg number	Chassis number	UK fleet number	UK reg number	Date reg in HK	Rebody date	Withdrawn
2D 4	AD7481	30076	6644	PBN662	01.07.73	24.08.77	28.04.83
2D 5	AD7482	30079	6647	PBN665	13.07.73	19.08.77	28.04.83
2D 6	AD7483	30081	6649	PBN667	27.07.73	23.02.76	25.01 83
2D 7	AD7484	30082	6650	PBN668	05.09.73	23.02.76	11.08.82

> **NOTE:**
> 1) UK fleet numbers shown are for SELNEC; Bolton numbers were 144/7/9/50.
> 2) **SELNEC:-** South East Lancashire, North East Cheshire - the forerunners of Greater Manchester Passenger Transport Executive (GMPTE).

DAIMLER FLEETLINE CRL6 - New to London Transport.   Total 100   - 2D8-107

Fleet number	HK reg number	Chassis number	UK fleet number	UK reg number	Date reg in H.K.	Withdrawn	Disposal see notes	Remarks
2D 8	CM2856	67050	DMS 725	TGX725M	01.04.81	04.12.87	Speedybus - Guangzhou No 1 Bus Co	
2D 9	CM2857	67086	DMS 739	TGX739M	01.04.81	04.12.87	Speedybus - Guangzhou No 1 Bus Co	
2D10	CM3985	67104	DMS 763	TGX763M	03.04.81	17.10.86	Trg.	Sc by 10/90
2D11	CM4087	66998	DMS 685	MLK685L	08.04.81	04.12.87	Speedybus - Guangzhou No 1 Bus Co	
2D12	CM4727	67080	DMS1554	THM554M	08.04.81	07.10.86	Trg	
2D13	CM4782	66960	DMS 698	TGX698M	08.04.81	04.12 87	Speedybus - Guangzhou No 1 Bus Co	
2D14	CM4841	67108	DMS 758	TGX758M	08.04.81	03.89	Speedybus - Wuzhou	
2D15	CM4879	67115	DMS 766	TGX766M	08.04.81	07.10.86	Sc in HK	Accident victim
2D16	CM4939	67977	DMS 693	MLK693L	08.04.81	17.10 86	Trg.	Sc by 10/90
2D17	CM8337	67154	DMS1587	THM587M	05.05.81	03.89	Speedybus - Wuzhou	
2D18	CM8999	67072	DMS1551	THM551M	05.05.81	17.10.86	Trg.	
2D19	CM8371	67150	DMS1577	THM577M	07.05.81	03.89	Speedybus - Tianjin	
2D20	CM8751	67097	DMS 751	TGX751M	07.05.81	03.89	Speedybus - Wuzhou	
2D21	CM8752	67121	DMS 765	TGX765M	07.05.81	17.10.86	Trg.	
2D22	CN 672	67075	DMS 735	TGX735M	15.05.81	17.05.84	Trg. -	withdrawn by 5/93
2D23	CM1282	67172	DMS 779	TGX779M	15.05.81	09.89	Sc	
2D24	CN1527	67041	DMS1532	THM532M	15.05.81	03.89	Speedybus - Tianjin	
2D25	CN1528	67124	DMS1574	THM574M	15.05.81	03.89	Speedybus - for spare parts.	
2D26	CN1529	66992	DMS 694	MLK694L	15.05.81	04.89	Speedybus - Dalian	
2D27	CN 855	67056	DMS 727	TGX727M	19.05.81	04.89	Sc. 9/89	
2D28	CN1106	67174	DMS1599	THX599M	19.95.81	04.89	Sc. 9/89	
2D29	CN 311	67083	DMS1557	THM557M	22.05.81	25.10.84	Sc. -/84	
2D30	CN 315	67105	DMS 760	TGX760M	22.05.81	17.10.86	Trg.	
2D31	CN 476	67139	DMS1593	THM593M	22.05.81	03.89	Speedybus - Tianjin	
2D32	CN2661	67071	DMS1550	THM550M	26.05.81	03.89	Speedybus - for spare parts - then to Dalian	
2D33	CN2494	67058	DMS1541	THM541M	27.05.81	03.89	Speedybus - Tianjin	
2D34	CN2740	67070	DMS1549	THM549M	27.05.81	24.10.84	Trg.	Sc by 10/90
2D35	CN2741	67043	DMS1533	THM533M	27.05.81	04.90	Sc. 4/89	
2D36	CN3370	67114	DMS1567	THM567M	01.06.81	04.09.85	Trg.	Sc by 10/90
2D37	CN4510	67011	DMS 699	TGX699M	01.06.81	16.02.88	Sc. 2/88	
2D38	CN4588	67168	DMS1595	THM595M	01.06.81	22.07.85	Trg.	Sc by 10/90
2D39	CN4590	67036	DMS 724	TGX724M	01.06.81	26.07.85	Trg. -	withdrawn by 5/93
2D40	CN4746	67062	DMS1545	THM545M	01.06.81	20.10.87	Sc. 10/87	
2D41	CN3897	67176	DMS1600	THM600M	04.06.81	23.03.88	Sc. 3/88	
2D42	CN3967	67156	DMS1588	THM588M	04.06.81	16.02.88	Speedybus - Dalian	
2D43	CN4255	67149	DMS 741	TGX741M	04.06.81	04.89	Sc 4/89	
2D44	CN4275	67136	DMS1582	THM582M	04.06.81	06.89	Speedybus - Wuzhou	
2D45	CN4375	67173	DMS1598	THM958M	04.06.81	12.08.85	Trg.	
2D46	CN3966	67002	DMS 700	TGX700M	05.06.81	11.88?	Speedybus - Dalian	
2D47	CN3970	67175	DMS 780	TGX780M	05.06.81	06.88	Speedybus - Guilin	
2D48	CN4313	67126	DMS1575	THM575M	05.06.81	26.07.85	Trg.	
2D49	CN5075	67100	DMS 757	TGX757M	10.06.81	16.02.88	Sc.	
2D50	CN5028	67087	DMS 749	TGX749M	12.06.81	04.89	Sc.	
2D51	CN6296	67107	DMS 753	TGX753M	17.06.81	20.10.87	Sc.	
2D52	CN6349	67137	DMS1583	THM583M	17.06.81	18.03.88	Sc.	
2D53	CN6431	66947	DMS 695	MLK695L	17.06.81	09.88	Sc.	
2D54	CN6777	67079	DMS 733	TGX733M	17.06.81	06.89	Not traced.	
2D55	CN7013	66954	DMS 642	MLK642L	23.06.81	06.89	Speedybus - Dalian	
2D56	CN7060	67194	DMS1605	THM605M	23.06.81	06.89	Speedybus - For spares	
2D57	CN7269	67167	DMS1594	THM594M	23.06.81	04.89	Sc.	
2D58	CN8512	67054	DMS1539	THM539M	02.07.81	26.08.88	Sc.	
2D59	CN9490	67017	DMS 703	TGX703M	03.07.81	25.07.88	Sc.	
2D60	CN9493	67151	DMS1555	THM555M	03.07.81	29.04.88	Sc.	
2D61	CN9710	67096	DMS1562	THM562M	03.07.81	10.09.85	Trg.	
2D62	CP 174	67113	DMS1566	THM566M	03.07.81	01.90	Speedybus - Shengyang	
2D63	CP 182	67061	DMS1544	THM544M	03.07.81	29.12.86	Sc.	
2D64	CP 287	67063	DMS 726	TGX726M	03.07.81	22.04.88	Sc.	
2D65	CP 298	67193	DMS 795	TGX795M	03.07.81	10.09.85	Trg.	
2D66	CP 322	67082	DMS1556	THM556M	03.07.81	03.89	Speedybus - Wuzhou	
2D67	CP 325	67123	DMS1573	THM573M	03.07.81	09.04.84	Trg.	Sc 3/90
2D68	CP2398	67111	DMS1564	THM564M	20.07.81	23.06.88	Sc.	
2D69	CP2155	67159	DMS1585	THM585M	22.07.81	04.90	Speedybus - Dalian	
2D70	CP2325	67047	DMS 722	TGX722M	22.07.81	08.89	SB 6/10/89 -	Not traced.
2D71	CP2484	67132	DMS 777	TGX777M	22.07.81	08.89	Speedybus - Shengyang	
2D72	CP3040	67208	DMS 800	TGX800M	22.07.81	22.04.88	Sc.	
2D73	CP3090	67102	DMS 762	TGX762M	24.07.81	29.04.88	Sc.	
2D74	CP3134	67200	DMS 797	TGX797M	24.07.81	31.10.85	Trg. Sc by 10/90	
2D75	CP3237	67067	DMS 730	TGX730M	24.07.81	04.09.85	Trg.	
2D76	CP3240	67183	DMS 787	TGX787M	24.07.81	29.04.88	Sc.	
2D77	CP3337	67098	DMS 756	TGX756M	24.07.81	05.07.88	Sc.	
2D78	CP3506	67196	DMS1606	THM606M	24.07.81	08.89	Sc.	
2D79	CP3638	67186	DMS 791	TGX791M	24.07.81	25.07.88	Sc.	
2D80	CP3690	67161	DMS 770	TGX770M	24.07.81	29.04.88	Sc.	
2D81	CP3824	67192	DMS1609	THM609M	24.07.81	08.89	Speedybus - Shengyang	
2D82	CP3325	67064	DMS 728	TGX728M	20.07.81	08.89	Speedybus - Shengyang	

> **NOTE: Disposal abbreviations:-**
> **SB-China:** Sent for use in China by Speedybus (SB).
> **Sc.** : Scrapped - final withdrawal.
> **Trg.** : To training fleet.
> **SB** : Speedybus Services Ltd **or** Speedybus Enterprises Ltd -(HK dealer).
> **GZ1** : Guangzhou No.1 Bus Company.

DAIMLER FLEETLINE CRL6 - New to London Transport (Continued.)   Total 100   -   2D8-107

MCW or PARK ROYAL BODIES

Fleet number	HK reg	Chassis number	UK fleet number	UK reg number	Date reg in H.K.	Withdrawn	Disposal see notes	Remarks
2D83	CP4395	67165	DMS1592	THM592M	04.08.81	22.11.85	Trg. Sc by 10/90	Sc by 10/90
2D84	CP4597	67095	DMS1561	THM561M	04.08.81	25.07.88	Sc. by 10/90	
2D85	CP4690	67116	DMS1570	THM570M	04.08.81	09.89	Sc.	
2D86	CP4748	67190	DMS1602	THM602M	04.08.81	05.07.88	Sc.	
2D87	CP4715	67142	DMS 783	TGX783M	06.08.81	10.05.88	Sc.	
2D88	CP4903	67068	DMS1547	THM547M	06.08.81	20.10.87	Sc.	
2D89	CP4970	67122	DMS1572	THM572M	06.08.81	05.89	Sc.	
2D90	CP5026	67120	DMS1571	THM571M	11.08.81	09.89	Sc.	
2D91	CP5092	67032	DMS 716	TGX716X	11.08.81	09.04.04	Trg.	Sc by 10/90
2D92	CP5432	67046	DMS 718	TGX718M	11.08.81	08.89	Speedybus - Shengyang	
2D93	CP5478	67112	DMS1565	THM565M	11.08.81	08.89	SB 6/10/89 - Not traced.	
2D94	CP9429	66943	DMS1483	MLH483L	27.08.81	16.06.88	Sc.	
2D95	CR4733	67340	DMS1635	THM635M	21.09.81	05.07.88	Sc.	
2D96	CS3583	67195	DMS1607	THM607M	09.11.81	11.88	Sc.	
2D97	CT3692	66951	DMS 687	MLK687L	06.01.82	20.10.87	Sc.	
2D98	CT6138	67073	DMS 732	TGX732M	20.01.82	20.10.87	Sc.	
2D99	CT7108	67127	DMS1576	THM576M	01.02.82	04.12.87	SB-China - GZ1	
2D100	CV9173	67338	DMS1637	THM637M	21.06.82	17.10.86	Trg. Sc by 10/90	
2D101	CV9753	67188	DMS 794	TGX794M	21.06.82	17.10.86	Trg.	
2D102	CX1775	67171	DMS1597	THM597M	12.10.82	09.12.85	Sc.	
2D103	CY4971	67141	DMS 785	TGX785M	16.03.83	04.12.87	SB-China - GZ1	
2D104	CY8605	66890	DMS 605	MLK605L	02.05.83	17.10.86	Trg. Sc by 3/90	
2D105	CY8517	67179	DMS 784	TGX784M	12.05.83	03.89	Speedybus - Wuzhou.	
2D106	CZ1465	66925	DMS1497	MLH497L	06.06.83	06.89	Speedybus - Tianjin	
2D107	CZ5930	67088	DMS 740	TGX740M	11.07.83	06.89	Speedybus - Tianjin	

2-Axle Buses Purchased New in the 1980's

LEYLAND B45 OLYMPIAN   Fleet nos. BL1-3
ECW BODIES

Fleet number	Regn. number	Chassis number	Date first registered
BL1	CP3323	ON/3	24.07.81
BL2	CP3830	B45/09	24.07.81
BL3	CR2963	B45/08	11.09.81

LEYLAND OLYMPIAN ON6LXB/1R - 2AXLE
Fleet nos. BL4-120   ALEXANDER BODIES

Fleet number	Regn. number	Chassis number	Date first registered
BL4	CZ8666	ON764	25.00.03
BL5	CZ9684	ON782	25.08.83
BL6	CZ9891	ON754	25.08.83
BL7	CZ9906	ON753	25.08.83
BL8	DA 210	ON752	25.08.83
BL9	DA 316	ON763	25.08.83
BL10	CZ9123	ON781	01.09.83
BL11	CZ9564	ON755	01.09.83
BL12	CZ9667	ON756	01.09.83
BL13	CZ9917	ON751	01.09.83
BL14	DA 110	ON803	01.09.83
BL15	DA 144	ON746	01.09.83
BL16	DA 489	ON780	01.09.83
BL17	CZ9873	ON802	08.09.83
BL18	CZ9894	ON776	08.09.83
BL19	CZ9328	ON801	13.09.83
BL20	DA 365	ON779	13.09.83
BL21	DA 491	ON777	13.09.83
BL22	DA1790	ON747	03.10.83
BL23	DA1921	ON775	03.10.83
BL24	DB6686	ON1219	01.05.84
BL25	DB7311	ON1187	01.05.84
BL26	DB7333	ON1195	01.05.84
BL27	DB7753	ON1189	01.05.84
BL28	DB7782	ON1229	10.05.84
BL29	DB8234	ON1186	01.05.84
BL30	DB8328	ON1188	01.05.84
BL31	DB8334	ON1194	01.05.84
BL32	DB6920	ON1240	03.05.84
BL33	DB6989	ON1233	03.05.84
BL34	DB7198	ON1218	03.05.84
BL35	DB7250	ON1231	03.05.84
BL36	DB7290	ON1230	07.05.84
BL37	DB7872	ON1232	07.05.84
BL38	DB9175	ON1239	09.05.84
BL39	DB9908	ON1256	09.05.84
BL40	DB9253	ON1254	14.05.84
BL41	DB9635	ON1241	14.05.84
BL42	DB9881	ON1255	14.05.84
BL43	DC 294	ON1248	14.05.84
BL44	DB8809	ON1253	09.05.84
BL45	DB9629	ON1263	23.05.84
BL46	DB9740	ON1257	24.05.84
BL47	DC 983	ON1265	01.06.84
BL48	DC1500	ON1266	01.06.84
BL49	DC1611	ON1267	01.06.84
BL50	DC2270	ON1150	01.06.84
BL51	DC 726	ON1264	05.06.84
BL52	DC1407	ON1247	05.06.84
BL53	DD3221	ON1217	15.10.84
BL54	DE4044	ON1502	18.02.85
BL55	DE3824	ON1503	18.02.85
BL56	DE3814	ON1538	18.02.85
BL57	DE2763	ON1539	10.02.85
BL58	DE3742	ON1567	18.02.85
BL59	DE4314	ON1473	18.02.85
BL60	DE2817	ON1500	18.02.85

LEYLAND OLYMPIAN   ON6LXB/1R - 2-axle
Fleet nos. BL4-120 (Contd.)   ALEXANDER BODIES

Fleet number	Regn. number	Chassis number	Date first registered	Withdrawn	Disposal
BL61	DE4413	ON1550	18.02.85		
BL62	DE3528	ON1540	18.02.85		
BL63	DE2928	ON1566	18.02.85		
BL64	DE2897	ON1585	18.02.85		
BL65	DE3037	ON1465	18.02.85		
BL66	DE3271	ON1568	18.02.85		
BL67	DE4058	ON1580	18.02.85		
BL68	DE3414	ON1553	18.02.85		
BL69	DE4076	ON1577	02.02.05		
BL70	DE3580	ON1565	05.03.85		
BL71	DE4472	ON1551	05.03.85		
BL72	DE2569	ON1586	07.03.85		
BL73	DE2717	ON1554	07.03.85		
BL74	DE3018	ON1597	11.03.85		
BL75	DE4016	ON1552	11.03.85		
BL76	DE4089	ON1584	11.03.85		
BL77	DE2889	ON1588	12.03.85		
BL78	DE3078	ON1541	12.03.85		
BL79	DE3104	ON1578	12.03.85		
BL80	DE3522	ON1576	12.03.85		
BL81	DE5456	ON1569	14.03.85		
BL82	DE5565	ON1609	14.03.85		
BL83	DE5729	ON1589	14.03.85		
BL84	DE5764	ON1587	14.03.85		
BL85	DE6127	ON1589	14.03.85		
BL86	DE6317	ON1571	14.03.85		
BL87	DE5387	ON1603	19.03.85		
BL88	DE5569	ON1601	19.03.85		
BL89	DE5655	ON1611	19.03.85		
BL90	DE5854	ON1602	19.03.85		
BL91	DE6085	ON1610	19.03.85		
BL92	DE5313	ON1485	21.03.85		
BL93	DE5491	ON1599	21.03.85		
BL94	DG 660	ON1885	16.08.85		
BL95	DG2105	ON1909	16.08.85		
BL96	DG 825	ON1897	19.08.85		
BL97	DG1457	ON1896	19.08.85		
BL98	DG1627	ON1899	19.08.85		
BL99	DG1701	ON1909	28.08.85		
BL100	DG1815	ON1920	28.08.85		
BL101	DG1874	ON1911	28.08.85		
BL102	DG2249	ON1916	28.08.85		
BL103	DG2548	ON1918	28.08.85		
BL104	DG2596	ON1907	28.08.85		
BL105	DG2633	ON1910	05.09.85		
BL106	DG2509	ON1930	11.09.85		
BL107	DG2520	ON1941	11.09.85		
BL108	DG2529	ON1929	11.09.85		
BL109	DG2935	ON1932	11.09.85		
BL110	DG3065	ON1942	11.09.85		
BL111	DG3514	ON1919	11.09.85		
BL112	DG4273	ON1933	11.09.85		
BL113	DG2692	ON1943	17.09.85		
BL114	DG3001	ON1939	17.09.85		
BL115	DG3098	ON1917	17.09.85		
BL116	DG3155	ON1950	17.09.85		
BL117	DG3734	ON1952	17.09.85		
BL118	DG4112	ON1898	17.09.85		
BL119	DG4152	ON1951	17.09.85		
BL120	DG4335	ON1931	17.09.85		

KMB sub-class: Leyland Olympian(C)   ALEXANDER BODIES

Fleet number	Regn. number	Chassis number	Date first registered
BL121	DG2896	ON1949	17.09.85
BL122	DG3324	ON1953	17.09.85
BL123	DG4421	ON1940	17.09.85

MCW METROBUS - 9.67 metre Fleet nos. M1-88
KMB sub-class: Metrobus(A), or M59-60 (B)   MCW BODIES

Fleet number	Regn. number	Chassis number	Date first registered	Withdrawn	Disposal
M 1	CZ 745	MB7186	09.06.83		
M 2	CZ1014	MB7191	09.06.83		
M 3	CZ1914	MB7190	09.06.83		
M 4	CZ2039	MB7188	09.06.83		
M 5	CZ2611	MB7187	14.06.83		
M 6	CZ2864	MB7189	14.06.83		
M 7	CZ3119	MB7184	14.06.83		
M 8	CZ4434	MB7185	14.06.83		
M 9	CZ4133	MB7192	16.06.83		
M10	CZ3903	MB7193	27.06.83		
M11	CZ3920	MB7196	27.06.83		
M12	CZ4418	MB7195	27.06.83		
M13	CZ3621	MB7194	07.07.83		
M14	CZ5866	MB7199	07.07.83		
M15	CZ5944	MB7200	07.07.83		
M16	CZ4825	MB7198	11.07.83		
M17	CZ6450	MB7197	11.07.83		
M18	CZ6482	MB7201	11.07.83		
M19	CZ5049	MB7202	19.07.83		
M20	DA3079	MB7203	24.10.83		

Metrobus(C)   MCW BODIES

Fleet	Regn.	Chassis	Date first		
M21	DC9306	MB7738	07.09.83		

Metrobus(A)   MCW BODIES

Fleet	Regn.	Chassis	Date first	Notes	Disposal
M22	DC9683	MB7737	07.09.83		
M23	DD 364	MB7739	07.09.83		
M24	DC8707	MB7736	10.09.83		
M25	DC9288	MB7742	10.09.83		
M26	DC9418	MB7740	14.09.83		
M27	DC9587	MB7741	14.09.83		
M28	DC9961	MB7735	14.09.83		
M29	DC9077	MB7744	17.09.83		
M30	DD 148	MB7743	17.09.83		
M31	DE6718	MB8031	25.04.85		
M32	DE7101	MB8029	25.04.85		
M33	DE8232	MB8027	25.04.85		
M34	DE8287	MB8028	25.04.85		
M35	DE7739	MB8026	29.04.85		
M36	DE8234	MB8032	29.04.85		
M37	DE8819	MB8034	02.05.85		
M38	DE9730	MB8033	02.05.85		
M39	DE8723	MB8038	05.05.85		
M40	DE8858	MB8035	06.05.85		
M41	DE9520	MB8030	06.05.85		
M42	DE8563	MB8037	09.05.85		
M43	DE8745	MB8036	09.05.85		
M44	DE9936	MB8042	09.05.85		
M45	DE8779	MB8039	13.05.85		
M46	DE8929	MB8041	13.05.85		
M47	DE9975	MB8043	13.05.85		
M48	DF1402	MB8048	16.05.85		
M49	DF2119	MB8025	16.05.85		
M50	DF 781	MB8040	22.05.85		
M51	DF 956	MB8045	22.05.85		
M52	DF1103	MB8049	22.05.85		
M53	DF1769	MB8046	22.05.85		
M54	DF 570	MB8047	24.05.85		
M55	DF2472	MB8044	24.05.85		
M56	DF1845	MB8050	29.05.85		
M57	DF2140	MB8051	29.05.85		
M58	DF3520	MB8053	13.06.85		Converted to (D) 18.12.91
M59	DF4437	MB8054	13.06.85 (B)	04.06.91 Sc	
M60	DF4467	MB8052	13.06.85 (B)		Converted to (D) 02.12.91
M61	DG7079	MB8180	25.10.85		
M62	DG7558	MB8175	25.10.85		
M63	DG7614	MB8183	25.10.85		
M64	DG8084	MB8177	25.10.85		
M65	DG9281	MB8185	04.11.85		
M66	DG9319	MB8192	04.11.85		
M67	DG9377	MB8174	04.11.85		
M68	DG9531	MB8184	04.11.85		
M69	DG9670	MB8182	04.11.85		
M70	DG9721	MB8178	04.11.85		
M71	DH 145	MB8179	04.11.85		
M72	DG8953	MB8181	08.11.85		
M73	DG9851	MB8186	08.11.85		
M74	DH 508	MB8176	22.11.85		
M75	DH 699	MB8188	22.11.85		
M76	DH 749	MB8191	22.11.85		
M77	DH 921	MB8173	22.11.85		
M78	DH1170	MB8189	22.11.85		
M79	DH2408	MB8190	22.11.85		
M80	DH2436	MB8187	22.11.85		

KMB sub-class: Metrobus(D) - as built new.   MCW BODIES

Fleet	Regn.	Chassis	Date first		
M81	EH4891	MB10357	11.89		
M82	EH5154	MB10358	11.89		
M83	EH5692	MB10362	11.89		
M84	EH5383	MB10361	11.89		
M85	EH5919	MB10360	11.89		
M86	EH6775	MB10359	11.89		
M87	EH6783	MB10363	11.89		
M88	EH7728	MB10363	11.89		

DENNIS DOMINATOR 9.5 metre - 2-axle
Fleet nos. DM1-40   DUPLE METSEC BODIES

Fleet number	Regn. number	Chassis number	Date first registered	Withdrawn	Disposal
DM 1	DA7157	DDA166-647	23.12.83		
DM 2	DA9195	DDA166-653	09.01.84		
DM 3	DA9223	DDA166-652	09.01.84		
DM 4	DA9385	DDA166-640	09.01.84		
DM 5	DB 450	DDA166-642	09.01.84		

DENNIS DOMINATOR 9.5 metre - 2-axle
Fleet nos. DM1-40 (Contd.)   DUPLE METSEC BODIES

Fleet number	Regn. number	Chassis number	Date first registered	Withdrawn	Disposal
DM 6	DA8529	DDA166-635	12.01.84		
DM 7	DA9153	DDA166-654	12.01.84		
DM 8	DA9537	DDA166-639	12.01.84		
DM 9	DA9068	DDA166-643	16.01.84		
DM10	DA9155	DDA166-648	16.01.84		
DM11	DB 416	DDA166-641	16.01.84		
DM12	DA8758	DDA166-651	19.01.84		
DM13	DA9289	DDA166-638	19.01.84		
DM14	DB 472	DDA166-636	19.01.84		
DM15	DB1256	DDA166-649	24.01.84		
DM16	DB 914	DDA166-650	27.01.84		
DM17	DB1931	DDA166-644	27.01.84		
DM18	DB 951	DDA166-645	30.01.84		
DM19	DB1785	DDA166-646	30.01.84		
DM20	DB2409	DDA166-633	30.01.84		
DM21	DC5078	DDA175-727	01.08.84		
DM22	DC5159	DDA175-732	01.08.84		
DM23	DC7298	DDA175-736	06.08.84		
DM24	DC7349	DDA175-734	06.08.84		
DM25	DC7856	DDA175-735	06.08.84		
DM26	DC8429	DDA175-728	06.08.84		
DM27	DC7287	DDA175-726	08.08.84		
DM28	DC6635	DDA175-737	10.08.84		
DM29	DC7520	DDA175-730	10.08.84		
DM30	DC7456	DDA175-771	13.08.84		
DM31	DC8351	DDA175-776	13.08.84		
DM32	DC6683	DDA175-729	16.08.84		
DM33	DC7541	DDA175-778	16.08.84		
DM34	DC6812	DDA175-777	20.08.84		
DM35	DC8289	DDA175-775	20.08.84		
DM36	DC6630	DDA175-733	24.08.84		
DM37	DC8066	DDA175-731	24.08.84		
DM38	DC7484	DDA175-773	28.08.84		
DM39	DC8397	DDA175-772	28.08.84		
DM40	DC9328	DDA175-774	03.09.84		

MERCEDES-BENZ Model 0.303 11 metre - 2-axle -
Prototype bus. Fleet no: ME1.   ALEXANDER BODY

Fleet number	Regn. number	Chassis number	Date first registered	Withdrawn	Disposal
ME 1	CZ6686	307001-21-0304043-6	04.08.83		

MERCEDES-BENZ Model 0.303 11-metre - 2-axle
Fleet nos: ME2-41.   ALEXANDER BODIES

Fleet number	Regn. number	Chassis number	Date first registered	Withdrawn	Disposal
ME 2	DF7102	044121	19.07.85		
ME 3	DF6700	043982	22.07.85		
ME 4	DF7896	044266	22.07.85		
ME 5	DF6548	044368	24.07.85		
ME 6	DF6853	044385	24.07.85		
ME 7	DF6909	044597	24.07.85		
ME 8	DF8120	044302	24.07.85		
ME 9	DF7632	044621	30.07.85		
ME10	DF7840	044639	30.07.85		
ME11	DF8962	044656	08.08.85		
ME12	DF9321	044194	08.08.85		
ME13	DF9534	044570	08.08.85		
ME14	DF9565	044605	08.08.85		
ME15	DF9698	044230	08.08.85		
ME16	DG 316	044655	08.08.85		
ME17	DF8519	044638	12.08.85		
ME18	DF9149	044636	12.08.85		
ME19	DF9520	044096	12.08.85		
ME20	DF9705	044552	12.08.85		
ME21	DF9740	044158	12.08.85		
ME22	DG2612	044449	17.09.85		
ME23	DG2758	044500	17.09.85		
ME24	DG4601	044433	23.09.85		
ME25	DG4602	044465	23.09.85		
ME26	DG6240	044534	23.09.85		
ME27	DG6377	044499	23.09.85		
ME28	DG5010	044400	25.09.85		
ME29	DG5071	044464	25.09.85		
ME30	DG5344	044401	25.09.85		
ME31	DG6407	044417	25.09.85		
ME32	DG6437	044059	25.09.85		
ME33	DG4756	044674	02.10.85		
ME34	DG5362	044337	02.10.85		
ME35	DG5565	044432	02.10.85		
ME36	DG5708	044569	02.10.85		
ME37	DG5717	044482	02.10.85		
ME38	DG6239	044539	02.10.85		
ME39	DG5082	044517	07.10.85		
ME40	DG7877	044637	10.10.85		
ME41	DG8380	044604	10.10.85		

VOLVO B10MD - Demonstration Vehicle for Evaluation
Fleet no: VMD1   ALEXANDER BODY

Fleet number	Regn. number	Chassis number	Date first registered	Withdrawn/ scrapped	Disposal
VMD 1	DC4146	B10M6939	03.07.84	23.02.88	Fire damaged:chassis purchased by Ranger Roadways Ltd. and returned to the UK Nov 88.

3-Axle Double-deck Buses - non-air-conditioned.

MCW METROBUS - 12-metre - 3-axle.
Fleet nos: 3M1-3 — MCW BODIES

Fleet number	Regn. number	Chassis number	Date first registered	Withdrawn	Disposal
3M1	CN 870	MB6258	10.05.81		
3M2	CP5280	MB6378	11.08.81		
3M3	CS1204	MB6379	28.10.81		

MCW METROBUS - 11-metre - 3-axles.
Fleet nos: S3M1-254 — MCW BODIES

Fleet number	Regn. number	Chassis number	Date first registered	Withdraw	Disposal
S3M 1	DK1367	MB8526	26.05.86		
S3M 2	DL2842	MB8526	28.08.86		
S3M 3	DL2504	MB8518	05.09.86		
S3M 4	DL2712	MB8522	05.09.86		
S3M 5	DL3138	MB8521	05.09.86		
S3M 6	DL4387	MB8528	05.09.86		
S3M 7	DL2767	MB8525	09.09.86		
S3M 8	DL4182	MB8523	09.09.86		
S3M 9	DL4480	MB8533	09.09.86		
S3M10	DL4505	MB8531	11.09.86		
S3M11	DL4596	MB8524	11.09.86		
S3M12	DL4682	MB8536	16.09.86		
S3M13	DL4728	MB8530	16.09.86		
S3M14	DL5300	MB8537	16.09.86		
S3M15	DL5862	MB8520	16.09.86		
S3M16	DL4669	MB8644	17.09.86		
S3M17	DL5352	MB8527	17.09.86		
S3M18	DL1870 / EM 360	MB8535	22.09.86	Rebuilt to MCW(B) spec 05.90	
S3M19	DL4898	MB8654	24.09.86		
S3M20	DL5040	MB8534	24.09.86		
S3M21	DL5730	MB8641	24.09.86		
S3M22	DL6216	MB8651	24.09.86		
S3M23	DL6651	MB8654	25.09.86		
S3M24	DL7203	MB8532	25.09.86		
S3M25	DL6733	MB8642	26.09.86		
S3M26	DL7411	MB8661	26.09.86		
S3M27	DL7966	MB8662	26.09.86		
S3M28	DL6720	MB8648	30.09.86		
S3M29	DL7081	MB8646	30.09.86		
S3M30	DL7119	MB8647	30.09.86		
S3M31	DL7441	MB8657	30.09.86		
S3M32	DL7831	MB8656	30.09.86		
S3M33	DL8044	MB8659	30.09.86		
S3M34	DL8082	MB8653	30.09.86		
S3M35	DL7028	MB8649	03.10.86		
S3M36	DL7322	MB8660	03.10.86		
S3M37	DL6714	MB8652	08.10.86		
S3M38	DL6765	MB8655	08.10.86		
S3M39	DL6877	MB8658	08.10.86		
S3M40	DL8056	MB8643	08.10.86		
S3M41	DM1043	MB8650	29.10.86		
S3M42	DM1224	MB8664	29.10.86		
S3M43	DM1817	MB8529	03.11.86		
S3M44	DM2108	MB8665	03.11.86		
S3M45	DM4622	MB8663	01.12.86		

KMB Sub-class: Metrobus(A) — MCW BODIES

Fleet number	Regn. number	Chassis number	Date first registered	Withdraw	Disposal
S3M46	DN9074	MB8749	17.03.87		
S3M47	DN9223	MB8728	17.03.87		
S3M48	DN9426	MB8743	17.03.87		
S3M49	DN9759	MB8740	17.03.87		
S3M50	DN9879	MB8739	17.03.87		
S3M51	DN9624	MB8747	20.03.87		
S3M52	DN9789	MB8735	29.03.87		
S3M53	DP 889	MB8727	23.03.87		
S3M54	DP1084	MB8741	23.03.87		
S3M55	DP1237	MB8733	23.03.87		
S3M56	DP1490	MB8730	23.03.87		
S3M57	DP1573	MB8745	23.03.87		
S3M58	DP1848	MB8732	23.03.87		
S3M59	DP2378	MB8751	23.03.87		
S3M60	DP2392	MB8750	23.03.87		
S3M61	DP1522	MB8737	24.03.87		
S3M62	DP1611	MB8731	24.03.87		
S3M63	DP2065	MB8742	24.03.87		
S3M64	DP2157	MB8746	24.03.87		
S3M65	DP1185	MB8729	30.03.87		
S3M66	DP1430	MB8734	30.03.87		
S3M67	DP1742	MB8752	30.03.87		
S3M68	DP1789	MB8754	30.03.87		
S3M69	DP2106	MB8738	30.03.87		
S3M70	DP1330	MB8748	01.04.87		
S3M71	DP1454	MB8744	01.04.87		
S3M72	DP1542	MB8736	01.04.87		
S3M73	DP1164	MB8753	02.04.87		
*	DP1932	MB8640	02.04.87	??.04.88 Air-conditioned bus *	
S3M74	DR4807	MB8854	25.06.87		
S3M75	DR5480	MB8870	25.06.87		
S3M76	DR4513	MB8871	01.07.87		
S3M77	DR4731	MB8873	01.07.87		
S3M78	DR4760	MB8836	01.07.87		
S3M79	DR4834	MB8828	01.07.87		
S3M80	DR5312	MB8872	01.07.87		
S3M81	DR6131	MB8858	01.07.87		
S3M82	DR6773	MB8827	07.07.87		
S3M83	DR6798	MB8831	07.07.87		
S3M84	DR7099	MB8833	07.07.87		
S3M85	DR7500	MB8830	07.07.87		
S3M86	DR7891	MB8834	07.07.87		
S3M87	DR8339	MB8835	07.07.87		
S3M88	DR8903	MB8841	10.07.87		
S3M89	DR9007	MB8837	10.07.87		
S3M90	DR9036	MB8848	10.07.87		
S3M91	DR9116	MB8856	10.07.87		

*Air-conditioning removed and bus later reinstated as **S3M145** - MCW(A).

KMB Sub-class: Metrobus(A) (Contd);

Fleet number	Regn. number	Chassis number	Date first registered	Withdrawn	Disposal	MCW BODIES
S3M92	DR9120	MB8838	10.07.87			
S3M93	DR9194	MB8840	10.07.87			
S3M94	DR9264	MB8829	10.07.87			
S3M95	DR9800	MB8843	10.07.87			
S3M96	DR9973	MB8825	10.07.87			
S3M97	DS 495	MB8869	10.07.87			
S3M98	DR8741	MB8853	16.07.87			
S3M99	DR9798	MB8832	16.07.87			
S3M100	DR9156	MB8824	21.07.87			
S3M101	DS 655	MB8857	21.07.87			
S3M102	DS 738	MB8846	21.07.87			
S3M103	DS1244	MB8860	21.07.87			
S3M104	DS1341	MB8855	21.07.87			
S3M105	DS1549	MB8839	21.07.87			
S3M106	DS2165	MB8865	21.07.87			
S3M107	DS2191	MB8864	21.07.87			
S3M108	DS3065	MB8849	11.08.87			
S3M109	DS3391	MB8851	11.08.87			
S3M110	DS6784	MB8852	24.08.87			
S3M111	DS7187	MB8867	24.08.87			
S3M112	DS7341	MB8859	24.08.87			
S3M113	DS7705	MB8850	24.08.87			
S3M114	DS8150	MB8844	24.08.87			
S3M115	DS8160	MB8862	24.08.87			
S3M116	DS8266	MB8845	24.08.87			
S3M117	DT1288	MB8863	21.09.87			
S3M118	DT1374	MB8842	21.09.87			
S3M119	DT 716	MB8868	22.09.87			
S3M120	DT1084	MB8847	22.09.87			
S3M121	DT1891	MB8891	22.09.87			
S3M122	DT3960	MB8866	26.10.87			
S3M123	DT6881	MB8826	03.04.89			
S3M124	ED2773	MB9191	03.04.89			
S3M125	ED2788	MB9196	03.04.89			
S3M126	ED2829	MB9193	03.04.89			
S3M127	ED3280	MB9186	03.04.89			
S3M128	ED3536	MB9178	03.04.89			
S3M129	ED3641	MB9192	03.04.89			
S3M130	ED3845	MB9179	03.04.89			
S3M131	ED4027	MB9184	03.04.89			
S3M132	ED4164	MB9182	03.04.89			
S3M133	ED4189	MB9194	03.04.89			
S3M134	ED4209	MB9189	03.04.89			
S3M135	ED4218	MB9180	03.04.89			
S3M136	ED4246	MB9176	03.04.89			
S3M137	ED4333	MB9189	03.04.89			
S3M138	ED4452	MB9171	03.04.89			
S3M139	ED2527	MB9190	04.04.89			
S3M140	ED2587	MB9177	04.04.89			
S3M141	ED2830	MB9179	04.04.89			
S3M142	ED3221	MB9168	04.04.89			
S3M143	ED3720	MB9183	04.04.89			
S3M144	ED4956	MB9187	04.04.89			
S3M145	DP1932	MB8640	04.04.89			*Also listed after S3M73 - formerly Air-conditioned.
S3M146	EE8647	MB9203	03.07.89			
S3M147	EE8847	MB9204	03.07.89			
S3M148	EE9030	MB9202	03.07.89			
S3M149	EE9255	MB9255	03.07.89			
S3M150	EE9435	MB9200	03.07.89			
S3M151	EE9748	MB9195	03.07.89			
S3M152	EE9760	MB9185	03.07.89			
S3M153	EE9861	MB9199	03.07.89			
S3M154	EE9927	MB9198	03.07.89			
S3M155	EF 107	MB9201	03.07.89			
S3M157	EF6840	MB9127	16.08.89			
S3M159	EF6966	MB9175	16.08.89			
S3M162	EF7484	MB9173	16.08.89			
S3M163	EF8241	MB9167	16.08.89			
S3M166	EF8576	MB9174	21.08.89			
S3M170	EF9516	MB9166	21.08.89			
S3M171	EF9927	MB9181	21.08.89			
S3M174	EF9648	MB9165	22.08.89			
S3M180	EG1424	MB9188	01.09.89			

KMB Sub-class: Metrobus(B) — MCW BODIES

Fleet number	Regn. number	Chassis number	Date first registered	Withdrawn	Disposal
S3M156	EF6702	MB10219	16.08.89		
S3M158	EF6919	MB10185	16.08.89		
S3M160	EF7111	MB10205	16.08.89		
S3M161	EF7424	MB10209	16.08.89		
S3M164	EF8304	MB10225	16.08.89		
S3M165	EF8401	MB10216	16.08.89		
S3M167	EF8815	MB10221	21.08.89		
S3M168	EF8826	MB10184	21.08.89		
S3M169	EF8860	MB10218	21.08.89		
S3M172	EF8865	MB10191	22.08.89		
S3M173	EF9551	MB10223	22.08.89		
S3M175	EF9676	MB10211	22.08.89		
S3M176	EG 113	MB10202	22.08.89		
S3M177	EG 734	MB10210	01.09.89		
S3M178	EG 780	MB10195	01.09.89		
S3M179	EG1022	MB10186	01.09.89		
S3M181	EG1651	MB10189	01.09.89		
S3M182	EG1694	MB10192	01.09.89		
S3M183	EG1923	MB10212	01.09.89		
S3M184	EG2093	MB10214	01.09.89		
S3M185	EG2226	MB10220	01.09.89		
S3M186	EG2346	MB10196	01.09.89		
S3M187	EG2352	MB10217	01.09.89		
S3M188	EG2378	MB10232	01.09.89		
S3M189	EG4812	MB10200	25.09.89		
S3M190	EG4945	MB10231	25.09.89		
S3M191	EG5081	MB10188	25.09.89		
S3M192	EG5247	MB10194	25.09.89		
S3M193	EG5358	MB10201	25.09.89		
S3M194	EG5523	MB10229	25.09.89		

KMB Sub-class: Metrobus(B) (Contd); MCW BODIES

Fleet number	Regn. number	Chassis number	Date first registered	Withdrawn	Disposal
S3M195	EG5705	MB10197	25.09.89		
S3M196	EG6456	MB10198	25.09.89		
S3M197	EG5114	MB10230	02.10.89		
S3M198	EG5118	MB10183	02.10.89		
S3M199	EG5138	MB10199	02.10.89		
S3M200	EG5163	MB10210	02.10.89		
S3M201	EG5345	MB10222	02.10.89		
S3M202	EG5521	MB10206	02.10.89		
S3M203	EG5677	MB10203	02.10.89		
S3M204	EG5836	MB10204	02.10.89		
S3M205	EG6176	MB10207	02.10.89		
S3M206	EG6212	MB10190	02.10.89		
S3M207	EG6461	MB10224	02.10.89		
S3M208	EG6463	MB10227	02.10.89		
S3M209	EG6772	MB10215	05.10.89		
S3M210	EG7663	MB10226	05.10.89		
S3M211	EG8196	MB10208	05.10.89		
S3M212	EG8389	MB10228	05.10.89		
S3M213	EG8451	MB10193	05.10.89		
S3M214	EG6915	MB10284	10.10.89		
S3M215	EG7138	MB10283	10.10.89		
S3M216	EG7754	MB10285	10.10.89		
S3M217	EG7915	MB10291	10.10.89		
S3M218	EG8505	MB10303	12.10.89		
S3M219	EG8900	MB10287	12.10.89		
S3M220	EG9515	MB10280	12.10.89		
S3M221	EH 439	MB10289	12.10.89		
S3M222	EH2701	MB10294	01.11.89		
S3M223	EH2940	MB10311	01.11.89		
S3M224	EH2968	MB10286	01.11.89		
S3M225	EH3072	MB10313	01.11.89		
S3M226	EH3349	MB10305	01.11.89		
S3M227	EH3431	MB10290	01.11.89		
S3M228	EH3608	MB10308	01.11.89		
S3M229	EH3792	MB10309	01.11.89		
S3M230	EH3825	MB10310	01.11.89		
S3M231	EH4312	MB10304	01.11.89		
S3M232	EH4472	MB10307	01.11.89		
S3M233	EH8559	MB10314	01.12.89		
S3M234	EH9408	MB10281	01.12.89		
S3M235	EJ3809	MB10319	21.12.89		
S3M236	EK 706	MB10298	13.02.90		
S3M237	EK 995	MB10288	13.02.90		
S3M238	EK1037	MB10296	13.02.90		
S3M239	EK1480	MB10299	13.02.90		
S3M240	EK1557	MB10295	13.02.90		
S3M241	EK2246	MB10306	13.02.90		
S3M242	EK2409	MB10318	13.02.90		
S3M243	EK2431	MB10297	13.02.90		
S3M244	EL6749	MB10293	01.05.90		
S3M245	EL6774	MB10300	01.05.90		
S3M246	EL6857	MB10282	01.05.90		
S3M247	EL7295	MB10317	01.05.90		
S3M248	EL7383	MB10292	01.05.90		
S3M249	EL7659	MB10315	01.05.90		
S3M250	EL7793	MB10286	01.05.90		
S3M251	EL7853	MB10312	01.05.90		
S3M252	EL8029	MB10301	01.05.90		
S3M253	EL8061	MB10316	01.05.90		
S3M254	EL7065	MB10187	07.05.90		

LEYLAND OLYMPIAN 12-metre.

Fleet no: 3BL 1 ECW BODY

Fleet number	Regn. number	Chassis number	Date first registered	Withdrawn	Disposal
3BL 1	CV 184	ON/119	01.04.82		

Leyland demonstrator; purchased by KMB July 1982.

LEYLAND OLYMPIAN 12-metre;

Fleet nos: 3BL 2-163

KMB class: Olympian 12M(A). ALEXANDER BODIES

Fleet number	Regn. number	Chassis number	Date first registered	Withdrawn
3BL 2	DA 759	ON 824	29.09.83	
3BL 3	DA 837	ON 839	05.10.83	
3BL 4	DA3277	ON 825	10.10.83	
3BL 5	DA3469	ON 793	10.10.83	
3BL 6	DA4082	ON 849	10.10.83	
3BL 7	DA3149	ON 879	10.10.83	
3BL 8	DA3604	ON 838	18.10.83	
3BL 9	DA3912	ON 848	18.10.83	
3BL10	DA2745	ON 878	20.10.83	
3BL11	DA2689	ON 889	24.10.83	
3BL12	DA2775	ON 915	01.11.83	
3BL13	DA2892	ON 857	01.11.83	
3BL14	DA3278	ON 856	01.11.83	
3BL15	DA3606	ON 853	01.11.83	
3BL16	DA5893	ON 909	03.11.83	
3BL17	DA4604	ON 890	07.11.83	
3BL18	DA5449	ON 910	07.11.83	
3BL19	DA5090	ON 840	11.11.83	
3BL20	DA5183	ON 914	11.11.83	
3BL21	DA6505	ON 852	14.12.83	
3BL22	DA3562	ON1169	16.03.84	
3BL23	DB4161	ON1114	16.03.84	
3BL24	DB4647	ON1178	20.03.84	
3BL25	DB5233	ON1175	20.03.84	
3BL26	DB5051	ON1176	22.03.84	
3BL27	DB5551	ON1168	22.03.84	
3BL28	DB4846	ON1179	27.03.84	
3BL29	DB4951	ON1177	27.03.84	
3BL30	DB5564	ON1170	27.03.84	
3BL31	DB6080	ON1201	27.03.84	
3BL32	DB6283	ON1202	27.03.84	
3BL33	DB6483	ON1199	27.03.84	
3BL34	DB5932	ON1200	29.03.84	
3BL35	DB5966	ON1192	29.03.84	

KMB Sub-class: Olympian12M(A) (Contd); ALEXANDER BODIES

Fleet number	Regn. number	Chassis number	Date first registered	Withdrawn scrapped	Disposal
3BL36	DB5302	ON1203	05.04.84		
3BL37	DB5455	ON1193	05.04.84		
3BL38	DB4685	ON1212	11.04.84		
3BL39	DB4823	ON1213	11.04.84		
3BL40	DB5841	ON1210	11.04.84		
3BL41	DB5971	ON1211	11.04.84		

KMB Sub-class: Olympian12M(B) ALEXANDER BODIES

Fleet number	Regn. number	Chassis number	Date first registered
3BL42	DD2650	ON1439	22.10.84
3BL43	DD2997	ON1452	22.10.84
3BL44	DD3068	ON1443	22.10.84
3BL45	DD3999	ON1441	22.10.84
3BL46	DD4108	ON1437	22.10.84
3BL47	DD4166	ON1446	22.10.84
3BL48	DD4180	ON1448	22.10.84
3BL49	DD4277	ON1429	22.10.84
3BL50	DD4385	ON1400	22.10.84
3BL51	DD2523	ON1477	26.10.84
3BL52	DD2871	ON1450	26.10.84
3BL53	DD3272	ON1474	26.10.84
3BL54	DD3462	ON1469	26.10.84
3BL55	DD4071	ON1466	26.10.84
3BL56	DD4471	ON1486	26.10.84
3BL57	DD4407	ON1489	29.10.84
3BL58	DD3022	ON1480	01.11.84
3BL59	DD3601	ON1495	01.11.84
3BL60	DD3261	ON1492	05.11.84
3BL61	DD8352	ON1483	17.12.84
3BL62	DH4596	ON1969	20.12.85
3BL63	DH5011	ON1978	20.12.85
3BL64	DH5054	ON1960	20.12.85
3BL65	DH5212	ON1956	20.12.85
3BL66	DH5486	ON1989	20.12.85
3BL67	DH5573	ON1967	20.12.85
3BL68	DH5585	ON2008	20.12.85
3BL69	DH5610	ON1958	20.12.85
3BL70	DH5622	ON1964	20.12.85
3BL71	DH5849	ON1980	20.12.85
3BL72	DH5865	ON1935	20.12.85
3BL73	DH5935	ON1982	20.12.85
3BL74	DH5964	ON1971	20.12.85
3BL75	DH6170	ON1974	20.12.85
3BL76	DH4715	ON1996	03.01.86
3BL77	DH4767	ON2048	03.01.86
3BL78	DH4951	ON2062	03.01.86
3BL79	DH4984	ON2024	03.01.86
3BL80	DH5954	ON2004	03.01.86
3BL81	DH5057	ON1991	03.01.86
3BL82	DH5333	ON2028	03.01.86
3BL83	DH5399	ON2050	03.01.86
3BL84	DH5498	ON2030	03.01.86

KMB Sub-class: Olympian(C). 3BL85-163 ALEXANDER BODIES

Fleet number	Regn. number	Chassis number	Date first registered
3BL85	DH5703	ON1998	03.01.86
3BL86	DH5831	ON1993	03.01.86
3BL87	DH5877	ON2014	03.01.86
3BL88	DH6008	ON1985	03.01.86
3BL89	DH6050	ON2038	03.01.86
3BL90	DH6058	ON2044	03.01.86
3BL91	DH6200	ON1994	03.01.86
3BL92	DH6293	ON2046	03.01.86
3BL93	DH6570	ON1987	13.01.86
3BL94	DH6691	ON2005	13.01.86
3BL95	DH7030	ON2034	13.01.86
3BL96	DH7747	ON2010	13.01.86
3BL97	DH8160	ON2040	13.01.86
3BL98	DH8226	ON2064	13.01.86
3BL99	DH8474	ON2018	13.01.86
3BL100	DH6555	ON2052	15.01.86
3BL101	DH6631	ON1976	15.01.86
3BL102	DH7022	ON2056	15.01.86
3BL103	DH7247	ON2016	15.01.86
3BL104	DH7439	ON2054	15.01.86
3BL105	DH7912	ON1954	15.01.86
3BL106	DH8136	ON2058	15.01.86
3BL107	DH8275	ON2032	15.01.86
3BL108	DH6924	ON2230	23.01.86
3BL109	DH7026	ON2026	23.01.86
3BL110	DH7796	ON2195	23.01.86
3BL111	DH8201	ON2036	23.01.86
3BL112	DH6956	ON2233	27.01.86
3BL113	DH8346	ON2216	27.01.86
3BL114	DH7411	ON2060	27.01.86
3BL115	DH7459	ON2204	27.01.86
3BL116	DH7861	ON2202	27.01.86
3BL117	DH7958	ON2226	27.01.86
3BL118	DH8508	ON2222	04.02.86
3BL119	DH8525	ON2205	04.02.86
3BL120	DH8581	ON2211	04.02.86
3BL121	DH8666	ON2220	04.02.86
3BL122	DH8672	ON2210	04.02.86
3BL123	DH8689	ON2229	04.02.86
3BL124	DH8737	ON2234	04.02.86
3BL125	DH8990	ON2218	04.02.86
3BL126	DH9145	ON2214	04.02.86
3BL127	DH9152	ON2212	04.02.86
3BL128	DH9306	ON2208	04.02.86
3BL129	DH9398	ON2255	04.02.86
3BL130	DH9411	ON2228	04.02.86
3BL131	DH9487	ON2200	04.02.86
3BL132	DH9627	ON2231	04.02.86
3BL133	DH9707	ON2215	04.02.86
3BL134	DH9849	ON2224	04.02.86
3BL135	DH9864	ON2232	04.02.86
3BL136	DJ 349	ON2206	24.02.86
3BL137	DJ 714	ON2260	24.02.86
3BL138	DJ 805	ON2244	24.02.86
3BL139	DJ1012	ON2256	24.02.86
3BL140	DJ1039	ON2253	24.02.86
3BL141	DJ1322	ON2252	24.02.86
3BL142	DJ1848	ON2236	24.02.86

KMB sub-class: Olympian(C) S3BL85-163 (Contd);
ALEXANDER BODIES

Fleet number	Regn. number	Chassis number	Date first registered	Withdrawn
3BL143	DJ1959	ON2235	24.02.86	
3BL144	DJ2046	ON2263	24.02.86	
3BL145	DJ 735	ON2254	03.03.86	
3BL146	DJ1454	ON2250	03.03.86	
3BL147	DJ2788	ON2042	11.03.86	
3BL148	DJ2813	ON2245	11.03.86	
3BL149	DJ3052	ON2240	11.03.86	
3BL150	DJ3220	ON2257	11.03.86	
3BL151	DJ3644	ON2238	11.03.86	
3BL152	DJ3770	ON2242	11.03.86	
3BL153	DJ4058	ON2249	11.03.86	
3BL154	DJ3014	ON2247	18.03.86	
3BL155	DJ3037	ON2248	18.03.86	
3BL156	DJ3134	ON2258	18.03.86	
3BL157	DJ3742	ON2261	18.03.86	
3BL158	DJ3850	ON2246	18.03.86	
3BL159	DJ3851	ON2251	18.03.86	
3BL160	DJ3978	ON2264	18.03.86	
3BL161	DJ4406	ON2259	18.03.86	
3BL162	DJ3429	ON2012	01.04.86	
3BL163	DK7612	ON2262	15.07.86	

LEYLAND OLYMPIAN 3-axle - 11-metre

Fleet nos: S3BL 1-470   ALEXANDER BODIES

Fleet number	Regn. number	Chassis number	Date first registered	Withdrawn	Disposal
S3BL 1	DJ9765	ON10022	09.05.86		
S3BL 2	DK7698	ON10033	18.07.86		
S3BL 3	DK7897	ON10034	18.07.86		
S3BL 4	DK6817	ON10036	21.07.86		
S3BL 5	DK6353	ON10046	21.07.86		
S3BL 6	DK7157	ON10037	22.03.86		
S3BL 7	DK7889	ON10035	22.07.86		
S3BL 8	DK8065	ON10024	23.07.86		
S3BL 9	DK8242	ON10017	23.07.86		
S3BL10	DK8472	ON10043	23.07.86		
S3BL11	DK6870	ON10038	25.07.86		
S3BL12	DK6882	ON10032	25.07.86		
S3BL13	DK8460	ON10066	25.07.86		
S3BL14	DK8640	ON10040	29.07.86		
S3BL15	DK8765	ON10039	29.07.86		
S3BL16	DK8798	ON10025	29.07.86		
S3BL17	DK8865	ON10019	29.07.86		
S3BL18	DK9031	ON10047	29.07.86		
S3BL19	DK9502	ON10064	29.07.86		
S3BL40	DK9523	ON10028	29.07.86		
S3BL41	DL 257	ON10021	29.07.86		
S3BL42	DK8785	ON10031	01.08.86		
S3BL43	DK8975	ON10048	01.08.86		
S3BL44	DK9086	ON10027	01.08.86		
S3DL45	DK9167	ON10065	01.08.86		
S3BL46	DK9643	ON10130	01.08.86		
S3BL47	DK9710	ON10042	01.08.86		
S3BL48	DK8605	ON10063	07.08.86		
S3BL49	DK8708	ON10051	07.08.86		
S3BL30	DK8949	ON10052	07.08.86		
S3BL31	DK9205	ON10026	07.08.86		
S3BL32	DK9434	ON10018	07.08.86		
S3BL33	DK9480	ON10023	07.08.86		
S3BL34	DL 431	ON10163	07.08.86		
S3BL35	DK9052	ON10135	08.08.86		
S3BL36	DK9913	ON10136	08.08.86		
S3BL37	DL 864	ON10029	11.08.86		
S3BL38	DL1899	ON10177	11.08.86		
S3BL39	DL 811	ON10041	13.08.86		
S3BL40	DL1085	ON10020	13.08.86		
S3BL41	DL1142	ON10055	13.08.86		
S3BL42	DL1750	ON10059	13.08.86		
S3BL43	DL1873	ON10071	13.08.86		
S3BL44	DL2085	ON10162	13.08.86		
S3BL45	DL2099	ON10164	13.08.86		
S3BL46	DL2328	ON10176	13.08.86	25.03.91	Sc.
S3BL47	DL2264	ON10058	18.08.86		
S3BL48	DL 751	ON10161	20.08.86		
S3BL49	DL1557	ON10075	20.08.86		
S3BL50	DL2027	ON10165	20.08.86		
S3BL51	DL 705	ON10167	20.08.86		
S3BL52	DL 706	ON10166	20.08.86		
S3BL53	DL1200	ON10170	20.08.86		
S3BL54	DL1412	ON10204	20.08.86		
S3BL55	DL1533	ON10186	20.08.86		
S3BL56	DL1857	ON10179	20.08.86		
S3BL57	DL 863	ON10209	22.08.86		
S3BL58	DL1738	ON10174	22.08.86		
S3BL59	DL1767	ON10178	22.08.86		
S3BL60	DL2587	ON10181	28.08.86		
S3BL61	DL2752	ON10229	28.08.86		
S3BL62	DL2993	ON10239	28.08.86		
S3BL63	DL3101	ON10183	28.08.86		
S3BL64	DL3601	ON10211	28.08.86		
S3BL65	DL3713	ON10169	28.08.86		
S3BL66	DL4492	ON10185	28.08.86		
S3BL67	DL2875	ON10196	04.09.86		
S3BL68	DL3605	ON10224	04.09.86		
S3BL69	DL3689	ON10252	04.09.86		
S3BL70	DL4426	ON10249	04.09.86		
S3BL71	DL7783	ON10172	06.10.86		
S3BL72	DL7819	ON10168	08.10.86		
S3BL73	DL8126	ON10241	08.10.86		
S3BL74	DL9075	ON10197	10.10.86		
S3BL75	DL9206	ON10225	10.10.86		
S3BL76	DL9587	ON10188	14.10.86		
S3BL77	DM 211	ON10195	14.10.86		
S3BL78	DL8529	ON10215	20.10.86		
S3BL79	DL8617	ON10219	20.10.86		
S3BL80	DL8645	ON10240	20.10.86		
S3BL81	DL9056	ON10226	20.10.86		
S3BL82	DL9368	ON10220	20.10.86		
S3BL83	DL9397	ON10214	20.10.86		

KMB sub-class: Olympian11M - S3BL 1-470 (Contd);
ALEXANDER BODIES

Fleet number	Regn. number	Chassis number	Date first registered	Withdrawn	Disposal
S3BL84	DL9903	ON10173	20.10.86		
S3BL85	DM 608	ON10203	29.10.86		
S3BL86	DM 639	ON10193	29.10.86		
S3BL87	DM 924	ON10175	29.10.86		
S3BL88	DM1355	ON10171	29.10.86		
S3BL89	DM1671	ON10228	29.10.86		
S3BL90	DM1953	ON10223	29.10.86		
S3BL91	DM2146	ON10184	29.10.86		
S3BL92	DM2186	ON10227	29.10.86		
S3BL93	DM 970	ON10189	03.11.86		
S3BL94	DM1638	ON10200	04.11.86		
S3BL95	DM2034	ON10194	04.11.86		
S3BL96	DM 731	ON10183	05.11.86		
S3BL97	DM2575	ON10243	10.11.86		
S3BL98	DM3043	ON10187	10.11.86		
S3BL99	DM3822	ON10221	10.11.86		
S3BL100	DM4266	ON10269	24.11.86		
S3BL101	DM4781	ON10180	27.11.86		
S3BL102	DM4938	ON10278	27.11.86		
S3BL103	DM6105	ON10267	27.11.86		
S3BL104	DM5966	ON10268	01.12.86		
S3BL105	DM1249	ON10210	21.01.87		

KMB sub-class: Olympian11M(A)   ALEXANDER BODIES

Fleet number	Regn. number	Chassis number	Date first registered	Withdrawn	Disposal
S3BL106	DN 641	ON10290	22.01.87		
S3BL107	DN1831	ON10285	22.01.87		
S3BL108	DN2536	ON10281	22.01.87		
S3BL109	DN3433	ON10283	23.01.87		
S3BL110	DN3254	ON10286	26.01.87		
S3BL111	DN3426	ON10298	26.01.87		
S3BL112	DN3671	ON10284	26.01.87		
S3BL113	DN3832	ON10287	26.01.87		
S3BL114	DN4053	ON10289	26.01.87		
S3BL115	DN2774	ON10297	27.01.87		
S3BL116	DN3011	ON10299	27.01.87		
S3BL117	DN3119	ON10302	27.01.87		
S3BL118	DN3274	ON10303	27.01.87		
S3BL119	DN3662	ON10291	27.01.87		
S3BL120	DN3861	ON10279	27.01.87		
S3BL121	DN3904	ON10288	27.01.87		
S3BL122	DN4098	ON10280	04.02.87		
S3BL123	DN3681	ON10294	10.02.87		
S3BL124	DN3839	ON10292	10.02.87		
S3BL125	DN4042	ON10306	10.02.87		
S3BL126	DN2551	ON10287	12.02.87		
S3BL127	DN3034	ON10300	12.02.87		
S3BL128	DN3061	ON10296	12.02.87		
S3BL129	DN4528	ON10304	13.02.87		
S3BL130	DN6097	ON10308	13.02.87		
S3BL131	DN4902	ON10305	17.02.87		
S3BL132	DN5576	ON10307	17.02.87		
S3BL133	DN5031	ON10301	20.02.87		
S3BL134	DN5421	ON10312	23.02.87		
S3BL135	DN6354	ON10310	23.02.87		
S3BL136	DN4976	ON10314	25.02.87		
S3BL137	DN6610	ON10319	27.02.87		
S3BL138	DN7073	ON10328	27.02.87		
S3BL139	DN7354	ON10318	27.02.87		
S3BL140	DN7534	ON10317	27.02.87		
S3BL141	DN8059	ON10311	27.02.87		
S3BL142	DN6990	ON10326	03.03.87		
S3BL143	DN7027	ON10325	03.03.87		
S3BL144	DN7657	ON10295	03.03.87		
S3BL145	DN7665	ON10313	03.03.87		
S3BL146	DN7990	ON10327	03.03.87		
S3BL147	DN8096	ON10293	03.03.87		
S3BL148	DN8165	ON10329	03.03.87		
S3BL149	DN8447	ON10316	03.03.87		
S3BL150	DP7040	ON10376	07.05.87		
S3BL151	DP7415	ON10387	08.05.87		
S3BL152	DP7392	ON10324	11.05.87		
S3BL153	DP6746	ON10382	11.05.87		
S3BL154	DP7243	ON10375	11.05.87		
S3BL155	DP8784	ON10390	14.05.87		
S3BL156	DP8862	ON10384	14.05.87		
S3BL157	DP9184	ON10400	14.05.87		
S3BL158	DP9778	ON10406	14.05.87		
S3BL159	DP9840	ON10410	14.05.87		
S3BL160	DP9987	ON10388	14.05.87		
S3BL161	DR 259	ON10383	14.05.87		
S3BL162	DR 316	ON10411	14.05.87		
S3BL163	DP8576	ON10399	19.05.87		
S3BL164	DP8741	ON10412	19.05.87		
S3BL165	DP8749	ON10385	19.05.87		
S3BL166	DP8978	ON10409	19.05.87		
S3BL167	DP9217	ON10415	19.05.87		
S3BL168	DP9256	ON10407	19.05.87		
S3BL169	DP9642	ON10398	19.05.87		
S3BL170	DP9891	ON10386	19.05.87		
S3BL171	DR2106	ON10389	26.05.87		
S3BL172	DR2458	ON10395	26.05.87		
S3BL173	DR 864	ON10397	26.05.87		
S3BL174	DR 889	ON10405	26.05.87		
S3BL175	DR 622	ON10413	28.05.87		
S3BL176	DR 647	ON10422	28.05.87		
S3BL177	DR 859	ON10452	28.05.87		
S3BL178	DR1622	ON10453	28.05.87		
S3BL179	DR1666	ON10408	28.05.87		
S3BL180	DR2006	ON10394	28.05.87		
S3BL182	DR4014	ON10460	08.06.87		
S3BL183	DR4424	ON10454	08.06.87		
S3BL184	DR2859	ON10396	09.06.87		
S3BL185	DR3414	ON10461	09.06.87		
S3BL186	DR3474	ON10459	09.06.87		
S3BL187	DR3670	ON10463	09.06.87		
S3BL188	DR3711	ON10457	09.06.87		
S3BL189	DR3894	ON10467	09.06.87		
S3BL190	DR4258	ON10462	09.06.87	10.92	Sc Accident victim.
S3BL191	DR4355	ON10465	09.06.87		
S3BL192	DR4517	ON10466	18.06.87		

KMB sub-class: Olympian11M(A) S3BL 1-470 (Contd);

ALEXANDER BODIES

Fleet number	Regn. number	Chassis number	Date first registered	Withdrawn	Disposal
S3BL193	DR5325	ON10455	18.06.87		
S3BL194	DR6425	ON10458	18.06.87		
S3BL195	DR4582	ON10456	23.06.87		
S3BL196	DR4591	ON10464	23.06.87		
S3BL197	DR4648	ON10470	23.06.87		
S3BL198	DR4998	ON10471	23.06.87		
S3BL199	DR6008	ON10468	23.06.87		
S3BL200	DR6324	ON10469	23.06.87		
S3BL201	DT1280	ON10492	21.09.87		
S3BL202	DT1880	ON10542	21.09.87		
S3BL203	DT1579	ON10514	22.09.87		
S3BL204	DT1613	ON10548	22.09.87		
S3BL205	DT2703	ON10491	28.09.87		
S3BL206	DT3076	ON10494	28.09.87		
S3BL207	DT3229	ON10567	28.09.87		
S3BL208	DT3269	ON10568	28.09.87		
S3BL209	DT3329	ON10563	28.09.87		
S3BL210	DT4016	ON10540	28.09.87		
S3BL211	DT4763	ON10562	05.10.87		
S3BL212	DT5367	ON10528	05.10.87		
S3BL213	DT5392	ON10545	05.10.87		
S3BL214	DT5556	ON10544	05.10.87		
S3BL215	DT5693	ON10560	05.10.87		
S3BL216	DT5992	ON10565	05.10.87		
S3BL217	DT6376	ON10546	05.10.87		
S3BL218	DT5747	ON10557	13.10.87		
S3BL219	DT5807	ON10510	13.10.87		
S3BL220	DT5931	ON10511	13.10.87		
S3BL221	DT6115	ON10506	13.10.87		
S3BL222	DT6132	ON10522	13.10.87		
S3BL223	DT6293	ON10505	13.10.87		
S3BL224	DT6490	ON10550	13.10.87		
S3BL225	DT7216	ON10502	26.10.87		
S3BL226	DT7604	ON10551	26.10.87		
S3BL227	DT7911	ON10509	26.10.87		
S3BL228	DT8167	ON10504	26.10.87		
S3BL229	DT8317	ON10482	26.10.87		
S3BL230	DV8569	ON10560	01.03.88		
S3BL231	DV8675	ON10564	01.03.88		
S3BL232	DV9708	ON10496	01.03.88		
S3BL233	DV9852	ON10495	01.03.88		
S3BL234	DV9902	ON10508	01.03.88		
S3BL235	DW 104	ON10490	01.03.88		
S3BL236	DW2466	ON10530	06.04.88		
S3BL237	DW2671	ON10481	06.04.88		
S3BL238	DW2886	ON10527	06.04.88		
S3BL239	DW2955	ON10561	06.04.88		
S3BL240	DW3065	ON10501	06.04.88		
S3BL241	DW3126	ON10523	06.04.88		
S3BL242	DW3149	ON10555	06.04.88		
S3BL243	DW3417	ON10549	06.04.88		
S3BL244	DW3555	ON10479	06.04.88		
S3BL245	DW3614	ON10541	06.04.88		
S3BL246	DW3630	ON10503	06.04.88		
S3BL247	DW3683	ON10526	06.04.88		
S3BL248	DW3708	ON10547	06.04.88		
S3BL249	DW3733	ON10480	06.04.88		
S3BL250	DW3833	ON10558	06.04.88		
S3BL251	DW3913	ON10513	06.04.88		
S3BL252	DW3927	ON10524	06.04.88		
S3BL253	DW3990	ON10543	06.04.88		
S3BL254	DW4022	ON10483	06.04.88		
S3BL255	DW4108	ON10529	06.04.88		
S3BL256	DW4358	ON10559	06.04.88		
S3BL257	DW7035	ON10552	02.05.88		
S3BL258	DW7152	ON10507	02.05.88		
S3BL259	DW7274	ON10556	02.05.88		
S3BL260	DW7549	ON10525	02.05.88		
S3BL261	DW7793	ON10512	02.05.88		
S3BL262	DX3248	ON10596	01.08.88		
S3BL263	DX3844	ON10633	01.08.88		
S3BL264	DY2528	ON10681	01.08.88		
S3BL265	DY2576	ON10678	01.08.88		
S3BL266	DY2719	ON10676	01.08.88		
S3BL267	DY2773	ON10677	01.08.88		
S3BL268	DY3249	ON10683	01.08.88		
S3BL269	DY4043	ON10679	01.08.88		
S3BL270	DY4190	ON10682	01.08.88		
S3DL271	DY4315	ON10675	01.08.88		
S3BL272	DY4386	ON10680	01.08.88		
S3BL273	DY8537	ON10695	30.08.88		
S3BL274	DY8561	ON10584	30.08.88		
S3BL275	DY8603	ON10691	30.08.88		
S3BL276	DY8711	ON10626	30.08.88		
S3BL277	DY8749	ON10619	30.08.88		
S3BL278	DY8912	ON10591	30.08.88		
S3BL279	DY8923	ON10613	30.08.88		
S3BL280	DY9071	ON10696	30.08.88		
S3BL281	DY9118	ON10657	30.08.88		
S3BL282	DY9218	ON10651	30.08.88		
S3BL283	DY9228	ON10610	30.08.88		
S3BL284	DY9259	ON10685	30.08.88		
S3BL285	DY9291	ON10589	30.08.88		
S3BL286	DY9362	ON10611	30.08.88		
S3BL287	DY9363	ON10620	30.08.88		
S3BL288	DY9392	ON10628	30.08.88		
S3BL289	DY9516	ON10622	30.08.88		
S3BL290	DY9582	ON10606	30.08.88		
S3BL291	DY9758	ON10693	30.08.88		
S3BL292	DY9803	ON10688	30.08.88		
S3BL293	DY9913	ON10659	30.08.88		
S3BL294	DZ 110	ON10694	30.08.88		
S3BL295	DZ 176	ON10631	30.08.88		
S3BL296	DZ 251	ON10692	30.08.88		
S3BL297	DZ 265	ON10627	30.08.88		
S3BL298	DZ 296	ON10689	30.08.88		
S3BL299	DZ 483	ON10690	30.08.88		
S3BL300	EA2714	ON10598	01.11.88		
S3BL301	EA2744	ON10661	01.11.88		
S3BL302	EA2762	ON10580	01.11.88		
S3BL303	EA2786	ON10603	01.11.88		
S3BL304	EA2933	ON10650	01.11.88		
S3BL305	EA3002	ON10585	01.11.88		
S3BL306	EA3520	ON10656	01.11.88		
S3BL307	EA3580	ON10599	01.11.88		
S3BL308	EA4020	ON10655	01.11.88		
S3BL309	EA4086	ON10660	01.11.88		
S3BL310	EA4101	ON10600	01.11.88		
S3BL311	EA4113	ON10653	01.11.88		
S3BL312	EA4158	ON10654	01.11.88		
S3BL313	EA4282	ON10664	01.11.88		
S3BL314	EA4329	ON10609	01.11.88		
S3BL315	EA4350	ON10615	01.11.88		
S3BL316	EA4452	ON10635	01.11.88		
S3BL317	EB2620	ON10574	03.01.89		
S3BL318	EB2636	ON10636	03.01.89		
S3BL319	EB3087	ON10601	03.01.89		
S3BL320	EB3170	ON10632	03.01.89		
S3BL321	EB3210	ON10575	03.01.89		
S3BL322	EB3316	ON10630	03.01.89		
S3BL323	EB3682	ON10579	03.01.89		
S3BL324	EB3909	ON10658	03.01.89		
S3BL325	EB3934	ON10637	03.01.89		
S3BL326	EB4431	ON10652	03.01.89		
S3BL327	EC 625	ON10577	01.02.89		
S3BL328	EC 843	ON10576	01.02.89		
S3BL329	EC 880	ON10671	01.02.89		
S3BL330	EC 963	ON10673	01.02.89		
S3BL331	EC1106	ON10662	01.02.89		
S3BL332	EC1349	ON10672	01.02.89		
S3BL333	EC1588	ON10571	01.02.89		
S3BL334	EC1647	ON10573	01.02.89		
S3BL335	EC1662	ON10569	01.02.89		
S3BL336	EC1693	ON10583	01.02.89		
S3BL337	EC1733	ON10670	01.02.89		
S3BL338	EC1907	ON10570	01.02.89		
S3BL339	EC2079	ON10667	01.02.89		
S3BL340	EC2083	ON10663	01.02.89		
S3BL341	EC2408	ON10581	01.02.89		
S3BL342	EC2429	ON10665	01.02.89		
S3BL343	EC2456	ON10669	01.02.89		
S3BL344	EC6542	ON10623	01.03.89		
S3BL345	EC6554	ON10668	01.03.89		
S3BL346	EC6571	ON10625	01.03.89		
S3BL347	EC6836	ON10572	01.03.89		
S3BL348	EC7127	ON10616	01.03.89		
S3BL349	EC7163	ON10621	01.03.89		
S3BL350	EC7169	ON10590	01.03.89		
S3BL351	EC7226	ON10587	01.03.89		
S3BL352	EC7308	ON10586	01.03.89		
S3BL353	EC7483	ON10608	01.03.89		
S3BL354	EC7492	ON10578	01.03.89		
S3BL355	EC7920	ON10594	01.03.89		
S3BL356	EC7989	ON10607	01.03.89		
S3BL357	EC8117	ON10634	01.03.89		
S3BL358	EC8129	ON10674	01.03.89		
S3BL359	EC8215	ON10612	01.03.89		
S3BL360	EC8251	ON10666	01.03.89		
S3BL361	EC8262	ON10592	01.03.89		
S3BL362	EC8476	ON10629	01.03.89		
S3BL363	ED2557	ON10614	03.04.89		
S3BL364	ED2914	ON10579	03.04.89		
S3BL365	ED2940	ON10618	03.04.89		
S3BL366	ED3060	ON10602	03.04.89		
S3BL367	ED3201	ON10617	03.04.89		
S3BL368	ED3873	ON10588	03.04.89		
S3BL369	ED3902	ON10582	03.04.89		
S3BL370	ED3935	ON10593	03.04.89		
S3BL371	EH2627	ON11369	01.11.89		
S3BL372	EH2687	ON11378	01.11.89		
S3BL373	EH2688	ON11380	01.11.89		
S3BL374	EH2736	ON11372	01.11.89		
S3BL375	EH2765	ON11385	01.11.89		
S3BL376	EH2790	ON11374	01.11.89		
S3BL377	EH2936	ON11376	01.11.89		
S3BL378	EH3021	ON11388	01.11.89		
S3BL379	EH3080	ON11359	01.11.89		
S3BL380	EH3350	ON11387	01.11.89		
S3BL381	EH4005	ON11368	01.11.89		
S3BL382	EH4145	ON11360	01.11.89		
S3BL383	EH4323	ON11358	01.11.89		
S3BL384	EH4358	ON11371	01.11.89		
S3BL385	EH4420	ON11312	01.11.89		
S3BL386	EH8911	ON11357	01.12.89		
S3BL387	EH9067	ON11361	01.12.89		
S3BL388	EH9097	ON11373	01.12.89		
S3BL389	EH9302	ON11377	01.12.89		
S3BL390	EH9592	ON11375	01.12.89		
S3BL391	EH9697	ON11370	01.12.89		
S3BL392	EH9957	ON11354	01.12.89		
S3BL393	EJ2843	ON11382	21.12.89		
S3BL394	EJ3115	ON11383	21.12.89		
S3BL395	EJ3741	ON11386	21.12.89		
S3BL396	EJ4025	ON11381	21.12.89		
S3BL397	EJ4147	ON11379	21.12.89		
S3BL398	EJ4474	ON11355	21.12.89		
S3BL399	EK 734	ON11356	13.02.90		
S3BL400	EK 867	ON11384	13.02.90		
S3BL401	EU4823	ON11448	06.05.91		
S3BL402	EU4965	ON11442	06.05.91		
S3BL403	EU5295	ON11434	06.05.91		
S3BL404	EU5472	ON11453	06.05.91		
S3BL405	EU5923	ON11435	06.05.91		
S3BL406	EU6010	ON11451	06.05.91		
S3BL407	EU6267	ON11441	06.05.91		
S3BL408	FA8930	ON11437	19.02.92		
S3BL409	FA9279	ON11438	19.02.92		
S3BL410	FA9288	ON11445	19.02.92		
S3BL411	FA9671	ON11452	19.02.92		
S3BL412	FB4748	ON11439	02.03.92		
S3BL413	FB4822	ON11440	02.03.92		
S3BL414	FB5764	ON11450	02.03.92		

KMB sub-class: Olympian11M(A) S3BL 1-470 (Contd);

Fleet number	Regn. number	Chassis number	Date first registered	Withdrawn	Disposal
S3BL415	FB8568	ON11446	11.03.92		
S3BL416	FB8617	ON11449	11.03.92		
S3BL417	FB8663	ON11436	11.03.92		
S3BL418	FB8886	ON11437	11.03.92		
S3BL419	FB8955	ON11443	11.03.92		
S3BL420	FB9469	ON11444	11.03.92		

KMB sub-class: Olympian11M(B)   ALEXANDER BODIES

Fleet number	Regn. number	Chassis number	Date first registered	Withdrawn	Disposal
S3BL421	FP6583	ON20593	11.06.93		
S3BL422	FP8167	ON20558	11.06.93		
S3BL423	FP7359	ON20596	17.06.93		
S3BL424	FP8329	ON20595	17.06.93		
S3BL425	FT7052	ON20592	29.09.93		
S3BL426	FT7111	ON20594	29.09.93		
S3BL427	FT8575	ON20598	04.10.93		
S3BL428	FT9910	ON20589	04.10.93		
S3BL429	FU4794	ON20561	25.10.93		
S3BL430	FU5476	ON20559	25.10.93		
S3BL431	FU5381	ON20599	27.10.93		
S3BL432	FU5593	ON20562	27.10.93		
S3BL433	FU6970	ON20597	28.10.93		
S3BL434	FU8336	ON20590	28.10.93		
S3BL435	FU7371	ON20591	01.11.93		
S3BL436	FU7902	ON20716	02.11.93		
S3BL437	FU8048	ON20560	02.11.93		
S3BL438	FU6538	ON20711	03.11.93		
S3BL439	FU9273	ON20715	05.11.93		
S3BL440	FU9572	ON20714	05.11.93		
S3BL441	FV 342	ON20714	05.11.93		
S3BL442	FU8538	ON20718	05.11.93		
S3BL443	FU9319	ON20710	09.11.93		
S3BL444	FV 682	ON20713	11.11.93		
S3BL445	FV2298	ON20600	11.11.93		
S3BL446	FV5261	ON20712	25.11.93		
S3BL447	FV5827	ON20717	25.11.93		
S3BL448	FV5004	ON20652	01.12.93		
S3BL449	FV5139	ON20649	01.12.93		
S3BL450	FV5383	ON20602	01.12.93		
S3BL451	FV5579	ON20635	01.12.93		
S3BL452	FV5668	ON20603	01.12.93		
S3BL453	FV6099	ON20639	01.12.93		
S3BL454	FV6105	ON20574	01.12.93		
S3BL455	FV6212	ON20651	01.12.93		
S3BL456	FV6840	ON20648	02.12.93		
S3BL457	FV7258	ON20637	02.12.93		
S3BL458	FV7886	ON20601	06.12.93		
S3BL459	FV8281	ON20653	06.12.93		
S3BL460	FW1440	ON20669	06.12.93		
S3BL461	FW2149	ON20657	16.12.93		
S3BL462	FW2243	ON20636	16.12.93		
S3BL463	FW2410	ON20650	16.12.93		
S3BL464	FW 614	ON20670	20.12.93		
S3BL465	FW1321	ON20671	20.12.93		
S3BL466	FW2096	ON20672	20.12.93		
S3BL467	FW7968	ON20676	13.03.94		
S3BL468	FY8389	ON20573	13.03.94	(B)	
S3BL469	FZ5653	ON20554	28.04.94		
S3BL470	GB2444	ON20638	21.06.94	(B)	

DENNIS DRAGON 12-metre; 3-axle Prototypes

Fleet nos: 3N1-3.   ALEXDANDER BODIES

Fleet number	Regn. number	Chassis number	Date first registered	Withdrawn	Disposal
3N1	CU5437	DDA601-101	12.03.82		
3N2	CV2718	DDA601-102	22.04.82		

KMB sub-class: Dragon 12M.   ALEXANDER BODY

3N3	CW5424	DDA-601-104	12.08.82		

DENNIS DRAGON 12-metre - 3-axle

Fleet nos: 3N4-63.

KMB sub-class: Dragon(B)12M.   DUPLE METSEC BODIES

Fleet number	Regn. number	Chassis number	Date first registered	Withdrawn	Disposal
3N 4	DA6422	DDA603-106	21.11.83		
3N 5	DA5634	DDA603-114	25.11.83		
3N 6	DA7099	DDA603-116	01.12.83		
3N 7	DA7269	DDA603-113	01.12.83		
3N 8	DA7746	DDA603-117	05.12.83		
3N 9	DA6922	DDA603-107	12.12.83		
3N10	DA7745	DDA603-115	12.12.83		
3N11	DA8218	DDA603-108	12.12.83		
3N12	DA7543	DDA603-111	14.12.83		
3N13	DA7194	DDA603-109	19.12.83		
3N14	DA8494	DDA603-118	19.12.83		
3N15	DA7462	DDA603-122	21.12.83		
3N16	DA8409	DDA603-110	21.12.83		
3N17	DA7355	DDA603-119	23.12.83		
3N18	DA9543	DDA603-112	28.12.83		
3N19	DA9594	DDA603-123	28.12.83		
3N20	DA9697	DDA603-120	28.12.83		
3N21	DA8566	DDA603-124	04.01.84		
3N22	DA8751	DDA603-125	04.01.84		
3N23	DA8876	DDA603-121	04.01.84		
3N24	DC1148	DDA604-127	07.06.84		
3N25	DC2201	DDA604-128	07.06.84		
3N26	DC1774	DDA604-130	14.06.84		
3N27	DC1953	DDA604-134	14.06.84		
3N28	DC2903	DDA604-129	21.06.84		
3N29	DC3906	DDA604-136	21.06.84		
3N30	DC3612	DDA604-133	26.06.84		
3N31	DC3740	DDA604-138	26.06.84		

KMB sub-class: Dragon(B)12M.   DUPLE METSEC BODIES

Fleet number	Regn. number	Chassis number	Date first registered	Withdrawn	Disposal
3N32	DC4078	DDA604-137	26.06.84		
3N33	DC4106	DDA604-139	26.06.84		
3N34	DC2759	DDA604-142	02.07.84		
3N35	DC2894	DDA604-126	02.07.84		
3N36	DC3207	DDA604-135	02.07.84		
3N37	DC4304	DDA604-141	02.07.84		
3N38	DC3283	DDA604-140	10.07.84		
3N39	DC3301	DDA604-144	10.07.84		
3N40	DC3938	DDA604-132	10.07.84		
3N41	DC6235	DDA604-145	12.07.84		
3N42	DC6323	DDA604-131	12.07.84		
3N43	DC5022	DDA604-143	17.07.84		
3N44	DD4565	DDA605-152	15.11.84		
3N45	DD4864	DDA605-158	15.11.84		
3N46	DD4728	DDA605-159	19.11.84		
3N47	DD5126	DDA605-160	19.11.84		
3N48	DD5377	DDA605-161	21.11.84		
3N49	DD5323	DDA605-148	26.11.84		
3N50	DD6266	DDA605-156	26.11.84		
3N51	DD4555	DDA605-146	28.11.84		
3N52	DD4571	DDA605-162	28.11.84		
3N53	DD4883	DDA605-164	28.11.84		
3N54	DD5228	DDA605-163	28.11.84		
3N55	DD5492	DDA605-157	28.11.84		
3N56	DD6267	DDA605-165	28.11.84		
3N57	DD6378	DDA605-149	28.11.84		
3N58	DD7012	DDA605-168	04.12.84		
3N59	DD7119	DDA605-155	04.12.84		
3N60	DD7178	DDA605-150	04.12.84		
3N61	DD7546	DDA605-151	04.12.84		
3N62	DD8146	DDA605-147	04.12.84		
3N63	DD6740	DDA605-167	06.12.84		
3N64	DD6933	DDA605-144	06.12.84		
3N65	DD6577	DDA605-169	10.12.84		
3N66	DD8285	DDA605-166	10.12.84		
3N67	DD8445	DDA605-172	10.12.84		
3N68	DD6683	DDA605-179	17.12.84		
3N69	DD7412	DDA605-170	17.12.84		
3N70	DD8167	DDA605-173	17.12.84		
3N71	DD8239	DDA605-153	17.12.84		
3N72	DD6509	DDA605-175	20.12.84		
3N73	DD7751	DDA605-171	20.12.84		
3N74	DD7946	DDA605-177	20.12.84		
3N75	DD6780	DDA605-192	21.12.84		
3N76	DD7011	DDA605-176	21.12.84		
3N77	DD7382	DDA605-184	21.12.84		
3N78	DD7523	DDA605-178	21.12.84		
3N79	DD8360	DDA605-183	21.12.84		
3N80	DD9774	DDA605-174	02.01.85		
3N81	DD9408	DDA606-186	04.01.85		
3N82	DD9416	DDA605-181	04.01.85		
3N83	DE 146	DDA605-194	04.01.85		
3N84	DD9052	DDA605-189	07.01.85		
3N85	DD9251	DDA606-187	07.01.85		
3N86	DD9821	DDA605-193	07.01.85		
3N87	DE 360	DDA605-182	07.01.85		
3N88	DD8555	DDA605-180	11.01.85		
3N89	DD8572	DDA605-195	11.01.85		
3N90	DD9094	DDA605-190	11.01.85		
3N91	DD9715	DDA605-188	11.01.85		
3N92	DD9956	DDA605-185	21.01.85		
3N93	DE 169	DDA605-191	21.01.85		

KMB sub-class: Dragon(C)12M.   DUPLE METSEC BODIES

3N94	DJ6702	DDA607-203	23.04.86		
3N95	DJ6711	DDA607-198	23.04.86		
3N96	DJ6962	DDA607-200	23.04.86		
3N97	DJ7005	DDA607-204	23.04.86		
3N98	DJ7550	DDA607-197	23.04.86		
3N99	DJ7586	DDA607-201	23.04.86		
3N100	DJ7630	DDA607-207	23.04.86		
3N101	DJ7889	DDA607-199	23.04.86		
3N102	DJ7970	DDA607-205	23.04.86		
3N103	DJ8465	DDA607-209	23.04.86		
3N104	DJ6533	DDA607-215	02.05.86		
3N105	DJ6638	DDA607-221	02.05.86		
3N106	DJ6989	DDA607-219	02.05.86		
3N107	DJ7068	DDA607-208	02.05.86		
3N108	DJ7290	DDA607-218	02.05.86		
3N109	DJ7524	DDA607-214	02.05.86		
3N110	DJ7599	DDA607-216	02.05.86		
3N111	DJ8066	DDA607-202	02.05.86		
3N112	DJ8292	DDA607-206	02.05.86		
3N113	DJ8315	DDA607-213	02.05.86		
3N114	DJ8341	DDA607-196	02.05.86		
3N115	DJ8496	DDA607-210	02.05.86		
3N116	DJ9117	DDA607-211	09.05.86		
3N117	DJ9349	DDA607-226	09.05.86		
3N118	DJ9553	DDA607-210	09.05.86		
3N119	DK 280	DDA607-227	09.05.86		
3N120	DK 289	DDA607-223	09.05.86		
3N121	DJ9546	DDA607-228	13.05.86		
3N122	DJ9751	DDA607-234	13.05.86		
3N123	DJ9757	DDA607-229	13.05.86		
3N124	DK 340	DDA607-222	13.05.86		
3N125	DJ8697	DDA607-220	16.05.86		
3N126	DJ9860	DDA607-225	16.05.86		
3N127	DJ8688	DDA607-230	19.05.86		
3N128	DJ9021	DDA607-235	19.05.86		
3N129	DJ9186	DDA607-232	19.05.86		
3N130	DJ9247	DDA607-236	19.05.86		
3N131	DJ9796	DDA607-240	19.05.86		
3N132	DK 319	DDA607-233	19.05.86		
3N133	DK 704	DDA607-242	26.05.86		
3N134	DK 810	DDA607-239	26.05.86		
3N135	DK1494	DDA607-241	26.05.86		
3N136	DK1127	DDA607-243	26.05.86		
3N137	DK1750	DDA607-245	26.05.86		
3N138	DK2191	DDA607-237	26.05.86		
3N139	DK2267	DDA607-224	26.05.86		
3N140	DK2327	DDA607-231	26.05.86		

KMB sub-class: Dragon12M(C) — DUPLE METSEC BODIES

Fleet number	Regn. number	Chassis number	Date first registered	Withdrawn	Disposal
3N141	DK 691	DDA1701-248	29.05.86		
3N142	DK1121	DDA1701-247	29.05.86		
3N143	DK1992	DDA1701-246	09.06.86		
3N144	DK2774	DDA1701-262	09.06.86		
3N145	DK2799	DDA1701-263	09.06.86		
3N146	DK3665	DDA1701-250	09.06.86		
3N147	DK3914	DDA1701-256	09.06.86		
3N148	DK4248	DDA1701-260	09.06.86		
3N149	DK4356	DDA1701-249	09.06.86		
3N150	DK2715	DDA1701-254	18.06.86		
3N151	DK3111	DDA1701-259	18.06.86		
3N152	DK3482	DDA1701-257	18.06.86		
3N153	DK3530	DDA1701-255	18.06.86		
3N154	DK4447	DDA1701-265	18.06.86		

KMB sub-class: Dragon12M(D) — DUPLE METSEC BODIES

Fleet number	Regn. number	Chassis number	Date first registered	Withdrawn	Disposal
3N155	DK2555	DDA607-238	18.06.86		
3N156	DK2680	DDA607-244	18.06.86		
3N157	DK3696	DDA607-217	18.06.86		

KMB sub-class: Dragon12M(C) (resumed) — DUPLE METSEC BODIES

Fleet number	Regn. number	Chassis number	Date first registered	Withdrawn	Disposal
3N158	DK2508	DDA1701-273	20.06.86		
3N159	DK3084	DDA1701-270	20.06.86		
3N160	DK3149	DDA1701-258	20.06.86		
3N161	DK3592	DDA1701-266	20.06.86		
3N162	DK3645	DDA1701-269	20.06.86		
3N163	DK3957	DDA1701-272	20.06.86		
3N164	DK4158	DDA1701-271	20.06.86		
3N165	DK5224	DDA1701-277	25.06.86		
3N166	DK5626	DDA1701-264	25.06.86		
3N167	DK4553	DDA1701-281	07.07.86		
3N168	DK4632	DDA1701-283	07.07.86		
3N169	DK4862	DDA1701-276	07.07.86		
3N170	DK5051	DDA1701-286	07.07.86		
3N171	DK5310	DDA1701-274	07.07.86		
3N172	DK5387	DDA1701-282	07.07.86		
3N173	DK5401	DDA1701-279	07.07.86		
3N174	DK5416	DDA1701-287	07.07.86		
3N175	DK5471	DDA1701-275	07.07.86		
3N176	DK5496	DDA1701-289	07.07.86		
3N177	DK5719	DDA1701-290	07.07.86		
3N178	DK5902	DDA1701-278	07.07.86		
3N179	DK6027	DDA1701-268	07.07.86		
3N180	DK6109	DDA1701-280	07.07.86		
3N181	DK6142	DDA1701-261	07.07.86		
3N182	DK6286	DDA1701-267	07.07.86		
3N183	DK6342	DDA1701-293	07.07.86		
3N184	DK6505	DDA1701-292	15.07.86		
3N185	DK6613	DDA1701-252	15.07.86		
3N186	DK7450	DDA1701-291	15.07.86		
3N187	DK8238	DDA1701-284	15.07.86		
3N188	DK8440	DDA1701-253	15.07.86		
3N189	DK8491	DDA1701-285	15.07.86		
3N190	DK8500	DDA1701-285	15.07.86		
3N191	DM3103	DDA1701-251	10.11.86		

DENNIS DRAGON - 11-metre - 3-axle.
Fleet nos: S3N1-270

Fleet number	Regn. number	Chassis number	Date first registered	Withdrawn	Disposal
S3N 1	DM6519	1801-229	16.12.86		
S3N 2	DM6533	1801-309	16.12.86		
S3N 3	DM6609	1801-303	16.12.86		
S3N 4	DM6723	1801-297	16.12.86		
S3N 5	DM6944	1801-298	16.12.86		
S3N 6	DM7134	1801-327	16.12.86		
S3N 7	DM7366	1801-305	16.12.86		
S3N 8	DM7366	1801-316	16.12.86		
S3N 9	DM7495	1801-313	16.12.86		
S3N10	DM7889	1801-307	16.12.86		
S3N11	DM7993	1801-314	16.12.86		
S3N12	DM8097	1801-295	16.12.86		
S3N13	DM8253	1801-296	16.12.86		
S3N14	DM8365	1801-306	16.12.86		
S3N15	DM8445	1801-304	16.12.86		
S3N16	DM6738	1801-300	18.12.86		
S3N17	DM6804	1801-301	18.12.86		
S3N18	DM6950	1801-321	18.12.86	18.12.86	
S3N19	DM6962	1801-312	18.12.86		
S3N30	DM7035	1801-317	18.12.86		
S3N31	DM7621	1801-302	18.12.86		
S3N32	DM8070	1801-311	18.12.86		
S3N33	DM8186	1801-308	18.12.86		
S3N34	DM8256	1801-310	18.12.86		
S3N35	DM8328	1801-325	18.12.86		
S3N36	DM6505	1801-315	19.12.86		
S3N37	DM7818	1801-294	19.12.86		
S3N38	DM8133	1801-319	19.12.86		

KMB sub-class: Dragon 11M

Fleet number	Regn. number	Chassis number	Date first registered	Withdrawn	Disposal
S3N39	DM8553	1801-320	02.01.87		
S3N30	DM8746	1801-318	02.01.87		
S3N31	DM8817	1801-322	02.01.87		
S3N32	DM9261	1801-323	02.01.87		
S3N33	DM9365	1801-328	02.01.87		
S3N34	DM9878	1801-326	02.01.87		
S3N35	DM8768	1801-340	07.01.87		
S3N36	DM8812	1801-333	07.01.87		
S3N37	DM9110	1801-353	07.01.87		
S3N38	DM9147	1801-336	07.01.87		
S3N39	DM9492	1801-324	07.01.87		
S3N40	DM9680	1801-343	07.01.87		
S3N41	DM9732	1801-329	07.01.87		
S3N42	DM9832	1801-332	07.01.87		
S3N43	DM9890	1801-341	07.01.87		
S3N44	DN 143	1801-338	07.01.87		
S3N45	DM8514	1801-339	09.01.87		
S3N46	DM8943	1801-330	09.01.87		

Dennis Dragon 11M. — DUPLE METSEC BODIES

Fleet number	Regn. number	Chassis number	Date first registered	Withdrawn	Disposal
S3N47	DM9031	1801-342	09.01.87		
S3N48	DM9418	1801-337	09.01.87		
S3N49	DN 342	1801-344	09.01.87		
S3N50	DN 366	1801-334	09.01.87		

KMB sub-class: Dragon 11M - Cummins L10 engine.

Fleet number	Regn. number	Chassis number	Date first registered	Withdrawn	Disposal
S3N51	DM8714	1802-348	12.01.87		
S3N52	DM8906	1802-347	12.01.87		
S3N53	DM9172	1802-331	12.01.87		
S3N54	DM9588	1802-346	12.01.87		
S3N55	DM9623	1802-345	12.01.87		

KMB sub-class: Dragon(B) 11M. — DUPLE METSEC BODIES

Fleet number	Regn. number	Chassis number	Date first registered	Withdrawn	Disposal
S3N56	DN 602	DDA1802-357	19.01.87		
S3N57	DN1230	DDA1802-358	19.01.87		
S3N58	DN1784	DDA1802-356	19.01.87		
S3N59	DN2408	DDA1802-352	19.01.87		
S3N60	DN2444	DDA1802-350	19.01.87		
S3N61	DN 510	DDA1802-363	21.01.87		
S3N62	DN 605	DDA1802-360	21.01.87		
S3N63	DN 645	DDA1803-373	21.01.87		
S3N64	DN 678	DDA1803-361	21.01.87		
S3N65	DN 806	DDA1803-349	21.01.87		
S3N66	DN1018	DDA1803-371	21.01.87		
S3N67	DN1195	DDA1803-351	21.01.87		
S3N68	DN1246	DDA1803-364	21.01.87		
S3N69	DN1251	DDA1803-354	21.01.87		
S3N70	DN1317	DDA1803-368	21.01.87		
S3N71	DN1471	DDA1803-353	21.01.87		
S3N72	DN1674	DDA1803-362	21.01.87		
S3N73	DN1720	DDA1803-366	21.01.87		
S3N74	DN1987	DDA1803-355	21.01.87		
S3N75	DN2006	DDA1803-365	21.01.87		
S3N76	DN 860	DDA1803-376	22.01.87		
S3N77	DN1262	DDA1803-374	22.01.87		
S3N78	DN1453	DDA1803-367	22.01.87		
S3N79	DN1732	DDA1803-370	22.01.87		
S3N80	DN1776	DDA1803-375	22.01.87		
S3N81	DN1866	DDA1803-372	22.01.87		
S3N82	DN1984	DDA1803-359	22.01.87		
S3N83	DN2027	DDA1803-369	22.01.87		
S3N84	DN2545	DDA1803-382	23.01.87		
S3N85	DN2861	DDA1803-379	23.01.87		
S3N86	DN3748	DDA1803-377	23.01.87		
S3N87	DN3770	DDA1803-381	23.01.87		
S3N88	DN3435	DDA1803-380	26.01.87		
S3N89	DN3586	DDA1803-378	26.01.87		
S3N90	DN3616	DDA1803-383	26.01.87		

KMB sub-class: Dragon(C) 11M. — DUPLE METSEC BODIES

Fleet number	Regn. number	Chassis number	Date first registered	Withdrawn	Disposal
S3N91	DP3097	DDA1803-387	13.04.87		
S3N92	DP3551	DDA1803-388	13.04.87		
S3N93	DP4134	DDA1803-390	13.04.87		
S3N94	DP4140	DDA1803-384	13.04.87		
S3N95	DP4652	DDA1803-391	15.04.87		
S3N96	DP4689	DDA1803-386	15.04.87		
S3N97	DP4712	DDA1803-395	15.04.87		
S3N98	DP4718	DDA1803-400	15.04.87		
S3N99	DP4946	DDA1803-399	15.04.87		
S3N100	DP4974	DDA1803-401	15.04.87		
S3N101	DP5088	DDA1803-385	15.04.87		
S3N102	DP5170	DDA1803-389	15.04.87		
S3N103	DP5208	DDA1803-393	15.04.87		
S3N104	DP6239	DDA1803-398	15.04.87		
S3N105	DP4767	DDA1803-405	21.04.87		
S3N106	DP5081	DDA1803-403	21.04.87		
S3N107	DP5794	DDA1803-402	21.04.87		
S3N108	DP6439	DDA1803-398	21.04.87		
S3N109	DP4707	DDA1803-394	23.04.87		
S3N110	DP4964	DDA1803-409	23.04.87		
S3N111	DP5569	DDA1803-396	23.04.87		
S3N112	DP5882	DDA1803-404	23.04.87		
S3N113	DP4537	DDA1803-408	27.04.87		
S3N114	DP4555	DDA1803-420	27.04.87		
S3N115	DP4618	DDA1803-407	27.04.87		
S3N116	DP5009	DDA1803-421	27.04.87		
S3N117	DP5277	DDA1803-412	27.04.87		
S3N118	DP5295	DDA1803-392	27.04.87		
S3N119	DP5392	DDA1803-416	27.04.87		
S3N120	DP5490	DDA1803-413	27.04.87		
S3N121	DP5782	DDA1803-411	27.04.87		
S3N122	DP5888	DDA1803-406	27.04.87		
S3N123	DP6029	DDA1803-414	27.04.87		
S3N124	DP4678	DDA1803-410	28.04.87		
S3N125	DP4806	DDA1803-425	28.04.87		
S3N126	DP5001	DDA1803-424	28.04.87		
S3N127	DP5060	DDA1803-426	28.04.87		
S3N128	DP5222	DDA1803-419	28.04.87		
S3N129	DP5480	DDA1803-418	28.04.87		
S3N130	DP5512	DDA1803-427	28.04.87		
S3N131	DP5587	DDA1803-422	28.04.87		
S3N132	DP6316	DDA1803-415	28.04.87		
S3N133	DP7290	DDA1803-428	01.05.87		
S3N134	DP7678	DDA1803-431	01.05.87		
S3N135	DP8219	DDA1803-430	01.05.87		
S3N136	DP6702	DDA1803-417	07.05.87		
S3N137	DP6841	DDA1803-433	07.05.87		
S3N138	DP7389	DDA1803-432	07.05.87		
S3N139	DP7701	DDA1803-423	07.05.87		
S3N140	DP7866	DDA1803-429	07.05.87		

KMB sub-class: Dragon(D) 11M. — DUPLE METSEC BODIES

Fleet number	Regn. number	Chassis number	Date first registered	Withdrawn	Disposal
S3N141	DS6749	DDA1804-439	02.09.87		
S3N142	DS6791	DDA1804-439	02.09.87		
S3N143	DS6890	DDA1804-438	02.09.87		
S3N144	DS7114	DDA1804-440	02.09.87		
S3N145	DS8046	DDA1804-434	02.09.87		
S3N146	DS8167	DDA1804-435	02.09.87		
S3N147	DS8634	DDA1804-449	07.09.87		
S3N148	DS8659	DDA1804-450	07.09.87		
S3N149	DS8748	DDA1804-452	07.09.87		

KMB sub-class: Dragon(D) 11M.   DUPLE METSEC BODIES

Fleet number	Regn. number	Chassis number	Date first registered	Withdrawn	Disposal
S3N150	DS8799	DDA1804-444	07.09.87		
S3N151	DS9320	DDA1804-436	07.09.87		
S3N152	DS9318	DDA1804-442	07.09.87		
S3N153	DS9326	DDA1804-443	07.09.87		
S3N154	DT 380	DDA1804-437	07.09.87		
S3N155	DS9063	DDA1804-448	09.09.87		
S3N156	DS9507	DDA1804-451	09.09.87		
S3N157	DT 137	DDA1804-453	09.09.87		
S3N158	DT 249	DDA1804-447	09.09.87		
S3N159	DT 265	DDA1804-445	09.09.87		
S3N160	DT 349	DDA1804-454	09.09.87		
S3N161	DT1821	DDA1804-446	21.09.87		
S3N162	DW6538	DDA1805-477	02.05.88		
S3N163	DW6724	DDA1805-472	02.05.88		
S3N164	DW6745	DDA1805-460	02.05.88		
S3N165	DW6755	DDA1805-491	02.05.88		
S3N166	DW6766	DDA1805-480	02.05.88		
S3N167	DW6791	DDA1805-462	02.05.88		
S3N168	DW6816	DDA1805-455	02.05.88		
S3N169	DW6899	DDA1805-469	02.05.88		
S3N170	DW6940	DDA1805-473	02.05.88		
S3N171	DW6978	DDA1805-457	02.05.88		
S3N172	DW7005	DDA1805-467	02.05.88		
S3N173	DW7110	DDA1805-458	02.05.88		
S3N174	DW7114	DDA1805-484	02.05.88		
S3N175	DW7123	DDA1805-464	02.05.88		
S3N176	DW7149	DDA1805-471	02.05.88		
S3N177	DW7174	DDA1805-461	02.05.88		
S3N178	DW7427	DDA1805-468	02.05.88		
S3N179	DW7163	DDA1805-478	02.05.88		
S3N180	DW7625	DDA1805-466	02.05.88		
S3N181	DW7716	DDA1805-470	02.05.88		
S3N182	DW7841	DDA1805-459	02.05.88		
S3N183	DW8242	DDA1805-479	02.05.88		
S3N184	DW8262	DDA1805-487	02.05.88		
S3N185	DW8306	DDA1805-485	02.05.88		
S3N186	DW8351	DDA1805-463	02.05.88		
S3N187	DX2531	DDA1805-512	01.06.88		
S3N188	DX2600	DDA1805-500	01.06.88		
S3N189	DX2618	DDA1805-492	01.06.88		
S3N190	DX2619	DDA1805-489	01.06.88		
S3N191	DX2643	DDA1805-475	01.06.88		
S3N192	DX2771	DDA1805-476	01.06.88		
S3N193	DX2813	DDA1805-509	01.06.88		
S3N194	DX2824	DDA1805-514	01.06.88		
S3N195	DX2920	DDA1805-504	01.06.88		
S3N196	DX3175	DDA1805-483	01.06.88		
S3N197	DX3197	DDA1805-465	01.06.88		
S3N198	DX3258	DDA1805-506	01.06.88		
S3N199	DX3290	DDA1805-501	01.06.88		
S3N200	DX3335	DDA1805-503	01.06.88		
S3N201	DX3371	DDA1805-481	01.06.88		
S3N202	DX3464	DDA1805-486	01.06.88		
S3N203	DX3514	DDA1805-474	01.06.88		
S3N204	DX3519	DDA1805-513	01.06.88		
S3N205	DX3523	DDA1805-502	01.06.88		
S3N206	DX3723	DDA1805-508	01.06.88		
S3N207	DX3816	DDA1805-488	01.06.88		
S3N208	DX3931	DDA1805-510	01.06.88		
S3N209	DX4029	DDA1805-482	01.06.88		
S3N210	DX4120	DDA1805-511	01.06.88		
S3N211	DX4333	DDA1805-493	01.06.88		
S3N212	DY8699	DDA1805-456	30.08.88		
S3N213	DY9452	DDA1805-505	30.08.88		
S3N214	DY9699	DDA1805-507	30.08.88		
S3N215	EA 684	DDA1805-494	01.11.88		
S3N216	EA1495	DDA1805-496	01.11.88		
S3N217	EA1725	DDA1805-499	01.11.88		
S3N218	EA1742	DDA1805-495	01.11.88		
S3N219	EA1965	DDA1805-497	01.11.88		
S3N220	EA2065	DDA1805-490	01.11.88		
S3N221	EA2263	DDA1805-498	01.11.88		

KMB sub-class: Dragon(E) 11M.
49 bodies; 50 chassis; 1 chassis destroyed;   DUPLE METSEC BODIES
Total complete vehicles: 49.

Fleet number	Regn. number	Chassis number	Date first registered	Withdrawn	Disposal
-	-	DDA1808-568		Chassis destroyed prior to or during shipping.	
S3N222	ES2845	DDA1808-586	19.12.90		
S3N223	ES3678	DDA1808-574	19.12.90		
S3N224	ES3833	DDA1808-570	19.12.90		
S3N225	ES2621	DDA1808-572	20.12.90		
S3N226	ES3433	DDA1808-569	20.12.90		
S3N227	ES3559	DDA1808-579	20.12.90		
S3N228	ES3578	DDA1808-583	20.12.90		
S3N229	ES4192	DDA1808-589	20.12.90		
S3N230	ES5532	DDA1808-602	02.01.91		
S3N231	ES4540	DDA1808-591	08.01.91		
S3N232	ES4644	DDA1808-588	08.01.91		
S3N233	ES4799	DDA1808-584	08.01.91		
S3N234	ES5024	DDA1808-592	08.01.91		
S3N235	ES5512	DDA1808-577	08.01.91		
S3N236	ES5348	DDA1808-585	08.01.91		
S3N237	ES5377	DDA1808-590	08.01.91		
S3N238	ES5485	DDA1808-594	08.01.91		
S3N239	ES5571	DDA1808-596	08.01.91		
S3N240	ES5382	DDA1808-564	08.01.91		
S3N241	ES5816	DDA1808-562	08.01.91		
S3N242	ES5859	DDA1808-598	08.01.91		
S3N243	ES5915	DDA1808-581	08.01.91		
S3N244	ES6048	DDA1808-587	08.01.91		
S3N245	ES6049	DDA1808-573	08.01.91		
S3N246	ES6240	DDA1808-563	08.01.91		
S3N247	ES6310	DDA1808-580	08.01.91		
S3N248	EU5242	DDA1808-571	06.05.91		
S3N249	EU8629	DDA1808-600	23.05.91		
S3N250	EU8914	DDA1808-565	23.05.91		
S3N251	EU9041	DDA1808-566	25.05.91		
S3N252	EU9286	DDA1808-575	23.05.91		
S3N253	EU9417	DDA1808-582	23.05.91		
S3N254	EU9505	DDA1808-603	23.05.91		

KMB sub-class: Dragon(E) 11M.   DUPLE METSEC BODIES

Fleet number	Regn. number	Chassis number	Date first registered	Withdrawn	Disposal
S3N255	EU9551	DDA1808-582	23.05.91		
S3N256	EU9848	DDA1808-603	23.05.91		
S3N257	EV 129	DDA1808-576	23.05.91		
S3N258	EV 806	DDA1808-567	03.06.91		
S3N259	EV1244	DDA1808-595	03.06.91		
S3N260	EV1582	DDA1808-604	03.06.91		
S3N261	EV1820	DDA1808-607	03.06.91		
S3N262	EV2259	DDA1808-593	03.06.91		
S3N263	EV2448	DDA1808-599	03.06.91		
S3N264	EV4019	DDA1808-610	20.06.91		
S3N265	EV4763	DDA1808-609	24.06.91		
S3N266	EV6424	DDA1808-611	24.06.91		
S3N267	EV4543	DDA1808-601	01.07.91		
S3N268	EV5641	DDA1808-608	01.07.91		
S3N269	EV5774	DDA1808-605	01.07.91		
S3N270	EY5022	DDA1808-606	12.11.91		

DENNIS DRAGON - ducted but not air-conditioned.
Fleet nos: S3N271-320

KMB sub-class: Dragon(F)11M.   DUPLE METSEC BODIES

Fleet number	Regn. number	Chassis number	Date first registered	Withdrawn	Disposal
S3N271	FX6545	DDA1816-875	22.02.94		
S3N272	FX6582	DDA1816-872	22.02.94		
S3N273	FX6774	DDA1816-913	22.02.94		
S3N274	FX6852	DDA1816-909	22.02.94		
S3N275	FX7103	DDA1816-908	22.02.94		
S3N276	FX7476	DDA1816-905	22.02.94		
S3N277	FX7611	DDA1816-910	22.02.94		
S3N278	FX7816	DDA1816-874	22.02.94		
S3N279	FX7994	DDA1816-914	22.02.94		
S3N280	FX8032	DDA1816-912	22.02.94		
S3N281	FX8173	DDA1816-873	22.02.94		
S3N282	FX8270	DDA1816-906	22.02.94		
S3N283	FX6953	DDA1816-891	24.02.94		
S3N284	FX7082	DDA1816-844	24.02.94		
S3N285	FX7484	DDA1816-907	24.02.94		
S3N286	FX7494	DDA1816-841	24.02.94		
S3N287	FX8116	DDA1816-869	24.02.94		
S3N288	FX8567	DDA1816-843	25.02.94		
S3N289	FX8719	DDA1816-866	25.02.94		
S3N290	FX9426	DDA1816-911	25.02.94		
S3N291	FX9694	DDA1816-869	25.02.94		
S3N292	FX9995	DDA1816-870	25.02.94		
S3N293	FY 714	DDA1815-818	01.03.94		
S3N294	FY1915	DDA1815-833	01.03.94		
S3N295	FY1970	DDA1815-826	01.03.94		
S3N296	FY2206	DDA1815-823	01.03.94		
S3N297	FY2213	DDA1815-831	01.03.94		
S3N298	FY2350	DDA1815-825	01.03.94		
S3N299	FY7679	DDA1815-830	04.03.94		
S3N300	FY2726	DDA1815-817	04.03.94		
S3N301	FY3145	DDA1815-032	04.03.94		
S3N302	FY3401	DDA1815-836	04.03.94		
S3N303	FY3686	DDA1815-816	04.03.94		
S3N304	FY3404	DDA1815-815	04.03.94		
S3N305	FY4417	DDA1816-845	04.03.94		
S3N306	FY2519	DDA1815-835	07.03.94		
S3N307	FY3418	DDA1815-834	07.03.94		
S3N308	FY3674	DDA1815-838	07.03.94		
S3N309	FY3128	DDA1815-822	08.03.94		
S3N310	FY4125	DDA1815-821	08.03.94		
S3N311	FY4528	DDA1815-839	10.03.94		
S3N312	FY5061	DDA1815-828	10.03.94		
S3N313	FY5774	DDA1815-837	10.03.94		
S3N314	FY6490	DDA1815-820	10.03.94		
S3N315	FY5814	DDA1815-829	18.03.94		
S3N316	FY5856	DDA1815-827	18.03.94		
S3N317	FY5867	DDA1816-842	18.03.94		
S3N318	FY6264	DDA1816-840	18.03.94		
S3N319	FY6705	DDA1815-824	22.03.94		
S3N370	GD 605	DDA1815-819	25.08.94		

DENNIS DRAGON (G) not air-conditioned
Fleet nos: S3N321-370.

KMB sub-class: Dragon(G)11M.   DUPLE METSEC BODIES

Fleet number	Regn. number	Chassis number	Date first registered	Withdrawn	Disposal
S3N320	FZ4577	DDA1818-948	28.04.94		
S3N321	FZ5289	DDA1818-951	28.04.94		
S3N322	FZ4602	DDA1818-988	29.04.94		
S3N323	FZ5135	DDA1818-934	29.04.94		
S3N324	FZ7937	DDA1818-920	02.05.94		
S3N325	FZ8420	DDA1818-964	02.05.94		
S3N326	FZ6681	DDA1818-962	03.05.94		
S3N327	FZ6940	DDA1818-959	03.05.94		
S3N328	FZ7116	DDA1818-950	03.05.94		
S3N329	FZ7161	DDA1818-944	03.05.94		
S3N330	FZ7219	DDA1818-963	03.05.94		
S3N331	FZ7423	DDA1818-961	03.05.94		
S3N332	FZ8028	DDA1818-952	03.05.94		
S3N333	FZ8258	DDA1818-949	03.05.94		
S3N334	FZ7068	DDA1818-936	05.05.94		
S3N335	FZ7458	DDA1818-935	05.05.94		
S3N336	FZ7591	DDA1818-942	05.05.94		
S3N337	FZ7602	DDA1818-943	05.05.94		
S3N338	FZ7935	DDA1818-946	05.05.94		
S3N339	FZ8084	DDA1818-929	05.05.94		
S3N340	FZ6723	DDA1818-922	09.05.94		
S3N341	FZ6737	DDA1818-919	09.05.94		
S3N342	FZ7574	DDA1818-945	09.05.94		
S3N343	FZ8046	DDA1818-918	09.05.94		
S3N344	FZ8158	DDA1818-926	09.05.94		

KMB sub-class: Dragon(G)11M — DUPLE METSEC BODIES

Fleet number	Regn. number	Chassis number	Date first registered	Withdrawn	Disposal
S3N345	FZ8217	DDA1818-921	09.05.94		
S3N346	GA 665	DDA1818-938	17.05.94		
S3N347	GA1252	DDA1818-923	17.05.94		
S3N348	GA1429	DDA1818-940	17.05.94		
S3N349	GA1468	DDA1818-933	17.05.94		
S3N350	GA1570	DDA1818-917	17.05.94		
S3N351	GA1614	DDA1818-947	17.05.94		
S3N352	GA1948	DDA1818-937	17.05.94		
S3N353	GA2116	DDA1818-924	17.05.94		
S3N354	GA2387	DDA1818-927	17.05.94		
S3N355	GA4943	DDA1818-943	01.06.94		
S3N356	GA5144	DDA1818-956	01.06.94		
S3N357	GA5145	DDA1818-957	01.06.94		
S3N358	GA5505	DDA1818-960	01.06.94		
S3N359	GA6027	DDA1818-915	01.06.94		
S3N360	GA6394	DDA1818-939	01.06.94		
S3N361	GA6479	DDA1818-958	01.06.94		
S3N362	GA4829	DDA1818-954	02.06.94		
S3N363	GA5311	DDA1818-941	02.06.94		
S3N364	GA5396	DDA1818-928	02.06.94		
S3N365	GA5685	DDA1818-916	02.06.94		
S3N366	GA6072	DDA1818-925	02.06.94		
S3N367	GA6107	DDA1818-930	02.06.94		
S3N368	GA6126	DDA1818-932	02.06.94		
S3N369	GA6324	DDA1818-931	02.06.94		

S3N370 appears after S3N319 on the previous page.

Air-conditioned Double-deck Buses.

SCANIA - N113 - Air-conditioned - 11.5-metre.
Prototypes - Fleet nos: AS1 & 2   Total: 2.   ALEXANDER BODIES

Fleet number	Regn. number	Chassis number	Date first registered	Withdrawn	Disposal
AS 1	FU 482	1820465	07.10.93		
AS 2	FU2948	1820466	18.10.93		

LEYLAND OLYMPIAN - Air-conditioned - 11-metre
Fleet no: AL1.   ALEXANDER BODY

Fleet number	Regn. number	Chassis number	Date first registered	Withdrawn	Disposal
AL 1	DX2437	ON10743	12.05.88	Fleet number not initially carried.	

LEYLAND OLYMPIAN - Air-conditioned -11-metre.
Fleet nos: AL2-150   ALEXANDER BODIES
KMB sub-class: Ley.Oly(A)11M AC

Fleet number	Regn. number	Chassis number	Date first registered	Withdrawn	Disposal
AL 2	ER4295	ON11483	29.10.90		
AL 3	ER2645	ON11475	01.11.90		
AL 4	ER2796	ON11500	01.11.90		
AL 5	ER2990	ON11495	01.11.90		
AL 6	ER3201	ON11507	01.11.90		
AL 7	ER3931	ON11473	01.11.90		
AL 8	ER4299	ON11478	01.11.90		
AL 9	ER4721	ON11478	05.11.90		
AL10	ER6572	ON11499	21.11.90		
AL11	ER6834	ON11480	21.11.90		
AL12	ER7707	ON11496	21.11.90		
AL13	ER7526	ON11477	21.11.90		
AL14	ER9171	ON11484	28.11.90		
AL15	ER9386	ON11476	28.11.90		
AL16	ES 271	ON11506	28.11.90		
AL17	ER8639	ON11481	28.11.90		
AL18	ES2803	ON11479	17.12.90		
AL19	ES3248	ON11474	17.11.90		
AL20	ES3246	ON11497	19.11.90		
AL21	ES3304	ON11482	20.11.90		
AL22	ET 576	ON11721	04.02.91		
AL23	ET 856	ON11708	04.02.91		
AL24	ET1282	ON11707	04.02.91		
AL25	ET1295	ON11742	04.02.91		
AL26	ET1836	ON11706	04.02.91		
AL27	ET1852	ON11711	04.02.91		
AL28	ET2117	ON11722	04.02.91		
AL29	ET2191	ON11709	04.02.91		
AL30	ET1076	ON11710	07.02.91		
AL31	ET1967	ON11723	08.02.91		
AL32	ET2032	ON11759	11.02.91		
AL33	ET2624	ON11761	12.02.91		
AL34	ET4946	ON11712	05.03.91		
AL35	ET5603	ON11760	12.03.91		
AL36	ET7228	ON11762	12.03.91		
AL37	ET7360	ON11713	20.03.91		
AL38	ET7856	ON11724	20.03.91		
AL39	EU 260	ON11745	25.03.91		
AL40	EU6705	ON11743	10.05.91		
AL41	EU8479	ON11744	10.05.91		
AL42	EX2907	ON20068	19.09.91		
AL43	EX3585	ON20082	19.09.91		
AL44	EX5154	ON20069	26.09.91		
AL45	EX5396	ON20083	26.09.91		
AL46	EX5627	ON20076	26.09.91		
AL47	EX7407	ON20061	02.10.91		

KMB sub-class: Ley.Oly(A)11M AC AL2-150   ALEXANDER BODIES

Fleet number	Regn. number	Chassis number	Date first registered	Withdrawn	Disposal
AL48	EX8076	ON20064	02.10.91		
AL49	EX6915	ON20073	03.10.91		
AL50	EX7190	ON20075	09.10.91		
AL51	EY 176	ON20065	09.10.91		
AL52	EX9092	ON20077	17.10.91		
AL53	EY 285	ON20058	17.10.91		
AL54	EY 607	ON20060	22.10.91		
AL55	EY1792	ON20081	22.10.91		
AL56	EY3231	ON20074	29.10.91		
AL57	EY4088	ON20063	29.10.91		
AL58	EY3211	ON20068	01.11.91		
AL59	EY4052	ON20066	01.11.91		
AL60	EY4921	ON20067	11.11.91		
AL61	EY6211	ON20059	11.11.91		
AL62	EY6501	ON20148	19.11.91		
AL63	EY7455	ON20186	19.11.91		
AL64	EY8244	ON20154	19.11.91		
AL65	EZ 521	ON20156	02.12.91		
AL66	EZ 933	ON20183	02.12.91		
AL67	EZ 945	ON20185	02.12.91		
AL68	EZ1252	ON20182	02.12.91		
AL69	EZ1951	ON20178	02.12.91		
AL70	EZ2818	ON20184	10.12.91		
AL71	EZ4416	ON20149	10.12.91		
AL72	EZ7241	ON20150	19.12.91		
AL73	EZ7739	ON20152	19.12.91		
AL74	EZ7970	ON20179	19.12.91		
AL75	EZ8310	ON20123	19.12.91		
AL76	EZ8538	ON20181	02.01.92		
AL77	EZ8701	ON20180	02.01.92		
AL78	FA3436	ON20155	17.01.92		
AL79	FA8827	ON20151	19.02.92		
AL80	FA9827	ON20147	19.02.92		
AL81	FB4642	ON20153	03.03.92		
AL82	FB5674	ON20177	02.03.92		
AL83	FD6512	ON20226	03.06.92		
AL84	FD7303	ON20228	03.06.92		
AL85	FD7935	ON20253	03.06.92		
AL86	FD8502	ON20241	10.06.92		
AL87	FD8592	ON20244	10.06.92		
AL88	FD8794	ON20225	10.06.92		
AL89	FD8900	ON20245	10.06.92		
AL90	FD9160	ON20223	10.06.92		
AL91	FD9754	ON20242	10.06.92		
AL92	FD9758	ON20248	10.06.92		
AL93	FD9808	ON20227	10.06.92		
AL94	FE 324	ON20229	10.06.92		
AL95	FD8977	ON20238	16.06.92		
AL96	FD9560	ON20240	16.06.92		
AL97	FE1022	ON20239	18.06.92		
AL98	FE1975	ON20237	18.06.92		
AL99	FE2747	ON20243	24.06.92		
AL100	FE4085	ON20231	24.06.92		
AL101	FE3173	ON20224	25.06.92		
AL102	FE3626	ON20246	25.06.92		
AL103	FE6927	ON20334	09.07.92		
AL104	FE7263	ON20247	09.07.92		
AL105	FE7384	ON20337	09.07.92		
AL106	FE7632	ON20230	09.07.92		
AL107	FE7686	ON20236	09.07.92		
AL108	FE7245	ON20390	10.07.92		
AL109	FE8498	ON20399	10.07.92		
AL110	FE8874	ON20366	17.07.92		
AL111	FE9169	ON20391	17.07.92		
AL112	FE9366	ON20396	17.07.92		
AL113	FE9690	ON20368	17.07.92		
AL114	FF 418	ON20397	17.07.92		
AL115	FF1149	ON20338	27.07.92		
AL116	FF1304	ON20336	27.07.92		
AL117	FF1526	ON20339	27.07.92		

KMB sub-class: Ley.Oly(A)11M AC AL2-150   ALEXANDER BODIES

Fleet number	Regn. number	Chassis number	Date first registered	Withdrawn	Disposal
AL118	FF1819	ON20335	27.07.92		
AL119	FF2295	ON20340	27.07.92		
AL120	FF2711	ON20367	04.08.92		
AL121	FF3078	ON20388	04.08.92		
AL122	FF3449	ON20389	04.08.92		
AL123	FF4377	ON20389	04.08.92		
AL124	FF4495	ON20398	04.08.92		
AL125	FF5186	ON20342	13.08.92		
AL126	FF6443	ON20365	13.08.92		
AL127	FF7526	ON20346	20.08.92		
AL128	FF7590	ON20349	20.08.92		
AL129	FF7842	ON20347	20.08.92		
AL130	FF8539	ON20343	24.08.92		
AL131	FG5039	ON20350	14.09.92		
AL132	FG6325	ON20345	14.09.92		
AL133	FH 866	ON20344	06.10.92		
AL134	FH1058	ON20331	06.10.92		
AL135	FH1887	ON20348	06.10.92		
AL136	FM4502	ON20468	08.03.93		
AL137	FN6884	ON20460	03.05.93		
AL138	FN7122	ON20462	03.05.93		
AL139	FN7710	ON20413	03.05.93		
AL140	FN7831	ON20457	03.05.93		
AL141	FP5351	ON20458	01.06.93		
AL142	FP5449	ON20464	01.06.93		
AL143	FP5469	ON20467	01.06.93		
AL144	FP5789	ON20465	01.06.93		
AL145	FP6071	ON20463	01.06.93		
AL146	FP6351	ON20415	01.06.93		
AL147	FP8644	ON20461	21.06.93		
AL148	FP9061	ON20414	21.06.93		
AL149	FP9563	ON20459	21.06.93		
AL150	FP9847	ON20466	21.06.93		

Air-conditioned DENNIS 'DRAGON' - 11-metre.
Fleet no: AD 1.   DUPLE METSEC BODY

Fleet number	Regn. number	Chassis number	Date first registered	Withdrawn	Disposal
AD 1	EL5113	DDA1806/515	20.04.90		

Air-conditioned DENNIS 'DRAGON' 11 metre.
Fleet nos: AD2-130.   DUPLE METSEC BODIES
KMB class: Dragon (A) 11M AC

Fleet number	Regn. number	Chassis number	Date first registered	Withdrawn	Disposal
AD 2	EU7058	DDA1811-655	10.05.91		
AD 3	EU7248	DDA1811-664	09.05.91		
AD 4	EU7509	DDA1811-668	09.05.91		
AD 5	EU7649	DDA1811-658	10.05.91		
AD 6	EU7767	DDA1811-669	09.05.91		
AD 7	EU7919	DDA1811-654	16.05.91		
AD 8	EU9201	DDA1811-670	21.05.91		
AD 9	EU9513	DDA1811-666	21.05.91		
AD10	EU9446	DDA1811-662	24.05.91		
AD11	EU9067	DDA1811-663	29.05.91		
AD12	EV2021	DDA1811-660	03.06.91		
AD13	EV3286	DDA1811-657	12.06.91		
AD14	EV4156	DDA1811-671	12.06.91		
AD15	EV4297	DDA1811-661	12.06.91		
AD16	EV3060	DDA1811-665	14.06.91		
AD17	EV3201	DDA1811-667	13.06.91		
AD18	EV3767	DDA1811-659	14.06.91		
AD19	EV5814	DDA1811-656	25.06.91		
AD20	EV4864	DDA1811-652	01.07.91		
AD21	EV5507	DDA1811-653	01.07.91		
AD22	EW 699	DDA1811-700	25.07.91		
AD23	EW1005	DDA1811-697	25.07.91		
AD24	EW1306	DDA1811-698	25.07.91		
AD25	EW1694	DDA1811-699	25.07.91		
AD26	EW2459	DDA1811-696	25.07.91		
AD27	EW3309	DDA1811-701	01.08.91		
AD28	EW3508	DDA1811-704	01.08.91		
AD29	EW3806	DDA1811-708	01.08.91		
AD30	EW3074	DDA1811-702	06.08.91		
AD31	EW3979	DDA1811-714	06.08.91		
AD32	EW5598	DDA1811-711	13.08.91		
AD33	EW6304	DDA1811-709	13.08.91		
AD34	EW5608	DDA1811-705	16.08.91		
AD35	EW6225	DDA1811-707	16.08.91		
AD36	EW7507	DDA1811-715	23.08.91		
AD37	EW8437	DDA1811-706	23.08.91		
AD38	EW9016	DDA1811-703	29.08.91		
AD39	EW9809	DDA1811-710	29.08.91		
AD40	EX2116	DDA1811-713	05.09.91		
AD41	EX2369	DDA1811-712	05.09.91		
AD42	FB8780	DDA1811-725	11.03.92		
AD43	FB9527	DDA1811-721	11.03.92		
AD44	FB9226	DDA1811-726	16.03.92		
AD45	FB9728	DDA1811-722	16.03.92		
AD46	FB9897	DDA1811-732	16.03.92		
AD47	FC 321	DDA1811-717	16.03.92		
AD48	FC4888	DDA1811-716	13.04.92		
AD49	FC5018	DDA1811-727	13.04.92		
AD50	FC5718	DDA1811-718	13.04.92		
AD51	FC5735	DDA1811-729	13.04.92		
AD52	FC5772	DDA1811-724	13.04.92		
AD53	FC6037	DDA1812-735	13.04.92		Gardner LG1200 engine
AD54	FC6282	DDA1811-723	13.04.92		
AD55	FC6330	DDA1811-730	13.04.92		
AD56	FC6469	DDA1812-736	13.04.92		Gardner LG1200 engine

KMB class: Dragon (A) 11M AC   DUPLE METSEC BODIES

Fleet number	Regn. number	Chassis number	Date first registered	Withdrawn	Disposal
AD57	FD1179	DDA1811-733	11.05.92		
AD58	FD1232	DDA1811-728	11.05.92		
AD59	FD1436	DDA1811-734	11.05.92		
AD60	FD1603	DDA1811-720	11.05.92		
AD61	FD1879	DDA1811-719	11.05.92		
AD62	FD2497	DDA1811-731	11.05.92		
AD63	FD1091	DDA1811-741	11.05.92		
AD64	FD1146	DDA1811-737	11.05.92		
AD65	FD1755	DDA1811-740	11.05.92		
AD66	FD2180	DDA1811-739	11.05.92		
AD67	FD4612	DDA1811-747	25.05.92		
AD68	FD4715	DDA1811-746	25.05.92		
AD69	FD4750	DDA1811-755	25.05.92		
AD70	FD5407	DDA1811-754	25.05.92		
AD71	FD5410	DDA1811-744	25.05.92		
AD72	FD5520	DDA1811-758	25.05.92		
AD73	FD5587	DDA1811-743	25.05.92		
AD74	FD5735	DDA1811-748	25.05.92		
AD75	FD5793	DDA1811-756	25.05.92		
AD76	FD5801	DDA1811-750	25.05.92		
AD77	FD5806	DDA1811-751	25.05.92		
AD78	FD5872	DDA1811-749	25.05.92		
AD79	FD5906	DDA1811-738	25.05.92		
AD80	FD5962	DDA1811-742	25.05.92		
AD81	FD5993	DDA1811-753	25.05.92		
AD82	FD6118	DDA1811-745	25.05.92		
AD83	FD6160	DDA1811-757	25.05.92		
AD84	FD6420	DDA1811-752	25.05.92		
AD85	FD6725	DDA1811-759	03.06.92		
AD86	FH 722	DDA1813-769	06.10.92		
AD87	FH1393	DDA1813-775	06.10.92		
AD88	FH2110	DDA1813-787	06.10.92		
AD89	FH2119	DDA1813-766	06.10.92		
AD90	FH2171	DDA1813-772	06.10.92		
AD91	FH2202	DDA1813-768	06.10.92		
AD92	FH4920	DDA1813-794	21.10.92		
AD93	FH5548	DDA1813-771	21.10.92		
AD94	FH8543	DDA1813-813	02.11.93		
AD95	FH9512	DDA1813-797	02.11.92		
AD96	FH9577	DDA1813-770	02.11.92		
AD97	FH8597	DDA1813-814	09.12.92		
AD98	FH8627	DDA1813-784	09.12.92		
AD99	FJ8921	DDA1813-773	09.12.92		
AD100	FJ8931	DDA1813-788	09.12.92		
AD101	FJ9070	DDA1813-789	09.12.92		
AD102	FJ9228	DDA1813-785	09.12.92		
AD103	FK 455	DDA1813-782	09.12.92		
AD104	FK2988	DDA1813-762	22.12.92		
AD105	FK3292	DDA1813-767	22.12.92		
AD106	FK4038	DDA1813-776	22.12.92		
AD107	FK4563	DDA1813-761	04.01.93		
AD108	FK4875	DDA1813-765	04.01.93		
AD109	FK5072	DDA1813-764	04.01.93		
AD110	FK5141	DDA1813-763	04.01.93		
AD111	FK6019	DDA1813-760	04.01.93		
AD112	FL 562	DDA1813-781	01.02.93		
AD113	FL 590	DDA1813-793	01.02.93		
AD114	FL 652	DDA1813-783	01.02.93		
AD115	FL 724	DDA1813-788	01.02.93		
AD116	FL 954	DDA1813-808	01.02.93		
AD117	FL1282	DDA1813-777	01.02.93		
AD118	FL1437	DDA1813-774	01.02.93		
AD119	FL1504	DDA1813-812	01.02.93		
AD120	FL1978	DDA1813-780	01.02.93		
AD121	FL2085	DDA1813-790	01.02.93		
AD122	FL2434	DDA1813-786	01.02.93		
AD123	FL2498	DDA1813-779	01.02.93		
AD124	FM4674	DDA1813-796	08.03.93		
AD125	FM4824	DDA1813-811	08.03.93		
AD126	FM5360	DDA1813-791	08.03.93		
AD127	FM5444	DDA1813-792	08.03.93		
AD128	FM5652	DDA1813-810	08.03.93		
AD129	FM5897	DDA1813-795	08.03.93		
AD130	FM5987	DDA1813-687	08.03.93		

> A further 115 AD-class 11m Dragons were on order late 1994

Air-conditioned SHORT (10.4M) DENNIS DRAGON.
Fleet nos: ADS 1-30.   DUPLE METSEC BODIES

Fleet number	Regn. number	Chassis number	Date first registered	Withdrawn	Disposal
ADS 1	FS7340	DDA2201-882	26.08.93		
ADS 2	FS8396	DDA2201-885	26.08.93		
ADS 3	FS8419	DDA2201-883	26.08.93		
ADS 4	FS9108	DDA2201-881	27.08.93		
ADS 5	FS9649	DDA2201-889	27.08.93		
ADS 6	FT 144	DDA2201-878	27.08.93		
ADS 7	FS9772	DDA2201-886	01.09.93		
ADS 8	FT 182	DDA2201-877	01.09.93		
ADS 9	FT1039	DDA2201-887	06.09.93		
ADS10	FT2125	DDA2201-884	06.09.93		
ADS11	FT3273	DDA2201-888	13.09.93		
ADS12	FT3339	DDA2201-891	13.09.93		
ADS13	FT3867	DDA2201-890	13.09.93		
ADS14	FT4173	DDA2201-894	14.09.93		
ADS15	FT4465	DDA2201-893	14.09.93		
ADS16	FT4764	DDA2201-900	20.09.93		
ADS17	FT5712	DDA2201-902	20.09.93		
ADS18	FT6673	DDA2201-897	24.09.93		
ADS19	FT7214	DDA2201-903	24.09.93		
ADS20	FT8209	DDA2201-901	24.09.93		
ADS21	FT9054	DDA2201-899	05.10.93		
ADS22	FT9056	DDA2201-895	05.10.93		
ADS23	FT8994	DDA2201-892	07.10.93		
ADS24	FT9178	DDA2201-876	07.10.93		
ADS25	FU1098(?)	DDA2201-896	13.10.93		
ADS26	FU1694	DDA2201-879	13.10.93		
ADS27	FU2195	DDA2201-867	13.10.93		
ASD28	FU2776	DDA2201-904	20.10.93		
ADS29	FU4387	DDA2201-890	20.10.93		
ADS30	FV7835	DDA2201-880	09.12.93		

323

VOLVO OLYMPIAN - Air-conditioned - 11.3 metre.
Fleet nos: AV 1-85. ALEXANDER BUS BODIES
KMB sub-class: Volvo Oly: 11M a/c

Fleet number	Regn. number	Chassis number	Date first registered	Withdrawn	Disposal
AV 1	FW5572	025009	10.01.94		
AV 2	GD1404	025487	30.08.94		
AV 3	GD1673	025501	30.08.94		
AV 4	GD1757	025505	30.08.94		
AV 5	GD2390	025464	30.08.94		
AV 6	GD 568	025465	31.08.94		
AV 7	GD 830	025489	31.08.94		
AV 8	GD1092	025466	31.08.94		
AV 9	GD1570	025502	31.08.94		
AV10	GD2187	025468	31.08.94		
AV11	GD2665	025488	02.09.94		
AV12	GD2923	025508	02.09.94		
AV13	GD3163	025486	05.09.94		
AV14	GD4055	025428	05.09.94		
AV15	GD2535	025425	07.09.94		
AV16	GD3191	025429	07.09.94		
AV17	GD3865	025506	07.09.94		
AV18	GD2896	025399	08.09.94		
AV19	GD3159	025467	08.09.94		
AV20	GD3383	25341	09.09.94		
AV21	GD4331	25340	09.09.94		
AV22	GD4816	25537	19.09.94		
AV23	GD5733	25534	19.09.94		
AV24	GD6641	25507	22.09.94		
AV25	GD7147	25447	22.09.94		
AV26	GD7748	25337	22.09.94		
AV27	GD8073	25398	22.09.94		
AV28	GD7765	25426	28.09.94		
AV29	GD8840	25556	03.10.94		
AV30	GD8856	25554	03.10.94		
AV31	GD9501	25504	03.10.94		
AV32	GE 506	25521	07.10.94		
AV33	GE1148	25553	07.10.94		
AV34	GE1943	25594	07.10.94		
AV35	GE 750	25396	12.10.94		
AV36	GE2449	25485	12.10.94		
AV37	GE2756	25552	14.10.94		
AV38	GE2641	25488	17.10.94		
AV39	GE4410	25450	17.10.94		
AV40	GE2852	25446	18.10.94		
AV41	GE2878	25449	18.10.94		
AV42	GE2915	25427	20.10.94		
AV43	GE4240	25397	20.10.94		
AV44	GE4005	25510	20.10.94		
AV45	GE4577	25557	24.10.94		
AV46	GE4660	25509	24.10.94		
AV47	GE4861	25566	26.10.94		
AV48	GE5513	25400	26.10.94		
AV49	GE5395	25568	11.94		
AV50	GE6531	25339	11.94		
AV51	GE6827	25533	11.94		
AV52	GE6864	25555	11.94		
AV53	GE6919	25338	11.94		
AV54	GE7832	25339	11.94		
AV55	GE8007	25338	11.94		
AV56	GE6568	25691	11.94		
AV57	GE6760	25535	11.94		
AV58	GE7672	25540	11.94		
AV59	GE7990	25551	11.94		
AV60	GF 227	25692	11.94		
AV61	GE9355	25683	11.94		
AV62	GE9596	25685	11.94		
AV63	GE9531	25688	11.94		
AV64	GF1242	25525	11.94		
AV65	GF2069	25518	11.94		
AV66	GF 786	25532	11.94		
AV67	GF1912	25667	11.94		
AV68	GF1110	25536	11.94		
AV69	GF2128	25690	11.94		
AV70	GF 557	25519	11.94		
AV71	GF2543	25522	11.94		
AV72	GF3087	25524	11.94		

KMB sub-class: Volvo Oly: 11M a/c (Contd) ALEXANDER BUS BODIES

Fleet number	Regn. number	Chassis number	Date first registered	Withdrawn	Disposal
AV73	GF3727	25523	11.94		
AV74	GF4035	25526	11.94		
AV75	GF3766	25686	11.94		
AV76	GF4070	25520	11.94		
AV77	GF4681	-?-	12.94		
AV78	GF5012	25684	12.94		
AV79	GF6351	25682	12.94		
AV80	GG8690	25687	01.95		
AV81	GG8734	25680	01.95		
AV82	GG9294	25681	01.95		
AV83	GG9302	25503	01.95		
AV84	GG9879	25678	01.95		
AV85	GH 377	25500	01.95		
AV86	GH 418	-?-	01.95		

VOLVO OLYMPIAN - Air-conditioned - 12-metre.
Fleet nos: 3AV 1-50. ALEXANDER BUS BODIES
KMB sub-class: Volvo Oly: 12M a/c

Fleet number	Regn. number	Chassis number	Date first registered	Withdrawn	Disposal
3AV 1	GB2607	25118	01.07.94		
3AV 2	GB3493	25092	01.07.94		
3AV 3	GB3693	25156	01.07.94		
3AV 4	GB3968	25094	01.07.94		
3AV 5	GB2585	25115	04.07.94		
3AV 6	GB3485	25093	04.07.94		
3AV 7	GB5018	25119	05.07.94		
3AV 8	GB5360	25114	05.07.94		
3AV 9	GB5045	25112	06.07.94		
3AV10	GB5483	25007	06.07.94		
3AV11	GB4860	25150	07.07.94		
3AV12	GB5890	25136	07.07.94		
3AV13	GB4603	25133	08.07.94		
3AV14	GB4681	25135	08.07.94		
3AV15	GB6156	25143	11.07.94		
3AV16	GB6251	25006	11.07.94		
3AV17	GB4995	25140	12.07.94		
3AV18	GB5976	25163	12.07.94		
3AV19	GB6517	25147	13.07.94		
3AV20	GB8469	25155	13.07.94		
3AV21	GB6839	25148	15.07.94		
3AV22	GB8218	25142	15.07.94		
3AV23	GB6673	25166	18.07.94		
3AV24	GB8235	25137	18.07.94		
3AV25	GB9206	25134	19.07.94		
3AV26	GB9549	25126	19.07.94		
3AV27	GB9437	25116	21.07.94		
3AV28	GC 431	25113	21.07.94		
3AV29	GB9487	25139	22.07.94		
3AV30	GB9848	25138	22.07.94		
3AV31	GC1189	25141	25.07.94		
3AV32	GC1425	25117	25.07.94		
3AV33	GC 629	25167	26.07.94		
3AV34	GC1659	25160	26.07.94		
3AV35	GC2329	25146	27.07.94		
3AV36	GC2488	25159	27.07.94		
3AV37	GC4401	25149	28.07.94		
3AV38	GC2805	25161	29.07.94		
3AV39	GC2875	25157	29.07.94		
3AV40	GC6682	25144	11.08.94		
3AV41	GC6720	25164	11.08.94		
3AV42	GC7236	25158	11.08.94		
3AV43	GC7353	25165	11.08.94		
3AV44	GC7464	25145	11.08.94		
3AV45	GC7522	25152	11.08.94		
3AV46	GC7591	25095	11.08.94		
3AV47	GC8095	25162	11.08.94		
3AV48	GC8099	25154	11.08.94		
3AV49	GC8229	25151	11.08.94		
3AV50	GC8347	25168	11.08.94		

DENNIS DRAGON - 11m A/C - Additional vehicles
Fleet nos: AD131-190 DUPLE-METSEC BUS BODIES
KMB sub-class: Dragon

Fleet number	Regn. number	Chassis number	Date first registered	Disposal
AD131	GG8659	DDA1819/1009	01.95	
AD132	GG9877	1065	01.95	
AD133	GH5169		95	
AD134	GH5221		95	
AD135	GH5398		95	
AD136	GH5813		95	
AD137	GH5874		95	
AD138	GH -?-		95	
AD139	GH5996		95	
AD140	GH6112		95	
AD141	GH6231	DDA1819-983	95	
AD142	GH6239		95	
AD143	GH6283		95	
AD144	GJ2541		95	
AD145	GJ2779		95	
AD146	GJ2783		95	
AD147	GJ2795		95	
AD148	GJ3073	DDA1819-1114	95	
AD149	GJ3078		95	
AD150	GJ3112		95	
AD151	GJ3184	DDA1819-1011	95	
AD152	GJ2319		95	
AD153	GJ3303		95	
AD154	GJ3318		95	
AD155	GJ3609		95	
AD156	GJ3659	DDA1819-1128	95	
AD157	GJ3750		95	
AD158	GJ3815	DDA1819-978	95	
AD159	GJ3964		95	
AD160	GJ3965	DDA1819-976	95	
AD161	GJ3975		95	
AD162	GJ4084	DDA1819-974	95	
AD163	GJ4150		95	
AD164	GJ4309		95	
AD165	GJ4369		95	

Dragon 11m A/C (Contd.) AD131-190

Fleet number	Regn. number	Chassis number	Date first registered	Disposal
AD166	GJ4402	DDA1819-986	95	
AD167	GJ4622		95	
AD168	GJ4721		95	
AD169	GJ4823		95	
AD170	GJ4960		95	
AD171	GJ4963		95	
AD172	GJ4985		95	
AD173	GJ5069		95	
AD174	GJ5216	DDA1819-1112	95	
AD175	GJ5319		95	
AD176	GJ5334		95	
AD177	-?-		95	
AD178	GJ5568		95	
AD179	GJ5727		95	
AD180	GJ5751		95	
AD181	GJ5792		95	
AD182	GJ5810		95	
AD183	GJ5913		95	
AD184	-?-		95	
AD185	GJ6017		95	
AD186	GJ6162		95	
AD187	GJ6209		95	
AD188	GJ6229		95	
AD189	GJ6426		95	
*	GJ8306			

* Indicates that this vehicle entered service without a fleet number.

Detailed information about these buses was incomplete at the time of going to press.

Corrections to the fleet list:
Page 301: Daimler(e) CS should be D490 (AD7396) - **not** AD7400.
Page 316: 3BL11 - registration number should be DA2698.
Page 320: After S3N19, the following ten fleet numbers are misprinted as S3N30-39; *these should be S3N20-29 (DM7035)*: registration numbers are all correct and in sequence. the numbers following are in correct sequence.

SINGLE-DECK POSTSCRIPT:
DENNIS 'DART' 98SDL - 9.8-metre.
Fleet nos: AA29-41. WADHAM STRINGER ASSEMBLED METSEC B43F BUS BODIES

Fleet number	Regn. number	Chassis number	Date first registered	Notes
AA29	GE7471	98SDL3039-	11.94	
AA30	GE7349		11.94	
AA31	GE8781		11.94	
AA32	GE8603		11.94	
AA33	GE9055		11.94	
AA34	GF1742		11.94	
AA35	GF2592		11.94	
AA36	GF6156		11.94	

WADHAM STRINGER ASSEMBLED METSEC B37F 'AIRBUS' BODIES

AA37	GF8477		12.94	
AA38	GF7823		12.94	
AA39	GF9466		12.94	Detailed information about these buses was incomplete at the time of going to press.
AA40	GG2513		12.94	
AA41	GG2513		01.95	
AA42	GG6508	98SDL3039-2044	01.95	

On order April 1995:
DENNIS DARTS - NORTHERN COUNTIES BODIES.

Correction to Dennis Darts on Page 297:

AA3-22 have DUPLE METSEC BODIES.
AA3-11, & 21 are service buses - total 10.
AA12-20 & 22 are 'Airbus' Dart(A) - total 10.

Correction to Dennis Lances on Page 297:

AN1-10, 12 & 13 are dual-doorway service buses - total 12.
AN11, 14-24 are single-doorway 'Airbus' Lance(A) - total 12

Training Fleet Buses. Listing is in registration number order.

Former fleet number	Regn. number	Old number	Make	New	In	Out
D 9	4210	-	Daimler(a)	1954	1/81	? to uniform store
D 11	4212		Daimler(a)	1954	1/81	? to preservation
D153	4216		Daimler(b)	1959	4/82	?
D153	4232	-	Daimler(b)	1959	3/82	?
D162	4235		Daimler(b)	1959	3/82	?
D163	4236		Daimler(b)	1959	1/82	?
D169	4244		Daimler(b)	1959	4/82	?
D180	4595		Daimler(b)	1959	4/82	?
D182	4597		Daimler(b)	1960	2/82	?
D184	4599		Daimler(b)	1960	11/81	?
D185	4600		Daimler(b)	1960	3/82	?
D188	4603		Daimler(b)	1960	1/82	?
D189	4604		Daimler(b)	1960	2.82	?
D192	4607		Daimler(b)	1960	4/82	?
D193	4608		Daimler(b)	1960	1/82	?
D196	4613		Daimler(b)	1960	2/82	?
D197	4614		Daimler(b)	1960	2/82	?
D200	4625		Daimler(b)	1960	3/82	?
D204	4665		Daimler(b)	1960	3/82	?
D206	4670		Daimler(b)	1960	3/82	?
D194	4690		Daimler(b)	1960	1/82	?
D 26	4961		Daimler(a)	1949	1/80	1/83 (note 1)
D 27	4962		Daimler(a)	1949	5/80	?
D 28	4963		Daimler(a)	1949	3/80	?
D 29	4964		Daimler(a)	1949	3/80	?
D 30	4965		Daimler(a)	1949	4/80	?
D 31	4966		Daimler(a)	1949	1/80	?
D210	4979		Daimler(b)	1960	3/82	?
D211	4980		Daimler(b)	1960	4/82	?
D212	4981		Daimler(b)	1960	4/82	?
D215	4988		Daimler(b)	1960	4/82	?
D218	4991		Daimler(b)	1960	2/82	?
D 40	HK4001		Daimler(a)	1950	3/80	?
D 44	HK4006		Daimler(a)	1950	2/80	?
D 45	HK4009		Daimler(a)	1950	9/80	?
D 46	HK4010		Daimler(a)	1950	4/80	?
D 61	HK4054		Daimler(a)	1951	9/80	?
D 97	HK4266		Daimler(a)	1955	12/80	?
D125	HK4326		Daimler(a)	1956	11/80	?
D150	HK4351		Daimler(a)	1957	12/71	? (note 2)
D233	HK4474		Daimler(b)	1960	12/81	?
D234	HK4475		Daimler(b)	1960	4/82	?
D237	HK4478		Daimler(b)	1961	4/82	?
D246	HK4487		Daimler(b)	1961	12.81	?
D250	HK4491		Daimler(b)	1961	3/82	?
D252	HK4493		Daimler(b)	1961	3/82	?
D255	HK4496		Daimler(b)	1961	4/82	?
D257	HK4498		Daimler(b)	1961	4/82	?
D258	HK4499		Daimler(b)	1961	1/82	?
L 11	HK4512		Albion VT17L	1961	11/80	?
L 19	HK4520		Albion VT17L	1961	1/81	?
L 22	HK4523		Albion VT17L	1961	2/81	?
L 26	HK4527		Albion VT17L	1961	2/81	?
L 27	HK4529		Albion VT17L	1961	1/81	?
L 31	HK4532		Albion VT17L	1961	1/81	?
L 32	HK4533		Albion VT17L	1961	1/80	?
L 33	HK4534		Albion VT17L	1961	1/80	?
L 35	HK4536		Albion VT17L	1961	2/80	?
L 36	HK4537		Albion VT17L	1961	3/80	?
L 38	HK4538		Albion VT17L	1961	2/80	?
L 39	HK4540		Albion VT17L	1961	3/81	?
L 56	HK4557		Albion VT17L	1961	11/79	?
L 61	HK4562		Albion VT17L	1961	11/79	?
L 62	HK4563		Albion VT17L	1961	1/80	?
L 74	HK4575		Albion VT17L	1961	12/80	?
L 76	HK4577		Albion VT17L	1961	11/79	?
-	AE6147	AD4710	Albion VT17L	1961	6/70	?
-	AJ7144	?	TSM K5LA7	1947/8	-/67	?
-	AJ7145	?	TSM K5LA7	1947/8	-/67	?
-	AJ7146	?	TSM K5LA7	1947/8	-/67	?
-	AP7879	?	TSM K5LA7	1947/8	-/69	?
-	AP7880	?	TSM K5LA7	1947/8	-/69	?
-	AP7881	?	TSM K5LA7	1947/8	-/69	?
-	AP7882	?	TSM K5LA7	1947/8	-/69	?
-	AP7883	?	TSM K5LA7	1947/8	-/69	?
-	AP7884	?	TSM K5LA7	1947/8	-/69	?
-	AP7885	?	TSM K5LA7	1947/8	-/69	?
-	AP7886	?	TSM K5LA7	1947/8	-/69	?
-	AP7887	?	TSM K5LA7	1947/8	-/69	?
-	AP7888	?	TSM K5LA7	1947/8	-/69	?
-	AP7889	?	TSM K5LA7	1947/8	-/69	?
-	AP7890	?	TSM K5LA7	1947/8	-/69	?
-	AP7891	?	?	?	?	?
L270	AS6125	AD4711	Albion VT17L	1961	7/70	11/73 #
-	AS6126	AD4712	Albion VT17L	1961	7/70	?
L276	AW5196	AD4709	Albion VT17L	1961	9/71	-/74 #
-	AW6340	AD4708	Albion VT17L	1961	10/71	?
-	AW7671	AD4707	Albion VT17L	1961	11/71	?
-	AW7672	AD4705	Albion VT17L	1961	11/71	?
L275	AW8727	AD4706	Albion VT17L	1961	11/71	8/74 #
-	AW8728	AD4703	Albion VT17L	1961	11/71	?
L274	AX1209	HK4588	Albion VT17L	1961	12/71	8/74 #
L273	AX1210	HK4590	Albion VT17L	1961	12/71	7/74 #
L278	AX1211	AD4701	Albion VT17L	1961	12/71	11/73 #
L271	AX1213	AD4704	Albion VT17L	1961	12/71	11/73 #
-	AX3283	HK4589	Albion VT17L	1961	1/72	?
D678	BG7693	-	Daimler CRG6LXB	1974	1/90	
D680	BG7695	-	Daimler CRG6LXB	1974	12/89	
D682	BG7697	-	Daimler CRG6LXB	1974	12/89	
CA 1	BH3002	-	Albion coach	1975	10/83	?
CA 2	BH3004	-	Albion coach	1975	10/83	?
CA 3	BH3005	-	Albion coach	1975	10/83	?
CA 5	BH3007	-	Albion coach	1975	10/83	?
CA 6	BH3008	-	Albion coach	1975	10/83	?
CA 7	BH3009	-	Albion coach	1975	11/83	?
CA 8	BH3010	-	Albion coach	1975	11/83	?
CA 9	BH3011	-	Albion coach	1975	11/83	?
CA11	BH3013	-	Albion coach	1975	11/83	?
D706	BH3536	-	Daimler CRG6LXB	1975	4/90	
D707	BH3537	-	Daimler CRG6LXB	1975	4/90	
CA20	BH3711	-	Albion coach	1975	3/87	?
CA63	BH7644	-	Albion coach	1975	5/83	
D739	BH7627	-	Daimler CRG6LXB	1975	7/89	
D746	BH7743	-	Daimler CRG6LXB	1975	7/89	
D751	BH8519	-	Daimler CRG6LXB	1975	7/89	
D752	BH8520	-	Daimler CRG6LXB	1975	7/89	
D759	BJ 687	-	Daimler CRG6LXB	1975	2/90	
D762	BJ 690	-	Daimler CRG6LXB	1975	6/89	
D813	BK1164	-	Daimler CRG6LXB	1976	6/90	
G 1	BJ9266	-	Guy Victory 'J'	1976	3/82	?
G 2	BJ9267	-	Guy Victory 'J'	1976	1/82	?
G 3	BJ9268	-	Guy Victory 'J'	1976	2/82	?
G 4	BJ9269	-	Guy Victory 'J'	1976	1/82	?
D894	BM4486	-	Leyland FE33AGR	1977	5/86	?
G 92	CC5179	-	Leyland Victory 2	1980	5/85	?
2D10	CM3985	-	Daimler 'DMS'	1973	10/86	?
2D12	CM4727	-	Daimler 'DMS'	1973	10/86	?
2D16	CM4939	-	Daimler 'DMS'	1973	10/86	?
2D21	CM8752	-	Daimler 'DMS'	1973	10/86	?
2D18	CM8999	-	Daimler 'DMS'	1973	10/86	?
2D30	CN 315	-	Daimler 'DMS'	1973	10/86	?
2D22	CN 672	-	Daimler 'DMS'	1974	5/84	?
2D34	CN2740	-	Daimler 'DMS'	1973	10/83	?
2D36	CN3370	-	Daimler 'DMS'	1973	9/85	?
2D48	CN4313	-	Daimler 'DMS'	1974	7/85	?
2D45	CN4375	-	Daimler 'DMS'	1974	8/85	?
2D38	CN4588	-	Daimler 'DMS'	1974	7/85	?
2D39	CN4590	-	Daimler 'DMS'	1973	7/85	?
2D61	CN9710	-	Daimler 'DMS'	1973	9/85	?
2D65	CP 298	-	Daimler 'DMS'	1974	9/85	?
2D67	CP 325	-	Daimler 'DMS'	1974	4/84	?
2D74	CP3134	-	Daimler 'DMS'	1974	10/85	?
2D75	CP3237	-	Daimler 'DMS'	1974	9/85	?
2D83	CP4395	-	Daimler 'DMS'	1974	11/85	?
2D91	CP5092	-	Daimler 'DMS'	1973	4/84	?
2D100	CV9173	-	Daimler 'DMS'	1974	10/86	?
2D101	CV9753	-	Daimler 'DMS'	1974	10/86	?
-	DL9739	SBS7003	Dennis Dominator	1982	-/85	ex-Singapore.
AT11	DV 647	-	Toyota	1988	13.11.91	
AT12	DV 699	-	Toyota	1988	13.11.91	

ABOVE: Still registered as a 'Public Vehicle', Tilling-Stevens 4636 was on driver training duties in Waterloo Road, Kowloon in August 1969. *(Julian Osborne*

RIGHT: Once the earlier Daimler(a)-type CVG5's were withdrawn from passenger service, they were drafted into the Driver Training Unit and could be seen in numbers on the road in the hands of novice drivers. Here 4210 was parked at Cha Kwo Ling parking area one weekend in May 1982. *(John Shearman*

LOWER RIGHT: AX3283 was originally HK4589 before its transfer to the training fleet in January 1972. Unlike other vehicles of this type, similarly transferred and re-registered at the same time, this bus was never to return to the passenger fleet. It was seen here in 1974 returning to its Kwai Chung base after a training session. *(Mike Davis*

Service Vehicles Based on Passenger Chassis.

Former fleet number	Regn. number	Old number	Make	Function	New	In	Out
D11	4212		Daimler(a)	Staff rest room	1954	-/82	
D179	4594		Daimler(b)	Tree cutter	1959	2/81	?
D187	4602		Daimler(b)	Tree cutter	1960	4/78	15.05.78sc
-	HK5406	?	Commer	Service car	?	?	?
-	HK5551	?	Thornycroft CD4	Crane	c1938	?	
-	HK5710	?	Commer	Service car	?	?	?
-	HK5747	?	Commer	Service car	?	?	?
-	HK5758	?	Commer	Service car	?	?	?
-	HK5759	?	Commer	Service car	?	?	?
-	HK5760	?	Commer	Service car	?	?	?
-	AA6147	?	Commer	Service car	?	?	?
L187	AD7204	-	Albion CH13AXL	Service truck	1965	8/82	?
L188	AD7205	-	Albion CH13AXL	Service truck	1965	8/81	?
L195	AD7212	-	Albion CH13AXL	Service truck	1965	7/82	?
L196	AD7213	-	Albion CH13AXL	Service truck	1965	6/82	?
L198	AD7215	-	Albion CH13AXL	Service truck	1965	9/81	?
L199	AD7216	-	Albion CH13AXL	Service truck	1965	6/82	?
L200	AD7217	-	Albion CH13AXL	Service truck	1965	9/81	?
L203	AD7220	-	Albion CH13AXL	Water tanker	1965	9/82	?
L205	AD7222	-	Albion CH13AXL	Service truck	1965	10/81	?
L216	AD7233	-	Albion CH13AXL	Service truck	1965	9/81	?
2D1	AD7478	-	Daimler ex-CCG6	Staff rest room	1964	?	?
-	AF4004	?	Commer	Service car	?	-/64	?
-	AF4963	?	Commer	Service car	?	-/64	?
-	AF5190	?	Commer	Service car	?	-/64	?
-	AG8023	?	Commer	Service car	?	-/65	?
-	AG8807	?	Commer	Service car	?	-/65	?
-	AG8896	?	Commer	Service car	?	-/65	?
-	AP7892	?	Dennis Pax	Service car	1952	?	?
-	AP7893	?	Dennis Pax	Service car		1952	?
-	AV5269	AD7202	Albion CH13AXL	Service truck	1965	6/71	?
-	BE7470	AR7639	Seddon Pennine 4	Water tanker	1970	12/73	
L277	BH3001		Albion VT23L	Service car	1964	1/82	
(CA19)	BH3710	-	Albion coach	Service lorry	1975	7/84	?
(CA26)	BH3717	-	Albion coach	Service lorry	1975	7/84	?
(CA29)	BH3720	-	Albion coach	Service lorry	1975	6/85	?
(CA34)	BH3725	-	Albion coach	Service lorry	1975	10/84	?
(CA38)	BH3729	-	Albion coach	Service lorry	1975	10/84	?
(CA45)	BH5089	-	Albion coach	Service lorry	1975	4/88	?
(CA47)	BH5091	-	Albion coach	Service lorry	1975	2/88	?
(CA52)	BH7633	-	Albion coach	Service lorry	1975	3/84	?
(CA59)	BH7640	-	Albion coach	Service lorry	1975	4/88	?
(CA60)	BH7641	-	Albion coach	Service lorry	1975	3/88	?
(CA62)	BH7643	-	Albion coach	Service lorry	1975	?	?
(CA66)	BH8196	-	Albion coach	Service lorry	1975	11/83	?
(CA68)	BJ1862	-	Albion coach	Service lorry	1975	11/83	
(CA70)	BJ1864	-	Albion coach	Water tanker	1975	2/84	
(CA71)	BJ1973	-	Albion coach	Service lorry	1975	2/89	?
(CA73)	BJ2629	-	Albion coach	Service lorry	1975	9/88	?
(CA80)	BJ4811	-	Albion coach	Service lorry	1975	3/88	
(CA83)	BJ4814	-	Albion coach	Service lorry	1975	2/88	
(CA84)	BJ4815	-	Albion coach	Service lorry	1975	2/88	
(CA93)	BJ1822	-	Albion coach	Cash van	1976	11/86	

Some photographs of service vehicles appear within the main text but a few interesting examples are shown here

LEFT: Perhaps the oldest commercially working Thornycroft is this pre-1941 example, formerly HK5551 (see page 15), now de-registered and allocated a KMB internal number ZZ202, the meaning of which reamains unclear. This interesting vehicle still had its Gardner 4LW engine when examined by the author in 1984. Until circa 1980, the front portion of its bus body was retained which resembled a post-war design widely used by KMB. *(John Shearman)*

BOTTOM LEFT: This Tilling-Stevens water tank was used entirely within Lai Chi Kok Depot until at least the mid-1980's. The original registration number as a bus cannot be traced but it was subsequently re-registered AA6147 in the goods vehicle series It was freshly repainted in this March 1981 view. *(John Shearman)*

BELOW: As referred to in the section describing the Commer Superpoise MkII buses, three of that type, AA5760, AF7493 & AG8968, were rebuilt whilst in use as service cars and acquired an angular bonnet and wing assembly. No photograph has come to light of one of these, except for this view from the background of another picture. *(V. J. Griffiths collection)*

ABOVE: KMB's very own 'duck & ducklings'. The Public Relations Department operates a pair of miniature buses, one two-axle and the other three-axle which are driven by small petrol engines and attend Shows attended or arranged by KMB. It is believed that they are sponsored by Dennis, whose badge they both carry. The bus, AD7336, is a Daimler(e) specially converted to carry the two 'little-ones'. *(From a KMB calendar illustration*

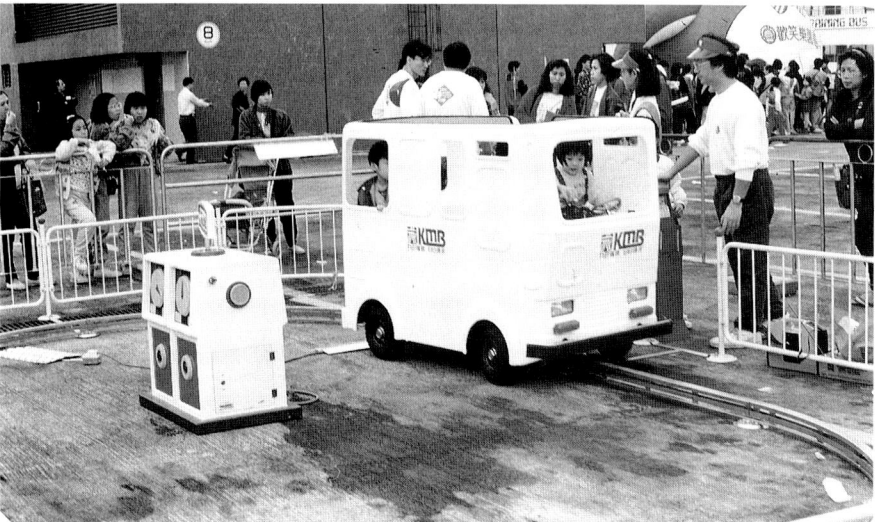

CENTRE LEFT: Also taken to events is an electric 'fun-bus', complete with its own guided circuit - could this be Hong Kong's answer to the 'GLT'? *(George Law*

LEFT: AD7202, the Albion Chieftain bus that was converted to a service truck prior to the 1973 fleet numbering after only a few years in service. It survived in its new role to receive the '1982' livery. *(Ron Phillips collection*

This Tilling-Stevens bus is one of many operated by the Kowloon Motor Bus Co. Ltd. of Hong Kong. Here, as in many other parts of the World, Tilling-Stevens vehicles have brought to Passenger Transport reliability, long life and travelling comfort which entirely depend upon fine design and precision workmanship.

TILLING-STEVENS
LIMITED
MAIDSTONE

An advertisement from the May 14th, 1948 edition of 'Passenger Transport' magazine. Note that the title of KMB is sign-written in full along the side panels in both English and Chinese.

Annex:
Bigger-Bus Development.

Buses Ever Larger.

Hong Kong's public transport operators have always, for one reason or another, had difficulty in satisfying every aspect of public demand at any one time.

In Kowloon, pre-1941, the increasing tide of refugees fleeing from the advance of the Japanese armies into South China resulted in ever increasing demands for bus travel but this demand was met by providing greater numbers of single-deck buses, seating a modest 35, or so, passengers.

With the the return of the population following the liberation of the Colony in 1945, a new wave of refugees threatened to flood Hong Kong and this time double-deck buses *were* seen to be the answer and KMB placed its first examples in service during April 1949. These early double-deckers were Daimler CVG5's, quite modest 56-seaters of traditional British appearance, with half-cabs and open rear platforms which soon became inadequate. The real answer lay in larger buses and KMB made an application to Government for permission to operate buses of very large dimensions by the standards of the day. Initially this permission was granted only to bring Hong Kong into line with the United Kingdom.

Daimler CVG6's, 30ft long and 8ft wide, were introduced in 1962 but KMB was still not able to meet traffic demands and continued its efforts to gain approval for oversized buses - oversized even by British standards. Faced with mounting public complaint about the inadequacy of the bus service, Government at last relented and a design was approved based on a 34ft long, 21ft wheelbase chassis by AEC Limited; an extended 'Regent Mk V' in fact, to carry 120 passengers.

Stillborn Plans for High-capacity Buses in the 1960's.

Following the success of the 34ft long AEC's introduced by KMB in 1963, the Company entered into discussions with a number of manufacturers and their Hong Kong agents, with a view to further increasing route capacity without increasing the numbers of either buses or crews.

ABOVE: 1960's 'Jumbo' - the KMB 34ft long 120 passenger AEC Regent Mk V. *(Mike Davis*

Numerous alternatives were examined, including very large rigid and articulated double-deck vehicles and similar single-deck options, not to mention a proposal by a German supplier for a large rigid 1-decked vehicle. A brief description of some of these interesting proposals is shown below.

Guy 'Arab' Mk VI - 3-axle.

According to Government records, during 1963, the KMB planners examined the specification for a Guy three-axle design based on the type being supplied to Johannesburg at that time. Drawings for this proposed type have become lost over the years but it is known that the turning or, more correctly, the swept circle was estimated as being 92ft, or 12 ft more than the 34ft AEC Regent Mk V, so one can suppose that the Guy would have been considerably longer than that. The laden weight would have been 22 or 23 tonnes and, together with the great turning circle required, was considered to be greater than felt desirable by the Commissioner for Police who, at the time, was the authority on transport matters. Seating, according to the manufacturers, would have been for 110, plus 82 standees. In a letter of 6th March 1964, replying to Far East Motors, the Guy agents in Hong Kong, KMB rejected the concept on two grounds, one, that, by their calculations, the capacity would have been 'only' 175 - divided 66/28+81, well below the capacity claimed by Guy Motors. The second objection was with regard to the restricted capacity of the rear doorway which would have interfered with orderly queue control (!).

ABOVE: Guy Arab Mk VI, based on the Sunbeam S7A trolleybus chassis. Vehicles of this type, proposed for KMB in 1963, were used in Johannasburg at that time. *(John Shearman collection*

Articulated Double-deck Concept.

Concurrently with the Guy project, KMB were studying a very ambitious scheme put forward forward for a 57ft (17.37m) long articulated double-deck bus to have a carrying capacity of approximately 220 passengers, but it is not known what ratio of standing to seating would have been provided. There is conflict in Government records as to who the manufacturer would have been, as one report in January 1964 refers to Leyland as being the designers, while the minutes of a later meeting of the Advisory Committee for Public Transport, dated 2nd March 1964 refers to AEC. It was at the January meeting that ACPT agreed to the articulated concept in principle but, at the March meeting, having discussed and examined the implications of such a large vehicle using the roads as they were at that time, the ACPT took the view that a bus of this size would take up too much kerb space at bus stops, resulting in less queue control and, accordingly, rejected the idea.

AEC 4-axle Proposal.

In May 1964, ACPT were presented with yet another new double-deck bus design of interest. This was to be rigid but on no less than four axles and eight wheels. Transport Equipment (Thornycroft) Limited presented a specification for an "AEC Special 8-wheel Bus Chassis".

As with other designs submitted at the time, the degree of manoeuvrability achieved by the 34ft long AEC Regent Mk V's was used as the acceptable norm, meaning that the 4-axle bus, with its 91ft turning circle, was 11ft outside the maximum permissible and thus the design was rejected.

Fortunately the specification remains intact and certain relevant items are as shown in the accompanying table.

SPECIFICATION: PROPOSED AEC 4-AXLE DOUBLE-DECK BUS.	
Engine:	AEC type-AH690, 6-cylinder, horizontal, 11,310cc, 192bhp.
Transmission:	AEC Monocontrol, incorporating fluid flywheel and an electro-pneumatically operated 5-speed epicyclic gearbox, driving both rear axles, having overhead worm reduction drive units and inter-axle differential.
Axle Ratio:	7.75:1
Steering:	Worm and nut, hydraulic power assistance with ratio of 40:1.
Brakes:	Air pressure.
Suspension:	Semi-eliptical leaf-springs, 3 in wide. Both springs on each side of rear bogie interconnected at shackles by a balance beam.
Overall length:	43ft 1in
Width:	7ft 11in
Height:	Not quoted
Wheelbase:	23ft from centre of front axle to mid-point of rear bogie.
Bogie centre to rear:	13ft 3in
Swept turning circle:	91ft.
Chassis weight:	15,500lbs
Body weight:	12,000lbs
Passenger load:	24,000lbs
Gross Vehicle Weight:	51,000lbs

The passenger load was calculated at 100 lbs per passenger, which would have given a total capacity of up to 240, including standees.

Estimated restart gradiant : 1 in 4 (approx).
Estimated maximum speed

1½-deck Proposal.

One idea that did not reach further than a letter from the local agents to the operator, KMB, was a proposal which would have gone some way to avoid the, then, disadvantage set by the Hong Kong Government against German, or other non-Commonwealth, manufacturers, the Germans being the strongest at the time in the local market, that the bus company franchise limitation which required, until 1984, that buses should be of British or Commonwealth origin.

In their letter of 18th January 1964, to the Transport Advisory Committee (TAC), 'ZF Garages' proposed a 1½-deck German-built crush-loading bus body fitted to a Leyland Leopard bus chassis, which was, of course, built in Britain.

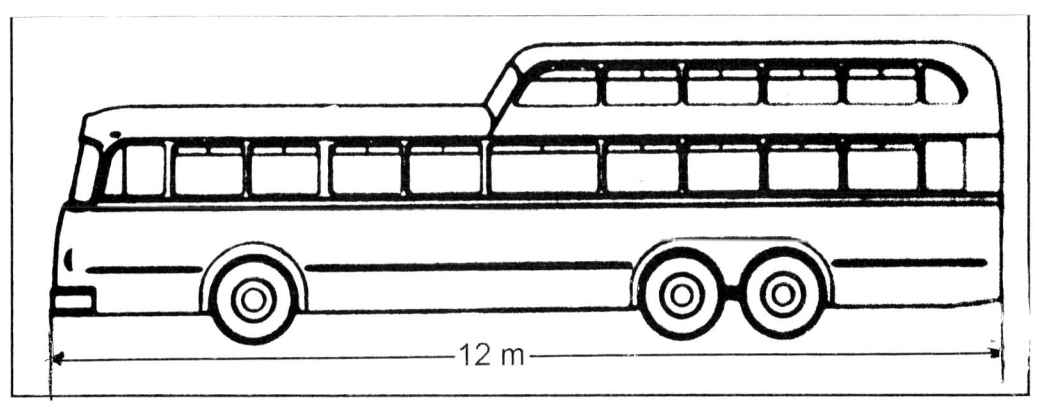

The body was to have been designed and built by 'Ludwig G.m.b.H.'. a German manufacturer. Intended for crush-loading and calculated by then current German law, which set ten passengers per m and, with an area available for standees of 8.1m , a possible capacity of 81 standees, plus 54 seated was indicated. Under Hong Kong law this would have been reduced to 62 standees but, if fifteen seats were removed, a total of 131 would have been theoretically possible. Nothing further came of this idea.

ABOVE: Outline drawing of a German proposal for a 12-metre one-and-a-half-deck bus promoted by the Mercedes dealer 'ZF Garages' in 1964.

Mercedes-Benz based Concepts.

Other designs put forward to relieve the pressure of passengers on the system included Mercedes-Benz based vehicle designs of both single-segment, rigid and two segment articulated configuration. While the capacity of these buses under continental European (pre-EC) conditions, was accepted by TAC, their application in Hong Kong was considered to be against the public interest, where the public would see the additional passengers as just so much more revenue for the not so hard-up bus company shareholders and without any balancing in additional passenger comfort. Also, the crush-load concept might not have been the best way to attract additional commuters onto buses at a time when every effort was being made to discourage the growing numbers of car owners from taking their vehicles into central areas.

The basis of the Mercedes-Benz concepts was, in all three cases - two rigid, one articulated - to have been the Daimler-Benz (Mercedes) O.317, mid-engined chassis, powered by the OM326/h, six-cylinder diesel-engine, delivering an optional 188bhp or 220bhp. Transmission alternatives to the standard, single-plate, dry clutch and DB 4-speed, synchromesh gearbox included the ZF Hydromedia 2-HM-70 or Voith Diwabus gearbox. The rigid versions would have been either two or three-door, multi-standee layout, capable of carrying, by weight, up to 160 passengers but, in reality, with seating taking up space, the actual number would have been in the order of 110 or 120, depending on body make and design, in a bus 11.925m long and having a wheelbase of 5.850m The proposed articulated version would have had the body and articulation built by Walter Vetter AG. Accommodation would have been provided for 40 seated passengers, plus a multitude of standees, quoted by the agents as 119, giving a total of 159 passengers and a crew of two, or, in the German notation. 2/159. Four doorways were specified, two each in tractor and trailer

BELOW: Mercedes-Benz articulated single-decker with 16.5 metre, high capacity, multi-standee layout

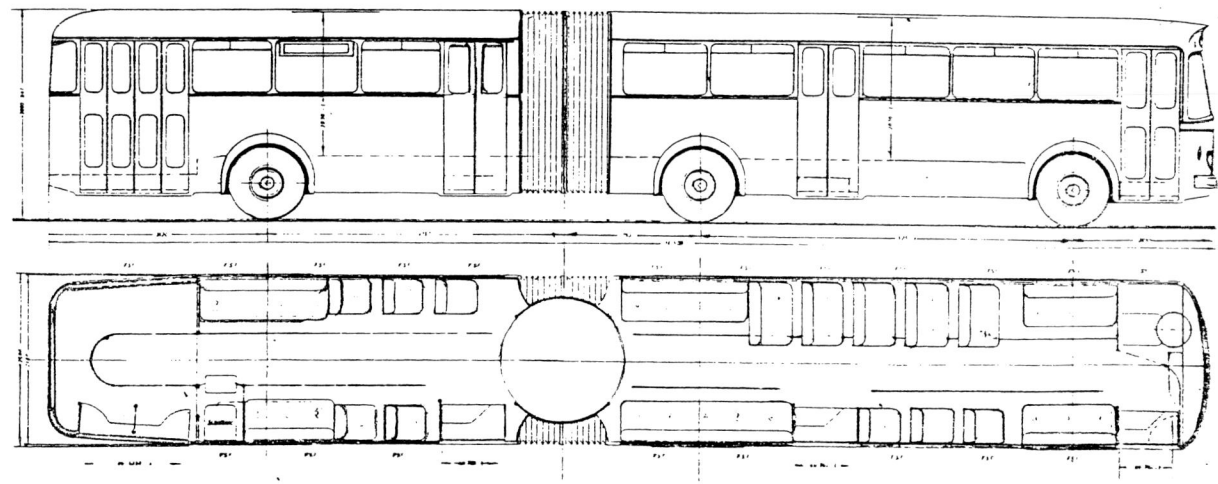

units. The overall length would have been 16.5m, with a wheelbase of 5.25m+6.08m. The same OM326/h engine would have been provided to drive the centre axle but would have been rated at 200bhp.

The manufacturer's agent in Hong Kong wrote to the TAC, saying that although the 0.317 had only an anticipated life of five years, he was sure that in Hong Kong conditions, double this could be expected! Nothing further is recorded on the subject.

AEC Regal Mk VI.

At a meeting of the TAC in June 1964, a discussion paper was presented, including the following:

It was suggested that consideration be given to the AEC Regal Mk VI, crush-loading, 139-passenger, single-deck bus. The same committee had so recently rejected the Mercedes-Benz proposals, without mention of the 'Buy British' requirement. It was put to the TAC that the existing double-deck AEC Regent Mk V could carry 120 passengers within both the manufacturer's gross maximum weight and the Government approved Gross Vehicle Weight but, nevertheless, was capable of safely loading 150 by crush-loading, of which 79 passengers had the relative comfort of a seat. With the single-deck Regal Mk VI, of the 139 crush-loaded passengers, only 33 would be seated, although loading and unloading would be quicker than with the double-decker.

The TAC's recommendation was that KMB be encouraged to buy a trial dozen Regal VI's for comparison with the Regent V's. KMB took no further action in this direction.

Vehicle length would have been 36ft and the design gross weight 15.75 tonnes. The engine would have been 11.31 litres, 186bhp.

Leyland Single-deck for Hilly Routes.

In order to increase speeds on steep hills - not vehicle capacity - Leyland carried-out trials in England with an unspecified two-axle vehicle, laden to simulate 30 seated and 10 standing passengers, plus a crew of two, making a total of 37° hundredweight. Body and chassis weight would have given a gross weight of 9 tonnes. Other specifications included a 185bhp engine and 5-speed gearbox, with 9.00 x 20 tyres. The vehicles eventually chosen for Route TWSK were to be Albion Viking single-deckers, from the Leyland stable, which later operated the route until being replaced by double-deck MCW Metrobuses in the late 1980's

Addendum to Page 31:
Seddon Mk17.

As so often happens, no sooner has a publisher gone to print, than a major new item worthy of inclusion comes to light and the following is such an item.

Quite by chance, during a visit to a rally in Manchester, a major contributor to this book, John Shearman came upon a piece of paper which turned-out to be a page from a 1962 edition of Motor Transport. On that page was an article about the then transport scene in Hong Kong, including a photograph of a full-fronted Seddon single-decker of the type already described on page 31 - except, that is for its having a traditional, exposed radiator.

This new information was quickly followed-up by John, together with Ron Phillips who had recorded in an earlier publication (*PSV Circle Publication OP2*) that there was a single Seddon with a Metal Sections body which included a Seddon produced scuttle (front dash panels) as opposed to the 99 others which had concealed radiators with Seddon-style ornamental surrounds.

More extraordinary still, it is recorded in the Seddon records that the chassis, number 32298, was a frustrated export order placed by Ashworth (agent) of Costa Rica.

This bus became the first of the 100 for KMB and was registered HK4352. Perhaps, as none of our contributors had even heard of this variant, it might have been rebuilt with a front, standard to the remainder of the type by the time detailed recording of the various bus types commenced c1964/5.

RIGHT: The recently unearthed illustration of the lone exposed-radiator Seddon Mk17 with Kowloon Motor Bus. Buses with this style of front were supplied to Singapore but this is the only example that can be accredited to Hong Kong.
(*Motor Transport, December 1st, 1962.*)